Peter Lemke · Hans-Werner J
Editors

Arctic Climate Change

The ACSYS Decade and Beyond

 Springer

Editors
Peter Lemke
Alfred-Wegener-Institute
for Polar and Marine Research
Postfach 12 01 61
27515 Bremerhaven
Germany
Peter.Lemke@awi.de

Hans-Werner Jacobi
Université Joseph Fourier-Grenoble 1/ CNRS
Laboratoire de Glaciologie et Géophysique
de l'Environnement UMR 5183
54, rue Molière
38041 Saint-Martin d'Hères cedex
France
jacobi@lgge.obs.ujf-grenoble.fr

ISSN 1383-8601
ISBN 978-94-017-8306-4   ISBN 978-94-007-2027-5 (eBook)
DOI 10.1007/978-94-007-2027-5
Springer Dordrecht Heidelberg London New York

Springer is part of Springer Science+Business Media (www.springer.com)

Dedication

Dr. Victor Savtchenko (22.07.1937–15.08.2008)

This book is dedicated to the memory of Victor (Gavrilovitch) Savtchenko, a scientist and coordinator of international climate science, who, in turn, dedicated a significant part of his life, knowledge and energy towards formulating the founding ideas and setting the stage for two major research projects of the World Climate Research Programme (WCRP), namely the Arctic Climate System Study (ACSYS) and the Climate and Cryosphere (CliC) Project.

Victor was born on 22 July 1937 in Poltava in the former Soviet Union. Now this city and the region surrounding it is a part of Ukraine. In 1969 Victor completed a postgraduate course of the Leningrad Hydrometeorological Institute (at present the Russian State Hydrometeorological University), and joined the Arctic and Antarctic Research Institute (AARI) in Saint Petersburg. In 1970 in his successfully defended dissertation for the academic degree of the Candidate of Science, he elegantly used sophisticated mathematical apparatus for a study of internal waves in fluids. This project solidified the characteristic style of his research that combined an accurate mathematical description of the simplified but essential features of complex phenomena with a deep and all-embracing understanding of their overarching complexity. Victor has kept this style of scientific investigation for his entire life and it proved to be particularly fruitful in theoretical studies of ocean and atmosphere interactions. Because of his affiliation with AARI, all scientific work of Victor had a strong focus on the role of the Arctic and Southern Oceans in climate variability and change.

For more than a decade Victor was the head of the Modelling Laboratory in the AARI's Department of Ocean – Atmosphere Interactions. He participated in several field expeditions to the Arctic and Antarctic. Main results of his studies are summarised in the Doctor of Science thesis, defended in 1989, in a well-known monograph 'Impact of heat fluxes from the ocean on oscillations of climate in high latitudes' (co-authored with close colleague Andrey Nagurny), and in over 60 research articles in leading scientific journals.

In 1978, Victor started to work for the World Meteorological Organization in Geneva, serving as senior scientific officer of the Global Atmospheric Research Programme, until 1981, and, in 1989–1999, as senior scientific officer of WCRP. His scientific insight and managerial support were decisive in the establishment of the WCRP Arctic Climate System Study (ACSYS). In the 1990s, while ACSYS was still underway, Victor started to work on a new global project to address the role of frozen water in the climate system. He was the one who recalled the now famous term 'cryosphere' and insisted on including it in the title of the new project, the Climate and Cryosphere (CliC).

Victor was a kind, sensitive, and modest person. Even for those who knew him well, it may come as a surprise that this very serious scientist and seasoned program manager had a charming hobby of collecting toy owls. Victor Savtchenko passed away on 15 August 2008 in Divonne-Les-Bains, France, at the age of 71. We are very grateful to Victor for his devoted service to climate and polar science and for his very significant contributions to founding ACSYS and CliC, two successful projects of the WCRP.

Geneva   Vladimir Ryabinin

Foreword

The Arctic is a region of rapid changes caused by global warming, resulting in rising atmospheric and oceanic temperatures, and declining sea ice cover and thickness, with significant impacts on ecosystems and human settlements. Therefore, a better understanding of climate processes in the Arctic is fundamental to assess major impacts of these changes in the Arctic and within the entire Earth system due to a variety of feedback processes.

The ACSYS project was fundamental in raising the awareness of the role of the Arctic in the global climate system. It has greatly advanced our understanding of the processes acting in the Arctic due to development of improved observing systems, using both in situ and remote sensing techniques, and numerical models describing the various components of the climate system and their interaction. ACSYS activated nearly 250 scientists from 19 countries for its final conference, with many more researchers, who participated in the respective field work and modelling activities.

ACSYS was designed as a ten-year project, which started in 1994 and finished at the end of 2003. It has provided a valuable legacy of data sets, model components and understanding, which is used as a basis for ongoing research within the framework of the bi-polar Climate and Cryosphere (CliC) project of the WCRP, as well as for many projects during the International Polar Year 2007–2009. This book represents an account of this legacy and its relevance for current Arctic climate research.

We express our thanks to the impressive number of scientists that have contributed to the success of the ACSYS project. This book could not have been realised without the huge support of the authors who submitted their chapters for this volume. We further acknowledge the following scientists for their help in reviewing the individual chapters of this book: Sabine Attinger (Helmholtz Centre for Environmental Research, Leipzig, Germany), Jens Hesselbjerg Christensen (Danish Climate Centre, Copenhagen, Denmark), Klaus Dethloff (Alfred Wegener Institute for Polar and Marine Research, Potsdam, Germany), Wolfgang Dierking (Alfred Wegener Institute for Polar and Marine Research, Bremerhaven, Germany), Peter S. Guest (Naval Postgraduate School, Monterey, USA), Stefan Hagemann (Max Planck Institute for Meteorology, Hamburg, Germany), Peter M. Haugan (University of Bergen, Norway), Daniela Jacob (Max Planck Institute for Meteorology, Hamburg,

Germany), Thomas Jung (Alfred Wegener Institute for Polar and Marine Research, Bremerhaven, Germany), Johann Jungclaus (Max Planck Institute for Meteorology, Hamburg, Germany), Wolfram Mauser (Ludwig-Maximilians-Universität Munich, Germany), Jens Meincke (University of Hamburg, Germany), Michael Tjernström (Arrhenius Laboratory, Stockholm, Sweden), Leif Toudal Pedersen (Technical University of Denmark, Lyngby, Denmark), Joachim Reuder (University of Bergen, Norway), Erich Roeckner (Max Planck Institute for Meteorology, Hamburg, Germany), Valery Vuglinsky (State Hydrological Institute of Roshydromet, St. Petersburg, Russia) and Michael Winton (Princeton University, USA).

Peter Lemke
Hans-Werner Jacobi

Contents

Contributors

G.V. Alekseev AARI, St. Petersburg, Russia, alexgv@aari.nw.ru

Leif Anderson Department of Chemistry, University of Gothenburg, SE-41296 Göteborg, Sweden, leifand@chalmers.se

Ghassem Asrar World Climate Research Programme, c/o WMO Secretariat, 7bis, Avenue de la Paix, CP2300, Geneva 2, CH-1211, Switzerland, GAsrar@wmo.int

Gerit Birnbaum Alfred Wegener Institute for Polar and Marine Research, Postfach 120161, D-27515 Bremerhaven, Germany, gerit.birnbaum@awi.de

Cecilia M. Bitz Atmospheric Sciences, University of Washington, Seattle, WA, USA, bitz@atmos.washington.edu

Howard Cattle National Oceanography Centre, Southampton, UK, hyc@noc.soton.ac.uk

K. Dethloff Alfred Wegener Institute for Polar and Marine Research, Research Unit Potsdam, Telegrafenberg A43, D-14473 Potsdam, Germany, Klaus.Dethloff@awi.de

Silke Dierer Meteorological Institute, Centre for Marine and Climate Research, University of Hamburg, Bundesstr. 55, D-20146 Hamburg, Germany, sdierer@web.de

W. Dorn Alfred Wegener Institute for Polar and Marine Research, Research Unit Potsdam, Telegrafenberg A43, D-14473 Potsdam, Germany, Wolfgang.Dorn@awi.de

Patrick Eriksson Finnish Meteorological Institute, P.O. Box 503, FI-00101 Helsinki, Finland, Patrick.Eriksson@fmi.fi

Eberhard Fahrbach Alfred-Wegener-Institute for Polar and Marine Research, P.O. Box 120161, D-27515 Bremerhaven, Germany, Eberhard.Fahrbach@awi.de

Eirik J. Førland The Norwegian Meteorological Institute, P.O. Box 43, Blindern 0313 Oslo, Norway, eirikjf@met.no

Thomas Garbrecht OPTIMARE Sensorsysteme GmbH & Co. KG, Am Luneort 15a, D-27572 Bremerhaven, Germany, thomas.garbrecht@optimare.de

Rüdiger Gerdes Sea Ice Physics, Alfred Wegener Institute for Polar and Marine Research, Bussestr. 24, 27570 Bremerhaven, Ruediger.Gerdes@awi.de

Vladimir M. Gryanik Alfred Wegener Institute for Polar and Marine Research, Postfach 120161, D-27515 Bremerhaven, Germany, vladimir.gryanik@awi.de

Micha Gryschka Institute of Meteorology and Climatology, Leibniz University of Hannover, D-30419 Hannover, Germany, gryschka@muk.uni-hannover.de

Stefan Hagemann Max-Planck-Institute for Meteorology, Bundesstraße 53, D-20146 Hamburg, Germany, Stefan.Hagemann@zmaw.de

Reinhard Hagenbrock WetterOnline Meteorologische Dienstleistungen GmbH, Am Rheindorfer Ufer 2, D-53117 Bonn, Germany, ReinhardHagenbrock@wetteronline.de

D. Handorf Alfred Wegener Institute for Polar and Marine Research, Research Unit Potsdam, Telegrafenberg A43, D-14473 Potsdam, Germany, Doerthe.Handorf@awi.de

Jörg Hartmann Alfred Wegener Institute for Polar and Marine Research, Postfach 120161, D-27515 Bremerhaven, Germany, jorg.hartmann@awi.de

Günther Heinemann Department of Environmental Meteorology, University Trier, D-54286 Trier, Germany, heinemann@uni-trier.de

Marika Holland National Center for Atmospheric Research, Boulder, CO, USA, mholland@ucar.edu

Martin Jakobsson Department of Geological Sciences, Stockholm University, SE-10691 Stockholm, Sweden, martin.jakobsson@geo.su.se

O.M. Johannessen NERSC, 5059 Bergen, Norway, ola.johannessen@nersc.no

E. Peter Jones Department of Fisheries and Oceans, Bedford Institute of Oceanography, P.O. Box 1006, Dartmouth NS B2Y 4A2, Canada, Peter.Jones@dfo-mpo.gc.ca

Lars Kaleschke Institute of Oceanography, University of Hamburg, Bundesstr. 55, D-20146 Hamburg, Germany, lars.kaleschke@zmaw.de

Lev Kitaev Institute of Geography RAS, Staromonetniy 29, 109017 Moscow, Russia, lkitaev@mail.ru

Peter Lemke Alfred Wegener Institute for Polar and Marine Research, Bremerhaven, Germany, Peter.Lemke@awi.de

Dennis P. Lettenmaier Department of Civil and Environmental Engineering, University of Washington, 164 Wilcox Hall, Box 352700, Seattle, WA 98195-2700, USA, dennisl@u.washington.edu

Christof Lüpkes Alfred Wegener Institute for Polar and Marine Research, Postfach 120161, D-27515 Bremerhaven, Germany, christof.luepkes@awi.de

A. Lynch School of Geography and Environmental Sciences, Monash University, Melbourne, VIC 3800, Australia, Amanda.Lynch@arts.monash.edu.au

Hermann Mächel German Weatherservice, Frankfurter Str. 135, D-63067 Offenbach, Germany, hermann.maechel@dwd.de

Thomas Maurer Federal Institute of Hydrology (BfG), Am Mainzer Tor 1, D-56002 Koblenz, Germany, Thomas.Maurer@bafg.de

T.A. McClimans SINTEF Fisheries and Aquaculture, 7465 Trondheim, Norway, thomas.mcclimans@sintef.no

Humfrey Melling Institute of Ocean Sciences, Fisheries and Oceans Canada, Sidney, BC, Canada, V8L 4B2, Humfrey.Melling@dfo-mpo.gc.ca

M.W. Miles Bjerknes Centre for Climate Research, 5007 Bergen, Norway and Environmental Systems Analysis Research Center, Boulder, USA

ESARC, Boulder, CO, USA, martin.miles@geo.uib.no

James E. Overland NOAA/Pacific Marine Environmental Laboratory, Seattle, WA 98115-6349, USA, james.e.overland@noaa.gov

Simon Prinsenberg Department of Fisheries and Oceans, Bedford Institute of Oceanography, P.O. Box 1006, Dartmouth NS B2Y 4A2, Canada, Simon.Prinsenberg@dfo-mpo.gc.ca

Siegfried Raasch Institute of Meteorology and Climatology, Leibniz University of Hannover, D-30419 Hannover, Germany, raasch@muk.uni-hannover.de

Vjacheslav Rasuvaev All-Russian Scientific Research Institute of the Hydrological and Meteorological Information – World Data Center, Koroleva 6, Obninsk 249020, Russia, razuvaev@meteo.ru

Jeff K. Ridley Hadley Centre for Climate Prediction, Met Office, Exeter, UK, jeff.ridley@metoffice.gov.uk

A. Rinke Alfred Wegener Institute for Polar and Marine Research, Research Unit Potsdam, Telegrafenberg A43, D-14473 Potsdam, Germany, Annette. Rinke@awi.de

Bert Rudels Department of Physics, University of Helsinki, P.O. Box 64, FI-00014 Helsinki, Finland

Finnish Meteorological Institute, P.O. Box 503, FI-00101 Helsinki, Finland, bert.rudels@helsinki.fi; bert.rudels@fmi.fi

Bruno Rudolf German Weatherservice, Frankfurter Str. 135, Offenbach D-63067, Germany, bruno.rudolf@dwd.de

Vladimir Ryabinin World Climate Research Programme, c/o WMO Secretariat, 7bis, Avenue de la Paix, CP2300, Geneva 2, CH-1211, Switzerland, VRyabinin@wmo.int

S. Saha C & H Division, Indian Institute of Tropical Meteorology, Pune 411008, India, Subodh@tropmet.res.in

Hannu Savijärvi Department of Physics, University of Helsinki, P.O. Box 64, 00014 Helsinki, Finland, hannu.sarvijarvi@helsinki.fi

Victor Savtchenko Deceased 15 August 2008

Ursula Schauer Alfred-Wegener-Institute for Polar and Marine Research, P.O. Box 120161, D-27515 Bremerhaven, Germany, Ursula.schauer@awi.de

K. Heinke Schlünzen Meteorological Institute, Centre for Marine and Climate Research, University of Hamburg, Bundesstr. 55, D-20146 Hamburg, Germany, heinke.schluenzen@zmaw.de

Mark C. Serreze CIRES/University of Colorado, Boulder, CO 80309-0449, USA

Konrad Steffen University of Colorado, CIRES, Campus Box 216, Boulder, CO 80309, USA, Konrad.steffen@colorado.edu

Fengge Su Institute of Tibetan Plateau Research, Chinese Academy of Sciences, No.18, Shuangqing Rd., Beijing 100085, China, fgsu@itpcas.ac.cn

Ole Einar Tveito The Norwegian Meteorological Institute, P.O. Box 43, Blindern 0313 Oslo, Norway, Ole.Einar.Tveito@met.no

Timo Vihma Finnish Meteorological Institute, P.O. Box 503, FIN-00101 Helsinki, Finland, timo.vihma@fmi.fi

Ulrike Wacker Alfred Wegener Institute for Polar and Marine Research, Postfach 120161, D-27515 Bremerhaven, Germany, ulrike.wacker@awi.de

Daqing Yang CliC International Project Office, Norwegian Polar Institute, Polarmiljøsenteret, NO-9296, Tromso, Norway, daqing.yang@npolar.no

Tom Yao 3989 W 18th Ave, Vancouver, BC, Canada V6S 1B6, tomyao@telus.net

Chapter 1
The Origins of ACSYS

Victor Savtchenko[†]

Abstract This chapter briefly reviews, in a historical perspective, the science background for establishing an internationally coordinated Arctic Climate System Study (ACSYS) under the auspices of the World Climate Research Programme (WCRP).

Keywords Arctic • ACSYS • Climate • WCRP • Sea ice • Ocean • Polar • Snow • Atmosphere • Ocean

The First Implementation Plan (FIP) of the WCRP was published in November 1985 (WCRP 1985). Based on the Scientific Plan (WCRP 1984), the FIP proposed activities needed to achieve the goals of the Programme and to efficiently develop the work among the WCRP projects according to their areas of expertise.

The scientific objectives (or 'streams') of the Programme were defined as follows:

- Establishing the physical basis of long-range weather prediction
- Understanding the predictable aspects of global climate variations over periods of several months to several years
- Assessing the response of climate to natural or man-made influences over periods of several decades

The first and second objectives were considered to be the necessary 'stepping stones' towards achieving the third objective. The WCRP activities were organized through six sub-programmes, namely:

- Atmospheric Climate Prediction Research (ACP)
- Coupled Atmosphere-Ocean Boundary Layer Research (CAOB)
- Cryosphere Research

V. Savtchenko
[†]Deceased 15 August 2008

P. Lemke and H.-W. Jacobi (eds.), *Arctic Climate Change: The ACSYS Decade and Beyond,* Atmospheric and Oceanographic Sciences Library 43, DOI 10.1007/978-94-007-2027-5_1, © Springer Science+Business Media B.V. 2012

- Tropical Ocean and Global Atmosphere (TOGA) project
- World Ocean Circulation Experiment (WOCE)
- Climate Sensitivity Assessment (CSA)

The ACP and the CAOB research activities constituted the basis for achieving the first objective of the WCRP. The Cryosphere Research and TOGA projects were designed to achieve the second objective. The WOCE and the CSA projects, together with the activities planned for the first and second 'streams', constituted elements of a scientific strategy for the achievement of the third objective of the WCRP.

The Cryosphere Research sub-programme was a forefather of both the ACSYS and the WCRP Climate and Cryosphere (CliC) projects. The sub-programme was focused on modelling the large-scale interaction between the atmosphere and sea ice. It concentrated on the study of seasonal and interannual variations of sea ice and on the determination of oceanic and atmospheric forcings which control them, with a view to incorporate the scientific advances in the knowledge of sea-ice processes into fully interactive atmosphere–ocean–ice models required for the assessment of long-term climate variations and impacts. The term 'cryosphere' was treated as the combination of the sea-ice cover of the polar oceans and the ice sheets over Greenland and the Antarctic continent. The consideration of snow was included in the study of hydrological processes in relation to the development of atmospheric climate models, i.e. in the Atmospheric Climate Prediction Research sub-programme (the first 'stream' of the WCRP).

The World Meteorological Organization (WMO)/International Council of Scientific Unions (ICSU) Joint Scientific Committee (JSC) for the WCRP recognized, at an early stage, the significant climate impact of sea ice as it affects the interaction between the ocean and the atmosphere as well as the surface albedo. The JSC-IV (1983) pointed out that since the extent of sea ice shows noticeable interannual as well as seasonal variations, sea-ice processes could be important for all three 'streams' of climate research. The observed extent of sea ice is, in particular, an important boundary condition to be specified for long-range weather prediction. The JSC-IV agreed that satellite observations of sea ice should be one key component of a comprehensive strategy for monitoring climate and the factors controlling its long-term variations.

It was underlined in the FIP that sea ice played a significant role in the fluctuations of the global ocean–atmosphere system on seasonal and longer time scales and constituted, therefore, an important interactive component of the climate system. It was stressed that the major obstacle, which hampered the development of interactive global atmosphere–ocean–ice models, was the lack of adequate observations to describe or to infer:

- Forcing fields, such as the ocean surface current and heat flux; the surface wind and air temperature and the incoming radiation flux at the surface
- Sea-ice state variables, such as the ice concentration, velocity, thickness and internal stress
- Physical processes, which determine mechanical properties of natural sea-ice fields

Existing or planned multi-national programmes (such as the Marginal Ice Zone Experiment and the Greenland Sea Project) addressed the above-indicated problems only partially. Information about the extent, concentration and age of sea ice on the global scale was expected to be provided by planned satellite missions with passive and/or active microwave sensors. Nevertheless, it was absolutely clear that substantial augmentation of existing programmes was needed to meet the observational requirements for large-scale sea-ice modelling in the Arctic Ocean basin and around the Antarctic continent. Therefore, the JSC-VI (1985) agreed on establishing an international research programme on sea ice and climate as an integral part of WCRP. A JSC Working Group on Sea Ice and Climate (WGSI) was constituted to design a strategy for global sea-ice research and to advice on a scientific strategy for the development of the fully interactive atmosphere–ocean–ice models. Prof. N. Untersteiner (Polar Science Centre, University of Washington, Seattle, USA) chaired the Group.

The WGSI suggested that ocean circulation studies in the ice-covered Southern Ocean be brought into the framework of the WOCE, while a specific Arctic Basin project might eventually be needed to improve predictions of climatic changes of the air–sea ice system. These recommendations of the Group were endorsed by the JSC-VII (1986). In 1988, the WGSI completed its assessment of sea-ice science issues relevant to the climate problem and put forward a programme of proposed activities that, it considered, should be carried out in the WCRP framework (WCP-18). The JSC-X (1989) expressed its agreement with the assessment done by the Group of the state of affairs in international polar sciences and outstanding priorities from the point of view of climate research. The Committee agreed that WCRP efforts should be focused on promoting the implementation of the following specific polar climate research projects identified by the Group:

- Polar radiation reference stations (to acquire standard downward shortwave and longwave radiation flux data)
- Coordinated global sea-ice modelling studies and intercomparison of results
- Arctic and Antarctic Ice-Thickness Monitoring Projects (based on upward-looking sonar measurements from under-ice moorings)
- Arctic Ice-Ocean Station Network (to acquire long-term series of systematic hydrographic, ocean current and sea-ice observations from a small number of sites in the Arctic basin)
- Arctic and Antarctic Drifting Ocean Data Buoys Programmes

The Sea-Ice Numerical Experimentation Group (SINEG) was established by the JSC-X in order to promote (by means of coordinated numerical experiments) the development of sea-ice models and coupled atmosphere–ice–ocean models for climate research, and the assessment of the sensitivity of models to internal parameterizations and external forcing fields. Prof. P. Lemke (Alfred Wegener Institute for Polar and Marine Research (AWI), Bremerhaven, Germany) was appointed the convener of the SINEG.

The fourth session of the WGSI held in November 1989 (WCP-41) reviewed various aspects of the role of sea ice in the climate system, corresponding WCRP

sea-ice research priorities, and the status of implementation of WCRP sea-ice observation and modelling projects. The Group noted that polar oceanic processes play a
dominant role in the production of the deep and bottom water masses of the world
ocean and therefore in the formation of the global thermohaline circulation, which
is responsible for planetary-scale redistribution of heat and freshwater. The Group
further noted that studies of the Antarctic deepwater formation and corresponding
sea-ice processes were part of the WOCE but that international large-scale Arctic
Ocean circulation studies had not been undertaken yet. In order to lay the foundation for an Arctic Ocean monitoring programme and to begin collection of long-
term series of hydrographic and tracer data in the Arctic basin, the Group agreed to
promote international cooperation for the implementation of deep ocean measurements of temperature, salinity, currents and tracer concentrations by exploiting
opportunities afforded by sea-ice studies in the Arctic.

To reflect the new research priorities, the Group was reconstituted under its previous name – Working Group on Sea Ice and Climate – as a joint body of the JSC
and the Committee on Climatic Changes and the Ocean (CCCO) formed by the
Intergovernmental Oceanographic Commission (IOC)/Scientific Committee on
Oceanic Research (SCOR). The Working Group on Sea Ice and Climate was charged
with the tasks of promoting the implementation of specific WCRP sea-ice research
projects and developing appropriate activities for improving the understanding of
the freshwater balance of polar regions and its relationship to deepwater formation
in the ocean (JSC-XI 1990). Prof. E. Augstein (AWI, Bremerhaven, Germany)
chaired the Group.

The International Conference on the Role of the Polar Regions in Global
Change was held in June 1990 at the University of Alaska, Fairbanks, USA. Many
papers presented at the conference emphasized the potentially significant and
complex influence of the Arctic Ocean sea-ice processes on the dynamics of the
global climate system, in particular the relation between circulation, salinity
structure and freshwater budget of the Arctic Ocean and the North Atlantic deep-
water production (International Conference 1991).

It was evident that the role of the Arctic Ocean processes in the maintenance and
change of the global climate system would be much better understood if observational data required for refining parameterizations of physical processes and for
validation of coupled atmosphere–ice–ocean models could be collected in the Arctic
and made available for the international climate research community. It was also
evident that consideration of the impact of the Arctic Ocean on the climate represented a gap in WCRP activities. At the same time, there was no doubt that the
WCRP was the most adequate international body to address the appropriate Arctic
topics. Informal discussions on the subject held by the chairman of the WGSI with
some leading climate scientists showed that a special Arctic component of WCRP
with strong hydrographic, sea-ice and atmospheric elements was highly desirable
from a scientific point of view and feasible, taking into account the changes that
happened in the international political climate. As a first step towards outlining an
Arctic experiment within the WCRP framework, an ad hoc workshop was organized

in Mainz, Germany, in December 1990 to prepare an appropriate proposal for consideration by the JSC.

The workshop developed the ACSYS proposal (Position Paper 1991), which was considered by the JSC-XII (1991). The main scientific goals of the ACSYS were defined as follows:

- To provide a scientific basis for a realistic representation of the Arctic region in coupled global climate models
- To suggest and possibly implement an effective climate monitoring scheme in the Arctic
- To develop regional coupled ocean–ice models and to carry out sensitivity studies and climate change computations for specified atmospheric and lateral boundary conditions
- To improve the treatment of Arctic clouds and radiation transfer in atmospheric and coupled climate models

Responding to the ideas in the proposal, the JSC-XII agreed to consider a coordinated research project to study the large-scale dynamics of the Arctic Ocean and sea ice, and the energy and freshwater budgets of the Arctic basin and adjacent regions. The Committee believed that a comprehensive statement of the scientific concept of the proposed Arctic Climate System Study should be prepared, which could convince the scientific community to participate and science funding agencies to support corresponding activities. For this purpose, the Committee established an ad hoc Study Group to consider the scientific goals and conditions for implementation of ACSYS, including logistic and organizational aspects. Prof. E. Augstein was appointed the chairman of the Study Group. The Study Group prepared the ACSYS concept document (WCRP-72) which was approved by the fifth session of the WGSI (WCRP-65). It was proposed that ACSYS comprises the following four main elements:

- Study of the structure and circulation of the Arctic Ocean
- Observation and modelling of sea ice
- Air–sea interaction
- Freshwater cycle in the Arctic region

The JSC-XIII (1992) welcomed the comprehensive approach developed by the Study Group for a multi-disciplinary study of the Arctic climate system to investigate the large-scale dynamics of the Arctic Ocean and sea ice, and the energy and freshwater budgets of the Arctic basin and adjacent regions. The Committee decided to undertake the ACSYS as a new project of the WCRP. To lead the implementation of the project, the JSC established an ACSYS Scientific Steering Group (SSG). Consequently, the Committee disbanded the WGSI. It was agreed that the ACSYS SSG should also assume the responsibility for oversight of sea-ice research and monitoring projects previously taken by the WGSI. Prof. K. Aagaard (NOAA Pacific Marine Environmental Laboratory, USA) was appointed the chair of the ACSYS SSG. The JSC-XIII further agreed that the SINEG would have an important role in supporting the ACSYS modelling efforts and should continue with appropriately modified terms of reference. The Committee emphasized that, in proposing the establishment of ACSYS, the importance of Antarctic science was not being overlooked or considered a lower scientific priority. The reason for focusing on an Arctic study in the WCRP was to take advantage of the then available and planned facilities in the Arctic basin. The JSC-XIII requested the ACSYS SSG to keep Antarctic science in mind and, when appropriate, to apprise the Committee of the possible need for WCRP initiatives in the Antarctic.[1]

The JSC-XIII underlined that an urgent task for the ACSYS SSG was the preparation of an internationally agreed implementation plan for the experiment. Responding to this request and building on the Scientific Concept of ACSYS, the SSG prepared the Initial Implementation Plan (IIP) for the ACSYS (WCRP-85). The main observational phase of the experiment began formally on 1 January 1994 and has been scheduled for a 10-year period, i.e. until 31 December 2003. The scientific goal of ACSYS was to ascertain the role of the Arctic in global climate. To attain this goal, ACSYS sought to develop and coordinate national and international Arctic science activities aimed at three main objectives:

- Understanding the interaction between the Arctic Ocean circulation, ice cover and the hydrological cycle
- Initiating long-term climate research and monitoring programmes for the Arctic
- Providing a scientific basis for an accurate representation of Arctic processes in global climate models

The IIP outlined the following five main areas of the ACSYS activity:

- Arctic Ocean circulation programme
- Arctic sea-ice programme
- Arctic atmosphere programme
- Study of the hydrology of the Arctic region
- Modelling of basin-wide ocean-ice processes and ocean–ice–atmosphere interaction

The first session of the ACSYS SSG (SSG-I 1993) reviewed the activities and terms of reference of the SINEG. In view of the orientation of SINEG's efforts towards coupled sea-ice/ocean modelling, both in the Arctic and the Antarctic, the SSG agreed that the SINEG should be renamed as the ACSYS Sea-Ice/Ocean Modelling Panel (SIOM) with appropriately modified terms of reference. Prof. P. Lemke was nominated to serve as the convener of the Panel. The SIOM Panel was converted into the ACSYS Numerical Experimentation Group in 1998, with expanded terms of reference. In 1999, Prof. P. Lemke stepped down as chair of the ACSYS NEG, and Dr. Gregory M. Flato from the Canadian Centre for Climate Modelling and Analysis became the new chair.

---

[1] The ACSYS SSG followed this advice and developed (in the late 1990s) a proposal focused on global cryospheric studies within the WCRP. The proposal was approved by the JSC as a core WCRP CliC project.

The ACSYS Conference on the Dynamics of the Arctic Climate System (WCRP-94) in Gothenburg, Sweden, in November 1994 reviewed the ACSYS IIP. The Conference fully concurred on the importance of the ACSYS goal and found ACSYS to be a well-focused and technically feasible project. The polar science community and science funding agencies of countries active in Arctic climate research and global change investigations gave strong support to the ACSYS. A number of additions and refinements to the IIP were proposed.

Subsequently, the ACSYS SSG reviewed and evolved the IIP, taking into account national contribution to the project and logistic coordination with other international programmes. The end product, which is regularly updated, is available on the ACSYS Web site (http://acsys.npolar.no) under the title 'ACSYS Implementation and Achievements'.

In a paper published in 1994 when ACSYS was launched as the only WCRP regional experiment, K. Aagaard and E. Carmack, discussing effects of the Arctic Ocean on global climate, put the following two questions:

- Is the Arctic Ocean, in fact, an important cog in the global climate machinery?
- Can realistic changes in the density structure and circulation of the Arctic Ocean effect significant changes on a much larger spatial scale?

They foresaw that getting answers to the fundamental questions would require a decade or two.

References

Aagaard K, Carmack EC (1944) The Arctic ocean and climate: a perspective. In: Johannessen OM, Muench RD, Overland JE (eds) The Polar oceans and their role in shaping the global environment; the Nansen centennial volume, vol 85, Geophysical monograph. American Geophysical Union, Washington, DC, pp 5–20

International Conference (1991) International conference on the role of the polar regions in global change. In: Proceedings of a conference held at the University of Alaska, Fairbanks, 11–15 June 1990, vols I and II

JSC-IV (1983) Report of the fourth session of the Joint Scientific Committee, Venice, 1–8 Mar 1983

JSC-VI (1985) Report of the sixth session of the Joint Scientific Committee, London, 26 Feb–5 Mar 1985. WMO/TD- No.54

JSC-VII (1986) Report of the seventh session of the Joint Scientific Committee, Lisbon, 12–18 Mar 1986. WMO/TD-121

JSC-X (1989) Report of the tenth session of the Joint Scientific Committee, Villefranche-sur-Mer, France, 13–18 Mar 1989. WMO/TD-No.314

JSC-XI (1990) Report of the eleventh session of the Joint Scientific Committee, Tokyo, Japan, 5–10 Mar 1990. WMO/TD- No.375

JSC-XII (1991) Report of the twelfth session of the Joint Scientific Committee, Bremen, Germany, 18–23 Mar 1991. WMO/TD-No.432

JSC-XIII (1992) Report of the thirteenth session of the Joint Scientific Committee, Victoria, BC, Canada, 23–28 Mar 1992. WMO/TD-No.497

Position paper (1991) Proposal for an Arctic Climate System Study (ACSYS). A position paper evolved from a WCRP ad hoc workshop, held on 3–4 Dec 1990, Mainz

SSG-I (1993) Report of the first session of the ACSYS scientific steering group, Seattle, 2–6 Nov 1992

WCP-18 (1988) Report of the third session of the working group on sea ice and climate, Oslo, 31 May–3 June 1988. WMO/TD-No.272

WCP-41 (1990) Report of the fourth session of the working group on sea ice and climate, Rome, 20–23 Nov 1989. WMO/TD-No.377

WCRP (1984) Scientific plan for the WCRP. WCRP Publications Series, No.2. WMO/TD-No.6

WCRP (1985) First implementation plan for the world climate research programme. WCRP Publications Series, No.5. WMO/TD-No.80

WCRP-65 (1992) Report of the fifth session of the working group on sea ice and climate, Bremerhaven, 13–15 June 1991. WMO/TD-No.459

WCRP-72 (1992) Scientific concept of the Arctic Climate System Study (ACSYS). Report of the JSC study group on ACSYS, Bremerhaven, 10–12 June 1991 and London, 18–19 Nov 1991. WMO/TD-No.486

WCRP-85 (1994) Arctic Climate System Study (ACSYS) initial implementation plan.WMO/TD-No.627

WCRP-94 (1996) Proceedings of the ACSYS conference on the dynamics of the arctic climate system, Göteborg, 7–10 Nov 1994. WMO/TD No.760

Part I
Observations

Chapter 2
Advances in Arctic Atmospheric Research

James E. Overland and Mark C. Serreze

Abstract The previous decade and a half saw major advances in understanding of the Arctic atmosphere and the ability to project future climate states based on reanalysis datasets, field studies, and climate models. Limitations continue to be the lack of direct observations of the Arctic troposphere. The balance of evidence now argues for an anthropogenic component to Arctic change. Today, we see positive Arctic-wide temperature trends in all seasons with an Arctic amplification relative to lower latitude changes, but with strong regional modulations from natural variability. These include a positive index of the Arctic Oscillation (AO) in the early 1990s, a record negative phase of the AO during the winter of 2009/2010, and increased prominence of an Arctic Dipole (AD) climate pattern. The negative AO period showed linkages between Arctic and subarctic weather. Despite deficiencies in climate models used for the International Panel of Climate Change (IPCC), all models project increased temperatures and sea ice loss by mid-century, amplified through Arctic feedback processes.

Keywords Arctic atmosphere • Energy balance • Climate change • Arctic Oscillation

2.1 Introduction

The previous decade and a half (1995–2010) was an active period for Arctic research based on retrospective analyses, field studies, and the development and application of improved coupled atmosphere–sea ice–ocean global climate models for change

J.E. Overland (✉)
NOAA/Pacific Marine Environmental Laboratory, Seattle, WA 98115-6349, USA
e-mail: james.e.overland@noaa.gov

M.C. Serreze
CIRES/University of Colorado, Boulder, CO 80309-0449, USA

P. Lemke and H.-W. Jacobi (eds.), *Arctic Climate Change: The ACSYS Decade and Beyond,* Atmospheric and Oceanographic Sciences Library 43, DOI 10.1007/978-94-007-2027-5_2, © Springer (outside the USA) 2012

detection and projections. The past 15 years also saw major shifts in atmospheric climate patterns, with strengthening of the polar vortex between the 1980s and early 1990s linked to a positive trend in the index of the Arctic Oscillation (AO) (Serreze et al. 2000; Overland et al. 2004a), a record negative phase of the AO during the winter of 2009/2010, and increased prominence of the Arctic Dipole climate pattern (Wu et al. 2006). Along with increased recognition of the role of atmospheric variability and change in driving extreme seasonal minima in sea ice extent, such as the record September minimum observed in 2007 (Stroeve et al. 2008), there was increased awareness of impacts of sea ice loss on emerging Arctic amplification of air temperature changes and that sea ice loss may be affecting patterns of atmospheric circulation with potential impacts beyond the Arctic (Serreze et al. 2008; Francis et al. 2009; Budikova 2009; Screen and Simmonds 2010; Kumar et al. 2010; Deser et al. 2010; Overland and Wang 2010; Petoukhov and Semenov 2010). In this chapter, we discuss several trends in climate research from the ACSYS period to the present, rather than claiming to be comprehensive in all areas of Arctic meteorology. We refer readers to the primary journal sources for in-depth review.

One fascinating story of the last two decades was the rise and fall of the Arctic Oscillation as the key organizing principle for understanding variability in Arctic atmospheric circulation (Kerr 1999); the AO was defined as the first principal component of variability in monthly winter sea level pressure fields with the positive phase associated with anomalously low Arctic surface pressures and a strong cyclonic polar vortex. The AO turned from a predominantly negative phase in the 1970s to a strong positive phase in the first half of the 1990s, with some authors considering it to be a manifestation of global warming (Palmer 1999; Feldstein 2002), but since the late 1990s, the AO index has been variable, and it has often been replaced by another climate pattern, the Arctic Dipole (AD). Winter 2009–2010 had an extreme negative phase of the AO associated with a meridional flow pattern; evidence arose of linkages between Arctic climate and midlatitude weather (Cattiaux et al. 2010; Seager et al. 2010; Overland et al. 2011). A shortcoming of the AO framework is that it represents a single geographic center of variability, while Arctic circulation variability often shows more than one center of action at a given time.

From 1982 to1999, the Arctic Ocean became cloudier and warmed in spring, but cooled and become less cloudy in winter. By contrast, since the dawn of the new millennium, autumn has become the season with the strongest positive temperature anomalies. This is in large part a result of a retreat of summer sea ice extent, resulting in increased heat uptake in the ocean mixed layer in newly sea-ice-free ocean areas; this heat is then transferred back to the atmosphere in autumn (Serreze et al. 2008; Screen and Simmonds 2010; Deser et al. 2010). This process of Arctic amplification has long been predicted by global climate models. This warming, while strongest at the surface, has impacted atmospheric temperatures and geopotential heights well up into the troposphere (Overland and Wang 2010).

A yearlong surface energy budget and boundary-layer climate assessment over the sea ice of the western Arctic was established using observed values during the SHEBA (Surface Heat Budget of the Arctic) experiment from October 1997 to October 1998. Many of the SHEBA results were summarized in a special issue of *Journal of Geophysical Research.*

Reanalysis fields of atmospheric variables became standard tools of climate research. Reanalysis represents blending of a short-term model forecast with observations through a data assimilation process. Examples are tropospheric pressure heights and winds. Modeled or forecast variables, typically those at the surface (e.g., precipitation, radiation, and turbulent fluxes), are not directly influenced by observations of that variable. Research through previous decade focused primarily on the Reanalysis of the National Centers for Environmental Prediction/National Center for Atmospheric Research (NCEP/NCAR) and the European Center for Medium-Range Weather Forecasts (ECMWF). A shortcoming of these reanalysis products is the lack of direct tropospheric observations over the Arctic. Next-generation reanalyses include ERA-Interim and MERRA, the Modern Era Retrospective-Analysis for Research and Applications, a product of the National Aeronautics and Space Administration.

A major question as recently as 5 years ago was whether the recent Arctic warming was unique in the instrumental record and whether such a trend will continue. Arctic temperature anomalies in the 1930s and 1940s were as positive as those in the 1990s, and the events of the early 1990s are associated with a positive persistence of the AO, considered a manifestation of natural variability. Attribution of Arctic change is difficult because one is assessing changes relative to large natural variability in an energetic atmospheric circulation. In the last 5 years, however, there are both observational and model evidence for systematic changes in the Arctic. An observational surprise was the 39% reduction in sea ice extent in summer 2007 relative to climatology, with sea ice extents for 2007 through 2010 all below the two standard deviation level based on a 1979–2000 reference level. Other surprises include breakdown of the strong polar vortex (positive AO) in recent years replaced by more meridional (north-south) flow as indicated by an AD climate pattern in spring and summer, and the record negative phase of the AO and NAO value (a regional manifestation of the AO) in winter 20009–2010. Persistent trends in many Arctic variables, including sea ice extent, the timing of spring snow melt, increased shrubbiness in tundra regions, and ocean temperatures, as well as Arctic-wide increases in air temperatures, can no longer be associated only with dominant climate variability patterns. Arctic-wide warming is occurring twice as fast as most other regions on the planet (Fig. 2.1). The next sections discuss these results in more depth.

2.2 Energy Fluxes

Differential solar heating drives an equator to pole gradient in atmospheric temperature and hence pressure heights, driving the general circulation of the atmosphere which transports heat poleward. Differential heating also results in a poleward oceanic heat transport. The Arctic can be viewed as the Northern Hemisphere heat sink as part of the planetary heat engine. The large-scale energy budget of the Arctic was reviewed by Serreze et al. (2007), and the surface energy budget observed at SHEBA was described by Persson et al. (2002). The January

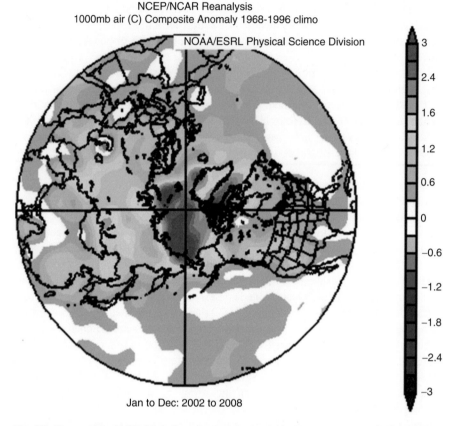

NCEP/NCAR Reanalysis
1000mb air (C) Composite Anomaly 1968-1996 climo

Jan to Dec: 2002 to 2008

Fig. 2.1 Near surface (1,000 hPa) air temperature anomaly multiyear composites (°C) for 2002–2009. Anomalies are relative to 1968–1996 mean and show a strong Arctic amplification of recent temperature trends (Data are from the NCEP–NCAR Reanalysis through the NOAA/Earth Systems Research Laboratory, generated online at www.cdc.noaa.gov)

and July energy budget of the polar cap, taken as the region poleward of 70°N, is shown in Fig. 2.2, using ERA-40 reanalysis data. At the top of the atmosphere, the longwave energy (LW) loss radiation is 230 W m⁻² in summer and 180 W m⁻² in winter, while the summer net shortwave (SW) radiation flux is 240 W m⁻² downward. The transfer of energy from the south into the Arctic (given by ∇•F) is about the same for summer and winter, 80–90 W m⁻². These radiative and advective transports give a net increase in energy storage of the Arctic Ocean mixed layer (the net surface heat flux) in July of 100 W m⁻² associated with sea ice melt and sensible heat gain, while in January, the surface has a net loss of 60 W m⁻² linked to ice growth and oceanic sensible heat loss. In winter, the radiative loss at the top of the atmosphere is about equal to the sum of the contributions from radiative loss at the surface and horizontal advection.

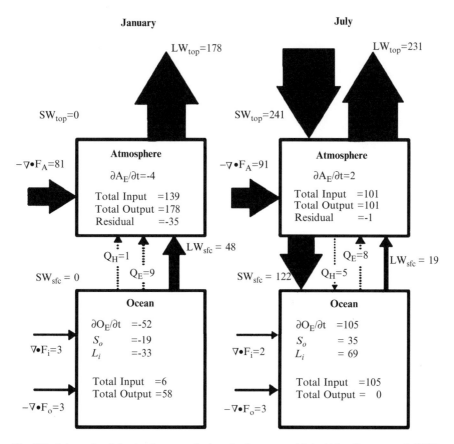

Fig. 2.2 Schematic of the Arctic energy budget for January and July (After Serreze et al. 2007). Symbols are defined in the text. Units are in W m^{-2}. The width of the arrows is proportional to the size of the transports

Measured energy fluxes at the surface in the Beaufort Sea for the SHEBA year are shown in Fig. 2.3. In Fig. 2.3a, F_{tot} is the total vertical heat flux and Q* is the sum of the longwave and shortwave contributions, separated out in Fig. 2.3b as Qs and Ql. Albedo is α, Hs and Hlb are sensible and latent heat flux, and C is the conductivity flux from the ocean through the sea ice to the surface. At the SHEBA site, the total surface flux varied from −25 to +12 W m^{-2} in winter to +37 to +129 W m^{-2} in July (Persson et al. 2002). Most variability is from changes in cloud cover. These mean observed flux magnitudes are considerably smaller than the Arctic-wide estimates by Serreze et al. (2007) in Fig. 2.2, but the Arctic-wide estimates include additional sensible heat flux from open water areas in the Atlantic sector north of 70°N. As in the Arctic-wide estimates, radiative terms dominate the surface energy budget in

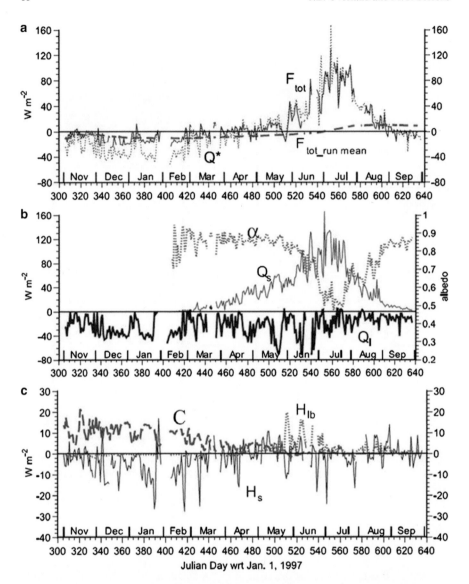

Fig. 2.3 Surface energy budget at the SHEBA camp site (After Persson et al. 2002). (**a**) Ftot is the total vertical heat flux, and Q* is total radiative flux. (**b**) Qs is the shortwave, and Ql is the long-wave contribution. Albedo is a. (**c**) Hs and Hlb are sensible and latent heat flux, and C is the conductivity flux from the ocean through the sea ice to the surface. The cumulative mean for Ftot starts on 1 November 1997

winter and summer; turbulent fluxes are five to ten times smaller in magnitude and are generally of opposing sign to the net radiation. By October, accumulation of excess energy, Ftot-run_mean, was equivalent to melting of nearly 1 m of sea ice.

2.3   The Arctic Oscillation, the PNA-Like, and Arctic Dipole Climate Patterns

The Arctic Oscillation (AO), also referred to as the northern annular mode (NAM), has traditionally been defined as the leading empirical orthogonal function of monthly values of wintertime sea level pressure poleward of 20°N (Thompson and Wallace 1998). This surface field is coupled to fluctuations at the 50 hPa level and can be interpreted as the surface signature of modulations in the strength of the polar vortex aloft. There was a positive trend for the AO for the 30 years prior to 1997 (Feldstein 2002). The AO pattern was first identified by Kutzbach (1970) and is closely related to the phase of the North Atlantic Oscillation (NAO) (Ambaum et al. 2001; Hurrell et al. 2003), the strength of the polar night stratospheric jet (PNJ) (Kuroda 2002), and the zonal index (ZI) of upper atmospheric winds (Li and Wang 2003). The latter two indices are more directly related to polar vortex winds, while the NAO is based on the difference in surface pressure over Iceland and the Azores. The monthly time series of the winter AO correlates with both a zonal index (50 hPa zonal wind at 65°N) at 0.6 and the NAO at 0.8.

A number of papers have addressed the physics of the AO and links with the NAO in detail. Comparing the winter average sea level pressure in the 1990s to the 1980s, Overland and Adams (2001) noted that the decadal differences were more symmetric with respect to the pole than suggested by the formal definition of the AO, and the negative lobe of the difference field was over the Mediterranean rather than over the Atlantic; they interpret these differences and the high AO–NAO correlation as due to the dominance of large intraseasonal variability in the Atlantic sector in the formal definition of the AO relative to interannual variability. Using rotated principal component analysis, Rogers and McHugh (2002) noted separate NAO- and AO-like modes in spring through fall with the AO-like mode having high sea level pressure variability over the Kara and Laptev Seas, while in winter, there was one combined mode associated with the strength of southeast winds over the north Atlantic. Monahan et al. (2003) used a nonlinear principal component technique which relaxes the assumption that the spatial distributions of positive and negative versions of the same mode are exact opposites; thus, while the positive AO simply represents a strong polar vortex, the negative value represents a breakdown of the polar vortex with more meridional and spatially variable flow patterns, *not a mirror image of the positive AO phase*. A striking example is December 2009 which saw the most negative NAO phase of the 145-year winter (DJF) record (Fig. 2.4). Following contours of 850 mb geopotential height (approximate streamlines), we see a link between southward flow over the Beaufort Sea and low temperatures over the central United States. There are also northerly winds over eastern Asia, and the North Atlantic wind jet is far south of its normal winter position allowing cold temperatures to develop in northern Europe. Thus, the single AO index may not be representative of the full range of changes occurring in the Arctic.

An important paper is Quadrelli and Wallace (2004) which considered projections of major climate patterns onto the phase space formed by the first two principal

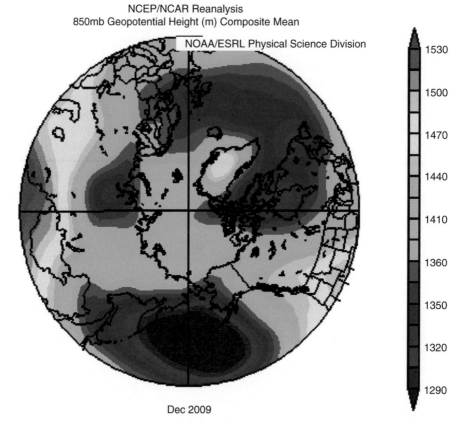

NCEP/NCAR Reanalysis
850mb Geopotential Height (m) Composite Mean

Dec 2009

Fig. 2.4 The observed 850 mb geopotential height field in December 2009. Note the near reversal of the normal pattern with the heights over the Arctic Ocean exceeding those at lower latitudes. Air streamlines follow the height contours, showing a connection between the Beaufort Sea region and the eastern United States and that the westerly flow into Europe is strongly displaced to the south. The normal region of the Icelandic Low show high heights (Data are from the NCEP–NCAR Reanalysis through the NOAA/Earth Systems Research Laboratory, generated online at www.cdc. noaa.gov)

components of sea level pressure. The first component (X-axis) is the AO by definition, and the orthogonal second component represents variability in the North Pacific which the authors refer to as a Pacific North American (PNA)-like pattern. These capture the two major patterns of variability for the Northern Hemisphere and represent about 35% of the month-to-month winter variance.

Other related patterns such as the NAO and stratospheric flow also represent pairs of basis functions, rotated slightly from the AO and PNA-like axes. The key result from their study is that there are primarily two patterns of variability for Northern Hemispheric flow. The discouraging conclusion is that over 50% of the Northern Hemispheric variability can be considered primarily chaotic in nature.

The nature of Arctic flow patterns in spring and summer appear to have changed in the last decade (Overland and Wang 2005, 2010; Wu et al. 2006; Maslanik et al. 2007; L'Heureux et al. 2008; Zhang et al. 2008). There is a persistent appearance of the Arctic Dipole (AD) pattern with pressure anomaly centers over the northern Beaufort Sea and north central Eurasia. Skeie (2000) also discussed this meridional flow pattern which he named the Barents Oscillation. The AD relates to the third principal component pattern in the Quadrelli/Wallace system based on data north of 20°N, or the second principal component based on data north of 70°N where the PNA climate pattern does not project. The AD pattern shifted from primarily small interannual variability to a persistent phase during spring (April through June) beginning in 1997 (except for 2006) and extending to summer (July through September) beginning in 2005. Climatologically, the sea level pressure pattern in summer shows a weak mean low over the central Arctic Ocean, but from year 2007 to 2010, the AD pattern has been prominent, contributing to sea ice loss in the northern Chukchi Sea.

Conceptual interpretations of the AO become even more complex when one begins to consider possible physical mechanisms for interannual to decadal variability. Based on the polar vortex interpretation of the AO, there are feedback interactions between the time mean flow (zonal wind) and transient eddies (De Weaver and Nigam 2000). This interpretation argues for a large climate noise paradigm for the process. Further mechanisms involve the effect of stratospheric flow on the refraction of planetary waves dispersed upward from the troposphere (Limpasuvan and Hartmann 2000; Ambaum and Hoskins 2002). Particularly with respect to the NAO, sea temperatures in the North Atlantic are thought to play a role, as deep mixing can provide multiyear memory to the climate system (Peng et al. 2003). Loss of sea ice is mentioned (Seierstad and Bader 2008). North–south differences in hemispheric radiative processes have been shown to couple the Arctic to the subtropics, particularly through stratospheric connections. This is supported both by data (Robock and Mao 1992) and modeling studies (Stenchikov et al. 2002; Gillett et al. 2003). Hoerling et al. (2001) argue for a tropical SST–NAO connection.

The negative AO/NAO phase during winter 2009/2010 was so extreme that one might speculate that it was not just a random occurrence but also involved recent shifts in Arctic conditions as a forcing to the overall atmospheric circulation pattern.

2.4 Warm Temperatures of the 1930s and 2000s

There was considerable discussion of the warm temperature anomalies that occurred in the Arctic in the 1930s and 1940s for placing the strong warming since the 1990s into context. The early twentieth century warm period, while reflected in the global average air temperature record, did not appear at midlatitudes nor on the Pacific side of the Arctic. By contrast, recent warm anomalies include midlatitudes and are Arctic-wide. Polyakov et al. (2003) argued that the Arctic record manifests a natural 50-year cycle in Arctic temperatures, that the warming of the 1990s is not unique, and that the

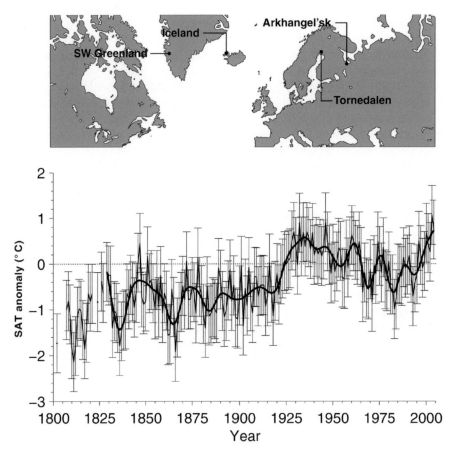

Fig. 2.5 Extended annual mean SAT record for the Atlantic–Arctic boundary region (T_{NA}). Confidence limits of 95% are shown. Decadal-scale variations are emphasized with a two-way Butterworth low-pass filter constructed to remove frequencies higher than 0.1 cycles per year (*black line*). The general regions represented by station-based composite SAT records used to compute T_{NA} are indicated in the map (Wood et al. 2010)

Arctic climate should swing toward a cool regime over the next decade. Bengtsson et al. (2004), Overland et al. (2004b), Grant et al. (2009), Wood and Overland (2010), and Wood et al. (2010) instead argue for the uniqueness of the 1930s event based on natural variability in North Atlantic atmospheric circulation (Fig. 2.5).

Of interest is that the strong Arctic warming linked to summer sea ice loss, especially since 2007, is influencing midtropospheric air temperatures in the following autumn (Schweiger et al. 2008; Serreze et al. 2008). Increased Arctic temperatures has the effects of weakening polar vortex through the thermal wind effect (Overland and Wang 2010). Seierstad and Bader (2008) and Francis et al. (2009) provide evidence that autumns following low summer sea ice extent tend to exhibit atmospheric anomalies resembling the negative phase of the NAO.

2.5 Recent Climate Change

Climate models are not able to reproduce the recent Arctic warming without the inclusion of anthropogenic radiative forcing (Stott et al. 2001; Johannessen et al. 2004; IPCC 2007). Instrumental and proxy data from the 1800s also do not show a strong 50- to 80-year cycle, although the available data are sparse (Wood et al. 2010). Thus, we have two competing views for the future of the Arctic. One is the cyclic view based on data fitting, and the other is a continued warming based on physical reasoning and model projections. While natural variability will always be a large part of the picture, the balance of the evidence argues for a major anthropogenic component to Arctic change based on the application of methodological, evidentiary, and performance scientific standards (Overland 2009). The most viable interpretation is that we are beginning to see a warming signal emerging from the background of large natural variability in the Arctic. There are large changes occurring in the Arctic across physical and biological systems. Latest conditions are summarized in the NOAA Arctic Report Card (http://www.arctic.noaa.gov/reportcard/). While some large climate signals going back to the early twentieth century can be explained by natural variations in climate patterns such as the AO, the impact of these patterns is regional. Today, we see Arctic-wide trends, with a continued superposition of influences from these regional patterns.

Global climate models project that the Arctic will warm at a greater rate over the coming decades and century compared with other regions of the planet (Arctic amplification), but there is still much uncertainty. Fourteen global climate models were used in IPCC AR-4 (2007) to project warming for the Arctic beyond the year 2030 for each of three widely adopted emission scenarios (labeled B1, A1B, and A2 according to IPCC terminology), although interannual variability of the surface air temperatures in the Arctic is large especially over land (Chapman and Walsh 2007). By the end of the twenty-first century, annual mean temperature changes projected by different models for the B1 emission scenario range from +1°C to +5.5°C, the A1B range is from +2.5°C to +7.0°C, while the A2 range is from +4.0°C to +9.0°C. Differences between the three different emission scenarios are small in the first half of the twenty-first century, but increased toward the end of the twenty-first century. Unlike midlatitudes, warming rates in the Arctic are smaller over land and largest over the Arctic Ocean, the latter linked strongly to sea ice loss. The greatest annual warming over land is near the Barents Sea (~6.5°C for the middle-of-the-road A1B scenario). All models project the largest warming in autumn and winter, although the rates vary considerably among the models and across emission scenarios. Walsh et al. (2008) evaluated the performance of 15 IPCC (2007) models and found a tendency for models with smaller errors to simulate larger warming and larger increases of precipitation over the Arctic (60–90°N).

Composite mean changes in seasonal temperature for the late twenty-first century from the IPCC AR-4 models are shown in Fig. 2.6. Temperature increases dominate in all seasons, and the warming is amplified with respect to the globe as a whole in all seasons except in summer when the melting "ice bath" of the Arctic

Projected changes of temperature: 2070-2090

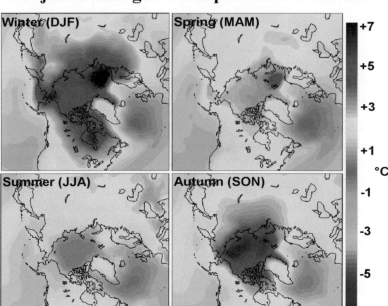

Fig. 2.6 Projected changes of temperature by season for the 2070–2090 time slice. Changes are composited over models participating in the IPCC-ARC forced by the A1B scenario. Outlier models were excluded (After Chapman and Walsh 2007)

Ocean constrains the surface air temperature. The warming is largest over the Arctic in autumn and winter (3–5°C by 2080), consistent with a loss of summer sea ice, greater absorption of solar radiation during summer, and release of this additional heat to the atmosphere during the cold season. Especially during winter, the spatial pattern bears the signature of sea ice loss, with local maxima of 6–7°C. Over much of the northern land areas, the warming is between 2°C and 3°C. The projections in Fig. 2.6 are based on the A1B, a middle-of-the-road scenario of greenhouse gas forcing. The corresponding spatial patterns for the A2 and B1 scenarios are similar, but the magnitudes of the warming are 30–50% larger (smaller) in the A2 (B1) scenario.

It is difficult to sort out cause and effect between loss of sea ice and increased air temperatures. Wang and Overland (2009) show that the better-performing models show a greater reduction rate in summer sea ice loss. Further, the observed extreme September sea ice minima of 2007–2010 argues that the faster sea ice loss ensemble members should be favored (Holland et al. 2008), giving a nearly sea-ice-free September by midcentury or perhaps even sooner.

Kattsov et al. (2007) analyzed precipitation changes for 2041–2060 and 2080–2099 relative to 1980–1999 data for a subset (13) of 21 of the IPCC AR4 models forced by emission scenarios B1, A1B, and A2. In all models and scenarios, precipitation increases throughout the Arctic through the twenty-first century, showing larger percentage increases than in the global mean precipitation, and with distinct regional patterns. Percentage increases are most pronounced over northeast Greenland followed by coastal Siberia and the Canadian Arctic Archipelago. Percentage increases projected by 2080–2099 vary across the Arctic from 5–40% (B1 scenario) to 5–70% (A2 scenario), with the largest increases projected for northeast Greenland. The Arctic precipitation changes have a pronounced seasonality, with the largest relative increases in winter and fall and the smallest in summer. It should be emphasized that the projected increases of precipitation, which are typically 1–2 cm per season in the Arctic, do not preclude drying of Arctic terrestrial regions. As shown by the IPCC (2007, Figure 10.3), increases of precipitation over Arctic land areas are accompanied by increases in evapotranspiration and runoff under greenhouse scenarios, resulting in decreases of soil moisture.

Serreze and Francis (2006) and Miller et al. (2010) argue that sea ice–albedo insolation feedback and sea ice insulation feedback lead to greater sensitivity of the response of the Arctic to greenhouse gases compared to lower latitudes, especially in autumn and winter. The Arctic is impacted by both with-Arctic feedbacks and forcing from lower latitudes (Döscher et al. 2009). The wide range in model projections for the next few decades points to an uncertain future, with natural and forced changes in atmospheric circulation operating together (Wang et al. 2009).

References

Ambaum MHP, Hoskins BJ (2002) The NAO troposphere-stratosphere connection. J Clim 15:1969–1978

Ambaum MHP, Hoskins BJ, Stephenson DB (2001) Arctic oscillation or north Atlantic oscillation? J Clim 14:3495–3507

Bengtsson L, Semenov V, Johannessen OM (2004) The early 20th century warming in the Arctic—a possible mechanism. J Clim 17:4045–4057

Budikova D (2009) Role of Arctic sea ice in global atmospheric circulation: a review. Glob Planet Chang 68:149–163

Cattiaux J, Vautard R, Cassou C, Yiou P, Masson-Delmotte V, Codron F (2010) Winter 2010 in Europe: a cold extreme in a warming climate. Geophys Res Lett 37:L20704. doi:10.1029/2010GL044613

Chapman WL, Walsh JE (2007) Simulations of Arctic temperature and pressure by global coupled models. J Clim 20:609–632

De Weaver E, Nigam S (2000) Zonal-eddy dynamics of the NAO. J Clim 13:3893–3914

Deser C, Tomas R, Alexander M, Lawrence D (2010) The seasonal atmospheric response to projected Arctic sea ice loss in the late 21st century. J Clim 23:333–351. doi:10.1175/2009JCLI3053.1

Döscher R, Wyser K, Meier M, Qian R, Redler G (2009) Quantifying Arctic contributions to climate predictability in a regional coupled ocean-ice-atmosphere model. Clim Dyn 34:1157–1176

Feldstein SB (2002) The recent trend and variance increases of the annular mode. J Clim 15:88–94

Francis JA, Chan W-H, Leathers DJ, Miller JR, Veron DE (2009) Winter northern hemisphere weather patterns remember summer. Geophys Res Lett 36:L07503. doi:10.1029/2009GL037274

Gillett NP, Zwiers FW, Weaver AJ, Stott PA (2003) Detection of human influence on sea-level pressure. Nature 422:292–294

Grant AN, Brönnimann S, Ewen T, Griesser T, Stickler A (2009) The early twentieth century warming period in the European Arctic. Meteorol Z 18:425–432

Hoerling MP, Hurrell JW, Xu T (2001) Tropical origins for recent north Atlantic climate change. Science 292:90–92

Holland MM, Bitz CM, Tremblay B, Bailey DA (2008) The role of natural versus forced change in future rapid summer Arctic ice loss. In: Arctic sea ice decline: observations, projections, mechanisms, and implications, Geophysical monograph Series 180. AGU, Washington, DC

Hurrell JW, Kushnir Y, Ottersen G, Visbeck M (2003) The north Atlantic oscillation: climate significance and environmental impact, Geophysical monograph series 134. AGU, Washington, DC, 279 pp

IPCC (2007) Climate change 2007: the physical science basis. In: Solomon S et al (eds) Contribution of working group I to the fourth assessment report of the Intergovernmental Panel on Climate Change. Cambridge University Press, Cambridge/New York, 996 pp

Johannessen OM, Bengtsson L, Miles MW, Kuzmina SI, Semenov VA, Alekseev GV, Nagurnyi AP, Zakharov VF, Bobylev LP, Pettersson LH, Hasselmann K, Cattle HP (2004) Arctic climate change: observed and modeled temperature and sea-ice variability. Tellus 56A:328–341

Kattsov VM, Walsh JE, Chapman WL, Govorkova VA, Pavlova TV, Zhang X (2007) Simulation and projection of Arctic freshwater budget components by the IPCC AR4 global climate models. J Hydrometeorol 8:571–589

Kerr RA (1999) A new force in high-latitude climate. Science 284:241–242

Kumar A, Perlwitz J, Eischeid J, Quan X, Xu T, Zhang T, Hoerling M, Jha B, Wang W (2010) Contribution of sea ice loss to Arctic amplification. Geophys Res Lett 37:L21701. doi:10.1029/2010GL045022

Kuroda Y (2002) Relationship between the polar-night jet oscillation and the annular mode. Geophys Res Lett 29:1240. doi:10.1029/2001GL013933

Kutzbach J (1970) Large-scale features of monthly mean Northern Hemisphere anomaly maps of sea-level pressure. Mon Weather Rev 98:708–716

L'Heureux ML, Kumar A, Bell GD, Halpert MS, Higgins RW (2008) Role of the Pacific-North American (PNA) pattern in the 2007 Arctic sea ice decline. Geophys Res Lett 35:L20701. doi:10.1029/2008GL035205

Li J, Wang JXL (2003) A modified zonal index and its physical sense. Geophys Res Lett 30(12):1632. doi:10.1029/2003GL017130

Limpasuvan V, Hartmann DL (2000) Wave maintained annular modes of climate variability. J Clim 13:4414–4429

Maslanik J, Drobot S, Fowler C, Emery W, Barry R (2007) On the Arctic climate paradox and the continuing role of atmospheric circulation in affecting sea ice conditions. Geophys Res Lett 34:L03711. doi:10.1029/2006GL028269

Miller GH, Alley RB, Brigham-Grette J, Fitzpatrick JJ, Polyak L, Serreze M, White JWC (2010) Arctic amplification: can the past constrain the future? Quat Sci Rev 29:1779–1790

Monohan AH, Fyfe JC, Pandolfo L (2003) The vertical structure of wintertime climate regimes of the northern hemisphere extratropical atmosphere. J Clim 16:2005–2021

Overland JE (2009) The case for global warming in the Arctic. In: Nihoul JCJ, Kostianoy AG (eds) Influence of climate change on the changing Arctic and Sub-Arctic conditions. Springer, Dordrecht, pp 13–23

Overland JE, Adams JM (2001) On the temporal character and regionality of the Arctic oscillation. Geophys Res Lett 28:2811–2814

Overland JE, Wang M (2005) The third Arctic climate pattern: 1930s and early 2000s. Geophys Res Lett 32:L23808. doi:10.1029/2005GL024254

Overland JE, Wang M (2010) Large-scale atmospheric circulation changes are associated with the recent loss of Arctic sea ice. Tellus 62A:1–9

Overland JE, Spillane MC, Soreide NN (2004a) Integrated analysis of physical and biological pan-Arctic change. Clim Chang 63:291–322

Overland JE, Spillane MC, Percival DB, Wang M, Mofjeld HO (2004b) Seasonal and regional variation of pan-Arctic surface air temperature over the instrumental record. J Clim 17:3263–3282

Overland JE, Wood KR, Wang M (2011) Warm Arctic-cold continents: climate impacts of the newly open Arctic Sea. Polar Res (in press)

Palmer TN (1999) A nonlinear dynamical perspective on climate prediction. J Clim 12:575–591

Peng MS, Robinson WA, Li S (2003) Mechanism for the NAO responses to the North Atlantic SST tripole. J Clim 16:1987–2004

Persson POG, Fairall CW, Andreas EL, Guest PS, Perovich DK (2002) Measurements near the atmospheric surface flux group tower at SHEBA: near-shore conditions and surface energy budget. J Geophys Res 107(C10):8045. doi:10.1029/2000JC000705

Petoukhov V, Semenov VA (2010) A link between reduced Barents-Kara sea ice and cold winter extremes over northern continents. J Geophys Res 115:D21111. doi:10.1029/2009JD013568

Polyakov IV, Bekryaev RV, Alekseev GV, Bhatt U, Colony RL, Johnson MA, Makshtas AP, Walsh D (2003) Variability and trends of air temperature and pressure in the maritime Arctic. J Clim 16:2067–2077

Quadrelli R, Wallace JM (2004) A simplified linear framework for interpreting patterns of northern Hemisphere wintertime climate variability. J Clim 17:3728–3744

Robock A, Mao J (1992) The volcanic signal in surface temperature observations. J Clim 8:1086–1103

Rogers JC, McHugh MJ (2002) On the separability of the north Atlantic oscillation and the Arctic oscillation. Clim Dyn 19:599–608

Schweiger AJ, Lindsay RW, Vavrus S, Francis JA (2008) Relationships between Arctic sea ice and clouds during autumn. J Clim 21:4799–4810

Screen JA, Simmonds I (2010) The central role of diminishing sea ice in recent Arctic temperature amplification. Nature 464:1334–1337

Seager R, Kushnir Y, Nakamura J, Ting M, Naik N (2010) Northern Hemisphere winter snow anomalies: ENSO, NAO and the winter of 2009/10. Geophys Res Lett 37:L14703. doi:10.1029/2010GL043830

Seierstad IA, Bader J (2008) Impact of a projected future Arctic sea ice reduction on extratropical storminess and the NAO. Clim Dyn. doi:10.1007/s00382-008-0463-x

Serreze MC, Francis JA (2006) The Arctic amplification debate. Clim Chang 76:241–264. doi:1007/s10584-005-9017-y

Serreze MC, Walsh JE, Chapin FS III, Osterkamp T, Dyurgerov M, Romanovsky V, Oechel WC, Morison J, Zhang T, Barry RG (2000) Observational evidence of recent change in the northern high-latitude environment. Clim Chang 46(1–2):159–207

Serreze MC, Barrett AP, Slater AG, Steele M, Zhang J, Trenberth KE (2007) The large-scale energy budget of the Arctic. J Geophys Res 112:D11122. doi:10.1029/2006JD008230

Serreze MC, Barrett AP, Stroeve JC, Kindig DN, Holland MM (2008) The emergence of surface-based Arctic amplification. The Cryosphere 2:601–622

Skeie P (2000) Meridional flow variability over the Nordic seas in the Arctic oscillation framework. Geophys Res Lett 27:2569–2572

Stenchikov G, Robock A, Ramaswamy V, Schwarzkopf MD, Hamilton K, Ramachandran S (2002) Arctic Oscillation response to the 1991 Mount Pinatubo eruption: effects of the volcanic aerosols and ozone depletion. J Geophys Res 107. doi:10.1029/2002JD002090

Stott PA, Tett SFB, Jones GS, Allen MR, Ingram WJ, Mitchell JFB (2001) Attribution of twentieth century temperature change to natural and anthropogenic causes. Clim Dyn 17:1–21

Stroeve J, Serreze M, Drobot S, Gearheard S, Holland M, Maslanik J, Meier W, Scambos T (2008) Arctic sea ice extent plummets in 2007. EOS Trans Am Geophys Union 89:13–14

Thompson DWJ, Wallace JM (1998) The Arctic oscillation signature in the wintertime geopotential height and temperature fields. Geophys Res Lett 25:1297–1300

Walsh JE, Chapman WL, Romanovsky V, Christensen JH, Stendel M (2008) Global climate model performance over Alaska and Greenland. J Clim 21:6156–6174

Wang M, Overland JE (2009) A sea ice free summer Arctic within 30 years? Geophys Res Lett 36:L07502. doi:10.1029/2009GL037820

Wang J, Zhang J, Watanabe E, Ikeda M, Mizobata K, Walsh JE, Bai X, Wu B (2009) Is the dipole anomaly a major driver to record lows in Arctic summer sea ice extent? Geophys Res Lett 36:L05706. doi:10.1029/2008GL036706

Wood KR, Overland JE (2010) Early 20th century Arctic warming in retrospect. Int J Climatol. doi:10.1002/joc.1973

Wood KR, Overland JE, Jónsson T, Smoliak BV (2010) Air temperature variations on the Atlantic–Arctic boundary since 1802. Geophys Res Lett 3:L17708. doi:10.1029/2010GL044176

Wu B, Wang J, Walsh JE (2006) Dipole anomaly in the winter Arctic atmosphere and its association with sea ice motion. J Clim 19:210–225

Zhang X, Sorteberg A, Zhang J, Gerdes R, Comiso J (2008) Recent radical shifts in atmospheric circulations and rapid changes in Arctic climate system. Geophys Res Lett 35:L22701. doi:10.1029/2008GL035607

Chapter 3
Sea-Ice Observation: Advances and Challenges

Humfrey Melling

Abstract The presence and characteristics of the relatively thin ice cover of Earth's oceans at high latitude are important determinants of the polar climate and of the polar role in the global climate system. The focus of this chapter is the observation of sea ice. Advances in polar ocean science have been critically dependent on technology to provide the means for frequent and detailed observation of sea ice. The Arctic Climate System Study was an opportunity for a productive coordinated application of existing technology to the study of the marine cryosphere and provided impetus for the development and evaluation of new techniques. The chapter begins with a brief summary of the knowledge of global sea ice in the early 1990s and of the contemporary capabilities for sea ice observation. This introductory section is as a back drop against which to view the continued improvements in observational capability and knowledge associated with ACSYS during the 1990s. The chapter includes discussion of new developments in observational technology, of new activity in ice research and reconnaissance and of new understanding of sea ice. It has been organized around the structure of the original ACSYS science plan which had six process-oriented elements – sea ice extent and concentration, drift, thickness, export to temperate oceans, atmosphere-ice-ocean interaction, sea-ice mechanics – and one geographic element – sea ice of the southern hemisphere. The chapter concludes with a list tangible deliverables from the Arctic Climate System Study in relation to sea ice and a discussion of some key tasks for the future. These topics remain highly relevant in view of the present continued rapid rate of change in polar climate.

Keywords Ablation • Antarctic • Arctic • Climate • Drift • Extent • Freezing • Ice • Leads • Ocean • Ridges • Thickness

H. Melling (✉)
Institute of Ocean Sciences, Fisheries and Oceans Canada, Sidney, BC, Canada, V8L 4B2
e-mail: Humfrey.Melling@dfo-mpo.gc.ca

P. Lemke and H.-W. Jacobi (eds.), *Arctic Climate Change: The ACSYS Decade and Beyond,* Atmospheric and Oceanographic Sciences Library 43, DOI 10.1007/978-94-007-2027-5_3, © All Rights Reserved 2012

3.1   The Arctic Marine Cryosphere in the Context of Climate

The permanent cover of pack ice is a remarkable feature of the central Arctic Ocean. This thin solid layer dramatically modifies the characteristics of the ocean surface, thereby influencing the absorption and reflection of sunlight and the exchanges of heat, moisture and momentum between the ocean and the atmosphere. Sea ice influences the global heat balance via these mechanisms. The annual cycle of sea-ice growth and ablation drives a downward flux of salt, causing freshening of surface waters. The resulting low salinity of Arctic outflows has impact on the thermohaline circulation of the global ocean. The presence and character of sea ice is also an important determinant of ice-adapted oceanic ecosystems and of human activities in northern seas.

The ice cover represents the net balance, perhaps delicate, of fluxes of energy through the ocean's surface. Change in other components of the climate system, atmospheric temperature for example, could alter the present equilibrium in high-latitude oceans so as to establish a different sea ice regime. Thus, sea ice is sensitive to change in local climate as well as being an element in the mediation of such change globally.

There is potential for positive feedback via sea ice in the global budget of atmospheric heat: higher air temperature might plausibly cause a reduction in the area covered by sea ice, which would in turn allow increased absorption of solar radiation within the polar seas; warmer polar seas might further increase the melting of ice via an unstable interaction known as the albedo/temperature feedback loop. In climate models, this loop underlies the prediction that the largest temperature change from increasing atmospheric CO_2 will occur at high latitude. This destabilizing feedback is counteracted to an unknown degree by stabilizing effects of high-albedo cloud and of infrared back-radiation from the ice (or sea) surface to space; according to the Stefan-Boltzmann equation, the rate of radiative heat loss from the surface increases very rapidly with increasing temperature.

In a warming Arctic, the albedo-temperature feedback becomes very strong as the mean surface air temperature approaches the freezing point, whereas the stabilizing infrared back-radiation has no special sensitivity to this temperature. Therefore, at some point during climate warming, a transition to an Arctic Ocean which is ice-free in summer, and perhaps even in the winter, could take place.

In the early 1990s prior to ACSYS, a conceptual appreciation of these issues focused scientific interest on the stability of the Arctic marine cryosphere. It was acknowledged that predictions of future Arctic ice conditions would be credible only if knowledge of pack ice and its interactions with ocean and atmosphere were improved. At this time, observations of the Arctic marine cryosphere suitable for the evaluation of predictions from computer models even over short time scales (seasons, years, decades) did not exist. For example, the available ice-thickness data in 1990 were barely adequate to sketch in the spatial distribution of ice volume within the central Arctic. They were completely inadequate to define the total volume, its inter-annual or decadal variations, annual ice production or annual export through connecting passages to the World Ocean. Comparable knowledge of Antarctic pack ice was merely a dream.

3.2   Objectives for the Arctic Climate System Study

The Arctic Climate System Study adopted two principal objectives pertaining to sea ice, namely to document the present 'state' of the Arctic pack and to study the feedback between sea ice and other elements of the climate system.

Four elements of the implementation strategy for ACSYS were dependent on observations of sea ice: (1) preparation of climatology for sea ice, (2) monitoring of the export of sea ice to temperate oceans, (3) study of ice-ocean-atmosphere interaction and (4) study of the physical processes that determine the spatial variation in ice thickness through freezing, deformation and melting.

Although ACSYS was primarily an Arctic study, the sea-ice component also encompassed the Southern Ocean through its coordination of two Antarctic studies, the International Programme for Antarctic Buoys (IPAB) and the Antarctic Sea Ice Thickness Programme (AnSITP).

Long-term observations of sea-ice extent, concentration, thickness and velocity are needed to study natural (seasonal, inter-annual and decadal) variation in the marine cryosphere and to detect anthropogenic influence masked by this variation. Such data provide benchmarks for the evaluation and improvement of sea-ice models. ACSYS has fostered the compilation of a global climatology for sea ice based upon historical observations and new observing initiatives.

The flux of pack ice into warmer waters is an important climatic variable, as a measure of the net production of sea ice and as a component of the freshwater supply to the convective gyres of the North Atlantic. The export of ice from polar seas is an effective constraint on model simulations of high-latitude climate. Ice leaves the Arctic via openings in Fram Strait, the Barents Sea, the Canadian Arctic Archipelago and Bering Strait. ACSYS has encouraged initiatives to measure the export of pack ice from the Arctic.

Clouds, the atmospheric and oceanic boundary layers and sea ice interact within a complex coupled system. A much-improved understanding of this system's adjustment to external perturbation is a prerequisite for its accurate representation in models. Study of the processes by which pack ice, ocean and atmosphere interact, and the parameterization of these processes at the spatial resolution (100 km) of present global climate models, has been fundamental to the legacy of ACSYS.

A theoretical framework for the redistribution of ice thickness during plastic deformation, freezing and melting and for the relationship between stress and strain in deforming pack ice was developed in the late 1970s. An accurate representation of pack ice in global climate models awaits substantiation of this theory. Observations that elucidate the evolution of the ice-thickness distribution are critically needed. ACSYS has promoted the study of the sea-ice mechanics at spatial scales (greater than 1 km) relevant to its rheological behaviour in a geophysical context.

3.3   Knowledge Base for the Arctic Climate System Study

3.3.1   *Extent and Concentration*

Encounters with sea ice have been recorded by mariners since the earliest days of polar exploration in the seventeenth century. Because the ships were generally involved in the pursuit of fish and marine mammals, the information in logbooks varies geographically and over time, in response to changing abundance of the resource and changing economic conditions. Mariner's reports provide information on the location of the ice edge, but little concerning conditions within the interior of the pack. Since logbooks are widely dispersed, few had been compiled when ACSYS was being planned.

With the increased pace of exploration, commercial marine traffic and global military activity in the mid-twentieth century, reconnaissance of ice from aircraft in tactical support of navigation became routine. Charts of ice extent were prepared during the shipping season, but again the information primarily concerns the pack-ice margin.

Only with the advent of meteorological satellites in the late 1960s was ice reconnaissance practical year-round and in remote areas. Navigational support remained the primary purpose of detailed ice charting, but satellites permitted the compilation of information over wide expanses of pack ice far from the ice edge. Prior to ACSYS, national collections of charts from the early satellite era were available only on paper or microfilm. Their scientific use was tedious and labour intensive.

Fifteen years before the start of ACSYS, NASA launched the Nimbus-5 satellite carrying the first microwave imaging sensor for routine sea-ice surveillance, the electrically scanning microwave radiometer (ESMR). The all-weather, wide-swath sensor permitted the estimation of sea-ice type and extent throughout both hemispheres at roughly 3-day intervals. Successive generations of this new technology (ESMR, SMMR, SSM/I, AMSR) have revolutionized knowledge of the marine cryosphere.

Measurements of emission in the microwave band permit sea-surface classification as open water (less than a few percent of ice by WMO definition), first-year ice and old ice using a combination of polarization and spectral gradient information. However, second-order effects influence the retrieval of ice parameters (Oelke 1997). For example, the 'open water' fraction determined by microwave emission can actually be a mixture of ice-free water and thin ice, or be thick ice dotted with meltwater ponds. Variation in wave breaking with wind blurs the open-water tie point. Snow on first-year ice can bias the spectral gradient between the high and low frequency bands to indicate an erroneous presence of multi-year ice. Emission from water vapour and droplets in the atmosphere affects the retrieval of both ice concentration and ice type information. Such errors in the algorithms for ice concentration and type can be reduced if microwave analysis is constrained from other sources (e.g. visual and thermal band scanners on satellites, radar on satellites and aircraft, etc.). National ice services have used this multi-sensor approach to charting in varying degrees since 1978.

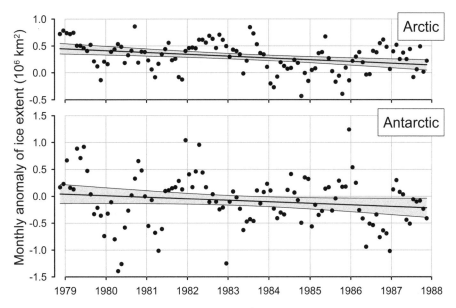

Fig. 3.1 Variations in the extent of sea ice in the northern and southern hemispheres determined by NASA's scanning multi-channel microwave radiometer (SMMR) during the 1980s. The 2.4% decline in the Arctic between late 1978 and late 1987 relative to 12.4×106 km² is statistically significant at the 95% level. That in the Antarctic (annual mean 11.9×106 km²) is indistinguishable from zero with such confidence (Update of Gloersen and Campbell (1991) using ftp://sidads.colorado.edu/DATASETS/NOAA/G02135)

Over the annual cycle, the area of Arctic ice fluctuates between about 6 and 14×10^6 km² (Gloersen and Campbell 1991). The inter-annual variability in the maximum and minimum extent is about 1×10^6 km². Parkinson and Cavalieri (1989) have estimated that the positional uncertainty in the ice edge from SMMR is about 30 km (1 pixel). Steffen and Schweiger (1991) estimate the accuracy of Arctic total ice-covered area based on microwave emission to be about 0.3×10^6 km² (3%). But this accuracy varies seasonally with the temperature and wetness of the ice surface, snow cover, atmospheric moisture, etc. The precision of estimates of ice-covered area was thought to be better, perhaps 0.1×10^6 km² (Gloersen and Campbell 1991). These numbers imply that the seasonal cycle in the extent of sea ice can be measured quite accurately, and that year-to-year variations can be resolved to one significant figure. This capability was the best available for monitoring polar ice fields in 1990.

By the late 1980s, the time series from passive microwave surveillance was long enough for a meaningful analysis of seasonal, regional and inter-annual variability of Arctic ice extent (Parkinson and Cavalieri 1989; Gloersen and Campbell 1991; see Fig. 3.1). Appreciable differences in variability between regions were noted (Parkinson 1991). The data set also lent itself to the study of variation in other measures of the marine cryosphere, such as the length of the ice season (Parkinson 1992b).

3.3.2   Drift

Early understanding of the drift of Arctic ice emerged from the discovery of identifiable flotsam on northern beaches and at sea: Belcher's ship *HMS Resolute*, which was abandoned in Viscount Melville Sound in the spring of 1854 and discovered adrift in Davis Strait; wreckage found off the southern tip of Greenland which was traced to the *Jeannette*, trapped in ice near Wrangell Island in 1881; driftwood on the beaches of Svalbard, Greenland and the Canadian Arctic Archipelago, traced to the mouths of rivers draining Siberia and Canada. The principal features of Arctic circulation, the Trans-polar drift exiting at Fram Strait and the Beaufort gyre, were identified from the tracking of drifting scientific expeditions, the ships *Fram* in 1893–1896, St Anna in 1912–1914, *Karluk* in 1913–1914 and *Sedov* in 1937–1938, Russian *Severnyi Polyus* (SP) camps starting in 1937 and US camps T3, ARLIS and AIDJEX between 1958 and 1976. There is a map of tracks in Colony and Thorndike (1984a). The historical data compiled by these authors can be acquired at http://nsidc.org/data/g01358.html (Colony and Thorndike 1984b).

Ice leaves the Arctic via the Canadian Archipelago as well as via Fram Strait. Fragments of ice island T1 were tracked across the Sverdrup Basin and through Byam Martin Channel in 1962 (Black 1965). Ice island WH-5 passed through Nares Strait during 1963–1964 (Nutt 1966). Melling (2002) summarizes other incidental information on ice drift across the Canadian polar continental shelf.

The use of radio-tracked beacons to measure the drift of Arctic ice was initiated by Russian scientists with drifting automatic radio meteorological stations (DARMS) in 1953 (Gudkovich and Nikolaeva 1961; Gorbunov and Moroz 1972). Locations were determined to a best accuracy of 10–15 km by radio direction finders located at the coast. With this tracking method, the error in drift exceeds the actual drift over intervals shorter than a few days.

The tracking of DARMS from coastal stations and aircraft was superseded in the mid-1970s by satellites using Doppler tracking. An array of such beacons was first deployed in association with AIDJEX during 1975–1976. The maturation of this technology permitted the inception of the International Arctic Buoy Programme (IABP) in 1979. The long-term objective of this multi-national project is the measurement of air pressure for operational meteorology. The systematic tracking of sea-ice circulation has been a fortuitous outcome.

By the mid-1980s, the general circulation of Arctic ice had been clearly delineated. Thorndike and Colony (1982) used early data from the IABP to investigate the dominant influences on ice drift. They demonstrated that ice drift within the central Arctic over seasonal or longer time intervals was attributable equally to forcing by geostrophic wind and by ocean current. On shorter time scales, velocity fluctuations were largely caused by varying wind. The forcing by both atmosphere and ocean in the central Arctic is correlated over large distances (more than 500 km). This large spatial scale facilitates objective analysis to determine the mean field of ice motion from observed trajectories. Colony and Thorndike (1984a) have used objective analysis to map the average drift pattern of Arctic pack ice, using all

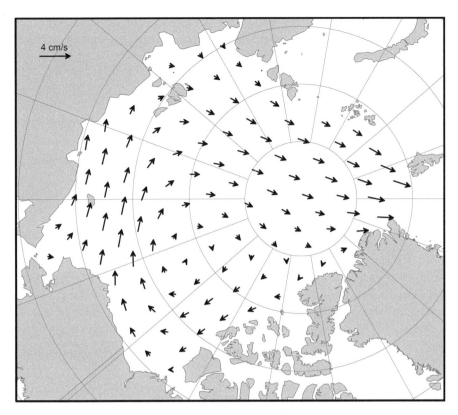

Fig. 3.2 The mean field of sea-ice motion in the central Arctic, as known at the start of ACSYS. The map is derived from an objective analysis of the trajectories of tracked ships and buoys in ice by Colony and Thorndike (1984a)

observations acquired between 1893 and 1982. Their analysis reveals a large Beaufort gyre, which dominates the circulation pattern over much the central Arctic, bounded on the Eurasian side by a Trans-polar drift carrying ice from the New Siberian Islands to Fram Strait (Fig. 3.2).

In a subsequent paper, Colony and Thorndike (1985) used the results of objective analysis, fields of mean drift and 'error' variance, to constrain a random walk for sea-ice motion. The random-walk model permits calculation of a number of interesting parameters as a function of location – the average age of ice, the future lifetime of ice before exit to peripheral seas (including Fram Strait) and the probabilities of ice leaving the central Arctic after a specified interval via various routes, or entering the central region from various source areas. The calculations revealed that the oldest ice in the Arctic, more than 6 years, is found against the north-western coast of the Canadian Archipelago; they indicated that this ice will not leave the central Arctic for another 5 years. In contrast, it appears that ice in the Trans-polar drift will typically leave the Arctic within 3 years.

McLaren et al. (1987) examined the seasonal variations of sea-ice motion in the Canada Basin using data from the first 6 years of the IABP. They identified a recurring tendency for the mean clockwise gyre of the Beaufort Sea to reverse in late summer, leading to eastward ice drift north of the Alaskan coast. They attributed the reversal to the establishment of a cyclonic (low pressure) atmospheric regime in late summer (August–September). Clearly, there is appreciable variance in the spectrum of ice velocity at periods as long as a year. Local drift trajectories will be strongly dependent on season as well as location.

3.3.3   Thickness

Capability to map the thickness of pack ice has lagged far behind that to map its extent, concentration and drift. In part, this is attributable to the apparent impracticality of suitable space-borne sensors; sea ice is opaque to electromagnetic radiation over a wide spectrum of spatially appropriate wave lengths. In 1990, there were only two well-established techniques, direct measurement of distance following drilling and remote measurement of draft by sonar mounted on a sub-sea platform. It was well known that single-point measurements were poorly representative of most pack-ice environments because of the wide diversity of ice forms and the short correlation scale of ice-thickness variation.

Measurements based on drilling are best suited to the study of level ice that annually covers wide areas in sheltered coastal environments. Weekly measurements of coastal land-fast ice became a routine at Arctic weather stations in Siberia in the late 1930s (Polyakov et al. 2003) and in northern Canada in the late 1940s (Bilello 1980; Ice Centre 1992). On the eve of ACSYS, Brown and Coté (1992) published a study of inter-annual variations in ice thickness based on data from selected stations in northern Canada. They found trends in ice thickness at each site, but no spatially coherent pattern of trend. The principal determinant of inter-annual variation in ice thickness was not variation in air temperature, but variation in the amount and timing of snow accumulation (Fig. 3.3).

Upward-looking sonar has been standard equipment on the nuclear submarines operating beneath Arctic pack ice since 1958 (Lyon 1984). The sonar was installed for its value in avoiding ice obstacles and in identifying 'sky lights', areas of relatively thin ice where a submarine might surface to launch missiles. The sonar focussed sound on a surface target less than 10 m across and measured the echo time. This measurement was converted to draft through knowledge of the submarine's depth and upper-ocean sound speed. With pings launched every second, a spatial transect of ice draft is acquired as the boat progresses at known speed.

As early as 1980, some observations from UK and US submarines had been subject to scientific analysis (Kozo and Tucker 1974; Williams et al. 1975; Wadhams 1977, 1978; Rothrock and Thorndike 1980; Wadhams and Horne 1980). Submarine navigational challenges and the cruise tracks of early voyages are summarized by Lyon (1984).

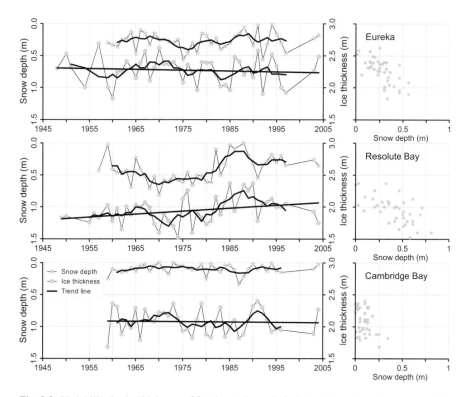

Fig. 3.3 Variability in the thickness of fast ice at the end of winter at three locations on a north-south transect of the Canadian Archipelago. Variability and trends are inconsistent between stations and most highly correlated with snow depth (Update of Brown and Coté (1992))

During the next decade, researchers published data by cruise, filling in blank areas of geography on the implicit assumption that the broad characteristics of the polar pack were unchangeable (Wadhams 1981, 1983b, 1989; McLaren et al. 1984; Wadhams et al. 1985). At the same time, attempts were made to understand the errors in sampling and measurement associated with instruments, operating protocol and analysis whose details were shrouded in military secrecy (Wadhams 1983a, 1988). The relationship between ice draft (derived from the sonar) and ice thickness was also a topic of interest (Wadhams and Lowry 1977; Bourke and Paquette 1989).

A landmark of the late 1980s was the publication of the first map of Arctic ice thickness by Bourke and Garrett (1987; see Fig. 3.4). This provided a dramatic update of Koerner's (1973) ice-thickness transect, acquired via drilling during the British Trans-Arctic Expedition in 1968–1969. By compositing seasonally segregated data, Bourke and Garrett (1987) were able to delineate the now familiar accumulation of very thick (7 m) ice against the Canadian Archipelago, which forms a reservoir feeding a south-westward drift across the southern Canada Basin and a south-eastward drift through the Canadian Arctic. They also mapped the progressive

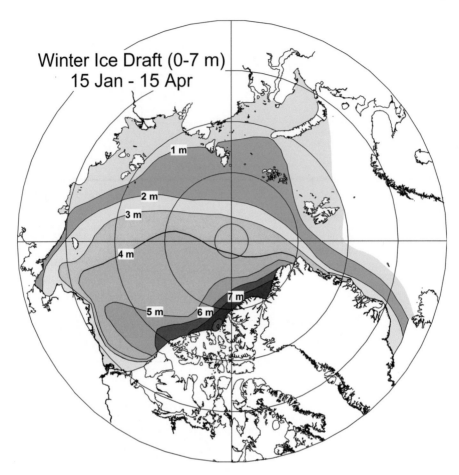

Winter Ice Draft (0-7 m)
15 Jan - 15 Apr

Fig. 3.4 Contour map of wintertime ice draft in the central Arctic prepared from observations by sonar on US naval submarines. This map summarizes the imperfect knowledge of this variable at the start of ACSYS (After Bourke and Garrett 1987)

thinning of ice across the Arctic towards Siberia. A second paper based on the same data presented maps of ice roughness within the central Arctic (Bourke and McLaren 1992).

The almost complete lack of information beyond the deep basins of the Arctic Ocean is not obvious from these maps, but is clear in the plotting of submarine cruise tracks by Lyon (1984). The thickness of pack ice within the ten million km^2 of the seasonal ice zone, which is two thirds of the northern marine cryosphere, was essentially unknown at this time.

Two developments at the end of the 1980s set the stage for ACSYS. Wadhams (1990) and McLaren et al. (1990) published the first discussions of possible progressive change in the thickness of Arctic pack ice. And upward-looking sonar designed to measure ice draft from sub-sea moorings began to provide data specifically for scientific studies (Vinje and Berge 1989; Hudson 1990; Pilkington and Wright 1991; Moritz 1992). Such instruments, measuring the draft of the drifting ice at intervals from seconds to minutes, survey several thousand kilometres of the pack during a 12-month deployment.

Three other techniques provided information on the thickness of offshore pack ice prior to ACSYS: measurements in bore holes, aircraft-mounted laser to measure ice-elevation profiles and an electromagnetic induction sensor to measure ice thickness. Drilling has provided precise data over small areas, with information on snow thickness, ice elevation, draft and thickness, but it is laborious and therefore unsuited to routine monitoring of highly inhomogeneous pack ice (Eicken and Lange 1989; Eicken et al. 1995). Laser has advantages in ease of deployment and high resolution, but careful analysis is necessary to estimate ice elevation and subsequently ice thickness (Ketchum 1971; Hibler 1972). Because snow generally conceals the ice surface and small changes in aircraft altitude are difficult to track, the technique is better suited to measuring ice roughness and ridging than absolute thickness (Tucker and Westhall 1973; Wadhams 1977; Tucker et al. 1979; Weeks et al. 1989; Krabill et al. 1990; Lytle and Ackley 1991). The use of an electromagnetic induction sensor to measure the thickness of sea ice dates from the mid-1970s (Sinha 1976), when the apparatus was operated on the ice surface. By the late 1980s, a probe had been developed to carry the sensor for aircraft deployment, typically via helicopter. Initial results were promising for level first-year ice and multi-year floes (Kovacs et al. 1987).

3.3.4   Export

Pack ice drifts out of the Arctic via Fram Strait, via the passages to the Barents Sea between Novaya Zemlya, Franz Josef Land and Svalbard, via the Canadian Archipelago and via Bering Strait. The Fram Strait efflux has long been regarded as principal among these, although in fact the other outflows have not received much scientific attention.

First estimates of the Fram Strait outflow were in terms of area and calculated as the speed of drift integrated across the width of the ice stream. Values are centred on about 1×10^6 km^2 year^{-1}, with a range of $\pm 50\%$ (Gordienko and Karelin 1945; Gordienko and Laktionov 1960; Antonov 1968). The average annual export is equivalent to 12% of the area covered by ice in winter within the central Arctic and peripheral seas (excluding the Barents Sea and Canadian Archipelago).

Wadhams (1983b) generated a first estimate of ice-volume efflux (9,150 km^3 year^{-1}) by convolving an ice-draft transect from submarine sonar with the average speed across the ice stream in Fram Strait. Østlund and Hut (1984) used observations of the oxygen-isotope anomaly in Arctic surface waters, which bear an imprint of net ice growth, to derive an appreciably smaller value (5,200 km^3 year^{-1}). The outcome of a thorough study by Vinje and Finnekåsa (1986) that integrated data on wind, ice drift and ice draft was a mean annual export of 5,000 km^3 year^{-1} during 1976–1984. The study delineated a seasonal cycle with maximum flux in January (+20%) and a minimum in August (−30%).

Estimates for the ice flux through the Canadian Archipelago have been based on very few data and are therefore diverse. Vowinckel and Orvig (1962) calculated the outflow through Davis Strait to be 491 km^3 year^{-1}. Dunbar (1973) quoted 225 km^3 year^{-1} for the sum of fluxes through Smith, Jones and Lancaster Sounds. The estimate of Aagaard and Carmack (1989) was 155 km^3 year^{-1}.

All estimates are compromised by a paucity of information. A confident estimate of ice flux requires spatially and temporally detailed observations of ice draft and velocity across the entire ice stream because cross-flow variations in both ice thickness and velocity are appreciable (Wadhams 1983b; Vinje and Finnekåsa 1986). Moreover, instruments must be carefully located to avoid bias. For example, ice melting occurs between 80°N and 81°N, in response to mixing between the southbound cold Transpolar Current and the northbound warm West Spitsbergen Current (Östlund and Hut 1984; Vinje and Finnekåsa 1986).

3.3.5   Atmosphere-Ice-Ocean Interaction

The interaction of pack ice with the atmosphere and ocean engages a wide range of process-oriented research. Elements of particular importance to pack ice locally are the surface energy and mass budgets, the impacts of snow accumulation, melt ponds and leads, the flux of heat from the ocean and that of salt from the ice and the consequences of ice-sheet failure and ridging driven by wind and current. Regionally, pack ice participates in several climatic feedback loops, through the impact of its high albedo on energy absorption at the surface, through the impact of meltwater on diffusive fluxes in the upper ocean, through the impact of freezing on the temperature of the pycnocline and through the impact of lead opening on atmospheric humidity, cloud formation and regional albedo. The network of flaw leads and polynyas that encircle the Arctic are of notable significance in the latter context.

The study of atmosphere-ice-ocean interaction in the Arctic has an impressive history that can be traced to the installation of a drifting scientific camp (SP-1) near the North Pole by the Soviet Union in 1937. By 1991, when SP-31 was abandoned and the programme fell into disfavour, the Soviet Union had supported almost a half century of scientific observations in the central Arctic. The first US drifting station was established on an ice-shelf fragment (T-3) in 1952 (Sater 1964). This platform was occupied until 1954, abandoned and re-occupied during 1957–1961 for the IGY. Other shorter-lived camps were established on ice floes, Alpha in 1957, Charlie in 1959, ARLIS-1 in 1960, ARLIS-2 in 1961. A camp was re-established on T-3 in 1965 and maintained until 1974.

In the 1970s and 1980s, several ambitious field studies of atmosphere-ice-ocean interaction were sponsored by US agencies. The Arctic Ice Dynamics Joint Experiment (AIDJEX) was a multi-facetted project focussed on pack-ice drift and deformation, with significant activity in the study of pack-ice thermodynamics. The principal field phase of AIDJEX was a year-long drift experiment in the Beaufort gyre during 1975–1976. The Marginal Ice Zone Experiments (MIZEX) were multi-national efforts in 1983 and 1984 focussed on the mechanics and thermodynamics of pack ice near its edge in the Greenland Sea, where there are complex interactions across the polar front between cold Arctic outflow and the warm Atlantic inflow. The Coordinated Eastern Arctic Experiment (CEAREX) field programme in 1987–1989 investigated the processes for exchange of momentum, heat and freshwater between ice and ocean in the vicinity of Svalbard. This study bridged the ice edge by conducting operations from both drifting camps and ships. In 1992, the US Office of Naval Research sponsored LEADEX, a drifting ice camp with rapid-response logistics specifically to study of physical processes above and below the surface of new refreezing leads.

Early study of the thermal conditions that maintain perennial ice in the central Arctic was conducted by Untersteiner (1961, 1964) and Hansen (1965). The land-mark paper of Maykut and Untersteiner (1971) presents a numerical model of sea-ice thermodynamics and discussion of the seasonal cycle in Arctic ice and factors related to the development of the perennial pack. This paper revealed the complexity of the interactions between sea ice, snow cover, cloud, radiation, ocean and atmosphere, and the weakness of existing knowledge concerning them. Of particular interest were the sensitivities of perennial-ice thickness to heat flux from the ocean and to snow accumulation. A modest change in the oceanic heat flux, from 1.5 to 5 $W m^{-2}$, reduced the thickness of perennial ice from 3 m to 1 m (Fig. 3.5). A doubling of snow depth from present values allowed the formation of extremely thick (5 m) perennial ice, a phenomenon explored further by Walker and Wadhams (1979). The paper by Brown and Coté (1992) provides further evidence for the importance of snow cover. These authors demonstrate that inter-annual variation in the accumulation of snow on land-fast sea ice, not variation in air temperature, is the principal determinant of inter-annual variation in the thickness of the underlying ice.

In subsequent study, attention was focussed on better understanding of the individual components of the surface energy balance for pack ice. Solar radiation is the principal contributor to summertime melting. Papers by Grenfell and Maykut (1977), Grenfell (1983), Grenfell and Perovich (1984), Maykut and Perovich (1987) discussed the albedo and internal optical properties of snow and ice at visible and near-infrared wavelengths. The sensitivity of sea ice to cloudiness, through its impact on both short-wave and long-wave radiant exchanges of energy, was explored by Shine and Crane (1984). Sensible heat within the ocean's surface layer can have dramatic impact in marginal ice zones and within the polar pack in summer. The penetration of solar radiation through sea ice to warm the upper ocean has been investigated by Grenfell (1979) and Perovich and Maykut (1990). The ocean's surface layer may warm in consequence of diffusive heat flux from deeper waters. The first direct measurements of turbulence beneath moving pack ice were acquired by McPhee and Smith (1976). Studies of the diffusive fluxes of heat from the deeper ocean to the base of sea ice have been completed by McPhee (1980, 1983, 1987, 1992),

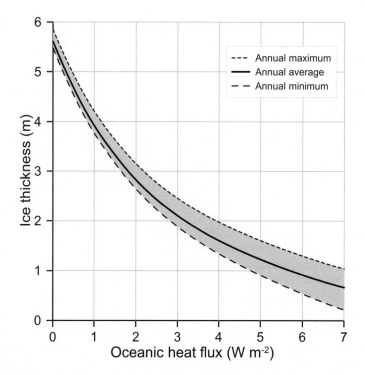

Fig. 3.5 High sensitivity of multi-year ice thickness to small changes in the net heat flux to the ice, here assumed to originate in the upper ocean. Incident short-wave flux in summer can larger by two orders of magnitude (After Maykut and Untersteiner 1971)

Mellor et al. (1986), McPhee et al. (1987) and Morison et al. (1987). Maykut (1978) discussed energy exchanges through thin sea ice, and Morison et al. (1992) summarized knowledge of freezing processes in leads. Estimates of atmosphere-ocean heat exchange for the entire central Arctic and of net ice production were published by Maykut (1982). This paper demonstrated that the net ice production through freezing and ablation is strongly dependent on assumptions regarding the probability distribution of ice thickness within the pack (Fig. 3.6). A review by Maykut (1986) summarized current understanding of sea-ice thermodynamics. Dynamical interactions between the growth and decay of sea ice and the ocean mixed layer were explored through simulations using a coupled one-dimensional sea-ice-ocean model by Lemke (1987) and Lemke et al. (1990).

The weak oceanic heat flux critical to the occurrence of perennial ice in the Arctic is attributable to the near-freezing temperature of the Arctic halocline. It is generally acknowledged that wintertime freezing in flaw leads and polynyas over the broad Arctic continental shelf produces the water masses necessary to maintain the cold halocline (Maykut and Untersteiner 1971; Aagaard et al. 1981; Melling et al. 1984). Much of the salt present in freezing seawater is rejected from the growing

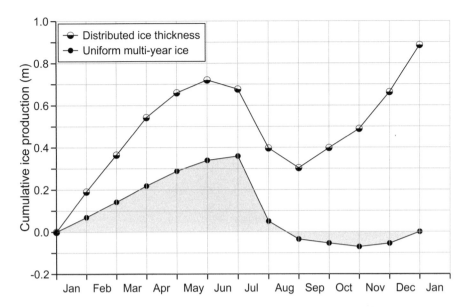

Fig. 3.6 Net ice production by month in the central Canada Basin. Calculations based on ice of uniform 3-m thickness are compared with those based on a more realistic distribution of ice thickness (After Maykut 1982). Note that there is a net annual production of ice in the latter circumstance

ice in concentrated brine, which increases the salinity (and density) of the ocean surface water. The cold water of shallow flaw leads may take up enough brine that it becomes sufficiently dense to ventilate and cool the halocline in adjacent deeper waters (Midttun 1985). Halocline ventilation by cold saline shelf waters has been observed at locations along the periphery of the Canada Basin by Garrison and Becker (1976) and by Melling and Lewis (1982), Melling (1993) and Melling and Moore (1995; see Fig. 3.7). The latter authors maintained that the amount of ice growth required to transform the shelf water of late summer to water that would ventilate the halocline was unrealistically large. They presented evidence indicative of shelf-water preconditioning to higher salinity prior to freeze-up. Storm winds are implicated both in the preconditioning of shelf waters and in the opening of flaw leads later during the winter.

Martin and Cavalieri (1989) used satellite data to map the opening and closing of flaw leads along the Siberian coast. Combining this information with calculations of ice growth, they estimated the total production of ice and of brine within the flaw leads, concluding that only 20–60% of the flux of water through the cold halocline could be supplied from this source. Killworth and Smith (1984), using a one-and-a-half dimensional box model of the Arctic Ocean, had earlier determined a corroborative result, that maintenance of a realistic cold halocline required a substantial interleaving of inflow from the Bering Strait.

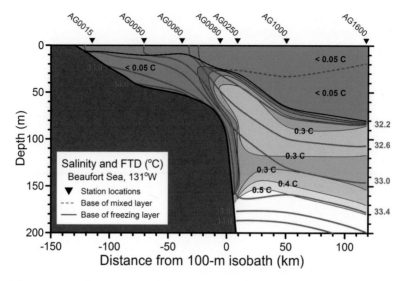

Fig. 3.7 Ventilation of the Arctic halocline over the continental shelf in the Beaufort Sea in late winter, as a consequence of brine accumulation from freezing in the flaw lead. Note the simultaneous upwelling of warm water and downwelling very cold water at the shelf edge (After Melling and Moore 1995)

3.3.6 Sea Ice Processes

Accurate simulation of the drift of sea ice requires realistic parameterization of the forces acting upon it: surface stresses exerted by wind and current, the Coriolis force, the component of gravitational force parallel to the sloping sea surface and the stress within the ice. Whereas reasonable parameterizations of the first four forces had been developed by 1970, this was not the case for the internal ice stress. At this time, the rheological behaviour of pack ice on a scale (100 km) appropriate to basin-wide computation of ice drift was virtually unknown. Moreover, it has been obvious from earliest times that the stress within pack ice frequently exceeds its strength, resulting in failure that creates ubiquitous leads, rafts and ridges. The former are zero-thickness lacunae that weaken the pack ice; the latter are thicker scars that strengthen it. Thus, there is a strong inter-dependence between the history of stresses applied to pack ice and its mechanical properties, which is mediated by the distribution of ice area as a function of thickness.

As a material, pack ice is clearly granular and highly inhomogeneous. Both its physical properties and its velocity can vary rather abruptly from one granule to another. In 1970, it was not known whether pack ice could be adequately modelled as a rheological continuum, and if so, on what spatial scale.

Internal stresses and the thickness distribution of pack ice are best studied in a Lagrangian sense by observing and modelling a parcel of pack ice as it evolves in response to thermal and mechanical forcing. The Arctic Ice Dynamics Joint Experiment (AIDJEX) was an ambitious field project based on this strategy which was conducted from drifting multi-year ice in the Beaufort Sea in the 1970s. The specific objective was the discovery of a quantitative relationship between internal stress and strain in pack ice on a regional scale.

A critical constraint on AIDJEX was lack of capability to measure stress in sea ice. As a consequence, the principal ice dynamics focus of the project was the accurate measurement of ice velocity and strain rate on a range of spatial scales. Progress in the understanding of pack-ice stress was developed through numerical experimentation with various forms of continuum mechanics. Success was measured in terms of capability to replicate the observed drift and deformation of pack ice subject to calculated wind and current stresses and Coriolis force. Two different rheological approaches were developed, one described by Pritchard (1975), Pritchard et al. (1977) and Coon and Pritchard (1979) and the other by Hibler (1977, 1979) and Hibler and Tucker (1979). This effort falls into the domain of Chap. 10, and is not discussed here.

An important step towards the objective of AIDJEX was the development of theory by Thorndike et al. (1975) to represent changes in the distribution of ice thickness. This theory was first incorporated into the strength parameterization of an ice dynamics model by Hibler (1980). Although this theory has now become part of many sea-ice models, it incorporates closure assumptions that remain little more than guesses.

One route towards the authentication of closure assumptions lies in study of the individual ridging events that are the localized response to pack-ice stress in terms of strain rate. The ice blocks that comprise ridges are represented as individual elements that move and re-orientate within a vertical plane in response to the forces acting upon them. Simulations of this type have been reported by Hopkins and Hibler (1990, 1991) and Hopkins et al. (1991); see Fig. 3.8.

Measurements of stress in pack ice were first practical in the early 1980s (Cox and Johnson 1983; Graham et al. 1983). Under impetus from the design of offshore oil-exploration structures, several experiments were conducted to measure pack-ice driving forces (Johnson et al. 1985; Croasdale et al. 1988; Coon et al. 1989; Comfort and Ritch 1990). Data contain signals indicative of stress induced by thermal expansion, tides and storms, but the quantitative interpretation of results remains enigmatic (Tucker et al. 1991).

3.4   Achievements of the Arctic Climate System Study

3.4.1   Extent and Concentration

3.4.1.1   Operational Ice Charting

Routine charting of the northern marine cryosphere by several national ice services continued throughout ACSYS decade. Operational ice charts are typically prepared to support navigation and are released daily, weekly or monthly, depending on shipping activity and season. Charts represent an analysis of ice conditions observed

Fig. 3.8 Detailed view of pressure-ridge development by flexural failure of thin lead ice at the edge of a thicker floe, as simulated numerically. The rheological behaviour of pack ice is the consequence of many local events of this type in response to pack-ice stress on a large scale (After Hopkins 1994)

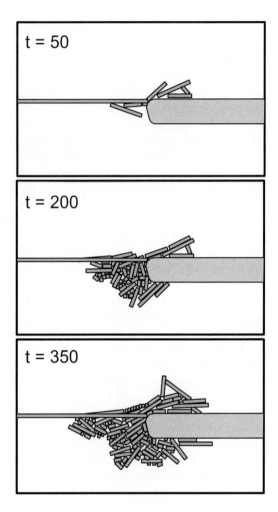

close to the time of issuance, but information used can be up to a few days old. In most countries, observations are complied from a variety of sources – observers on ships and aircraft, weather satellites in the visual and thermal bands, passive microwave sensors on satellites and imaging radar on aircraft and satellites. Ice analysts subjectively delineate distinct zones within the pack, identify the three dominant ice types and assign each a partial concentration and floe size. This information is presented for each polygonal zone on the chart using the WMO 'egg code'.

Countries with agencies preparing charts include the USA (National Ice Center), Russia (Arctic and Antarctic Research Institute), Canada (Canadian Ice Service), Finland, Sweden, Denmark, Norway, Germany, Poland, Iceland, Netherlands and Japan. US charts provide global coverage and Russian charts cover much of the Arctic.

Other countries emphasize national priorities. Because the focus of operational ice charting is navigation, detail is usually greater in seaways (marginal ice zones) than in the interior of ice fields.

There have been progressive improvements in the technology applied to ice reconnaissance, data communications and the preparation of ice charts since these activities began in the middle of the twentieth century. In general, ice services have access to more information with shorter delays as the years pass. Recent publications document the history of changing methods of ice charting in Canada (Ballicater Consulting 2000) and the USA (Dedrick et al. 2001). With the launch of satellites carrying synthetic aperture radar (SAR) during the ACSYS decade (ERS-1 in 1991, Radarsat in 1995, Envisat in 2002), the quality and availability of data for the preparation of ice charts have increased markedly, particularly where direct links to ground stations are possible. SAR on satellites has high resolution (25 m) and operates independently of cloud and darkness. However, there may be significant ambiguity in the interpretation images of pack ice from SAR, particularly in first-year pack and during the warm season. Thus, independent signals available via other sensors, even at lower resolution, provide a distinct advantage in ice-type classification (e.g. Hauser et al. 2002).

The disadvantage of reliance on a single data source is well illustrated by the study of Agnew and Howell (2003) in Canadian waters. Compared with the Canadian charts, which represent information gleaned from satellite images in the visible, thermal and microwave bands, from satellite radar and from visual observations from ships and aircraft, the NASA Team algorithm underestimates the ice-covered area by 20–33% during summer and by 7–44% during autumn. Overestimates vary widely because some Canadian regions are only partly within the marginal ice zone while others (e.g. Hudson Bay) are entirely so. The US hemispheric chart series (heavily reliant on passive microwave scanners) indicate 18% less ice in summer than corresponding Canadian charts. In other seasons, the underestimate decreases to about 8%. Ice concentration on US charts for the northern hemisphere is also on average higher than values from passive microwave by the Bootstrap algorithm. The average bias over the northern hemisphere increases from a minimum of about 4% in winter to a peak of 23% in early August (Partington et al. 2003).

National ice services have maintained their collections of historical ice charts on paper and on microfilm. Prior to ACSYS, these archives were little used for scientific research because of the difficulty of accessing and manipulating the information. During ACSYS, a joint US-Russia working group prepared digital versions of Russian and US ice charts of the Arctic on a 25-km grid, with coverage from 1950 to 1992 at 10-day intervals and 1972 to 1994 at weekly intervals, respectively. The data were published on CD as the *Sea Ice Atlas for the Arctic Ocean* (Arctic Climatology Project 2000). Charts for the Arctic prepared by the US NIC since 1994 are available in vector graphic form (E00 format: ArcInfo export) from http://www.natice.noaa.gov/. Russian charts prepared by AARI since 1997 are available in raster graphic form from http://www.aari.nw.ru/index_en.html. The vector graphic format has the advantage of preserving the boundaries of the polygons classified in the original ice analysis, including the ice edge.

Partington et al. (2003) have examined hemispheric modes of ice-concentration variability calculated from digitized charts in the *Sea Ice Atlas for the Arctic Ocean*. There are two principal modes. The first has centres of action in the northern Atlantic and Pacific, with variations out of phase across the Atlantic and between the Bering and Okhotsk Seas; variance is greatest in late winter. The second reflects out-of-phase anomalies on the Pacific (Beaufort, Chukchi, East Siberian and Laptev Seas) and Atlantic (Barents, Greenland, Baffin and Hudson Bay) sides of the Arctic; it has greatest amplitude in September. These patterns are consistent with variability in atmospheric circulation, particularly the Arctic Oscillation.

The Canadian Ice Service also completed the digitization of its weekly regional ice charts during ACSYS. A CD set covering the period 1968–1998, with charts in vector form (E00 format), was published late in 1999. Earlier charts (for 1956–1974) with less complete coverage were released in TIF format on CD a year later. Current charts may be obtained in raster or vector graphical formats from http://ice-glaces.ec.gc.ca/. Melling (2002) and Crocker and Carrieres (2000) discuss aspects of sea-ice variability revealed in the Canadian chart collection. In most regions of the Canadian exclusive economic zone, including the North-west Passage, there have been no statistically significant trends in ice conditions over the past 35 years. The exceptions are decreases in ice extent in Hudson Bay and the south-eastern Archipelago.

3.4.1.2   Historical Ice Charts

Notes from mariners prior to the era of operational ice charting can provide valuable data on historical variation in ice-edge locations. The availability of information with respect to geographic sector, time period and season reflects economic activity, primarily the hunting of marine mammals and inter-continental trade. ACSYS has promoted study of ice-edge variations over centuries based on the sources summarized by WMO (2003: Table 1). The North Atlantic sector is best served, with observations dating back to 1,100 in Icelandic waters, to the mid-sixteenth century in the Barents and Norwegian Seas (Fig. 3.9), the mid-eighteenth century in west Greenland waters and the nineteenth century off eastern Canada. In the Russian Far East (Okhotsk Sea), observations began the mid-seventeenth century, but we know little of conditions in the western Arctic (Chukchi and Beaufort Seas) until the late nineteenth century. A workshop on sea-ice charts of the Arctic was convened by ACSYS in the summer of 1998 (WMO 1999).

Vinje (2001a) has prepared an ice-edge time series for the Barents Sea north of Norway that covers more than 400 years. The latitude of the ice edge in summer has fluctuated over more than 3° on a wide range of time scales. Conditions of extreme retreat similar to today occurred in the late sixteenth century and for much of the eighteenth. Summertime ice was more extensive in the mid-seventeenth and early nineteenth centuries. Fluctuations in European air temperature since 1860 are well correlated with the ice-edge latitude in the Barents Sea.

On the other side of the Atlantic, a time series of ice extent in winter off eastern Canada has been complied from ship reports by Hill and Jones (1990). The series

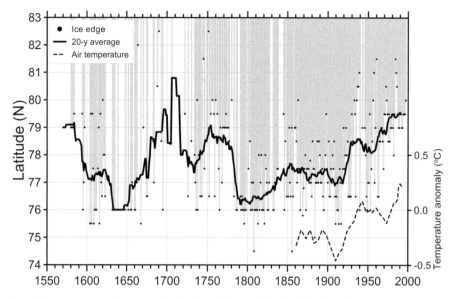

Fig. 3.9 Changes over four centuries in the springtime latitude of the ice edge in the Barents Sea based on reports from ships in the area. On a 20-year running average, the ice-edge position is correlated with the anomaly in mean air temperature for the northern hemisphere (After Vinje 2001a)

documents pack ice observed south of 55°N beginning in 1810. In early years, there is a single chart per winter, but after 1859, monthly (February, March, April) charting was practical. Inter-decadal variability dominates, but there is a slow variation evident from lighter to more extensive ice during 1810 to 1890, with a return to lighter conditions in the 1970s. The data and summaries to 2000 can be accessed at http://researchers.imd.nrc.ca/~hillb/icedb/ice/.

The value of many of these data in delineating past climatic variation remains to be exploited. To facilitate research, ACSYS has worked with the Norwegian Polar Institute to compile an ACSYS Historical Ice Chart Archive, which identifies and brings together ice information from a variety of nautical sources (particularly whaling ships) for the Arctic as far back as the seventeenth century.

3.4.1.3   Hemispheric Variability

Routine mapping of sea-ice extent through detection of natural microwave emission by satellite-borne radiometers (ESMR, SMMR, SSM/I) has revolutionized the study of pack-ice extent and its variations during ACSYS. Microwave data have been used to map pack ice since 1972 (ESMR), but data suited to long-term study date from 1978 (SMMR). The US NASA has maintained a routine acquisition and processing activity involving passive microwave imagery since this time. The effort has generated hemispheric information at 25-km resolution for ice concentration and (with some qualification) for ice type at 2-day intervals. The data are archived at World Data Center C (Glaciology: NSIDC) and available via http://nsidc.org/data/seaice_index/.

In general, ice targets viewed via the large pixel of microwave scanners are inhomogeneous. With the available combinations of frequency and polarization, each pixel may be classified as a mix of only three surface types, typically ice-free, first-year ice and multi-year ice (Rothrock et al. 1988). Analysts define end points for pure surface types on a mixing diagram and determine fractional composition from the position of data points within the triangle defined by pure types. This method of analysis gives ambiguous results where more than three types are present or where the ice surface or snow harbour liquid water, as is common during the melting season and in marginal ice zones year-round.

Prior to ACSYS, the preferred method for deriving pack-ice parameters from orbiting scanners was the NASA Team algorithm (Cavalieri et al. 1991). In the early 1990s, a new Bootstrap algorithm was developed for use with SSM/I data (Comiso 1995). Both algorithms produced results with seasonally varying bias relative to optical scanners at higher resolution. Comiso et al. (1997) evaluate a version of the Bootstrap algorithm with end points adjusted to match values of concentration for melting pack derived from orbiting SAR. The modified algorithm corresponds well with the NASA Team algorithm with respect to ice-edge location, but disagrees appreciably in the interior of Antarctic pack ice and in the seasonal sea ice zone of the Arctic (Cavalieri et al. 1997). Results from the two algorithms may differ by as much as ±25%. In comparison with independent mapping by optical and microwave sensors at higher resolution, the passive microwave yields lower values of concentration. This result is consistent with documented bias relative to operational ice charts (Agnew and Howell 2003) and to kinematic determination of ice-free area via SAR image analysis (Kwok 2002).

In areas where multi-year ice is not anticipated, the corresponding end point for ice classification can be re-assigned to the properties of microwave emission associated with new ice. The resulting pack-ice classification in terms of ice-free, new ice and first-year ice has value in the identification and monitoring of polynyas and flaw leads in the seasonal ice zone. The results of an inter-comparison with AVHRR imagery in the Bering Sea have been very encouraging (Cavalieri 1994), despite sub-optimal resolution for such narrow features.

The feasibility of mapping sea ice using space-borne scatterometer has also been investigated, at C-band from ERS-1/2 (Gohin and Cavanié 1994) and at Ku-band with NSCAT (Ezraty and Cavanié 1999; Remund and Long 1999). Resolution can be enhanced by combining information from images at different times, but with an inevitable increase in noise because of ice motion and changes in viewing geometry. Results comparable to those from SSM/I are achievable during the cold months. A dramatic decrease in backscatter accompanying thaw precludes use in the warm months.

Despite the unresolved issues of accuracy and bias, the 25-year time series from orbiting microwave radiometers is relatively homogeneous. It has provided a fertile ground for study of the interactions between pack ice and climate throughout the decade of ACSYS.

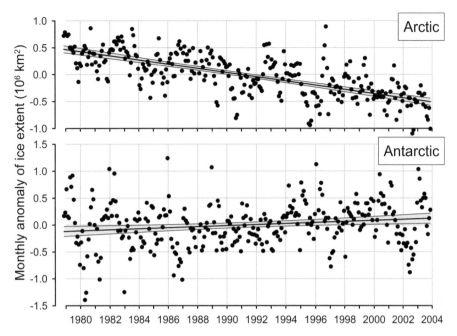

Fig. 3.10 Varying extent of sea ice in the northern and southern hemispheres over a quarter century 1978–2004 with linear trend lines indicated. The 8% decline in the Arctic is statistically significant at the 95% level, whereas the 2% increase in the Antarctic has marginal statistical significance (Update of Cavalieri et al. (2003) using ftp://sidads.colorado.edu/DATASETS/NOAA/G02135)

A first analysis of progressive change in the extent of polar sea ice was prepared using data from the 9-year lifetime of SMMR (Gloersen and Campbell 1988). The same data were presented with corrections for sensor drift by Gloersen and Campbell (1991). The change over 8 years was −2.1 ± 0.9% in the northern hemisphere and insignificant in the south. The length of the time series was considered too short to justify extensive speculation on significance and causation at that time.

An analysis of the data available by the mid-1990s provided evidence of continued decrease in ice extent and ice-covered area (Johannessen et al. 1995), although adjustments for the change in sensor from SMMR to SSM/I in 1987 had yet to be made. The time series were effectively merged by Bjorgo et al. (1997), yielding a 16.8-year record. Over this period, the extent and ice-covered area of the northern hemisphere decreased by 4.5% and 5.7%, respectively, while the corresponding trends in the southern hemisphere were less dramatic (−1.1%, −0.5%). Calculated trends in the 18-year series available by the end of 1996 were decreases of 5.3% in the extent and ice-covered area of the northern hemisphere and increases of about 2.7% in both measures for the southern (Cavalieri et al. 1997). Detailed discussions of methodology followed in papers by Cavalieri et al. (1999) and Gloersen et al. (1999). Recently updated information on trends in sea-ice extent has been provided by Vinnikov et al. (2002) and Cavalieri et al. (2003). The latter reference incorporates newly merged observations from the 1970s (Fig. 3.10).

Change in ice extent has not been uniform throughout the Arctic or invariant over season (Parkinson et al. 1999b). Decreases in Hudson Bay and the Okhotsk, Greenland and Barents Seas have been countered by increases in eastern Canadian waters and the Canadian Archipelago (see Parkinson et al. 1999a, Fig. 3.1 for delineation of areas). Trends have been largest in spring (AMJ) and summer (JAS) and smallest in autumn (OND). Partington et al. (2003) report that there is seasonal dependence also in trends derived from US NIC ice charts. However, the trends based on microwave data alone are larger, 43% more during summer (mid-June to mid-September) and 30% more in the annual average.

By the late 1990s, close examination of the passive microwave record had revealed dramatic change in the composition of Arctic pack ice. During the cold season, multi-year ice can be distinguished from first-year ice via its microwave emission. Through use of this signal, Johannessen et al. (1999) concluded that the area covered by multi-year ice decreased by 14% between 1978 and 1998, more than three times the overall decrease in ice-covered area. Because concern had been expressed that the micro-wave signature of multi-year ice is insufficiently stable for such an analysis (Kwok et al. 1996), Comiso (2002) took a different approach. He used the annual minima in ice extent and in ice-covered area as proxy for the area of multi-year ice. Between 1979 and 2000, the annual minimum ice extent decreased from 7.7 to 6.7 million square kilometres (13%) and the annual minimum ice-covered area from 6.5 to 5.5 (15%), changes consistent with Johannessen's analysis. During the 1996–1997 winter specifically, the loss of multi-year ice from the central Arctic can be accounted entirely by export, principally though Fram Strait (Kwok et al. 1999).

Both of Comiso's measures include some first-year ice since if no first-year ice were to survive the melt season, multi-year ice would soon disappear. The decrease mapped by Comiso (2002) is the combined result of change in the 'graduating class' of first-year ice at the end of summer and in the multi-year ice population itself. Since multi-year ice is thicker than its seasonal counterpart, progressive loss of this class from the Arctic represents a decrease in the average thickness of the polar pack.

Record minima in Arctic ice extent were reached several times during ACSYS. The record in 1990 was associated with unusual loss of ice from the Kara, East Siberian and Chukchi Seas. Serreze et al. (2003) concluded that unusual warm and windy conditions in May and June promoted early break-up and melting of ice, with feedback from radiation via the consequent lowered albedo in the early summer. The authors noted that similar atmospheric conditions occurred north of Alaska in 1990, but that ice conditions in that sector were normal. Maslanik et al. (1996) discussed ice-extent minima in 1990, 1993 and 1995. They proposed a linkage between these anomalies and the increase in the frequency of atmospheric lows over the central Arctic since 1989. The record in 1998 resulted from greatly reduced ice in the Beaufort Sea and the Canadian Archipelago. Maslanik et al. (1999) attributed the anomaly to unusual atmospheric circulation throughout the preceding autumn and winter, favouring south-easterly winds that prevented influx of old ice from the north and cleared the area of seasonal ice starting in late April. Ice on the Siberian shelves was heavy in 1998. The authors speculated that the strong Aleutian low during the 1997–1998 ENSO event may be implicated in the ice-cover anomaly. The record minimum in 2002 was established through unusual ice loss in the Beaufort, Chukchi, East Siberian, Laptev and Greenland Seas. Serreze et al. (2003) explained the event in terms of anomalous warm southerly winds in spring that forced ice polewards and persistent low pressure and warm air over the Arctic in summer that promoted the opening of leads and accelerated melting of ice. Drobot and Maslanik (2003) reinforced the argument that anomalies in summer ice extent are forced primarily by anomalies in the wind-driven circulation of ice, with feedback to enhanced melting via albedo, sea-surface temperature and air temperature. On a longer perspective, Vinje (2001a) discussed correlation between variation and trend in ice-extent and in atmospheric circulation in the Nordic Seas since 1864.

Much attention has been focussed on the central Arctic. An analysis of century-long record of ice extent in the Russian shelf seas, 1900–2000, has provided evidence that trends in extent over the past century have been small and generally not statistically significant. Dominant variability occurs in response to a low-frequency atmospheric oscillation of multi-decadal period (Polyakov et al. 2003), which complicates the accurate estimation of trend.

3.4.2   Drift

3.4.2.1   Satellite Tracking of Beacons

The International Arctic Buoy Programme has been the principal contributor to the ACSYS sea-ice motion climatology. Observations accumulated during its 25-years life have seen wide application in Arctic science. The programme has suffered intermittently from sparse coverage over the Arctic basin, related to the vagaries of ice drift, beacon failure and deployment logistics. In general, there has not been good coverage in peripheral seas. Here, pack ice is more dynamic, and losses of buoys to ridging and melting are high. In areas where ice persistently diverges at the coast (Beaufort, Chukchi and Laptev Seas), repeated seeding with new buoys would be necessary to maintain continuity since ice quickly vacates such zones. Frequent re-seeding is usually logistically and financially impractical. Frequent re-seeding would also be needed to maintain observations in marginal ice zones where southbound ice streams vanish through melting in temperate ocean waters (e.g. Bering, Labrador and Greenland Seas).

Continued tracking of beacons on drifting ice during ACSYS has occasionally revealed ice leaving the Arctic via the Canadian polar continental shelf. Three IABP buoys have passed from the area north of Ellesmere Island through Nares Strait and into Baffin, one in 1989–1990, one in 1991–1992 and one in 2000–2001 (Rigor 2002). The ice island Hobson's choice moved from the periphery of the Beaufort gyre into the Canadian Archipelago in October 1988, remained fast in pack ice until the autumn of 1991 and then completed a rapid transit across the Sverdrup Basin and through Penny Strait (Jeffries and Shaw 1993). An IABP buoy passed eastward through Parry Channel during 1993–1994. Successful transits of ice-tracking buoys from the Beaufort gyre to Baffin Bay are rare because their destruction is very likely within the region of severe ice deformation that borders the Canadian Archipelago.

The IABP has calculated the motion field of Arctic ice at half-day intervals since 1979, via optimal interpolation of tracking data from instrumented floes. Because ice drift is strongly correlated with geostrophic wind over short intervals and ocean current over long, the first-guess field is based on sea-level pressure and average current. This constraint generates better results in regions of the central Arctic where buoys are not present (Pfirman et al. 1997) but may create unrealistic vectors in coastal areas in the absence of tracking data. Figure 1 of Pfirman et al. (1997) displays the mean field of ice drift updated using 15 years of IABP data. Pfirman et al. (1997) and Tucker et al. (1999) have used the IABP data to determine where floes sampled in the Eurasian Basin actually first formed by back-calculating their trajectories.

Proshutinsky and Johnson (1997) have used a wind-driven ice-ocean model, with simulated ice drift corroborated by ice-floe tracking to delineate inter-decadal variation in the drift field of Arctic ice. The field alternates between anti-cyclonic and cyclonic patterns, each persisting for 5–7 years. The anti-cyclonic pattern is the classic picture of Colony and Thorndike (1984a, b). In the cyclonic pattern, the Beaufort gyre shrinks into the southern Canada Basin, the Trans-polar drift is directed towards the Canadian Archipelago rather than Fram Strait and a cyclonic gyre occupies much of the Eurasian Basin. During the era of the IABP, the Beaufort gyre was dominant during 1984–1988 (Proshutinsky and Johnson 1997) and perhaps since 1998 (Johnson et al. 1999). Polyakov et al. (1999) discuss oceanographic implications of difference in seasonal cycle between the two phases: ice circulation is anti-cyclonic in both winter and summer during the anti-cyclonic phase, whereas during the opposite phase, only the wintertime circulation is anti-cyclonic. Although the immediate driver of the oscillation is atmospheric circulation, the authors speculate on oceanographic links involving alternating storage of freshwater in the Beaufort gyre and release to the Greenland Sea.

Kwok (2000) has further explored the relationship between changing winds and patterns of ice drift revealed by drifting buoys. Patterns of winter ice drift (October–May), composited for the years 1978–1996 according to the index of the NAO, clearly reveal the changes in the Beaufort gyre and Trans-polar drift described by Proshutinsky and Johnson (1997). The difference in ice circulation between high and low NAO index is a cyclonic movement of ice eastward along the Siberian and Alaskan coasts towards to Canadian Archipelago, accompanied by increased export of ice via Fram and Davis Straits (Fig. 3.11). This result is further discussed by Rigor et al. (2002) and Rigor and Wallace (2004), who use IABP motion fields to argue that the recent high NAO values and a consequent increased export of sea ice may be in part the cause of recent thinning and shrinking of Arctic ice.

3.4.2.2 Kinematic Analysis of Satellite Imagery

The continued utilization of drifting buoys during ACSYS has been complemented by developing techniques for measuring pack-ice motion using satellite imagery.

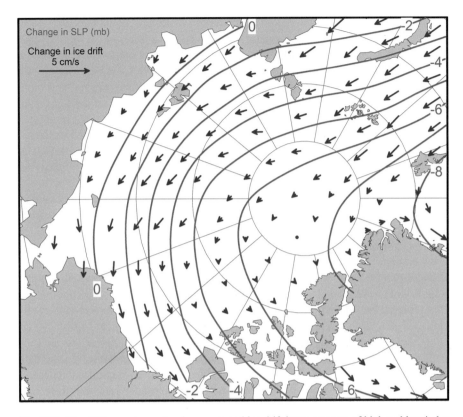

Fig. 3.11 The difference in sea-level pressure and ice drift between years of high and low index for the North Atlantic Oscillation. The lower Arctic sea-level pressure with high NAO creates a cyclonic anomaly in ice drift that decreases the size of the Beaufort gyre and pushes thick ice against the Canadian coast (After Kwok 2000)

Methods based on the cross-correlation of displaced patterns of leads in sequential scenes have modest beginnings using low-resolution, cloud-plagued AVHRR imagery (Ninnis et al. 1986; Fily and Rothrock 1987). Present techniques apply fully automated drift tracking to consecutive geo-referenced all-weather SSM/I and SAR images (Holt et al. 1992; Emery et al. 1997; Maslanik et al. 1998). In contrast to the sparse Lagrangian information from buoys, satellite images can provide basin-wide, quasi-Eulerian information on the velocity field of sea ice with high spatial resolution. However, the temporal resolution (days) is much longer than tracking interval for buoys (hours), and important variance in the ice-motion field at tidal, inertial and storm-related periods may be missed.

The most detailed and accurate motion fields have been determined from scenes acquired by satellite-borne synthetic aperture radar, ERS1/2 and Radarsat (Kwok et al. 1990). The Radarsat Geophysical Processing System (RGPS) was developed by NASA-JPL and has been operating since 1999. RGPS basic products are based on moving tie points that are tracked from week to week throughout the entire basin of the Arctic Ocean. The initial spacing of the tie points at freeze-up is 5 km. Output products are ice motion, age histogram, thickness histogram, multi-year ice fraction, open water fraction and dates of melt onset and freeze-up. The Lagrangian ice motion product from the RGPS can be as accurate as Argos-tracked buoys or perhaps more so (Lindsay and Stern 2003; Stern and Moritz 2002). However, there are significant tracking problems when using the RGPS in summer, at the edge of radar swaths, in coastal regions and in regions where deformation is large. There may be benefit in using wavelet analysis for tracking ice features in SAR imagery (Liu et al. 1997). The RGPS provides highly detailed information on ice-cover deformation, which has stimulated new initiatives in ice mechanics, to be discussed in a later section.

Passive microwave sensors (SSM/I) provide extensive and frequent coverage of pack ice at much lower cost than radar. However, ice-drift determination from SSM/I is less accurate and more prone to tracking errors because the smallest resolved feature is 10–1,000 times larger than for SAR. Despite greater interference from weather, the 85-Ghz channel is preferred over lower frequencies because its resolution is better (pixel size is 12 km). Various investigators have evaluated the accuracy of ice velocity derived via SSM/I against the movement of drifting buoys. Comparisons reveal that the error in positioning features (±12 km at 95% confidence) is comparable to the pixel size (Agnew et al. 1997; Kwok et al. 1998; Meier et al. 2000). With this large uncertainty, 3-day displacements are relatively inaccurate except in rapidly moving ice (e.g. East Greenland Current). However, the data are well-suited for delineation of long-term average circulation (Martin and Augstein 2000). Mean fields of motion for both Arctic and Antarctic pack ice were first derived from these data for the period 1988–1994 (Emery et al. 1997; see Fig. 3.12), and NSIDC (Fowler 2003) now supports an updated analysis (1978-present) of ice motion that uses SMMR/SSM/I, AVHRR and buoy data in combination. Ice-tracking using images is ideal for delineation of the complex patterns of ice drift near coastlines and marginal ice zones, which are rarely sampled by drifting buoys. Unfortunately, errors are also larger in such regions where increased rotation and deformation of ice features reduces the cross-correlation between views at different times (Kwok et al. 1998).

Images acquired using space-borne scatterometer such as NSCAT (25-km pixel) and QuikSCAT (12-km pixel) – e.g. http://www.ifremer.fr/cersat/en/data/overview/gridded/psidrift.htm – can also be analysed to yield motion fields for the marine cryosphere. Liu et al. (1999) have pioneered this approach, using the wavelet transform to enhance resolution. Accuracy is comparable to that achieved with SSM/I, but conditions for optimal performance differ (Zhao et al. 2002). There is thus advantage in computing ice-drift vectors from images in the microwave band acquired via both radiometer and scatterometer.

All microwave methods produce best results where there is strong contrast between multi-year floes and the surrounding matrix of first-year ice, or between thick first-year ice and new ice in leads. Results in summer have not been encouraging, in part because microwave contrast between ice types deteriorates as the surface become warmer and wetter. Marginal sea-ice zones experience similar conditions year-round.

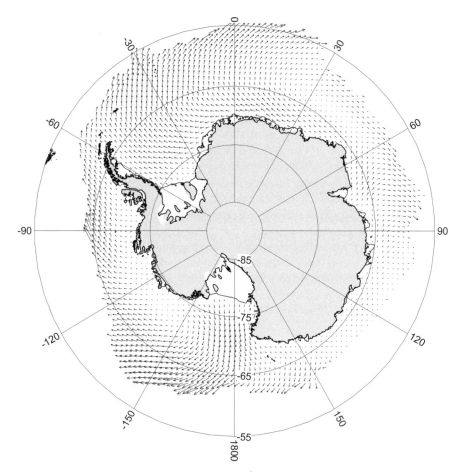

Fig. 3.12 Satellite view of the average circulation of pack ice around Antarctica, revealing three unbounded cyclonic gyres (Emery et al. 1997, Fowler 2003)

3.4.2.3   Doppler Sonar

As indicated earlier, long-term measurement of ice drift in the seasonal sea ice zone is challenging. Techniques based on tracking features in imagery are hindered by the high rate of pack-ice deformation and by the poor contrast between recognizable features and the background under melting conditions. Beacons deployed for tracking by satellite may quickly succumb to effects of ridging and break-up caused by high stresses and wave action or to melting as the pack drifts into warmer seas. During ACSYS, investigators have established Doppler sonar as an effective complement to other technologies for ice-drift measurement.

When measuring ice velocity, Doppler sonar is positioned upward-looking on a sub-sea mooring at a depth sufficient to minimize hazard from drifting ice. It provides an Eulerian measurement, velocity at a fixed point over time, not a trajectory or Lagrangian measurement such as obtained via beacon tracking. Integration of velocity provides a progressive vector or pseudo-trajectory, which may be a reasonable proxy for the actual trajectory of drift within a limited distance of the mooring.

The feasibility of Doppler sonar was established by Belliveau et al. (1990). The method was put to routine use for ice observation in the Beaufort Sea in 1990 (Melling et al. 1995), beginning a continuous record at several locations that now spans 14 years (Melling and Riedel 2004). Multi-year records have also been acquired in the Okhotsk, Pechora Seas and Greenland Seas, in Baffin Bay and in the Canadian Archipelago. Applications in climate research are reviewed by Mullison et al. (2004). Correlation sonar has been evaluated in the same application (Galloway and Melling 1997), but such instruments have not been widely used.

Because the underside of sea ice is extremely rough, the surface echo must be identified independently by the three or four pulsed sonar that comprise the instrument (e.g. RD Instruments); a specialized signal-processing algorithm is required. In open pack, there are a number of complications associated both with flat-calm water and surface waves (Melling et al. 1995; Visbeck and Fischer 1995; Zedel and Gordon 1996).

3.4.3   Thickness

3.4.3.1   In Situ Measurements

In situ methods of ice-thickness measurement include drilling for direct measurement, photography of blocks turned during ship transit for graphic analysis and installation of thermistor chains for temperature-gradient measurement.

Weekly measurements of land-fast ice thickness and on-ice snow depth from 195 sites across Canada were digitized by the Canadian Ice Service during ACSYS. Some of the Arctic series date from the late 1940s. Data are available at http:// ice-glaces.ec.gc.ca/. Flato and Brown (1996) have used time series from two locations with long records of snow depth on ice, Resolute Bay and Alert, to evaluate multi-year simulations of ice-thickness variation. With a reasonable assumption for oceanic heat storage, the model replicates seasonal and inter-annual variability in maximum ice thickness. Variation in the accumulation of snow on the ice is the dominant control of variation in end-of-season ice thickness.

An analysis of several half-century records in Siberian seas has provided evidence that trends in land-fast ice thickness over the past century have been small and generally not statistical significant. Brown and Coté (1992) reached the same conclusion from analysis of data from Arctic Canada. Dominant variability occurs in response to a low-frequency atmospheric oscillation of multi-decadal period (Polyakov et al. 2003). Unfortunately, many of the stations in Russia and Canada with long ice-thickness records were closed during the 1990s. Such stations have provided the only long records wherein the interplay between sea ice and snow cover could be studied.

Melling (2002) has examined ice-thickness data from 123,703 drill holes completed during the 1970s in the northern Canadian Archipelago. Here the ice pack is a mix of two populations: one is heavily ridged multi-year ice imported from the Beaufort gyre and the other a mix of relatively undeformed first-year, second-year and multi-year ice that grows and ages within the basin. The average thickness in late winter is 3.4 m, but sub-regional means reach 5.5 m. The ice thickness in this area is strongly influenced by heat flux that originates in the Atlantic-derived waters of the Arctic Ocean. The drift of ice through the basin is controlled by stable ice bridges that block channels in winter. Melling (2002) argues that a relaxation of this control in a warmer climate may bring worse ice conditions to the North-west Passage.

Tunik (1994) reported the first systematic determination of pack-ice thickness from ships in transit. Downward-looking cameras on the bridge wing of the *Rossya* photographed blocks of ice turned on edge by the action of the ship during a voyage across the Eurasian Basin in 1990. Photogrammetric analysis of suitably oriented blocks provided an ice-thickness distribution. Such sampling is biased by a number of factors, not the least of which is the mariners' search for the path of least resistance. However, the data provide a measure of ice conditions along such a path, directly related to the minimum difficulty of navigation. Average values mapped by Tunik (1994) are about 1 m smaller than values contoured by Bourke and Garrett (1987) from submarine sonar. Japanese scientists have used stereo photography for the same purpose in the Okhotsk Sea (Cho et al. 2002) and have added a forward-looking camera to determine ice concentration and floe size (Muramoto et al. 1994).

The thickness of sea ice during the freezing season can be deduced from measurements of temperature at depths spaced within the ice (Lewis 1967; Perovich and Tucker 1989). In typical installations, a chain of thermistors at 5–10 cm spacing is frozen into a small hole drilled through the ice. The chain extends from above the snow to well below the maximum anticipated depth of the ice. The upper and lower surfaces coincide with abrupt changes in the gradient of temperature during the cold season because the thermal conductivities of snow, ice and underlying water are greatly different. During the melting season, when the temperatures of the ice and the upper ocean are very close, the bottom of the ice cannot be identified using temperature; only the total bottom ablation between the beginning and end of the summer can be determined (Perovich et al. 1997). Ablation of the top surface can be followed intermittently throughout the summer since the lower atmosphere is frequently warmer than thaw temperature (Fig. 3.13).

Measurements of this type were made at nine locations with different ice forms during the year-long SHEBA drifting study (Perovich and Elder 2001). Despite very similar atmospheric and oceanic forcing, wintertime thickening and summertime ablation varied dramatically with site characteristics and snow cover.

Until the late 1990s, data were recorded locally for later retrieval. An installation as part of the North Pole Environmental Observatory in April 2000 was the first to exploit satellite communication links to relay data (Morison et al. 2002). Thermistor chains were augmented with an acoustic sounder to measure snow depth and with wind sensors, so that fluxes critical to the ice mass balance could be estimated. When data can be retrieved via satellite, the economics of Arctic logistics renders the instruments expendable.

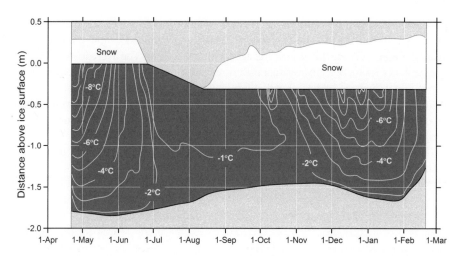

Fig. 3.13 Autonomous observations by an ice-mass balance buoy drifting in the central Arctic Ocean. The measurements of surface weather, internal ice temperature and snow depth elucidate the life progression of a multi-year ice floe in terms of thickness, draft and elevation of an ice floe (After Perovich and Elder 2001)

3.4.3.2   Top-Side Close Sensing

Instruments for 'close' sensing of ice thickness from above the ice include the laser profilometer mounted on aircraft and the electromagnetic induction radar mounted on aircraft, ships or sleds.

The laser profilometer was not used extensively for sea-ice measurement during ACSYS. Dierking (1995) reported the use of a laser operated from a ship-based helicopter to observe the height and spacing of ridges along a transect across the Weddell Sea in 1992. Hvidegaard and Forsberg (2002) operated an aircraft equipped with gravimeter, dual GPS receivers and an inertial measurement unit in addition to laser profilometer, to survey ice-covered waters north of Greenland. In this new approach, the first estimate of ice elevation was derived from the instrument data, with the geoid, accurate within a few decimetres, defined by the concurrent measurements of gravity. Elevation was adjusted such that minimum values were zero, on assumptions that the residuals have long wavelength and that minimum elevation is associated with ice-free water. Ice thickness was determined through multiplication by a factor of 7.84, based on climatological means for snow depth, ice density and water density derived by Wadhams et al. (1992). The surveys produced a map of average ice thickness (2–8 m) across the north of Greenland, with estimated 1-m standard error.

Prinsenberg et al. (1996) have completed an evaluation of the electromagnetic induction sensor, combined with laser profilometer and impulse radar in a probe suspended at 15–20 m altitude below a helicopter. In principal, the laser measures the range to the snow surface, the impulse radar that to the ice surface and the induction sensor measures the combined thickness of ice and snow. The application to sea ice is via a half-space model, wherein the ice (low salinity) is a layer of low electrical conductivity above a semi-infinite expanse of conducting seawater. The simplest version assigns zero conductivity and uniform thickness to the ice layer within the sensor's footprint. More elaborate specification of the geometry and conductivity is possible. The resulting thickness is dependent on the inversion model used.

In an evaluation conducted over seasonal pack in the Labrador Sea, Prinsenberg et al. (1996) demonstrated that the thickness of level floes can be measured remotely to an accuracy of 0.1 m. Over deformed ice, the electromagnetic induction sensor indicated greater and spatially more variable thickness, but it was not possible to establish the relation between indicated and geometric thickness in such features.

Topham and Bowen (1996) have compared measurements from the electromagnetic probe (with additional sensors) against a detailed ground survey of a large first-year ridge in the Beaufort Sea. They concur with the earlier assessment of the sensor for uniform ice. For ridged ice, they conclude that interpretation of the signal is seriously compromised by the large size of the footprint (radius equals twice the range) relative to the width of the ridge and by the loosely consolidated character of the ridge itself. Poor consolidation implies that a significant fraction of the ridge keel is flooded with electrically conductive seawater, so the dielectric transition from ice to ocean is gradual not abrupt. The thickness from the electromagnetic induction sensor was only about half of the 16-m geometric value for the feature surveyed.

Multala et al. (1996) have developed an electromagnetic induction sensor to be mounted in a Twin Otter aircraft and flown in conjunction with a laser profilometer. In trials over the Baltic Sea, they determined the horizontal resolution to be 100 m when flying 25 m above the ice. The thickness of uniform ice was accurate to ±0.2 m, but in deformed ice, with no independent knowledge of the ridge shape, the apparent thickness is only about half of the true value. The authors note that an assumption of zero bulk conductivity for deformed ice produces poor results.

Electromagnetic induction sensors can be suspended much closer to the ice from ships than from aircraft, with concomitant gain in horizontal resolution. Haas (1998) presents data from such a deployment from *Polarstern* in the Bellinghausen Sea. Results were good in level ice, but ridge thickness was underestimated. Better horizontal resolution was not achieved even within the sensor at a height of 4 m, apparently because the off-the-shelf instrument in use had low sensitivity. Haas comments that the results are biased not only by the mariner's choice of path through the pack but also because the action of icebreaking generates seawater-flooded cracks beneath the sensor, thereby decreasing the apparent ice thickness.

The 1990s saw a return to surface-based deployments of electromagnetic induction sensors, for improved spatial resolution. Worby et al. (1999) used an electromagnetic induction sensor at the surface of Antarctic sea ice in winter to determine its thickness. Accurate (±10%) values were obtained for undeformed ice, but data for deformed ice were much less good. Haas et al. (1997) report results obtained for Arctic ice under both freezing and thawing conditions, with results accurate within ±10% for level ice. A ground-based system mounted in a kayak was also used at

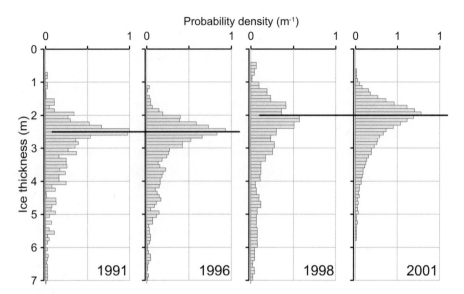

Fig. 3.14 Observations of thinning multi-year ice floes in the Trans Polar Drift during the 1990s, derived from direct bore-hole measurements across relatively level floes (After Haas 2004)

SHEBA to measure the summertime ablation of sea ice (Eicken et al. 2000). Measurements were sufficiently accurate (±0.05 m for 2-m ice) to reveal the progressive ablation along 13 repeated transects. The authors note that the presence of fresh meltwater in surface and sub-surface pools can influence the apparent thickness derived by the induction method. Surface ponds of typical depth and salinity bias thickness low by about 0.1 m and typical sub-surface ponds produce a comparable bias of opposite sign.

Haas and Eicken (2001) and Haas (2004) discuss the use of a ground-based electromagnetic induction sensor in a long-term project to examine variability in the thickness of floes in the Trans-Polar Drift (Fig. 3.14). During the 1990s, it was practical to survey 120 km of ice on 146 floes during four cruises because the method was simple and quick to use. The absence of ice-free and thin-ice fractions and under-representation of ridged ice (all for reasons of access) bias the results. However, the data are of value in representing inter-annual variability in the thickness of an identifiable ice class – large relatively undeformed floes that are safe to traverse in summer.

3.4.3.3 Bottom-Side Remote Sensing

Sonar provides the means for remotely sensing pack-ice thickness from below. Instruments in present use include fan-beam imaging sonar and narrow-beam profiling sonar. Platforms include naval submarines, sub-sea moorings and autonomous underwater vehicles.

Side-scan sonar has been operated from submarines under pack ice since the 1970s, providing plan-view images to complement the topographic sections acquired by ice-profiling sonar. Wadhams (1978, 1988) has reported that several ice types (open water, thin ice, first-year ice, multi-year ice and deformed ice) can be identified through subjective interpretation of side-scan imagery. Wadhams and Martin (1990) have attributed the blistered appearance of old floes in such imagery to refrozen under-ice melt ponds, which may play an important role in the thermal weathering of multi-year ice. An objective classification of a 140-km image swath by Sear and Wadhams (1992) provided quantitative data at high resolution on the fractional coverage by ice type and on the orientation and separation of quasi-linear features such as ridges and leads.

ACSYS heralded a new era of cooperation in the Arctic. Many naval data classified as secret were released for scientific study under the auspices of the joint US-Russia Environmental Working Group. The release included data acquired by ice-profiling sonar on US naval submarines within a zone of the central Arctic termed the Gore box. In addition, the USA sponsored six submarine-based Scientific Ice Expeditions (SCICEX) between 1993 and 1999, during which ice-draft data were acquired. Data from 14 cruises by US submarines between 1977 and 1993 were included in the Joint US-Russian Sea Ice Atlas (Arctic Climatology Project 2000). Since that time, observations from four SCICEX cruises (1993, 1996, 1997 and 1998) and from six earlier cruises (two by the UK) have been added to the archive (http://nsidc.org/data/g01360.html). Although the positions and times of data acquisition have been 'blurred' for reasons of national security, the scientific value is not compromised.

The US Navy introduced digital recording of sonar depth, target range and timing in 1987. This development greatly facilitated the retrieval and use of recent observations for climate research following declassification. Prior to 1987, echoes were recorded in analogue form on strip charts – 1800 rolls. Digitization and quality assessment of these records is currently (2004) being carried out at the Applied Physics Laboratory of the University of Washington.

An understanding of observational and sampling errors is essential to an assessment of significance for apparent secular change in ice draft. Wadhams (1997) examined mean values of ice draft derived from separate 50-km transects across nominally homogeneous pack ice in the central Arctic. He concluded that the standard error, representing the combined contribution from all sources (measurement, calibration, bias, sampling), was about 13% of the mean value, for means ranging from 3.7 to 4.9 m. Wadhams suggested that the total error scales with the mean value and decreases in inverse proportion to the square root of the transect length, but acknowledged that there was little empirical justification for these assumptions. An analysis of data from intersecting radial paths over the Chukchi Cap has provided support to the first assumption (Rothrock et al. 1999). The standard error over a 50-km transect with 1.2-m mean draft was 0.13 m (11% of the mean). Based on Wadhams (1997), the 95% bounds of confidence for the true mean over a 50-km section are about ±0.7 m at 3-m average draft and ±0.35 m at 1.5-m average draft.

During ACSYS, data from submarine-based sonar have been used almost exclusively to document the thinning of Arctic perennial ice. Wadhams (1991, 1992) documented a 15% decrease in the draft of ice (5.34–4.55 m) over a wide area north of Greenland between 1976 and 1987, and a progressive thinning of ice with southward drift within the East Greenland Current. Wadhams noted the sensitivity of regional ice thickness to variation in drift patterns. He concluded that attribution of the thinning to warming climate was not defensible at the time. An analysis of ice draft near the North Pole, measured six times between 1977 and 1990, led McLaren et al. (1992) to a similar conclusion – a trend could not be identified with confidence within appreciable inter-annual variation.

A re-examination of observations by Wadhams (1994) prompted the same conclusion regarding progressive changes. The outcome of an extended analysis of ice draft near the Pole, encompassing 12 submarine cruises between 1958 and 1992, was again a null result for thinning trend in the presence of inter-annual variation between 2.8 and 4.4 m (McLaren et al. 1994). Shy and Walsh (1996) examined the variability in the same data set in relation to ice-drift trajectories preceding the observation. They concluded that variations in average draft were correlated with the recent (2 weeks) direction of drift: prior motion towards Greenland (relatively close) was associated with higher average draft, whereas lower average draft was correlated with prior motion towards the Canada Basin.

By the late 1990s, a very different story was told. A paper by Rothrock et al. (1999) examined 50-km mean values of draft sampled at the same location in different years. Values measured in the mid-1990s were less than values measured between 1958 and 1977 at every crossing point of cruise tracks. The change was least (-0.9 m) in the southern Canada Basin, greatest (-1.7 m) in the Eurasian Basin and averaged about 42%. The study included very few data within the seasonal sea ice zone and none within 200 nautical miles of Canada or Greenland. The authors suggested that the observed decrease could be explained as the equilibrium response to a more than doubled oceanic heat flux (from 4 to 7 $W\,m^{-2}$), a 13% increase in poleward atmospheric heat transport or a similar increase in absorbed solar radiation. Additional data from 1976 to 1996 in the area between Fram Strait and the Pole revealed a comparable 43% reduction in average ice draft (Wadhams and Davis 2000).

Arguments presented in two papers published in 2001 suggest that the reduction in ice thickness has not been gradual, but occurred quite abruptly before 1991. Winsor (2001) examined data from six cruises along 150°W during 1991–1996, which provide no evidence of a thinning trend during this time. Winsor suggests that an understanding of natural variability in the thickness of Arctic ice may not be possible without time series much longer than those currently available. Tucker et al. (2001) examined a longer sequence of observations (1976–1994) along a single meridian (Barrow Alaska to the North Pole; see Fig. 3.15). They documented a dramatic drop in ice draft in the late 1980s and concurred with Winsor (2001) for the 1990s. The change in mean draft resulted from a decrease in the fraction of thick ice (more than 3.5-m draft) and an increase in the fraction of thin. Tucker et al. argue that the likely cause of thinner ice along 150°W in the 1990s is reduced storage of multi-year ice in a smaller Beaufort gyre and the export of 'surplus' via Fram Strait.

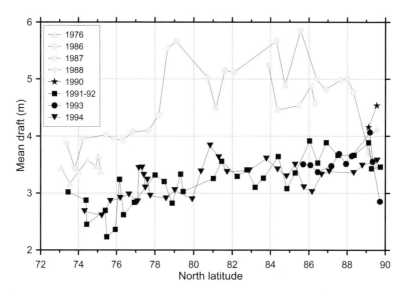

Fig. 3.15 Average pack-ice draft for 50-km segments of a transect along 150°W. Data acquired over an 18-year period via submarine sonar. Change in draft along this transect is manifest as an abrupt decrease by 1–1.5 m occurring between 1988 and 1990 (After Tucker et al. 2001)

Thus, a change in the circulation of the polar pack, driven by the Arctic Oscillation, was proposed as the primary cause of reduced ice thickness, with associated thermodynamic effects via the same forcing.

Rothrock et al. (2003) have rationalized the various reports of change in Arctic sea ice between 1987 and 1997. Their mathematical simulation of recent change is a plausible replication of the observations. The model suggests that ice thickness was about 0.6 m greater than the long-term mean in the mid-1960s and 0.8 m less in the mid-1990s. It indicates an increase in ice thickness since 1996.

Ice-draft measurement using specialized upward-looking sonar (viz. ice-profiling sonar, or IPS) on oceanographic moorings increased steadily during the ACSYS decade. The WCRP Arctic Ice Thickness Project (AITP), established in 1988 (WCRP-41 1990), had grown to encompass more than 20 Arctic installations during 2003–2004. Data have been acquired by this means continuously since 1990 in both the southern Beaufort Sea, by Canada, and in the western Greenland Sea, by Norway, Germany and USA. Although installations have been far too sparse to permit mapping the marine cryosphere, this new technology has been particularly useful in the monitoring of variability and change, in the measurement of ice flux, in observing areas not visited by submarines and in the study of ice processes. Thus, the use of submarines and moorings as platforms for sonar has been complementary in study of the marine cryosphere.

Ice-profiling sonar records the delay and amplitude of the surface echo, the total hydrostatic pressure and the pitch and roll angles of the sonar beam. Ice draft is calculated as the difference between the depth of the instrument and the vertical range to the target. The former is derived from the measurement of total pressure and estimates of sea-level pressure and seawater density; the latter is based on an estimate of depth-averaged sound speed. Instruments sample at intervals ranging between 1 and 300 s. The principal challenges to accurate observation are the uncertainties in sound speed and atmospheric pressure and the identification of spurious targets (bubble clouds, plankton, surface waves). Melling et al. (1995) provided a detailed technical discussion of these issues. They estimate 95% confidence limits for draft measured by moored IPS to be ±0.1 m under optimal conditions. Additional information on random and systematic errors, calibration and echo interpretation has been provided by Strass (1998) and Vinje et al. (1998).

Uncertainty in the range to the surface (and therefore in ice draft) is very sensitive to the varying speed of sound between the sonar and the target, which is rarely known. In standard practice, preliminary values of ice draft are 'calibrated' by subtracting the apparent draft of ice-free patches drifting over the instrument. Clearly, accuracy of draft is dependent on trustworthiness in open-water classification. The recognition of ice-free patches is difficult, especially during the freezing season when they are scarce and perhaps only a few metres in width (Melling et al. 1995). Incorrect classification of refrozen leads as open water can introduce several decimetres of negative bias, as well as random error, into time series of ice draft. Misclassification of refrozen leads is also a potential source of bias in ice-draft data acquired by submarine.

Does the amplitude of echoes provide guidance in identifying the narrow expanses of ice-free water that are needed for calibration? Bush et al. (1995) have reported that echoes from sea ice at 300 kHz to narrow-beam (2.6°) sonar are only 10% of the expectation for reflection. Melling (1998b) has since analysed echoes received from a variety of ice types during a 6-month deployment in the Beaufort Sea, and concurs with Bush et al. (1995) that echoes result from scattering at the ice-water interface, not reflection. The dominance of scattering over reflection precludes discrimination of targets from the amplitude of a single echo. Because scattering is a stochastic process, the strength of sequential echoes from a moving uniform target fluctuates over 10–20 dB, comparable to the average difference in scattering (13 dB) between an ice-free surface and thick first-year ice (Melling 1998b). Accurate calibration of ice draft is most practical when ice-free expanses can be resolved as features in a topographic profile of the ice. This typically requires that the ping interval be shorter than 10 s.

Melling et al. (1995) and Melling (2000b) have demonstrated that statistical measures of ice draft computed from time series at a fixed location frequently differ significantly from those computed from spatial profiles. The origin of difference is the variable speed of ice drift. The consequence is incompatibility between data acquired from submarines and from moorings. To facilitate blending of data from these complementary sources, Melling and Riedel (1995) recommend that the time series of draft be mapped to a pseudo-spatial coordinate and re-sampled at uniform spacing. The path surveyed by the IPS across the underside of the pack is the integral of drift velocity, which can be measured by Doppler sonar (Melling et al. 1995).

Many of the ACSYS-inspired scientific deployments of ice-profiling sonar in the 1990s were outside the area observed by submarine sonar (Fig. 3.4). Moreover, ACSYS coincided with a strong industrial interest in the use of ice-profiling sonar to identify ice hazards in coastal waters. Data collected by oil companies in the Okhotsk (1996–2002), Pechora (2001–2002) and Caspian Seas (2002) remain proprietary at this time, but those acquired for scientific purposes are now appearing in the public domain (Vinje and Berge 1989; Melling and Riedel 1993, 1994, 2004; http://nsidc.org/noaa/moored_uls/index.html). The release of data has been slow because of lingering concern about accuracy and bias in processed data. Workshops to discuss these issues were sponsored by ACSYS in Monterey USA in 1997 (WMO 2000) and in Tromso in 2002. The latter prepared guidelines for IPS data processing and documentation, which may be viewed via the URL for NSIDC given above.

The flux of ice through Fram Strait over a 6-year interval has been estimated by combining draft from sonar and drift from satellite-tracked beacons (Vinje et al. 1998). Melling and Riedel (1996a) have measured the ice exported from a flaw lead with ice-profiling and Doppler sonar in order to determine the production of new ice within it. Inter-regional differences in pack ice were illustrated by Moritz (1992) through comparison of annual draft-time series acquired by sonar in the north-eastern Chukchi, southern Beaufort and eastern Greenland Seas. Melling and Riedel (1995) documented the topographic properties of thick first-year pack ice in the Beaufort Sea. The development of Beaufort Sea ice through the cold season via both freezing and deformation has been described using IPS data by Melling and Riedel (1996a); see Fig. 3.16. These authors followed the development of distinct seasonal ice populations with different modal drafts, traceable to new ice formed during episodic openings of the Bathurst polynya. Their observations revealed that by late winter, ridges accounted for three-quarters of the volume of seasonal pack ice and half its area. Melling and Riedel (1996b) have described the properties of heavy multi-year ice during its incursion into Arctic shipping lanes in 1991. The thickness and ridging of pack ice at its wintertime southern limit in the Okhotsk Sea is the subject of a paper by Fukamachi et al. (2003). Ice-profiling sonar has contributed to studies of physical processes that accompany refreezing of flaw leads (Melling and Riedel 1996a; Melling 1998a; Drucker et al. 2003) and the formation of ice ridges (Amundrud et al. 2004). The latter authors have selected specific events wherein reciprocal motion of the pack enabled the sonar to measure pack-ice draft both before and after ice-field compression. The results approximate experiments in ice deformation since both the strain and the change in ridging can be measured.

Sonar mounted both on moorings and on small submersible vehicles has also been used for detailed study of the features in pack ice, principally ridges. Melling et al. (1993) and Bowen and Topham (1996) integrated information from sonar, theodolite survey and stereo photography to prepare three-dimensional maps of ridged first-year ice. Melling (1998a) combined top-side views ice pack ice via SAR on aircraft and satellite (ERS-1) ice-draft from moored sonar to evaluate the visibility of ridges from satellite. Goff (1995a, b) applied stochastic modelling to the analysis of sea-ice topographic profiles, and Hughes (2003) completed a statistical analysis of pack ice based on an ice-floe concept. Amundrud et al. (2004) have examined the distribution of ice volume by draft in ridge keels.

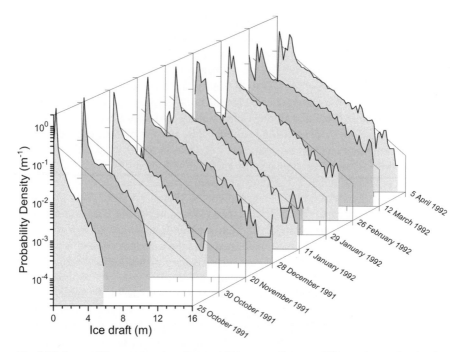

Fig. 3.16 Ice-profiling sonar on moorings provides an excellent capability to observe local variation of pack ice on time scales from days to decades. This cascade plot illustrates the progressive
development of the draft distribution within in seasonal pack ice of the Beaufort Sea during the
winter of 1991–1992 (After Melling and Riedel 1996a)

Connors et al. (1990) demonstrated the promise of autonomous underwater
vehicles (AUVs) for ice-draft measurement. There have been several initiatives in
the development of an AUV for long-range surveys under ice, with IPS in the complement of instruments (e.g. Autosub http://www.bodc.ac.uk/projects/uk/aui/ and
the ALTEX AUV http://www.mbari.org/auv/). The AUV has the potential to replace
manned submarines in long-range ice surveys. However, such platforms have yet to
provide data for Arctic cryospheric research.

3.4.3.4   Top-Side Remote Sensing

Two satellite-based methods for measuring thin ice were explored during ACSYS,
one based on thermal emission and one on kinematic analysis. In addition, there are
two developing technologies for the remote measurement of pack-ice thickness
from satellite – the microwave scatterometer and the laser.

Thin ice has disproportionate significance to climate relative to its fractional
coverage because it conducts heat more rapidly than does thick ice. The higher flux
is manifest in a higher surface temperature, which is measurable via thermal radiation.

Models of the surface energy balance can be solved to derive ice thickness, subject to assumptions about snow thickness and the temperature gradient in the atmospheric boundary layer (Groves and Stringer 1991). Lindsay and Rothrock (1995) used AVHRR images throughout 1989 to map the presence of thin ice and to investigate bias associated with the large pixel (1–4 km) of this sensor. In a companion paper, Yu et al. (1995) discussed the accuracy of ice surface temperature derived from AVHRR via three different algorithms. Miles and Barry (1998) have used imagery in both thermal and visual bands over a 5-year period to compile a climatology of lead density and orientation within the Beaufort gyre.

Yu and Rothrock (1996) used a one-dimensional thermodynamic sea-ice model and satellite-derived values of surface temperature and albedo to map ice less than 1 m thick. They compared results with measurements by ice-profiling sonar in the Beaufort and Greenland Seas, concluding that accuracies of 50% in thickness and one-tenth in concentration were achievable for ice up to 1 m thick. Yu and Lindsay (2003) reported a favourable comparison of ice-thickness distributions (less than 1 m only) derived from satellite AVHRR and from kinematic analysis of SAR images (discussed below). Where significant deformation occurred on scales less than 10 km, which is poorly resolved in AVHRR, the SAR analysis yielded more thin ice. The authors note a critical dependence of results on assumed snow accumulation.

Highly detailed (0.5 km) views of pack-ice deformation became a reality with the launch of Radarsat in 1995, using methods proposed by Stern et al. (1995). Automated tracking of the displacement of patterns (leads, floes, ridges) in the 3-day interval between images was implemented within the Radarsat Geophysical Processor System (R-GPS), developed at NASA's Jet Propulsion Laboratory (Kwok 1998). The quality of ice motion from the R-GPS has been assessed by Lindsay and Stern (2003) as a 323-m median difference in displacement relative to satellite-tracked buoys. A one-dimensional thermodynamic sea-ice model can be used to determine new ice growth in the divergent areas of the pack identified by the R-GPS. At the same time, the R-GPS follows the drift of Lagrangian elements. Kwok and Cunningham (2002) have exploited the approach to calculate the production of first-year ice over the Arctic Ocean during the winter of 1996–1997. Yu and Lindsay (2003) have compared the probability distributions of thin ice determined from the R-GPS and from thermal satellite imagery, as already noted. There is agreement within the large range of uncertainty of each method.

Ridges in seasonal pack ice represent up to three-quarters of the ice volume. A reliable means for counting ridges would provide useful information on the average thickness of such ice fields. Melling (1998a, b) has demonstrated that ridge counts on images acquired by aircraft-mounted SAR at shallow angle of incidence are tightly correlated with average ice draft from ice-profiling sonar. Unfortunately, satellite-based SAR was a poor performer in ridge detection even at high resolution (12.5-m pixel). The primary reason is poor contrast between level and ridged ice at steep viewing angles (20°).

The microwave altimeter has been used from Earth satellites to measure sea level for more than two decades. Early use in the study of sea ice was constrained by the low orbital inclination of existing satellites, which permitted only occasional coverage of the marine cryosphere. Progress in sea-ice applications prior to ACSYS has been summarized by Fetterer et al. (1992). Data from a space-borne microwave altimeter were first available at high latitude (to 81.5°) on the launch of the European Radar Satellite (ERS-1) in 1991. A successor, ERS-2, launched in 1995 operates to the present (2004).

The ERS microwave altimeter, orbiting at an altitude of 785 km, is used to measure the elevation above sea level of the top surface of the ice. Ice thickness can be estimated from elevation via scaling discussed in the next section. Since the scaling factor for thickness is typically between 5 and 10, the accuracy in measuring elevation must be much better than ±5 cm for data to have value in climate research. In addition to the usual altimetric requirements for precise orbital tracking, an accurate knowledge of the marine geoid and corrections for ionospheric delay, wet and dry tropospheric delays, ocean tides and the inverse barometer effect, there are two other sources of uncertainty. First, contributions to sea level that are dynamically coupled to the average and time variable components of ocean circulation must be determined. Second, the relationship between the apparent elevation of the radar target and a useful metric of sea-ice elevation within the footprint must be established. Since the footprint of the altimeter is not beam-limited, diffuse echoes may represent an area as much as 10 km in diameter, far greater than the decorrelation scale of pack-ice topography (approximately 30 m).

A first successful approach to processing backscatter to an altimeter from sea ice was described by Laxon (1994). An improved methodology has been developed by Peacock and Laxon (2004). The basis is largely empirical since theoretical understanding of microwave back-scattering from pack ice at normal incidence is limited. The researchers have identified two types of echo: a coherent or near coherent reflection at nadir, which is intense and short-lived, and a more diffuse weaker echo of long duration. It is evident from co-registered satellite imagery that the strong specular echoes are returned by calm ice-free water or new ice, whereas diffuse echoes signify either pack ice at very high concentration or wind-roughened open seas. Since specular echoes are dominant even if only a small fraction (1%) of the pack is calm water, only diffuse echoes can be used to determine ice elevation.

Peacock and Laxon (2004) have determined the mean sea level over the Arctic and its variability from echoes acquired during a 4-year period in the late 1990s. First, they sorted echoes into specular and diffuse types according to a pulse-shape parameter. Next, diffuse echoes were grouped into compact ice and open sea categories on the basis of synchronous SSM/I imagery. In ice-free areas, the diffuse echoes have been used to map sea-surface height by established techniques. Within pack ice, the specular returns provide this information, after correction (by up to a metre) for the vagaries of the ERS Ocean Mode tracking system and several stages of quality control and filtering. Peacock and Laxon (2004) have estimated the random uncertainty in sea-surface height from data along repeated orbital tracks and at crossover points to be about ±15 cm (95% bounds of confidence). The uncertainty combines the effects of measurement error and of ocean variability up to interannual period. Neglected sea-level variance associated with decadal oscillations in

Arctic Ocean circulation (Proshutinsky and Johnson 1997) might possibly be misinterpreted as variance and trend in ice thickness.

Onstott (1992) has reviewed understanding of microwave scattering from sea ice. The scattering of microwaves is strongly dependent on wavelength-scale surface roughness (either geometric or dielectric) and less so on bulk dielectric difference. Under cold conditions, the scattering from first-year ice is dominated by the influence of small-scale surface roughness because the high salinity of the near-surface layer causes abrupt attenuation of transmitted energy. In multi-year ice, the drainage of brine leaves the upper few decimetres almost fresh with numerous small pockets of air. Microwaves pass through the fresh upper layer with little attenuation, scattering strongly from the bubbles. The relative importance of volume and surface scattering, and therefore the relationship between the radar-derived and geometric elevations of multi-year ice, is not known.

The elevation of sea ice is determined from diffuse echoes to the altimeter. These incorporate backscatter from a wide area of pack ice with elevation ranging from near zero to as much as 6 m; the distribution of volume with elevation is not uniform, but exponential (e.g. Wadhams et al. 1992). It is not obvious whether the range to the target, evaluated at some specified amplitude threshold for the echo, corresponds to the mean, mode, median or some more complex statistic of surface topography.

At temperature below about $-10°C$, the snow on ice is relatively transparent to microwaves (Barber et al. 1994). Liquid water that appears within the snow at higher temperature causes strong attenuation of propagating microwaves. Even at relatively low fractions of liquid water, the penetration depth in snow is close to zero, and backscatter comes entirely from the snow surface. Because the ice surface is not detected under such conditions, the use of radar altimeter to measure sea-ice elevation must be restricted to the cold months, perhaps October to April in the central Arctic (Laxon et al. 2003). A tighter seasonal restriction may be appropriate in marginal ice zones.

Laxon et al. (2003) present estimates of Arctic ice thickness from radar altimeter, averaged over the cold months (October–March) and averaged over 1993–2001 (Fig. 3.17). The average excludes ice thinner than 0.5–1 m because ice in this range of thickness is thought to generate specular echoes which are excluded from analysis. Laxon et al. calculated thickness by multiplying elevation by 9.41, with adjustments for snow burden based on a climatology prepared by Warren et al. (1999). The measurements reveal a plausible geographic variation of thickness, with values increasing from about 2 m near Siberia to 4.5 m off the coasts of Canada and Greenland. Values derived independently by altimeter and by submarine sonar in the 1990s (excluding ice thinner than 0.5 m) are correlated and on average almost equal. Residuals of the linear regression span ±1 m.

A laser altimeter was launched on ICESat in 2003. It measures the delay for the surface reflection from targets 70 m in diameter, spaced at 170-m intervals. Kwok et al. (2004b) have reported preliminary results. Their derived map of ice-surface roughness on scales larger than 170 m is similar to a map of microwave backscatter derived via QuikSCAT. The spatial variation in nominal ice elevation and reflectivity from the laser along track lines is correlated with features in coincident images from

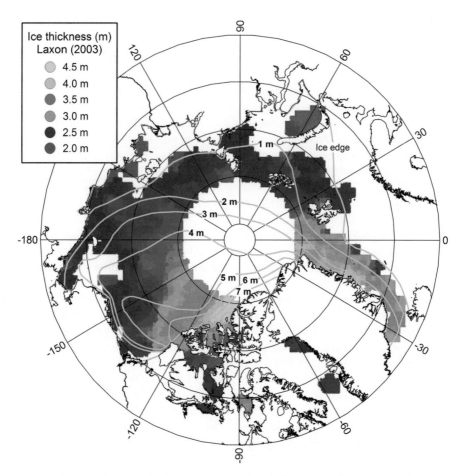

Fig. 3.17 Arctic ice thickness derived by radar altimeter from Earth satellite. The map depicts the average over all values exceeding about 0.5 m for the winter months between October 1993 and March 2001 (After Laxon et al. 2003). The overlay shows contours of wintertime ice draft derived by submarine prior to the mid-1980s (After Bourke and Garrett 1987)

Radarsat. Ice elevation fluctuates over tens of centimetres. The authors point out that uncertainty in the geoid and in sea-surface topography presents big challenges since thickness is nine times the apparent elevation. The unknown thickness of snow has similar impact. As with the microwave scatterometer, the statistic of ice elevation over the footprint (mean, mode, median, etc.) that corresponds to the detected range has yet to be determined.

3.4.3.5 Ice Thickness from Draft or Elevation

Although the dynamics and thermodynamics of sea ice are formulated in terms of ice thickness, most methods of measurement do not yield this parameter directly.

Moreover, it is clear from many detailed studies of deformed ice that the volume occupied by a ridge may only be 60–80% ice (e.g. Leppäranta et al. 1995). The remainder is air (in the sub-aerial portion) and water below the surface. These voids are certainly important in many aspects of pack ice dynamics. It is likely that ice thickness, defined as the distance between the most highly elevated and most deeply depressed ice at a point, is a necessary but not sufficient requirement of observations. A complementary estimate of ice mass per unit area might be of value.

Melling et al. (1993) and Bowen and Topham (1996) have examined the inter-relationship of ice elevation, draft and thickness via detailed three-dimensional surveys over sizeable areas of deformed first-year ice. They noted extensive rafted areas within deformation features, misalignment of sail and keel features, indications of porosity and appreciable local isostatic imbalance. The observations illustrate that constant factors for conversion of draft or elevation to thickness are inappropriate on scales at least as large as those investigated (235 m). For example, the 95% bounds on conversion factors to thickness at the largest scale were (1.1, 1.2) for draft and (5.9, 10) for elevation, representing thickness uncertainty of 9% and 70% if derived from draft or elevation, respectively. On a 10-m scale, the corresponding bounds were (1.05, 1.33) with draft and (4, 20) with elevation.

Bourke and Paquette (1989) acquired data on draft and thickness from 1,547 holes drilled in multi-year ice. The conversion factor to thickness based on linear regression was 1.130 for draft and 7.7 for elevation. The 95% bounds on residuals were ±0.54 m, substantially greater than the average difference between thickness and draft.

One study has been conducted from a statistical perspective, comparing sonar and laser data along nearby but not coincident survey tracks (Wadhams et al. 1991). Comiso et al. (1991) obtained congruent probability densities for elevation and draft when the former were scaled by 7.91. Wadhams et al. (1992) have used the data to prepare a seasonally varying ratio of draft to elevation, dependent on ice draft, estimated snow burden and seawater density. For 3-m ice, the ratio varies between 4.6 in early June and 8.0 in late August. The corresponding conversion factors to thickness on a 50-km scale are in the range (1.13, 1.22) for draft and (5.6, 9) for elevation, with snow burden being the dominant cause of variation.

There is presently an observational conundrum in that estimates of ice thickness from draft are much more accurate than those from elevation, but top-side surveillance via satellite is the only feasible means to acquiring hemispheric observational coverage.

3.4.4 Export

3.4.4.1 Fram Strait

The Norwegian Polar Institute has maintained moorings with sonar to measure ice draft in Fram Strait since 1990 (Vinje et al. 1998). In the early years of ACSYS, ice drift was determined via satellite image analysis, but Doppler sonar was added to some moorings in the late 1990s. The number of installations was increased with the European VEINS initiative in 1997. To date, the observational focus has been the East Greenland Current over the continental slope. Logistic constraints have precluded measurement of the ice flux that occurs over the broad East Greenland shelf.

Based on these observations, Vinje et al. (1998) reported a mean draft (open water excluded) during 1990–1996 of 2.88 m, and an annual cycle with a maximum in April–May and a minimum in September. They estimated the southward flux of ice area via an empirical relationship between monthly mean drift speed and along-strait geostrophic wind. The values range between 0.82 and 1.53×10^6 km^2 year^{-1}, with the minimum in 1990–1991 and the maximum in 1994–1995. They estimated the flux of ice volume via the ice-area flux and an estimate of the cross-strait profile of average ice draft. Amid large inter-annual variation, ranging between 2,046 km^3 in 1990–1991 and 4,687 km^3 in 1994–1995, the 6-year mean was 2,850 km^3 year^{-1}. Martin and Wadhams (1999) discuss ice flux in the East Greenland Current during 1993–1994, based on similar data and analysis. Meridional profiles of flux, illustrating the flux divergence associated with melting, are a unique aspect of this discussion.

Kwok and Rothrock (1999) have estimated the ice-area flux through Fram Strait directly, through analysis of SSM/I image pairs. Data were unavailable during the warm months (JJAS) when images are not suitable for analysis, but values were provided via a regression of flux against geostrophic wind. During 1978–1996, wintertime fluxes ranged between 0.45×10^6 km^2 in 1984–1985 and 0.91×10^6 km^2 in 1994–1995, with values strongly correlated to the NAO. Including the summer months, they calculated an annual mean of 0.92×10^6 km^2 year^{-1}. Incorporating the cross-stream profile of ice draft derived by Vinje et al. (1998), they derived an average ice-volume flux equal to 2,366 km^3 year^{-1}. A disproportionate fraction (76%) of this flux occurred during the eight winter months.

Kwok et al. (2004a) present an improved analysis of ice export through Fram Strait, updated for 1978–2002; their revised value for annual mean area flux is 0.866×10^6 km^2 year^{-1} (Fig. 3.18). Their analysis has revealed that the cumulative export of pack ice from the Arctic in the 1990s was greater than that in the 1980s by 0.4×10^6 km^2. The higher efflux in the 1990s may be at least part of the reason for decreasing ice area and thickness in the Arctic during the 1990s. During the 1996–1997 winter specifically, the flux of ice area through Fram Strait was sufficient to account for the entire observed shrinkage of the perennial pack in the central Arctic (Kwok et al. 1999).

Estimates of ice flux throughout the last half of the nineteenth century have been prepared by Vinje (2001b); ice-area flux was calculated via a regression on geostrophic wind and ice-volume flux using the cross-stream ice-thickness profile of Vinje et al. (1998). The mean annual export is 2,900 km^3, and the time series is without significant trend over 50 years. Decadal variability dominates, with anomalies reaching 40% of the mean. Efflux is strongly correlated with 1-year lagged change in multi-year ice area in the Arctic.

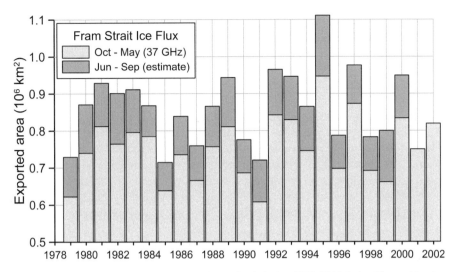

Fig. 3.18 Area of ice leaving the Arctic via Fram Strait during 1978–2002 derived by tracking ice features in sequential scenes acquired by the SSM/I. The analysis reveals an increased ice export from the Arctic during the 1990s (After Kwok et al. 2004a)

The principal weakness of ice-flux estimates in Fram Strait is the shortage of ice-thickness data. The ice stream extends between about 3°E and 15°W at 79°N, but there are observations only within 0–7°W, and only in certain years. These observations have been used to parameterize the variation in mean ice thickness across the entire outflow. Dependence of mean thickness on distance from the ice edge has not been incorporated, and possible variation in mean values over time has not been acknowledged in analyses that extend beyond the period of direct measurement, 1990-present.

3.4.4.2 Barents Sea

The mean drift of ice in the Barents Sea is about 6 cm s^{-1} to the southwest, in response to prevailing wind (Vinje and Kvambekk 1991). The average drift of ice determined via satellite image analysis, 1988–1994, is consistent with this result in the central Barents Sea, but indicates a slower net influx from the northeast, perhaps 1 cm s^{-1} and approximately zero net influx from the north (Emery et al. 1997). Drift through the opening from the Kara Sea, about 360 km wide, might therefore be as much as 0.1×10^6 km^2 year^{-1}. Vinje and Kwambekk (1991) estimate a loss of 500 km^3 year^{-1} from the Arctic via this path and a small gain of 50 km^3 year^{-1} via the opening to the west of Franz Josef Land.

Unfortunately, this issue has received little systematic study. The estimates here are at least 10% of the export through Fram Strait. Moreover, the melting of pack ice in the northern Barents Sea has been implicated in the formation of the cold, lower-halocline water mass found throughout the central Arctic (Steele et al. 1995). Ice flux into the Barents Sea merits closer attention.

3.4.4.3   Bering Sea

During the months (November–May) when ice is present in the Bering Sea, it drifts at an average 5 cm s^{-1} towards the southwest driven by prevailing wind (Emery et al. 1997). The movement constitutes the 'conveyor belt' described by Pease (1980), wherein ice forms rapidly in the polynyas of the northern Bering Sea and subsequent drift south to melt on contact with warmer sub-polar waters. A similar regime exists off the eastern coast of Canada and in the Okhotsk Sea.

Although ice drifts episodically in both directions through Bering Strait in winter, Kozo et al. (1987) examined 11 years of satellite images and concluded that the average flux of area is southward. In summer, ice may stream southward along the western side of the strait in some years, most recently 1998. Existing studies have been summarized by Melling (2000a).

There has been no long-term quantitative study of ice flux into the north Pacific. A rough calculation based on Emery et al. (1997), an assumption of five-tenths ice concentration and a 4-month ice season indicates that the ice-area export could be as large as 0.35×10^6 km^2 year^{-1}, a third of the export through Fram Strait. Assuming a 0.3-m average ice thickness, the corresponding volume flux is 100 km^3 year^{-1}. There is no systematic measurement of pack-ice thickness in the Bering Sea.

3.4.4.4   Canadian Archipelago

Melling (2000a, 2002) has reviewed existing knowledge of ice flux within the Canadian Archipelago. The flux is seasonal because all channels are blocked by land-fast ice for several months each year. Among the three principal pathways from the Arctic Ocean, Nares Strait has the most dynamic ice cover. Ice starts to move here between mid-July and mid-August and continues typically until some time between late December and late March. In some years, it never consolidates (Agnew 1998). Parry Channel becomes mobile at about the same time but freezes up earlier, between mid-November and mid-January. Pathways through the Sverdrup Basin are blocked by late November. The minimum combined width of contributing channels is 125 km, about one-third of the width of the ice stream in Fram Strait.

The short season for drift ice within the Archipelago is countered to some extent by the high average ice thickness, typically 3–5 m. However, there has been no long-term systematic study of fluxes through the channels. Kwok et al. (1999) have determined that about 0.034×10^6 km^2 of multi-year ice passed through Nares Strait during the winter of 1996–1997. Since a five-tenths concentration of multi-year ice is typical in Nares Strait, the total export may have been twice this value. Melling (2000a) provided a rough estimate of ice-volume flux through the Archipelago, 480 km^3 year^{-1}, which is about 20% of the Fram Strait value.

There is also a substantial production of ice in winter on the Atlantic side of the fast-ice barriers in the Archipelago. The ice-covered area is 1.2×10^6 km^2 at maximum extent in March. The volume of ice formed each winter is 2,400 km^3 if the average thickness is 2 m. Much of this ice drifts south to melt at the polar front in the Labrador Sea, forming a 'conveyor belt' between freezing and melting areas similar to that on the Pacific side of North America. Loder et al. (1998) have estimated the average annual ice flux through Davis Strait to be 1,100 km^3 year^{-1}, a large fraction of that which passes through Fram Strait.

At present, observations are too sparse to permit accurate estimation of ice fluxes through Canadian waters.

3.4.5 Atmosphere-Ice-Ocean Interaction

3.4.5.1 Principal Initiatives

The reasons for the existence of perennial sea ice in the Arctic Ocean and the identification of climatic changes that might cause its demise were central concerns of ACSYS. Ultimately, the understanding sought by ACSYS will emerge from numerical models, for only models have the potential for 'comprehending' the complex non-linear interactions that link the ocean, ice and atmosphere. However, because the interactions span wide ranges in space and in time, they cannot be fully resolved within numerical models. The impact of small-scale phenomena must be parameterized at the climate-modelling scale (nominally 100 km). The formulation of realistic algorithms may require detailed understanding of inhomogeneity and physical processes at smaller scales, which must be obtained through observations and their interpretation.

Important issues of parameterization on an aggregate scale are:

- Sensitivity of the surface energy balance to ice thickness, age, salinity and density, to melt pond area and depth and to lead fraction
- Sensitivity of the surface energy balance to snow cover
- Effect of leads on the fluxes of momentum, heat and moisture into the Arctic atmosphere
- Impact of the ice-surface characteristics on Arctic boundary-layer cloud and cyclogenesis
- Formation of open water areas as a function of pack-ice characteristics and wind forcing
- Processes in the oceanic mixed layer, including ice-bottom accretion and ablation, and the impact of brine rejection and meltwater on the rates of these processes
- Impact of brine and meltwater injection on the overall water-column stability of the Arctic Ocean

For obvious reasons, there is overlap on these issues with the atmosphere and ocean tasks of ACSYS. Only aspects most closely tied with ice will be reviewed here.

The chief activities addressing these objectives during ACSYS were the Sea Ice Monitoring and Modelling Site (SIMMS), maintained by Canadian scientists in the central Canadian Archipelago during 1990–1997 (e.g. LeDrew and Barber 1994; Barber et al. 1994) and the Surface Heat Budget of the Arctic (SHEBA) study, funded by the US National Science Foundation. The field phase of this project began in the Canada Basin in October 1997 and continued for 12 months. The Joint Ocean Ice Study (JOIS) was a Canadian component that focussed on biological and hydro-chemical issues.

Support to SHEBA encouraged retrospective analyses of data gathered during AIDJEX and other earlier drift-camp projects. McPhee (1992) re-examined mea-surements of oceanic turbulence from three such projects and proposed a simplified model for heat and mass flux at the freezing interface. Romanov (1995) prepared an atlas using systematic measurements of Arctic snow and ice thickness and mass balance collected during Russian expeditions over a half century. An analysis of the snow data was published by Warren et al. (1999). Maykut and McPhee (1995) used data from AIDJEX to conclude that the flux of oceanic heat to pack ice in the central Arctic is derived mainly from solar radiation absorbed beneath leads and not from deeper layers in the ocean. New parameterizations of radiative fluxes to sea ice were developed by Key at al. (1996), with an estimated accuracy of 4–21% depending on cloud cover. Podgorny and Grenfell (1996) developed a model to represent the partitioning of solar radiation within sea-ice melt ponds.

Three relevant review papers were published in the proceedings of a NATO Workshop on the freshwater budget of the Arctic in 1998 (Alekseev et al. 2000; Rothrock et al. 2000; Steele and Flato 2000).

3.4.5.2   Climate Sensitivity of Level Ice

There have been several initiatives in the use of one-dimensional thermodynamic models of sea ice to explore the sensitivity of ice thickness to varying climate. Flato and Brown (1996) used a model of this type to simulate the growth of seasonal land-fast ice and its inter-annual variation at coastal locations in Arctic Canada. They demonstrated that inter-annual fluctuations in the amount and timing of snow accu-mulation had greater impact on variability in ice thickness than did anomalies in air temperature. The model indicated that small (less than 3 W m^{-2}) changes in annual mean surface energy balance could force a transition from seasonal ice that melted in summer to much thicker permanent ice.

Dumas et al. (2003) used the same model, driven by meteorological observations at Russian drifting stations during 1954–1991, to simulate inter-annual variations in the thickness of multi-year ice in the central Arctic Ocean. They concluded that appreciable (±0.5 m) variation in multi-year ice thickness could be driven by varia-tion in thermodynamic forcing. Trend attributable to thermodynamic changes was negligible over the study period east of the data line and negative (−0.3 m per decade) during 1976–1991 north of 80°N. Ice thickness was anomalously low throughout the area between the late 1960s and the mid-1970s. Lindsay (1998)

worked with the same data and a similar model to define the average seasonal variation in the surface energy balance for sea ice. Net radiation was a significant forcing
factor year-round, dominating both heat loss from the ice surface in winter and melt
rate in summer. The flux of sensible heat from the atmosphere was negligible during
melting but a source of heat in winter, partially offsetting radiative losses.

The ice-albedo feedback mechanism was investigated using one-dimensional
models by Curry et al. (1995). They concluded that this feedback can operate via
the influence of surface moisture and melt ponds without the opening of leads.
The strength of feedback was reduced if a thickness distribution for sea ice was
included.

3.4.5.3 Freezing

New studies of freezing during the 1990s include an investigation of snow-ice formation by Maksym and Jeffries (2000). Heavy snow accumulation on young ice is
a common occurrence in the marginal sea ice zone. The cover of snow permits
warming of the ice and consequent widening of brine channels. If the weight of
snow is sufficient to depress floes below sea level, the surface floods and snow-ice
is formed. These authors investigated snow-ice formation using a one-dimensional
model that permitted the upward percolation of seawater. They discovered a feedback between the occurrence of flooding and future snow-ice formation, wherein
percolation from below increases brine volume and permeability of the ice, thus
facilitating future percolation.

Sturm et al. (2001) examined spatial inhomogeneity in the conductive heat loss
from sea ice in winter near SHEBA. They measured variations in conductive flux by
a factor of four over a distance of 10 m, attributable to variation in snow depth and
ice topography on the same scale. They concluded that heat conduction horizontally
in sea ice is an important aspect of its growth in winter (cf. Melling 1983). The issue
was further explored by Sturm et al. (2002), who reported a twofold discrepancy
between the effective thermal conductivity of snow derived from measurements of
ice growth in combination with a one-dimensional ice growth model and values
measured directly, which are lower. They use a two-dimensional finite-element
model to illustrate that half the discrepancy is caused by heat conduction horizontally to areas poorly insulated by thin snow cover. The cause of the remaining discrepancy remains uncertain, although convective heat transfer within the snow may
play a role.

Overland et al. (2000) used images from AVHRR to determine ice-surface temperature and surface energy balance under clear skies in winter. The distribution of
data values within a 100-km^2 region around the SHEBA site was skewed, with most
data clustered about SHEBA measurements but some forming a tail to higher value.
This result demonstrates that although the sensible heat flux at SHEBA (thick ice)
was downward, that over areas of thinner ice was upward, from ice to atmosphere.
The average flux of sensible heat over the region was close to zero. Measurements
of surface fluxes in the vicinity of leads reinforce the conclusions of this indirect

analysis. Ruffieux et al. (1995) reported a sensible heat flux of 150 Wm^{-2} near the downwind edge of a 1-km wide lead covered with 0.1-m ice. Downwind of another 100-m wide lead, the sensible heat flux was 20–50 Wm^{-2} upward, whereas remote from the lead the flux was downward and 0–10 Wm^{-2}. The net long-wave flux is heavily weighted towards radiation from thin ice because the intensity of upwelling radiation is proportional to the fourth power of surface temperature. In the analysis of Overland et al. (2000), the regional average of upwelling long-wave radiation was upward everywhere and 22% greater than local measurements at SHEBA.

Kwok and Cunningham (2002) have integrated measurements of ice drift and deformation from the R-GPS with models for ice growth based on freezing-degree days (snow depth is implicit) and ridging-rafting to estimate ice production in winter. They determined the total production of ice over the Amerasian Basin during the 1996–1997 winter to be 1,000 km^3, equivalent to a 0.4-m average thickening. There is no independent means of evaluating such an estimate at this time.

3.4.5.4   Ice-Bottom Ablation

Sensible heat that causes melting at the bottom of sea ice reaches the upper ocean from two main sources, the deeper ocean via upwelling and diffusion and the sun via absorption of short-wave radiation in leads. In traditional views of the Arctic Ocean, the flux of heat towards the ice from the deeper ocean (viz. the Atlantic inflow) is intercepted by the cold halocline. The maintenance of low temperature within the cold halocline has been attributed to intrusions of shelf water either chilled and salted via freezing (Aagaard et al. 1981; Melling and Lewis 1982) or chilled and freshened by melting (Steele et al. 1995), and to submergence of the polar mixed layer beneath shelf-water outflow (Rudels et al. 1996). Steele and Boyd (1998) presented evidence for an unprecedented retreat of the cold halocline from the Eurasian Basin during the early 1990s. Although thinning of ice in this sector is a likely consequence of halocline warming, a causal connection with the loss of ice in the 1990s has been difficult to establish. There were strong indications that a cold layer within the halocline had been re-established by the late 1990s (Bjork et al. 2002). Martinson and Steele (2001) have suggested that a vanishing cold halocline in the Arctic might herald a predominately seasonal ice regime similar to that of the Antarctic.

Maykut and McPhee (1995) have demonstrated that solar radiation absorbed in the upper ocean beneath leads is the dominant source of heat flux to ice on an annual average, even though the diffusive flux cannot be ignored during the dark months. Perovich and Elder (2002) have completed a similar study through the pycnocline at SHEBA, coming to the same conclusion. However, at SHEBA (1997–1998), the annual average oceanic heat flux was more than twice that measured at AIDJEX in 1975. Since a larger lead fraction at SHEBA was not the cause, the authors speculate that the thinner ice at SHEBA may have permitted a greater penetration of solar energy through snow-free ice. Perovich and Richter-Menge (2000) have calculated that in the instance of young ice in refreezing leads, 30% of the incident solar energy is absorbed in the ice and 25% in the underlying ocean. Skyllingstad et al. (2003) have integrated measurements of under-ice turbulence and a large-eddy simulation to examine the influence of ice topography on the oceanic heat flux. They established that flow disturbance by a 10-m ice keel caused fivefold enhancement of flux for several 100 m downstream. Clearly, under-ice topography has an important influence on the surface energy balance. Martin and Wadhams (1999), within the East Greenland drift, and Melling (2002), in the Canadian Archipelago, have documented the preferential thinning of thick ridged ice by oceanic heat flux. Theoretical attempts to understand these observations have been so far unsuccessful (Schramm et al. 2000).

3.4.5.5   Ice-Top Ablation

Ablation at the top surface of sea ice was an important element of ACSYS because of the significance of sea-ice-albedo feedback to global climate. Hanesiak et al. (1999) alerted researchers to the importance of the diurnal cycles in thermodynamic forcing of sea ice, especially during the melting season. They compared results of simulations using hourly and daily average meteorological data to drive a thermodynamic sea-ice model. With resolution of the diurnal cycle, the melting of snow began earlier, and the decay of ice was completed 13 days earlier. The principal contributor to differences was the non-linear response to snowmelt that occurred at peak daily insolation; this reduced snow albedo and increased solar absorption. Perovich et al. (2001) documented the seasonal changes in the morphology of sea ice – the break-up of floes, the persistence of ice-free leads, the smoothing of ridge sails to hummocks, the formation of melt ponds – and their impact on the disposition of solar radiation. Yackel et al. (2000) surveyed features associated with the ablation of land-fast ice in the Canadian Archipelago, identifying four surface classes – snow, saturated snow and light and dark coloured melt ponds. Aerial photographic surveys of the same type were completed by Perovich et al. (2002a) over multi-year sea ice near SHEBA. The observations have provided information on seasonal changes in the albedo of multi-year sea ice in the Arctic (Perovich et al. 2002b).

The reduction of sea-ice albedo that accompanies the melting of snow and formation of ponds on sea ice can be clearly seen from space (Lindsay and Rothrock 1994), when skies are clear. The rarity of cloud-free skies over the Arctic Ocean in summer has encouraged effort to identify proxy indicators of melting. An increase in microwave emission from seasonal ice in spring is indicative of the onset of melting (Anderson 1987). This signal has been exploited in a number of studies of multi-year ice, most recently by Anderson (1997) and Smith (1998a, b). Smith (1998a) has used the SMMR-SSM/I time series to detect a 10-day increase in the length of the melt season between 1978 and 1996 (Fig. 3.19). The method is constrained by the large spatial footprint of space-based microwave sensors, which may encompass several types of ice plus open water, and their sensitivity to atmospheric moisture.

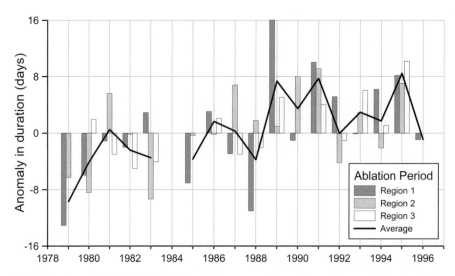

Fig. 3.19 Anomalies in the length of the ablation season for pack ice of the central Arctic during 1978–1996 deduced from change in the microwave emission from snow when wet. The duration of conditions conducive to ablation has increased by an average 5 days per decade (After Smith 1998a)

There is also a change in the backscatter of microwaves from sea ice at the onset of melting in the surficial snow. Whereas cold dry snow is almost transparent to microwaves, liquid water in warm snow typically causes a 2-dB reduction in transmitted energy (Winebrenner et al. 1994). The two-way path through a melting snow decreases backscattered flux from multi-year ice by about 4 dB even though the scattering cross-section of ice itself varies little with temperature. A larger (−10 dB) transient (several days) in backscatter can also be a convenient marker for the onset of melting (Winebrenner et al. 1998). Yackel et al. (2001) used the identified response in average backscatter to determine the onset of melt in the vicinity of the North Water polynya in Baffin Bay. Kwok et al. (2003) have used the change in average backscatter, plus two other metrics of the probability distribution of backscatter, as indicators of melting in springtime on pack of mixed ice types. The onset date of melt deduced from SAR was consistent with direct measurements of albedo at SHEBA in 1998 and matched within a few days the timing of 0°C measurements by drifting buoys. Discrepancies between dates derived using SAR and using SSM/I were larger, a 7-day bias and a 2.5-day standard error.

The scattering cross-section of first-year ice increases with the expansion of brine pockets as this ice type warms (Barber et al. 1995; Barber and Nghiem 1999). Since this increase counteracts the attenuation of wet snow, the backscatter from this ice type changes little with springtime warming. The different responses of first-year and multi-year ice to warming contribute to the much reduced microwave contrast between the two principal ice types during the summer months.

Yackel and Barber (2000) have examined change in microwave scattering from sea ice with the appearance of melt ponds later in the season. Backscatter increases at this time, primarily because of strong returns from the wind-roughened surface of ponds. The correlation between scattering cross-section and pond-covered area was high when wind was strong (5.3 m s^{-1}), but of little value otherwise.

3.4.5.6 Autonomous Monitoring of Surface Energy Balance

The thickness of sea ice represents the time integral of any small imbalance in energy fluxes at the surface. Ice thickness during the freezing season is readily measured using strings of thermistors passing through the ice. In the early 1990s, thermistor strings were integrated within satellite-reporting platforms to determine the growth and ablation of unmanned drifting ice floes. Additional sensors were added so that the reason for observed change could be established. The Ice-Ocean Environment Buoy (IOEB), first deployed in 1992, carried wind, temperature and pressure sensors for atmospheric measurement, ice stress sensors, thermistor strings and an echo sounder to detect the floe bottom.

The successor to the IOEB, the Compact Arctic Drifter (J-CAD) developed by the Japan Marine Science and Technology Centre, lacks ice sensors. However, an ice mass balance buoy (IMB) has been developed by the US Cold Regions Research Engineering Laboratory specifically for sea-ice measurement. For initial deployments of the IMB, thermistors were the only sensors (Perovich et al. 1997). An early installation on a snow-free 2.13-m thick floe at the end of the melting season revealed 0.45 m of accretion during the winter and 0.92 m of ablation during the summer, 73% from the top surface. Oceanic heat flux was calculated as 4 W m^{-2} on average.

Nine IMBs were installed in different ice environments near SHEBA during 1997–1998; one was supplemented with an echo sounder to measure snow depth (Perovich and Elder 2001). Despite very similar atmospheric and oceanic forcing, wintertime thickening and summertime ablation varied dramatically with site characteristics and snow cover. However, at every site, the net thickness change over 12 months from October 1997 was negative. Ablation started in mid-June and tapered off in mid-September; upper surface melting peaked in July and that at the lower surface in August (Perovich et al. 2003). Surface loss was highest in ponds and bottom loss at ridge keels.

Ice mass balance buoys have been deployed as a component of the North Pole Observatory since April 2000 (Morison et al. 2002). The floe instrumented in 2000 thinned progressively between late May and late November and after late January, as it drifted toward Fram Strait. Oceanic heat flux was an important contributor to melt during the second phase.

3.4.5.7 Flaw Leads and Polynyas

Leads are ubiquitous transient features in pack ice, important within the climate system because they facilitate strong coupling between the atmosphere and ocean during winter. Morison et al. (1992) have reviewed oceanographic aspects.

Coastal flaw leads, by virtue of their large size and geographic recurrence, are important producers of new ice, which is the raw material for the perennial pack and which is ultimately exported to temperate oceans. During ACSYS, studies of ice production and air-sea interaction within Arctic flaw leads have been pursued using conventional oceanographic techniques (Melling 1993; Melling and Moore 1995; Weingartner et al. 1998; Drucker et al. 2003), satellite imagery (Cavalieri and Martin 1994; Yu et al. 2001), sonar remote sensing (Melling and Riedel 1996a; Melling 1998; Drucker et al. 2003) and numerical models (Winsor and Bjork 2000; Winsor and Chapman 2002). These studies have provided estimates of the cumulative ice production in flaw leads, the corresponding production and characteristics of frigid saline shelf waters and the role of these waters in cooling the Arctic halocline via ventilation. By present estimates, maintenance of the cold halocline is impractical solely via the ventilation that occurs in flaw leads (Winsor and Bjork 2000).

The International North Water Polynya Project (1997–1999) in Baffin Bay has provided new understanding of the interactive dynamics of atmosphere, ice and ocean in polynyas (Melling et al. 2001; Ingram et al. 2002). Observations revealed that ice removal, oceanic upwelling and mixing were not independent processes, so that the traditional classification of polynyas in terms of latent-heat and sensible-heat forcing is inappropriate. In the Russian-German project System Laptev Sea, sea-ice studies have been carried out within the domain of the Laptev flaw lead. Time series of sea-ice circulation in the Laptev Sea were derived from passive-microwave satellite data and a large-scale sea-ice model (Eicken et al. 1997; Alexandrov et al. 2000). Variations in the rate of ice-area export at annual and decadal periods, extended back into the 1930s using Russian observations, were not strongly coupled to the atmospheric circulation. The summertime retreat of ice in the Laptev Sea co-varies with river discharge (Bareiss et al. 1999), but any direct impact of river discharge is confined to the coastal region. The dispersed brackish plume from the river does have important effects on ice forma-tion over the continental shelf, both in spring and in autumn (Dmitrenko et al. 1999; Golovin et al. 2000).

Wintertime ice growth in flaw leads creates reservoirs of cold saline water over Arctic continental shelves that are thought critical to maintenance of the cold layer within the halocline of the Arctic basins. The low temperature and stability of the halocline impede upward flux of sensible heat, so that perennial ice in the Arctic basins is protected. There is a feedback loop wherein decreased ice growth in flaw leads results in reduced cold-water ventilation of the halocline and increased oce-anic heat flux to perennial ice, further reducing Arctic ice volume. Sources of ven-tilating water have been identified in the Beaufort (Melling and Lewis 1982; Melling 1993), Chukchi (Weingartner et al. 1998) and Barents Seas (Midttun 1985; Steele et al. 1995).

Observations in the mid-1990s revealed that the cold layer within the halocline had disappeared over a large area of the Eurasian Basin (Steele and Boyd 1998); the stability of the upper ocean had weakened and the thermocline reached to the base of the surface layer, permitting entrainment of sensible heat into the under-ice layer. The authors suggested that the altered winds of the early 1990s had deflected

Siberian river run-off eastward, permitting the warm inflow from Fram Strait to interact directly with the ice. By 2001, there were indications that the cold halocline of the Amundsen Basin had been partially re-established through capping by river-freshened outflow from the Siberian shelves (Boyd et al. 2002; Bjork et al. 2002); the calculated consequence of renewed cooling of the halocline is about 0.25 m more ice growth during winter.

3.4.6 Sea-Ice Processes

3.4.6.1 Ice Thickness Redistribution

The redistribution of sea-ice thickness through the formation of ridges is central to understanding Arctic pack ice. However, although measurements of existing ice ridges are quite practical, measurements during the ridging process are not. A discrete element model of level ice developed by Hopkins (1994) has permitted realistic study of the process of pack-ice deformation. The basic geometrical units are two-dimensional blocks for which the model tracks position, orientation, velocity and imposed forces. Only flexural failure is permitted. The model simulations are realistic in terms of the occurrence of rafting, of the horizontal displacement of ridge sail and keel, of incomplete ridge consolidation, of the ratio of sail height to keel draft and of the ratio of keel draft to lead-ice thickness. The change in potential energy of the ice through ridging is less than 10% of the total energy dissipated; above-water friction is the dominant dissipation mechanism. The model has been used to generate an ensemble of ridges, from which the average redistribution of ice thickness can be calculated (Hopkins 1996a; see Fig. 3.20). Hopkins (1998) has identified four stages in the formation of a ridge: building of the sail to the maximum height consistent with the buckling strength of lead ice; building of the keel to the draft consistent with maximum sail height; broadening of the keel; redistribution of ice blocks under compression by thicker adjacent floes. The process can cease at any stage, if convergence ceases.

The ridging and rafting of synthetic ice sheets floating in a laboratory tank has been studied by Hopkins and Tuhkuri (1999) and Tuhkuri and Lensu (2002). Ridges never formed if the converging ice sheets were of uniform thickness; the outcome was rafting. Sheets assembled from floes of two different thickness values were used to study ridging. The initial deformation was rafting in all instances. The rafting was prolonged with thin sheets of near uniform thickness, while ridging was favoured with thick sheets containing disparate thicknesses. Ridging force had a maximum value related to horizontal ridge growth and the formation of a new ridge elsewhere. The ratio of work to change in potential energy was 15 for ridges and 35 for rafts.

Although rafting has long been recognized in young ice, its occurrence during deformation of thick ice is a relatively new discovery. Melling et al. (1993) and Bowen and Topham (1996) integrated information from sonar, theodolite survey

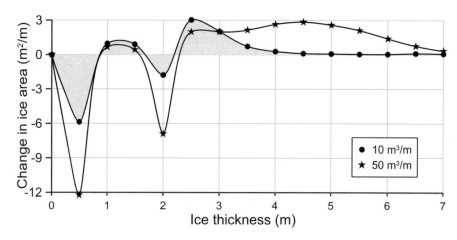

Fig. 3.20 A detailed representation of forces between ice blocks during ridge formation permits calculation of the redistribution of ice thickness during pack-ice deformation. Here the fracture of 0.5-m ice and the piling of rubble adjacent 2-m ice cause loss of ice in these two thickness categories and gain at greater thickness with the creation of ridges to 7-m maximum thickness (After Hopkins 1996b)

and stereo photography to prepare three-dimensional maps of deformation features built from thick first-year ice. They noted misalignment of sail and keel and extensive rafts of two and three thicknesses. Quasi-Lagrangian observations of change in the ice-thickness distribution of drifting pack ice were acquired in the Canada Basin through SCICEX and SHEBA. The study involved repeated under-ice profiling by submarine sonar of areas roughly 200 km in diameter, tracked by satellite. In sampling, the region was repeatedly traversed to obtain a reliable statistical representation of the under-ice topography. Patch A was surveyed in September 1996, October 1996 (40 days later) and in September 1997. Babko et al. (2002) compared ice-draft distributions for the first two surveys of this patch and concluded that rafting had been a significant component of the mechanical redistribution of thickness during deformation. Toyota et al. (2004) have discussed the role of rafting in the thickness redistribution of young ice in the Sea of Okhotsk.

Patch B, north-west of the SHEBA ship, was surveyed in September 1997, July 1998 and August 1998 (3 weeks later). Over the intervening months, ice deformation was observed using Radarsat and the surface heat budget through the SHEBA programme. The evolution of the ice-thickness distribution around the SHEBA site has been modelled using these data to drive thermodynamic and dynamic processes, with data assimilated into the thin end of the ice-thickness distribution (Lindsay 2003). The deformation of the ice was determined primarily from buoy tracking and R-GPS measurements. Assimilated observations were derived from aircraft surveys by camera and by infrared and microwave radiometers. The model indicates a 59% increase in mean ice thickness over the year, from 1.53 to 2.44 m, whereas Perovich et al. (2003) reported a net loss of ice at all mass-balance sites. The thickening was caused dynamically. A 40% divergence in the fall allowed rapid ice production, followed in the spring and summer by convergence that created much thick ridged ice. Evaluation of the simulation awaits availability of ice-draft data from the second SCICEX cruise in August 1998. The principal sources of uncertainty, exclusive of data available for this simulation from SHEBA, were snow depth and albedo for young ice, the thickness multiplier on ridging, the fraction of ice participating in ridging and the fraction of oceanic sensible heat utilized for lateral melting.

Sonar moored beneath statistically homogeneous seasonal ice can measure the evolution in the ice-thickness distribution in response to measured kinematic forcing. With a change in wind, ice that moved across the sonar beam during compression against the coast may move back along the same or a parallel path, permitting topographic surveys before and after deformation. Using events selected carefully from a 13-year record in the Beaufort Sea, Amundrud et al. (2004) have demonstrated that the draft of keels formed during pack-ice deformation is controlled both by the buckling strength of the thinnest level ice and by the amount of level ice available (floe size). They have also substantiated the earlier suggestion of Melling and Riedel (1995) that the distribution of ice with depth in a typical keel is not uniform (viz. triangular cross-section); the distribution is on average exponential, implying that keel slopes steepen towards the crest.

Ridge consolidation is important issue for ice-thickness climatology. There is appreciable (30%) void space within the sails and keels of new ridges (Leppäranta and Hakala 1992). Ice thickness is typically calculated from the remotely sensed elevation and draft of such features assuming solid ice. Melling and Riedel (1995) have estimated that ice volume calculated on this assumption may easily be biased high by 15–20%, about 0.5 m for 3-m ice. Although multi-year ridges contain few macroscopic voids, their evolutionary progression from porous first-year ridges is slow and poorly understood. Leppäranta et al. (1995) have recorded change in a Baltic ice ridge over 3.5 months in winter. Their ridge, with a maximum thickness of 6 m, formed from 0.15-m ice. Over the following months, the layer of consolidated ice at the waterline thickened to 1 m, the bulk porosity decreased from 28% to 18% and the thickness decreased by 1 m. Thickening of the consolidated layer was consistent with freezing, but consolidation deeper in the keel appeared to result from re-packing. Additional observations of similar consequence have been reported by Høyland and Løset (1999) and Høyland (2002).

3.4.6.2 Stress and Strain

The US Sea-Ice Mechanics Initiative (SIMI) during 1993–1994 and SHEBA during 1997–1998 were the principal field studies with focus on pack-ice rheology during ACSYS. The former generated a special issue of the Journal of Geophysical Research in September 1998 (103, C10) that contains 14 papers on pack-ice mechanics. An Arctic Sea Ice Dynamics Workshop was held in Seattle in March 2000.

There have been two complementary approaches to understanding pack-ice rheology. In the first, a rheological model is proposed and implemented in a realistic context. Results are subsequently tuned to match an observed variable, and the implications of that tuning are explored. Flato and Hibler (1995) took this path. They conducted realistic simulations of Arctic pack ice over a 6-year period in the early 1980s, using measured atmospheric forcing and calculated oceanic effects. They represented pack-ice thickness in 28 categories. The model was tuned such that monthly mean ice drift matched the observed drift of buoys. They determined the ratio of energy dissipation to change in potential energy to be 17, consistent with simulations of ridging by Hopkins (1994). They also identified two gaps in knowledge through comparison of observed and modelled ice-thickness distributions: the processes of consolidation that convert first-year ice to multi-year ice and those responsible for the known preferential melt of thick ridged ice.

The second approach has been to simulate the basic geometrical units of pack ice and their interactions. This has been the initiative of Hopkins (1996a). He has developed a granular model of pack ice in which individual multi-year floes and surrounding first-year ice are identifiable polygons. There is no assumed rheology. Rather, interactions between polygons are specified in terms of sliding and ridge building, with forces of the latter parameterized on the basis of Hopkins (1994). Changes in ice-thickness distribution are calculated by adding changes at each ridging site. An assumed critical thickness determines whether failure between interacting floes results in buckling (to form ridges) or crushing. Results may also be dependent on the assumed random orientation of leads and disposition of floe thickness. Hopkins has calculated the resulting relationship between stress and strain rate on the mesoscale of this simulated pack ice. The results indicate that the commonly used elliptical yield curve is a reasonable representation of plastic behaviour. It comes as no surprise that the yield curves are sensitive to the assumed critical thickness and to the sliding resistance between floes. Moreover, calculated strain rate vectors violate the commonly adopted normal flow rule. One interesting result is the approximate equality during shear of energy dissipation in ridging and in sliding. The latter process is commonly ignored in ice dynamics models.

Obvious discontinuities in the motion field of pack ice provide incentive to incorporate elements of Hopkins' approach into a pack-ice rheology. Coon et al. (1998) and Pritchard (1998) have formulated an anisotropic sea-ice mechanics model, wherein the formation and existence of leads are acknowledged. Their modelled ice pack comprises elements of two types: isotropic ice, which is itself made of interlocking rigid fragments 100–500 m across in clusters up to 20 km in diameter, and oriented ice, which may be open water, new and young ice or ridges when such occur in long narrow features. The model converts oriented ice into isotropic ice when its strength reaches the latter's.

Ukita and Moritz (1995) have developed a theoretical framework for pack-ice rheology that relates energy dissipation on the scale of in-plane sliding and ridging in pack ice to dynamics on the geophysical scale, described by internal ice stress as a function of strain rate. Their model is constrained by minimization of the maximum shear stress, which provides a yield curve, a flow rule and a directional relation between stress and strain rate. They demonstrate that various forms of pack-ice rheology – cavitating fluid, viscous-plastic, Mohr-Coulomb – correspond to different choices of kinematic model.

Hibler and Hutchings (2002) noted that the dependence of pack-ice strength on its average thickness provided the potential for distinct equilibrium states within the Arctic marine cryosphere. Ice is lost from the Arctic Ocean via a constriction in Fram Strait. The outflow is forced by wind. Because thick pack deforms less readily than thin, its flow through this aperture will be slower and may even stop. Such stoppage is a routine occurrence on other routes of exit (Melling 2002). Hibler and Hutchings (2002) identified two stable solutions for a slightly cooled climate, one with thick ice and slow outflow and the other with thinner ice and more rapid outflow. The thin ice moves easier through Fram Strait in response to wind, so that Arctic residence is short and the Arctic ice cover remains thin. Conversely, thick ice is constrained by ice stress and drifts out slowly, so that Arctic residence is long and the ice stays thick. The possibility of two disparate states under identical forcing has intriguing implications with respect to recent change in Arctic pack ice.

Patterns of fracture delineated by leads in pack ice, first discussed by Marko and Thomson (1977), provide valuable insight into pack-ice rheology. Schulson and Hibler (1991) argued that such patterns were indicative of common mechanisms for failure over a wide range of spatial scales. Access to all-weather, high-resolution SAR imagery of pack ice following the launch of Radarsat in July 1995 has renewed interest in this aspect of sea-ice mechanics. Slip lines are clearly visible at 30–150 km separation in the velocity fields derived by feature tracking (e.g. Kwok and Cunningham 2002; see Fig. 3.21). The apparent self-similarity of granules (floes or floe aggregations) in pack ice over a wide range of spatial scale suggests that it should behave as a aggregate material, with regional behaviour traceable to interactions between individual floes. However, large-scale lead patterns and their relationship to distant coastal boundaries evident in satellite images indicate that small-scale (10 km) processes can directly influence large-scale (500 km) dynamics at some times. There is evidence for a hierarchical structure in sea-ice mechanics with three natural scales, 100–300 km, 10–50 km and 1 km (Overland et al. 1995).

Overland et al. (1998) have examined the deformation of Beaufort pack ice during 1993–1994 based on the drift of 13 tracked buoys and fine-scale differential motions derived from ERS-1 SAR images. They deduced that a continuum approximation of pack ice is valid on scales greater than about 10 km, which is an order of magnitude larger than the scale of granules. Their inference regarding ice dynamics is consistent with the behaviour of a granular hardening plastic.

A review of fracture-based models of pack ice has been published by Hibler (2001). An interesting finding from fracture-based models is that substantial energy is dissipated by shearing stresses as floes slide past each other at flaws. It is likely that sliding friction has significance equivalent to ridging in dissipating energy during pack-ice deformation. Hibler and Schulson (2000) have examined the interaction of oriented flaws (viz. thinner ice) embedded in thick ice. They note that the failure of sea ice over a range of spatial scales is characterized by the propagation of oriented leads and cracks. Where the pack has flaws available at all orientations, the propagation of fractures proceeds along flaws oriented at 20° to the principal stress in the far-field.

Hutchings and Hibler (2002) have demonstrated that narrow linear zones of high deformation rate can be created within the pack ice of a viscous-plastic model using

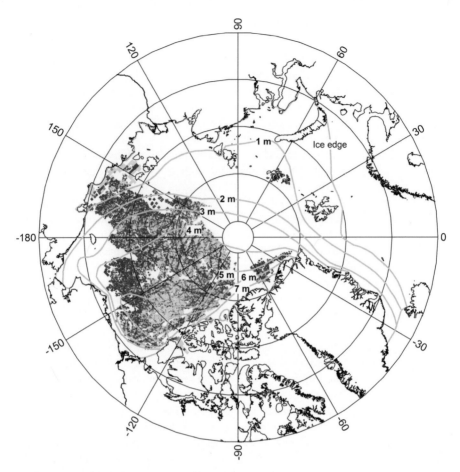

Fig. 3.21 The cumulative shear of cells within in pack ice of the central Arctic tracked using the Radarsat geophysical processing system (R-GPS) from early November 1996 to the end of the following April. Zones of persistent shearing (*slip lines*) can be seen clearly (After Kwok and Cunningham 2002). The overlay shows contours of wintertime ice draft derived by submarine prior to the mid-1980s (After Bourke and Garrett 1987)

an isotropic rheology, provided that ice is heterogeneous at the grid scale. They introduce heterogeneity as random perturbations of ice strength at grid points. The pack is allowed to weaken or strengthen in proportion to local divergence. Fracture zones in the model develop into linear features that extend across the model's domain.

Hopkins et al. (2004) have developed an Arctic sea-ice model with thousands of discrete interacting polygonal floes, 1–4 km across. The model is Lagrangian and models leads and ridges explicitly; there is specified large-scale rheology. Fractures develop naturally within the model domain, propagating along joints that delineate plates 10–100 km across that are aggregates of many floes. Deformation subsequent to aggregation occurs at plat boundaries.

Sub-diurnal motions in pack ice, which attracted scientific interest during AIDJEX (McPhee 1978), have been ignored in modelling pack ice, where the conventional water-drag formulation does not allow inertial motions, and tides are ignored.

Recent detailed measurement of ice deformation using R-GPS has revealed a widespread and persistent oscillatory motion and deformation of roughly 12-h period (Kwok et al. 2003). The associated divergence-convergence has 0.1–0.2% amplitude, peak-to-peak. Heil and Hibler (2002) have argued that the inertial dynamics associated with differential motion of pack ice at sub-diurnal frequency and its damping by internal ice stresses are necessary for accurate simulation of sea-ice motion. Moreover, inertial motions open and close leads roughly twice daily, permitting significantly enhanced ice growth on exposed open water and ridge formation by crushing young ice.

The measurement of internal ice stresses and their interpretation continues to be challenging. Richter-Menge and Elder (1998) measured stresses at a number of locations in a drifting floe during the 1993–1994 winter. Compressive stresses varied appreciably over time and between sites. Variations near the floe edge were typically of larger magnitude and higher frequency than near the floe centre. Variations in the minor principal stress were relatively uniform across sites and well correlated with ice temperature. Major principal stresses appeared correlated with ice motion as well as ice temperature. However, segregation of ice-motion-induced stresses from thermally induced stress is difficult since both variations occur within the same (meteorological) band of frequencies. Richter-Menge and Elder used the difference between primary and secondary principal stresses as an approximation to the ice-motion-induced stress. The data suggest that the strength on a geophysical scale is relatively low, only 30–150 kPa.

Lewis and Richter-Menge (1998) have compared the ice-motion-induced stress, computed as the difference between measured principal components, to the residual of forces computed from wind, current and ice motion. They grant correlation between observed and calculated stresses and approximate quantitative agreement if the former are distributed over tens of metres. The three main events of high stress all occurred during convergence. Lewis and Richter-Menge acknowledge some difficulty in understanding the results, and suggest that point measurements of stress may be more easily interpretable if the spatial distribution of stress within a floe were known.

During SHEBA, sensors were deployed to measure stress at several locations within an area 15 km across, to determine whether measured stress could be linked quantitatively to regional deformation (Richter-Menge et al. 2002). The estimate of ice-motion-induced stress was the difference between primary and secondary principal stresses, averaged over all sites. Four events were studied, providing qualitative evidence for a relationship between stress and strain. However, the relative magnitude and timing of peak activity varied appreciably between nearby sites.

3.5 Sea Ice of the Southern Hemisphere

Although an Arctic regional initiative of the WCRP, ACSYS adopted a coordinating role for marine cryospheric activity in the Southern Ocean on request of the Joint Scientific Committee of WCRP. The principal activities were the International

Programme for Antarctic Buoys (IPAB), an analogue of the International Arctic Buoy Programme (IABP) and the Antarctic Sea-Ice Thickness Project (AnSITP1).

The International Programme for Antarctic Buoys (http://www.ipab.aq/) worked to establish and maintain a network of drifting buoys in the southern marine cryosphere for the measurement of ice drift, atmospheric pressure and temperature. Through these activities, it provided meteorological data to operational weather centres, contextual environmental data to support regional marine research and an expanding base of climatological information. The operational area is south of 55°S, within the maximum domain of seasonal sea ice. The IPAB was already operating informally in 1993. Typically, there have been 9–14 buoys operating each month, with greater activity in March and April. Activity has been concentrated in the Weddell and Indian Ocean sectors. The systematic northward drift out of the sea-ice zone seriously abbreviates the operating lifetime of platforms on ice. This is a significant challenge to maintaining an optimal array.

The Antarctic Sea-Ice Thickness Project started in 1990 with six moorings (http://www.awi.de/en/research/research_divisions/climate_science/) carrying ice-profiling sonar. There have been 53 deployments, with records obtained from at least 27 and some installations still active. Germany (AWI), Australia (ACRC), USA (WHOI) and Norway (NPI) have been partners in the project. Most instruments have been deployed in the Weddell Sea, but Australian interest has been focussed near East Antarctica and US interest west of the Antarctic Peninsula.

Ice movement plays an important role in the interaction between ocean and atmosphere around Antarctica. The ice cover is generally divergent, forming in the south and melting in the north. Advection of ice increases the average albedo of more northerly seas and lowers their salinity through addition of meltwater. Early use of buoys to track ice drift in the Southern Ocean dates back to Ackley (1979) in the Weddell Sea and Allison (1989a, b) off East Antarctica. Satellite-tracked buoys were deployed again in the Weddell Sea in the late 1980s and analysed by Crane and Wadhams (1996). They concluded that ice drift was dominantly forced by wind and well correlated over distances of several 10 km. Massom (1992) incorporated passive microwave imagery into the analysis of drift trajectories from buoys to delineate the large-scale pattern of sea-ice advection in the Weddell Sea. There were additional deployments in this sector in 1990–1991 and 1991–1992 (Vihma and Launiainen 1993). Vihma et al. (1996) reported that 40–80% of the variance in the ice drift measured via these deployments was related to wind. They used this result, ECMWF-analysed winds and ice thickness from prior drillings in the northern Weddell Sea to estimate the net northward export of ice, 250–700 km^3 year^{-1}. Heil and Allison (1999) have analysed the drift of 39 buoys deployed between 1985 and 1996 off East Antarctica. The trajectories delineate a westward drift near the continent between 20°E and 160°E, a northward push near 70–80°E and a tendency for eastward drift north of 60°S. A map of average ice circulation around Antarctic was prepared by Emery et al. (1997). It depicts easterly drift near the continent deflected cyclonically northward in three places: at 70–80°E at Prydz Bay, at 140–160°E west of the Ross Sea and over 0–60°W in the Weddell Sea. None of the cyclonic ice streams show much inclination to turn south and complete a gyre. The possible exception was a weak indication of average southward drift near 130°E.

Most early data on the thickness of Antarctic sea ice came from drilling (Wadhams et al. 1987; Lange and Eicken 1990; Lange 1991). A review was prepared by Wadhams (1994), who noted the significant findings of this laborious and rather sparse sampling: Antarctic sea ice is generally very thin, less than 60 cm, a heavy snow burden frequently depresses floes below sea level, leading to snow-ice formation, and ridge keels deeper than 6 m are very rare.

In 1990, ice-profiling sonar was deployed on moorings in the Southern Ocean for the first time. The deployments and methods of processing have been described by Strass (1998). Data acquired during the first 2 years of operation, from six moorings spanning the northern Weddell Sea, have been presented and described by Strass and Fahrbach (1998). The mean draft of ice was 0.8 m in the central gyre, 2.2 m in the eastern inflow and 2.8 m near the tip of the Antarctic Peninsula. The authors note that these values are substantially higher than averages acquired via drilling. The higher means from the sonar records are attributable to more ice of deep draft since modal values are consistent with surveys by drilling. The authors discuss the possible 'contamination' of measured distributions of draft by the unwitting sampling of iceberg fragments within the record. Data on ice draft have been used in the estimation of sea-ice transport in the Weddell Sea by Harms et al. (2001). The difference between import across the Prime Meridian and export to the north was about 1,600 km^3 year^{-1}. Data were recovered from ice-profiling sonar in East Antarctic waters after a deployment in 1994–1995 (Worby et al. 2001). Monthly mean ice draft ranged between 0.6 and 1.2 m, with no drafts measured in excess of 4 m. These results are consistent with pack-ice drillings in the same area.

The electromagnetic induction method was also used to measure ice thickness in Antarctic waters during ACSYS. Haas (1998) measured ice using a probe suspended ahead of the R/V Polarstern. In the Bellinghausen Sea, the mean ice thickness was 1.3 m and the mode 0.9 m. Ice in the Amundsen Sea was more heavily deformed, with mean values of thickness between 2.3 and 3.1 m and modal values between 1.6 and 2.7 m. Worby et al. (1999) used a portable unit on the surface of land-fast ice with corroboration via drilling; average ice thickness ranged between 0.5 and 1.5 m including level sections and small ridges. The EM31 radar generated results within 10% over level ice but under-estimated ridge thickness by 50%.

Use of the airborne laser in the Antarctic has been reported by Weeks et al. (1989) in the Ross Sea and by Dierking (1995) and Granberg and Lepparanta (1999) in the Weddell Sea. Weeks et al. concluded that large ridges are significantly more likely in the Arctic than in the Ross Sea. Their mean ridge height was about 1.2 m and their maximum less than 3.5 m. Dierking's data for the Weddell Sea revealed similar mean (1.16 m using a 0.8-m detection threshold) and maximum values (3.0 m). The average height of ridges measured by Granberg and Lepparanta in the eastern Weddell Sea was 1.32 m (using a 0.91-m threshold) and the maximum greater than 4 m. In all studies, the distribution of ridge heights was exponential at low ridge density. Lytle and Ackley (1991) have used an acoustic echo sounder, downward-looking from a ship, to measure ridge height in the Weddell Sea. Their mean value was 1.1 m (for 0.75-m threshold) and their maximum was 3 m.

Conventional surveys of ice thickness by drilling were continued during ACSYS. During a cruise in 1993 in the Bellinghausen Sea, Worby et al. (1996) combined drilling sorties with a systematic routine of ice reports from the bridge, developed by Australian scientists. The study revealed that the thickening of Antarctic sea ice is highly dependent on deformation. Ice less than 0.3 m is extensively rafted, whereas floes thicker than 0.6 m frequently encompass ridges. The quarter of the top surface of pack ice that is ridged contains between half and three quarters of the total ice volume. A similar survey in the Ross Sea was completed by Jeffries and Adolphs (1997). Adolphs (1999a, b) has used both sets of data to develop a statistical model of ice and snow thickness in the Antarctic, based on equations that represent the relevant processes – freezing, lead opening, rafting and ridging. Systematic observations from a ship in transit have been integrated with other sources of data to document ice thickness and ridging and its impact on oceanic heat flux in the Ross Sea (Jeffries et al. 2001). In one study off East Antarctica, photogrammetric techniques have been used for detailed three-dimensional survey of the sub-aerial exposure of pressure ridges (Lytle et al. 1998).

Knowledge of sea-ice extent and its variation are less well known in the Antarctic than in the Arctic. Gloersen and Campbell (1988) published data from the first decade of continuous monitoring via passive microwave emission. In a later paper (Gloersen and Campbell 1991), they presented a trend analysis of ice extent in both hemispheres, concluding that Antarctic ice was neither expanding nor shrinking. Parkinson (1992a) followed with a discussion of inter-annual variability in monthly averaged ice distribution. Chapman and Walsh (1991) compiled information on southern ice extent from charts prepared by the US National Ice Center since 1973. The information at monthly intervals and 1-degree resolution is available via the Internet (http://www.natice.noaa.gov/), updated in 1996. A revision of the passive microwave analysis to the same date has been prepared by Gloersen et al. (1999). The apparent increased sea-ice extent in the Southern Ocean was considered statistically insignificant at the time.

Concern about inaccuracy in ice concentration deduced using the NASA algorithms led Hanna and Bamber (2001) to develop and test an alternative tuned to Antarctic conditions. The algorithm was adapted by implementing manual tuning of its end points (which represent specific surface types) for changes in season and environmental conditions in Antarctic seas. Following evaluation against AVHRR and Geosat altimeter, the algorithm was used to calculate a time series of ice extent in the Southern Ocean. The time series reveals a 3% increase in extent between 1987 and 1997. Simmonds and Jacka (1995) have investigated the connections between inter-annual variability in Antarctic sea ice and the Southern Oscillation.

There has been extensive study of atmosphere-ice-ocean interaction in the Antarctic during ACSYS. Knowledge of pack ice in the Southern Ocean at the end of the 1980s was summarized in a review paper by Lange et al. (1989), which focussed on the Weddell Sea. Complementary papers by Allison et al. (1993) and

Allison and Worby (1994) reviewed understanding of East Antarctic pack over the seasonal cycle. Much subsequent research has been concerned with the impacts of either snow cover or oceanic heat flux on ice thickness, growth, ablation and the seasonal cycle. Work has been conducted in the Weddell, Amundsen, Bellinghausen and Ross Seas and off the coast of East Antarctica. Studies of snow and its role in the formation of snow ice have been published by Jeffries et al. (1994a, b, c), Sturm et al. (1996, 1998), Massom et al. (1997) and Maksym and Jeffries (2000). In a number of projects, the crystal and oxygen isotope structure have been determined from cores to elucidate the processes involved in ice formation – rafting, flooding, frazil ice accumulation (Jeffries et al. 1994c; Kawamura et al. 1997). The oceanic heat flux has been the focus of research by Heil et al. (1996), Lytle and Ackley (1996), Ohshima et al. (1998) and Lytle et al. (2000). In many areas, the oceanic heat flux is 10–20 $W\,m^{-2}$, large enough to cause melting of thick (more than 0.6 m) snow-covered ice even under the coldest of winter conditions. An updated review of East Antarctic pack ice has been published by Worby et al. (1998).

Heil et al. (1998) have discussed the enhancement of ice growth by freezing that occurs in consequence of sea-ice deformation in the Antarctic environment, while related research by Eisen and Kottmeier (2000) explored the role of leads in promoting ice growth in the Weddell Sea.

3.6 Deliverables

The following might be considered the tangible deliverables of the Arctic Climate System Study in relation to sea ice. They have already, or will likely be made available to the scientific community through publications, public Internet sites and international data archives, as appropriate.

- Gridded data bases of Arctic ice extent, type and concentration at resolutions of 10–100 km and 1–10 days during at least the decade of ACSYS (1994–2003).
- Gridded motion fields of Arctic pack ice at resolutions of 10–100 km and 1–3 days during at least the decade of ACSYS. The IABP will yield the better temporal resolution (1 day), whereas imaging via satellite will yield the better spatial resolution (10 km via SAR analysis, 75–150 km via passive microwave).
- Statistical summaries of sea-ice draft, by location and date, and original processed temporal or spatial sequences of observations.
- A time series of the rate of ice-mass export from the Arctic through the Fram Strait of at least 10-year duration.
- Revised estimates of the average rate of ice-mass export through other connections between the Arctic and temperate oceans.
- Revised estimates of the rate of new ice production in Arctic flaw leads.
- Detailed observations of ice-ocean-atmosphere interaction over an annual cycle in the Arctic Ocean.

- Tested parameterizations of the processes involved in ice-ocean-atmosphere interaction, suitable for use in global climate models operating at a resolution on the order of 100 km.
- Tested parameterizations of the redistribution of pack-ice thickness in response to dynamic and thermodynamic forcing, suitable for use in global climate models operating at a resolution on the order of 100 km.
- Ice-thickness and ice-drift data sets from the two WCRP Antarctic programmes (AnSITP and IPAB).

3.7 Tasks for the Future

3.7.1 The Seasonal Sea Ice Zone

Perennial ice has been the dominant focus of scientific interest in the marine cryosphere. Knowledge of its circulation, dynamics and thermodynamics and response to changing climate far exceeds that of seasonal ice. However, perennial ice makes up less than one third of the northern marine cryosphere at its maximum extent and a much smaller fraction of the southern. The remainder is seasonal sea ice. In the northern hemisphere, seasonal sea ice does receive attention from the operational sector because of its impact on navigation and other economic activities. But this sector is generally content with mapping the extent and concentration according to a simple characterization by age class. We know comparatively little about the circulation of seasonal sea ice, its thickness and snow cover, its dynamics and thermodynamics and its interactions with the open ocean. This is particularly true for the vast areas of seasonal ice outside the Arctic Ocean, and in the Antarctic.

The much-publicized decrease in the extent of northern sea ice, detected via microwave remote sensing, is most pronounced in autumn, the time when seasonal ice is at minimum extent. Moreover in the southern hemisphere, where most of the pack ice is seasonal, the recent trend in ice extent is weak. Seasonal ice forms the margin of much of the marine cryosphere. Warm wet snow and flooding which are common on the surface of such pack have poorly understood effects on microwave emission. The result is lingering doubt concerning the accuracy of microwave-derived ice concentration and ice-edge position near the margin of the marine cryosphere.

Although the average thickness of pack ice in the Arctic Ocean has apparently decreased, data on ice thickness within the seasonal ice zone provide no evidence for progressive change during the last half century. Clearly, we must redress our tendency to neglect seasonal pack ice if knowledge of climate change impact on the marine cryosphere is to be balanced and complete. A greatly enhanced network for observing the thickness of seasonal ice in both hemispheres must have very high priority.

The boundary of the marine cryosphere, or ice edge, borders the open ocean in most areas and at most times. A close look reveals that the concept of an ice edge is actually poorly defined, with a variety of pragmatic definitions in use. The edge may be based variously upon the presence of ice at some concentration, upon the difficulty or hazard for navigation, upon the average thickness of ice, upon average short-wave albedo, upon average brightness temperature in the microwave band, etc.

We have a correspondingly poor understanding of what controls the position of the ice edge. Is atmospheric temperature the dominant control, as implied in many recent discussions? Is the edge determined by ice-albedo feedback, or by the location of thermal fronts in the ocean? Studies at various locations around the northern marine cryosphere (Okhotsk Sea, Bering Sea, Labrador Sea, Greenland Sea, Barents Sea) have revealed a 'conveyor-belt' circulation where ice melts as it drifts southward into warmer seas and the ice edge is the location where the rate of ice loss through melting equals that of delivery via drift. If this type of control is active on the time scales of climate, then ice flux may have a more direct significance than ice thickness. Yet we have virtually no knowledge of equator-ward ice fluxes and how they might change with climate. Since there are roles here for air temperature, prevailing wind and ocean current, there may be unexplored mechanisms of feedback between sea-ice extent and the climate system – albedo, sea-surface temperature, cyclogenesis, and storm tracks.

3.7.2 Global Sea Ice Volume

During the ACSYS decade, significant new knowledge of ice thickness has emerged from formerly proprietary data sets – observations by naval submarines in the central Arctic and by oil companies in peripheral seas. Geographical composites of submarine observations acquired at various locations over relatively long periods of time have brought a new awareness of changes in the thickness of the polar pack. However, uncertainties associated with poorly resolved seasonal, inter-annual and spatial variation are appreciable, and information is restricted to the central Arctic. In a complementary way, moored sonar has provided ice thickness data with poor spatial coverage but at high accuracy and temporal resolution.

Nonetheless, knowledge of the global volume of sea ice remains elusive. In part, this reflects the relative neglect of the seasonal sea ice zone as described in the preceding section. However, an important reason for slow progress is the lack of a suitable and suitably accurate method for ice-thickness measurement on a hemispheric scale. Typically, we look to satellite-borne sensors to provide such coverage. However, various characteristics of pack ice conspire to make the remote measurement of ice thickness extremely challenging: pack ice is relatively thin; it floats with 80–100% of its thickness below the surface; its thickness varies drastically over tens of metres; it is relatively opaque over a wide range of electromagnetic frequencies; its surface is obscured by snow with thickness comparable to the freeboard. The ACSYS decade has encompassed encouraging developments in the use of space-borne microwave and laser altimeters to measure the pack-ice freeboard. However, these techniques are not yet mature; there are needs for inter-comparison with better established methods in a variety of pack-ice environments and for mitigation of acknowledged shortcomings, such as the influence of snow. In time, they may provide much needed hemispheric coverage to useful accuracy when linked at various key sites to more accurate data from in situ methods.

There is need also for improvement in the application of sonar to the determination of global sea ice volume. The calibration of ice-draft observations by sonar is tedious and trouble-prone. Accuracy frequently falls short of the needs of climate research, particularly with sonar on moving platforms. Beyond the central Arctic and in the Antarctic, the challenge is more severe since the ice is thinner and the speed of sound in the upper ocean more variable.

Ridges contain more than 50% of the volume of pack ice. Analysts typically assume that the envelope of a ridge sail or keel measured remotely is solid ice. In fact, many ridges are poorly consolidated, with only 60–80% of the envelope filled with ice. Information comes from a few detailed studies of isolated features. We have little knowledge of the consolidation of pack ice on a regional scale. Clearly, the uncertain degree of consolidation is an important issue when considering global sea-ice volume.

An improved understanding of the redistribution of sea-ice thickness through deformation and melting is essential to modelling the impact of changing atmospheric and oceanic conditions on the marine cryosphere. Because ridges contain such a large fraction of pack-ice volume, the processes controlling ridge development and decay are key factors in the response of the marine cryosphere to changing climate. Since ice floes fracture and pile into ridges in response to dynamical, not thermal forcing, changes in storm climate may be influential in shaping change in the marine cryosphere.

3.7.3   Data Assimilation

Issues of technical feasibility and cost will continue to constrain the effective hemispheric monitoring of global pack ice, despite its apparent high significance in the global climate system. The variable providing greatest obvious challenge is ice thickness. Ice thickness varies significantly on a metre scale over an area 16 million square kilometres and has so far resisted efforts for accurate measurement using sensors on Earth satellites. A complementary technology, data assimilation, has the potential for 'intelligent' interpolation between sparse observations in such situations. It has been under active investigation in the context of sea ice for some time (Rothrock and Thomas 1992; Thomas et al. 1996). Data assimilation in the context of sea ice is a complex undertaking. First steps in relation to remotely sensed data have been proposed by Weaver et al. (2000).

The techniques of data assimilation are built around physical models, which ideally have the capability to compute values of a variable that is difficult to measure using other data that are physically related and more readily determined. A relevant example involves ice ridges. Ridges are important contributors to total ice thickness, probably containing more than half the total ice volume, but practically they cannot be measured throughout the Arctic. However, convergence in an ice field, which is accessible through analysis of high-resolution SAR imagery, is physically related to ice ridges through the redistribution of ice thickness that occurs with strain. If the redistribution function is well known, assimilation of ice motion into a model for ice thickness may permit realistic calculation of ice thickness. Any thickness data that are available here and there and from time to time can also be assimilated into the model where they can correct, or constrain, the results of calculations. Accurate estimates of the random and bias errors in observations are essential to the effective use of data assimilation. In many cases, such errors are not well known at the present time.

Three recent studies of Arctic sea ice have assimilated observations into models and evaluated the impact on simulated ice over and movement. Meier et al. (2000) assimilated ice motion, observed via feature tracking using SSM/I images during 1988–1993, into a dynamic-thermodynamic sea-ice model having four ice categories. Errors in ice velocity from both observations and modelling were comparable, but uncorrelated. Assimilation improved the correlation of simulated ice drift relative to IABP buoys, changed the average ice thickness by up to 0.3 m and influenced the ice flux through Fram Strait. The impacts of assimilation varied with both location and year.

Lindsay (2003) completed a regional simulation of pack-ice development in the vicinity of SHEBA during 1997–1998. The model contained components representing the thermodynamic growth and melt of ice and the redistribution of ice thickness through opening and ridging. Forcing for the model was derived from camp observations, and elements of ice thickness and concentration were assimilated into the model where available. Thickness observations at the thin end of the distribution were assimilated from aircraft surveys by infrared and microwave radiometer and by camera. Drifting buoys and the R-GPS provided data on ice drift and deformation. Unfortunately, data from the SCICEX cruise in August 1998 are not yet available to evaluate the simulation. The study benefited from detailed direct observations available through SHEBA. Lindsay identified snow cover, melt-pond coverage and parameters in the ridging model as significant unknowns and potential sources of discrepancy.

Zhang et al. (2003) worked with a thickness and enthalpy distribution model, coupled to an ocean model. Observations of ice drift from buoys and SSM/I (85 GHz) analysis were assimilated by optimal interpolation. The more accurate, but sparse, data from buoys were particularly effective in reducing error, but data from SSM/I were effective in data-sparse regions. With data assimilation, computed ice drift was faster and less frequently stopped and outflow through Fram Strait increased. With faster drift, there was greater ice deformation, a doubling of the production of ridged ice to 0.8 m year^{-1} and an inventory of half the total ice volume in ridges. The spatial distribution of volume corresponded better to submarine observations, although the standard error remained large (0.66 m). When drift data from both sources were assimilated, the ice near the Canadian Archipelago and north Greenland was thicker by 1 m and that in the central Arctic by 0.2–0.4 m.

References

Aagaard K, Carmack EC (1989) The role of sea ice and other fresh water in the Arctic circulation. J Geophys Res 94(C10):14485–14498

Aagaard K, Coachman LK, Carmack EC (1981) On the halocline of the Arctic Ocean. Deep Sea Res 28:529–545

Ackley SF (1979) Drifting buoy measurement on Weddell Sea pack ice. Antarct J USA 14(5):106–108

Adolphs U (1999a) Roughness variability of sea ice and snow cover thickness profiles in the Ross Amundsen and Bellinghausen Seas. J Geophys Res 104(C6):13577–13591

Adolphs U (1999b) Representativity analysis and statistical modelling of snow and ice thickness data sets from the southern polar Pacific Ocean. J Geophys Res 104(C6):13615–13625

Agnew TA (1998) Drainage of multi-year sea ice from the Lincoln Sea. CMOS Bull 26(4):101–103

Agnew TA, Howell S (2003) The use of operational ice charts for evaluating passive microwave ice concentration data. Atmos Ocean 41(4):317–331

Agnew TA, Le H, Hirose T (1997) Estimation of large-scale sea-ice motion from SSM/I 85.5 GHz imagery. Ann Glaciol 25:305–311

Alekseev GV, Bulatov LV, Zakharov VF (2000) Freshwater freezing/melting cycle in the Arctic Ocean. In: The freshwater budget of the Arctic Ocean. NATO/WCRP/AOSB Kluwer Academic, Amsterdam. Proceedings of WCRP/AOSB/NATO Adv Res Workshop, Tallinn, Estonia. April 1998, pp 589–608

Alexandrov VY, Martin T, Kolatschek J, Eicken H, Kteyscher M, Makshtas AP (2000) Sea ice circulation in the Laptev Sea and ice export to the Arctic Ocean: results from satellite remote sensing and numerical modeling. J Geophys Res 105(C7):17143–17159

Allison I (1989a) The East Antarctic sea ice zone: ice characteristics and drift. GeoJournal 18(1):103–115

Allison I (1989b) Pack ice drift off East Antarctica and some implications. Ann Glaciol 12:1–8

Allison I, Worby AP (1994) Seasonal changes of sea ice characteristics off East Antarctica. Ann Glaciol 20:195–201

Allison I, Brandt RE, Warren SG (1993) East Antarctic sea ice albedo thickness distribution and snow cover. J Geophys Res 98(C7):12417–12429

Amundrud T, Melling H, Ingram RG (2004) Geometric constraints on the evolution of ridged sea ice. J Geophys Res 109:C06005. doi:101029/2003JC002251

Anderson MR (1987) The onset of spring melt in first-year ice regions of the Arctic as determined from scanning multi-channel microwave radiometer data for 1979 and 1980. J Geophys Res 92:13153–13163

Anderson MR (1997) Determination of melt onset date for Arctic sea ice regions using passive microwave data. Ann Glaciol 33:74–78

Antonov VS (1968) The possible cause of water exchange variations between the Arctic and Atlantic Oceans. Probl Arkt Antarkt 29:12–18

Arctic Climatology Project (2000) In: Tanis F, Smolyanitsky V (eds) Environmental working group Joint US-Russian Sea ice atlas. National Snow and Ice Data Center, Boulder, CD-ROM

Babko O, Rothrock DA, Maykut GA (2002) Role of rafting in the mechanical redistribution of sea ice thickness. J Geophys Res C107(8):doi:101029/1999JC000190

Ballicater Consulting (2000) Documentation for the Canadian Ice Service digital sea ice database. Contract Report (unpublished) Canadian Ice Service, Ottawa (T Carrieres, sci auth)

Barber DG, Nghiem SV (1999) The role of snow on the thermal dependence of microwave backscatter over sea ice. J Geophys Res 104(C11):25789–25803

Barber DG, Papakyriakou TN, LeDrew EF (1994) On the relationship between energy fluxes dielectric properties and microwave scattering over snow-covered first-year sea ice during the spring transition period. J Geophys Res 99:22401–22411

Barber DG, Papakyriakou TN, LeDrew EF (1995) An examination of the relationship between spring period evolution of the scattering coefficient and radiative fluxes over land-fast sea ice. Int J Remote Sens 16:3343–3363

Bareiss J, Eicken H, Helbig A, Martin T (1999) Impact of river discharge and regional climatology on the decay of sea ice in the Laptev Sea during spring and early summer. Arct Antarct Alp Res 31(3):214–229

Belliveau DJ, Bugden GL, Eid BM, Calnan CJ (1990) Sea ice velocity measurements by bottom mounted Doppler current profilers. J Atmos Ocean Technol 7:596–602

Bilello MA (1980) Maximum thickness and subsequent decay of lake, river and fast sea ice in Canada and Alaska. CRREL Report 80–6. US Army Cold Regions Research and Engineering Laboratory, Hanover, p 160

Bjorgo E, Johannessen OM, Miles MW (1997) Analysis of merged SMMR-SSMI time series of Arctic and Antarctic sea ice parameters 1978–1995. Geophys Res Lett 24(4):413–416

Bjork G, Soderkvist S, Winsor R, Nikolopoulos A, Steele M, Morison J (2002) Return of the cold halocline layer to the Amundsen Basin of the Arctic Ocean: implications for the sea ice mass balance. Geophys Res Lett 29(11):doi:101029/2001GL014157

Black WA (1965) Sea Ice Survey Queen Elizabeth Islands Region Summer 1962. Geographical Paper No 39. Canada Dep Mines Tech Surv Geogr Branch. Queens Printer, Ottawa, 44 pp

Bourke RH, Garrett RP (1987) Sea ice thickness distribution in the Arctic Ocean. Cold Reg Sci Technol 13:259–280

Bourke RH, McLaren AS (1992) Contour mapping of Arctic Basin ice draft and roughness parameters. J Geophys Res 97:17715–17728

Bourke RH, Paquette RG (1989) Estimating the thickness of sea ice. J Geophys Res 94(C1):919–923

Bowen RG, Topham DR (1996) A study of the morphology of a discontinuous section of a first-year Arctic pressure ridge. Cold Reg Sci Technol 24:83–100

Boyd TJ, Steele M, Muench RD, Gunn JT (2002) Partial recovery of the Arctic cold halocline. Geophys Res Lett 29(14):doi:101029/2001GL014047

Brown RD, Coté P (1992) Inter-annual variability of land-fast ice thickness in the Canadian High Arctic 1950–89. Arctic 45(3):273–284

Bush GM, Duncan AJ, Penrose JD, Allison I (1995) Acoustic reflectivity of Antarctic sea ice at 300 kHz. In: Proceedings of oceans '95, 9–13 Oct 1995. IEEE and MTS 2 962, San Diego, CA

Cavalieri DJ (1994) A microwave technique for mapping thin sea ice. J Geophys Res 99(C6):12561–12572

Cavalieri DJ, Martin S (1994) The contribution of Alaskan Siberian and Canadian coastal polynyas to the cold halocline layer of the Arctic Ocean. J Geophys Res 99(C9):18343–18362

Cavalieri DJ, Crawford JP, Drinkwater MR, Emery WJ, Eppler DT, Farmer LD, Jentz RR, Wackerman CC (1991) Aircraft active and passive microwave validations of sea ice concentrations from DMSP SSM/I. J Geophys Res 96:21989–22008

Cavalieri DJ, Gloersen P, Parkinson CL, Comiso JC, Zwally HJ (1997) Observed hemispheric asymmetry in global sea ice changes. Science 278:1104–1106

Cavalieri DJ, Parkinson C, Gloersen P, Comiso JC, Zwally HJ (1999) Deriving long-term time series of sea-ice cover from satellite passive microwave multi-sensor data sets. J Geophys Res 104(C7):15803–15814

Cavalieri DJ, Parkinson CL, Vinnikov KY (2003) 30-year satellite record reveals contrasting Arctic and Antarctic decadal sea ice variability. Geophys Res Lett 30(18):1970. doi:101029/2003GL018031

Chapman WL, Walsh JE (1991, updated 1996) Arctic and Southern Ocean Sea ice concentrations. National Snow and Ice Data Center/WDC Glaciol digital media, Boulder

Cho K, Taniguchi Y, Nakayama M, Shimoda H, Sakata T (2002) Sea ice thickness measurement using stereo images. In: Proceedings of the 17th international symposium on Okhotsk Sea & sea ice, Japan, Feb 2002, pp 169–172

Colony R, Thorndike AS (1984a) An estimate of the mean field of Arctic sea ice motion. J Geophys Res 89:10623–10629

Colony R, Thorndike AS (1984b) Arctic Ocean drift tracks from ships buoys and manned research stations 1872–1973. National Snow and Ice Data Center Digital Media, Boulder

Colony R, Thorndike AS (1985) Sea ice motion as a drunkard's walk. J Geophys Res 90:965–974

Comfort G, Ritch R (1990) Field measurements of pack ice stresses. In: Ayorinde OA, Sinha NK, Sodhi DS (eds) Proceedings of the 9th international conference offshore mechanics and arctic engineering, vol 4, Houston, TX, 18–23 Feb 1990. ASME, New York, pp 177–181

Comiso JC (1995) SSM/I ice concentrations using the Bootstrap algorithm. NASA report 1380

Comiso JC (2002) A rapidly declining perennial sea ice cover in the Arctic. Geophys Res Lett 29(17):1–4

Comiso JC, Wadhams P, Krabill WB, Swift RN, Crawford JP, Tucker WB III (1991) Top/bottom multi-sensor remote sensing of Arctic sea ice. J Geophys Res 96:2693–2709

Comiso JC, Cavalieri D, Parkinson C, Gloersen P (1997) Passive microwave algorithms for sea ice concentrations: a comparison of two techniques. Remote Sens Environ 60(3):357–384

Connors DN, Levine ER, Shell RR (1990) A small-scale under-ice morphology study in the High Arctic. Sea ice properties and processes. In: Ackley SF, Weeks WF (eds) Proceedings WF weeks sea ice sympoisum, CRREL monograph no 90-1. CRREL, Hanover, pp 145–151

Coon MD, Pritchard RS (1979) Mechanical energy considerations in sea-ice dynamics. J Glaciol 24:377–389

Coon MD, Lau PA, Bailey SH, Taylor BJ (1989) Observations of ice floe stress in the eastern Arctic. In: Axelsson KBE, Fransson LA (eds) Proceedings of the 10th international conference port ocean engineering under arctic conditions, vol 1, Lulea, Finland, 12–16 June 1989. Tekniska Hogskolan I, Lulea, pp 44–53

Coon MD, Knoke GS, Echert DC, Pritchard RS (1998) The architecture of an anisotropic elastic-plastic sea-ice mechanics constitutive law. J Geophys Res 103:21915–21925

Cox GFN, Johnson JB (1983) Stress measurements in ice. CRREL Rep 83–23. US Army Cold Regions Research Engineering Laboratory, Hanover

Crane D, Wadhams P (1996) Sea-ice motion in the Weddell Sea from drifting buoy and AVHRR data. J Glaciol 42(141):249–254

Croasdale KR, Comfort G, Frederking RMW, Graham BW, Lewis EL (1988) A pilot experiment to measure Arctic pack ice driving forces. In: Sackinger WM, Jeffries MO (eds) Proceedings of 9th international conference on port and ocean engineering under Arctic conditions, vol 3. Geophysical Institute, University of Alaska, Fairbanks, pp 381–395

Crocker GB, Carrieres T (2000) The Canadian Ice Service digital database: History of data and procedures used in the preparation of regional ice charts. Contract Report No. 00-02, Ballicater Consulting Ltd Ottawa. 54 pp. (avail. Canadian Ice Service, Environment Canada, Ottawa K1A 0H3)

Curry JA, Schramm JL, Ebert EE (1995) On the ice albedo climate feedback mechanism. J Clim 8:240–247

Dedrick K, Partington K, Van Woert M, Bertoia C, Benner D (2001) US National/Naval Ice Center digital sea ice data and climatology. Can J Remote Sens 27(5):457–475

Dierking W (1995) Laser profiling of the ice surface topography during the Weddell Winter Gyre Study 1992. J Geophys Res 100(C3):4807–4820

Dmitrenko IA, Golovin PN, Gribanov VA, Kassens H (1999) The influence of the river runoff on the ice and hydrological conditions on the shelf zone of the Russian Arctic seas. Dokl Earth Sci 369(5):687–691

Drobot SD, Maslanik JA (2003) Inter-annual variability in summer Beaufort Sea ice conditions: relationship to winter and summer surface and atmospheric variability. J Geophys Res 108(C7):3233. doi:101029/2002JC001537

Drucker R, Martin S, Moritz R (2003) Observations of ice thickness and frazil ice in the St Lawrence Island polynya from satellite imagery upward looking sonar and salinity-temperature moorings. J Geophys Res 108(C5):3149. doi:101029/2001JC001213

Dumas JA, Flato GM, Weaver AJ (2003) The impact of varying atmospheric forcing on the thickness of Arctic multi-year sea ice. Geophys Res Lett 30(18):1918. doi:101029/2003GRL017433

Dunbar M (1973) Ice regime and ice breakup in Nares Strait. Arct 26(24):282–291

Eicken H, Lange MA (1989) Sea ice thickness data: the many vs the few. Geophys Res Lett 16:495–498

Eicken H, Lensu M, Leppäranta M, Tucker WB III, Gow AJ, Salmela O (1995) Thickness structure and properties of level summer multiyear ice in the Eurasian sector of the Arctic Ocean. J Geophys Res 100(C11):22697–22710

Eicken H, Reimnitz E, Alexancrov V, Martin T, Kassens H, Viehoff T (1997) Sea-ice processes in the Laptev Sea and their importance for sediment export. Cont Shelf Res 17(2):205–233

Eicken H, Tucker WB III, Perovich DK (2000) Indirect measurements of the mass balance of summer Arctic sea ice with an electromagnetic induction technique. Ann Glaciol 33:194–200

Eisen O, Kottmeier C (2000) On the importance of leads in sea ice to the energy balance and ice formation in the Weddell Sea. J Geophys Res 105(C6):14045–14060

Emery WJ, Fowler CW, Maslanik JA (1997) Satellite-derived maps of Arctic and Antarctic sea ice motion 1988 to 1994. Geophys Res Lett 24(8):897–900

Ezraty R, Cavanié A (1999) Inter-comparison of backscatter maps over Arctic sea ice from NSCAT and ERS scatterometer. J Geophys Res 104(C5):11471–11484

Fetterer FM, Drinkwater MR, Jezek KC, Laxon SWC, Onstott RG, Ulander LMH (1992) Sea ice altimetry. In: Carsey F (ed) Microwave remote sensing of sea ice, Geophysics monograph series. AGU, Washington, DC, pp 111–135

Fily M, Rothrock DA (1987) Sea ice tracking by nested correlations. IEEE Trans Geosci Rem Sens GE-25:570–80

Flato GM, Brown RD (1996) Variability and climate sensitivity of land-fast Arctic sea ice. J Geophys Res 101(C11):25767–25778

Flato GM, Hibler WD III (1995) Ridging and strength in modelling the thickness distribution of Arctic sea ice. J Geophys Res 100(C9):18611–18626

Fowler C (2003) Polar pathfinder daily 25-km EASE-grid sea ice motion vectors. http://nsidc.org/data/nsidc-0116.html. Digital media US National Snow and Ice Data Center, Boulder

Fukamachi Y, Mizuta G, Oshima KI, Melling H, Fissel D, Wakatsuchi M (2003) Variability of sea-ice draft off Hokkaido in the Sea of Okhotsk revealed by moored ice-profiling sonar in the winter of 1999. Geophys Res Lett 30(7):1376. doi:1010292002/GL016197

Galloway JL, Melling H (1997) Tracking the motion of sea ice by correlation sonar. J Atmos Ocean Technol 14(3):616–629

Garrison GR, Becker P (1976) The Barrow Submarine Canyon: a drain for the Chukchi Sea. J Geophys Res 81:4445–4453

Gloersen P, Campbell WJ (1988) Variations in Arctic Antarctic and global sea ice covers during 1978–1987 as observed with the Nimbus 7 scanning multi-channel microwave radiometer. J Geophys Res 93(C9):10666–10674

Gloersen P, Campbell WJ (1991) Recent variations in Arctic and Antarctic sea-ice covers. Nature 352:33–36

Gloersen P, Parkinson CL, Cavalieri DJ, Comiso JC, Zwally HJ (1999) Spatial distribution of trends and seasonality in the hemispheric sea ice covers: 1978–1996. J Geophys Res 104(C9):20827–20836

Goff JA (1995a) Quantitative analysis of sea ice draft: 1, methods for stochastic modelling. J Geophys Res 100(C4):6993–7004

Goff JA (1995b) Quantitative analysis of sea ice draft: 2, application of stochastic modelling to intersecting topographic profiles. J Geophys Res 100(C4):7005–7018

Gohin F, Cavanié A (1994) A first try at identification of sea ice using the three-beam scatterometer of ERS-1. Int J Remote Sens 15:1221–1228

Golovin PN, Dmitrenko IA, Kassens H, Holemann J (2000) Frazil ice formation along the upper boundary of seasonal pycnocline and its role in glacial marine sedimentation in the Arctic seas (using the Laptev Sea as an example). Dokl Earth Sci 370(2):237–242

Gorbunov YuA, Moroz VG (1972) The main results of the usage of DARM for ice drift observations and DARMS for study of ice drift in the Arctic Ocean. Probl Arkt Antark 39(1):33–39

Gordienko PA, Karelin DB (1945) Problems of the movement and distribution of ice in the Arctic Basin. Probl Arkt 3:25–38, Leningrad

Gordienko PA, Laktionov AF (1960) Principal results of the latest oceanographic research in the Arctic Basin. Izvest Akad Nauk SSSR Ser Geograf 5:22–33

Graham BW, Chabot LG, Pilkington GR (1983) Ice load sensors for offshore Arctic structures. In: Proceedings of the 7th international conference on port ocean engineering under Arctic Condition, vol 4, Helsinki, Finland, 5–9 Apr 1983. Technical Research Center Finland, Espoo, pp 547–562

Granberg HB, Lepparanta M (1999) Observations of sea ice ridging in the Weddell Sea. J Geophys Res 104(C11):25735–25746

Grenfell TC (1979) The effects of ice thickness on the exchange of solar radiation over the polar oceans. J Glaciol 22:305–320

Grenfell TC (1983) A theoretical model of the optical properties of sea ice in the visible and near infrared. J Geophys Res 88:9723–9735

Grenfell TC, Maykut GA (1977) The optical properties of ice and snow in the Arctic Basin. J Glaciol 18:445–463

Grenfell TC, Perovich DK (1984) Spectral albedos of sea ice and incident solar irradiance in the southern Beaufort Sea. J Geophys Res 89:3573–3580

Groves JE, Stringer WJ (1991) The use of AVHRR thermal imagery to determine sea ice thickness within the Chukchi polynya. Arctic 44:130–139

Gudkovich ZM, Nikolaeva AYa (1961) Some results of ice drift study obtained by DARM for ice drift observations. Probl Arkt Antarkt 8:11–18

Haas C (1998) Evaluation of ship-based electromagnetic inductive thickness measurements of summer sea ice in the Bellingshausen and Amundsen Seas Antarctica. Cold Reg Sci Technol 27:1–16

Haas C (2004) Late-summer sea ice thickness variability in the Arctic Transpolar Drift 1991–2001 derived from ground-based electromagnetic sounding. Geophys Res Lett 31:L09402. doi:101029/2003GL019394

Haas C, Eicken H (2001) Inter-annual variability of summer sea ice thickness in the Siberian and central Arctic under different atmospheric circulation regimes. J Geophys Res 106(C3):4449–4462

Haas C, Gerland S, Eicken H, Miller H (1997) Comparison of sea-ice thickness measurement under summer and winter conditions in the Arctic using a small electromagnetic induction device. Geophysics 62(3):749–757

Hanesiak JM, Barber DG, Flato GM (1999) Role of diurnal processes in the seasonal evolution of sea ice and its snow cover. J Geophys Res 104(C6):13593–13604

Hanna E, Bamber J (2001) Derivation and optimization of a new Antarctic sea-ice record. Int J Remote Sens 22(1):113–139

Hanson AN (1965) Studies of the mass budget of Arctic pack-ice floes. J Glaciol 5(41):701–709

Harms S, Fahrbach E, Strass VH (2001) Sea ice transports in the Weddell Sea. J Geophys Res 106(C5):9057–9073

Hauser A, Lythe M, Wendler G (2002) Sea-ice conditions in the Ross Sea during spring 1996 as observed on SAR. Atmos Ocean 40(3):281–292

Heil P, Allison I (1999) The pattern and variability of Antarctic sea-ice drift in the Indian Ocean and western Pacific sectors. J Geophys Res 104(C7):15789–15802

Heil P, Hibler WD III (2002) Modelling the high-frequency component of Arctic sea ice drift and deformation. J Phys Oceanogr 32(11):3039–3057

Heil P, Allison I, Lytle VI (1996) Seasonal and inter-annual variations of the oceanic heat flux under land-fast Antarctic sea ice cover. J Geophys Res 101(C11):25741–25753

Heil P, Lytle VI, Allison I (1998) Enhanced thermodynamic ice growth by sea-ice deformation. Ann Glaciol 27:433–437

Hibler WD III (1972) Removal of aircraft altitude variation from laser profiles of the Arctic ice pack. J Geophys Res 27(36):7190–7195

Hibler WD III (1977) A viscous sea ice law as a stochastic average of plasticity. J Geophys Res 82:3932–3938

Hibler WD III (1979) A dynamic thermodynamic sea ice model. J Phys Oceanogr 9(4):815–846

Hibler WD III (1980) Modelling a variable thickness sea ice cover. Mon Weather Rev 108:1943–1973

Hibler WD III (2001) Sea ice fracturing on the large scale. Eng Fract Mech 68:2013–2043

Hibler WD III, Hutchings JK (2002) Multiple equilibrium Arctic ice cover states induced by ice mechanics. In: Squire V, Langhorne P (eds) Ice in the environment: proceedings of the 16th IAHR international symposium ice 2, Dunedin, New Zealand, 2–6 Dec 2002. International Association of Hydraulic Engineering and Research, pp 228–238

Hibler WD III, Schulson EM (2000) On modelling the anisotropic failure and flow of flawed sea ice. J Geophys Res 105(C7):17105–17120

Hibler WD III, Tucker WB III (1979) Some results from a linear-viscous model of the Arctic ice cover. J Glaciol 22(87):293–304

Hill BT, Jones SJ (1990) The Newfoundland ice extent and the solar cycle from 1860 to 1988. J Geophys Res 95(C4):5385–5394

Holt B, Rothrock DA, Kwok R (1992) Determination of sea ice motion from satellite images. In: Carsey F (ed) Microwave remote sensing of sea ice. American Geophysical Union, Washington, DC, pp 343–354

Hopkins MA (1994) On the ridging of intact lead ice. J Geophys Res 99(C8):16351–16360

Hopkins MA (1996a) On the mesoscale interaction of lead ice and floes. J Geophys Res 101(C8):18315–18326

Hopkins MA (1996b) The effects of individual ridging events on the ice thickness distribution in Arctic ice pack. Cold Reg Sci Technol 2:75–82

Hopkins MA (1998) Four stages of pressure ridging. J Geophys Res 103(C10):21883–21891

Hopkins MA, Hibler WD III (1990) Simulation of the ice ridging process. In: Ackley SF, Weeks WF (eds) Sea ice properties and processes. Proceedings of WF weeks sea ice symposium, CRREL monograph no 90–1. CRREL, Hanover, pp 152–155

Hopkins MA, Hibler WD III (1991) On the ridging of a thin sheet of lead ice. Ann Glaciol 15:81–86

Hopkins MA, Tuhkuri J (1999) Compression of floating ice fields. J Geophys Res 104(C7):15815–15826

Hopkins MA, Hibler WD III, Flato GM (1991) On the numerical simulation of the sea ice ridging process. J Geophys Res 96:4809–4820

Hopkins MA, Frankenstein S, Thorndike AS (2004) Formation of an aggregate scale in Arctic sea ice. J Geophys Res 109:C1032. doi:101029/2003JC001855

Høyland KV (2002) Consolidation of first-year ridges. J Geophys Res 107(C6):3062. doi:10.1029/2000JC000526

Høyland KV, Løset S (1999) Measurements of temperature distribution consolidation and morphology of a first-year sea ice ridge. Cold Reg Sci Technol 29:59–74

Hudson RD (1990) Annual measurement of sea-ice thickness using upward-looking sonar. Nature 344:135–137

Hughes BA (2003) A statistical description of Arctic level-ice pans using the USS Gurnard ice-draft data set. Canadian Technical Reports of Hydrogrography and Ocean Science 229, Institute of Ocean Sciences, Sidney, iv, 45 pp

Hutchings JK, Hibler WD III (2002) Modelling sea ice deformation with a viscous-plastic isotropic rheology. In: Squire V, Langhorne P (eds) Ice in the environment. Proceedings of 16th IAHR international symposium on ice 2, Dunedin, New Zealand, 2–6 Dec 2002. International Association of Hydraulic Research, Madrid, pp 358–366

Hvidegaard SM, Forsberg R (2002) Sea-ice thickness from airborne laser altimetry over the Arctic Ocean north of Greenland. Geophys Res Lett 29(20):1952. doi:101029/2001GL014474

Ice Centre (1992) Ice thickness climatology 1961–1990 normals. Ministry of Supply and Services Canada report En57–28. Avail Canada Communication Group Publishing, Ottawa, 277 pp

Ingram RG, Bâcle J, Barber DG, Gratton Y, Melling H (2002) A review of physical processes in the North Water. Deep Sea Res II 49(22–23):4893–4906

Jeffries MO, Adolphs U (1997) Early winter ice and snow thickness distribution ice structure and development of the western Ross Sea pack ice between the ice edge and the Ross ice shelf. Antarct Sci 9(2):188–200

Jeffries MO, Shaw MA (1993) The drift of ice islands from the Arctic Ocean into the channels of the Canadian Arctic Archipelago: the history of Hobson's Choice ice island. Polar Rec 29(171):305–312

Jeffries MO, Morris K, Worby AP, Weeks WF (1994a) Late winter characteristics of the seasonal snow cover on sea ice floes in the Bellingshausen and Amundsen Seas. Antarc J USA 29(1):9–10

Jeffries MO, Morris K, Worby AP, Weeks WF (1994b) Late winter sea ice properties and growth processes in the Bellingshausen and Amundsen Seas. Antarc J USA 29(1):11–13

Jeffries MO, Shaw MA, Morris K, Veazey AL, Krouse HR (1994c) Crystal structure stable isotopes (d18O) and development of sea ice in the Ross Amundsen and Bellingshausen Seas Antarctica. J Geophys Res 99:985–995

Jeffries MO, Morris K, Maksym T, Kozlenko N, Tin T (2001) Autumn sea ice thickness ridging and heat flux variability in and adjacent to Terra Nova Bay Ross Sea Antarctica. J Geophys Res 106(C3):4437–4448

Johannessen OM, Miles M, Bjorgo E (1995) Arctic's shrinking ice cover. Nature 376:126–127

Johannessen OM, Shalina EV, Miles MW (1999) Satellite evidence for an Arctic sea ice cover in transformation. Science 286:1937–1939

Johnson JB, Cox GFN, Tucker WB III (1985) Kadluk ice stress measurement program. In: Proceedings of the 8th international conference port and ocean engineering under Arctic conditions, vol 1, Horsholm, Denmark, 7–14 Sept 1985. Danish Hydraulic Institute, Horsholm, pp 88–100

Johnson MA, Proshutinsky AY, Polyakov IV (1999) Atmospheric patterns forcing two regimes of Arctic circulation: a return to anti-cyclonic conditions? Geophys Res Lett 26(11):1621–1624

Kawamura T, Ohshima KI, Takizawa T, Ushio S (1997) Physical structural and isotopic characteristics and growth processes of fast ice in Lutzow-Holm Bay Antarctica. J Geophys Res 102(C2):3345–3356

Ketchum RD (1971) Airborne laser profiling of the Arctic pack ice. Remote Sens Environ 2:41–52

Key JR, Silcox RA, Stone RS (1996) Evaluation of surface radiative flux parameterizations for use in sea ice models. J Geophys Res 101(C2):3839–3850

Killworth PD, Smith JM (1984) A one-and-a-half dimensional model for the Arctic halocline. Deep Sea Res 31:271–293

Koerner RM (1973) The mass balance of the sea ice in the Arctic Ocean. J Glaciol 12:173–185

Kovacs A, Valleau N, Holladay JS (1987) Airborne electromagnetic sounding of sea ice thickness and sub-ice bathymetry. Cold Reg Sci Technol 14:289–311

Kozo TL, Tucker WB III (1974) Sea ice bottom-side features in the Denmark Strait. J Geophys Res 79:4505–4511

Kozo TL, Stringer WJ, Torgerson LJ (1987) Mesoscale nowcasting of sea ice movement through the Bering Strait with a description of major driving forces. Mon Weath Rev 115:193–207

Krabill WB, Swift RN, Tucker WB III (1990) Recent measurements of sea ice topography in the eastern Arctic. In: Ackley SF, Weeks WF (eds) Sea ice properties and processes. Proceedings of WF weeks sea ice symposium, CRREL monograph no 90-1. CRREL, Hanover, pp 132–136

Kwok R (2000) Recent changes in Arctic Ocean sea ice motion associated with the North Atlantic Oscillation. Geophys Res Lett 27(6):775–778

Kwok R (2002) Sea ice concentration estimates from satellite passive microwave radiometry and openings from SAR motion. Geophys Res Lett 29(9):doi:101029/2002GL014787

Kwok R, Cunningham GF (2002) Seasonal ice area and volume production of the Arctic Ocean: November 1996 through April 1997. J Geophys Res 107(C10):8038. doi:101029/2000JC000469

Kwok R, Rothrock DA (1999) Variability of Fram Strait ice flux and North Atlantic Oscillation. J Geophys Res 104(C3):5177–5189

Kwok R, Curlander JC, McConnell R, Pang SS (1990) An ice-motion tracking system at the Alaska SAR facility. IEEE J Ocean Eng 15:44–54

Kwok R, Comiso JC, Cunningham GF (1996) Seasonal characteristics of the perennial ice cover of the Beaufort Sea. J Geophys Res 101:28417–28439

Kwok R et al (1998a) The RADARSAT geophysical processor system. In: Analysis of SAR data of the polar oceans. Springer, New York, pp 235–257

Kwok R, Schweiger A, Rothrock DA, Pang S, Kottmeier C (1998b) Sea ice motion from passive microwave imagery assessed with ERS SAR and buoy motions. J Geophys Res 103(C4):8191–8214

Kwok R, Cunningham GF, Yueh S (1999) Area balance of the Arctic Ocean perennial ice zone: October 1996 to April 1997. J Geophys Res 104(C11):25747–25760

Kwok R, Cunningham GF, Ngiem SV (2003) A study of the onset of melt over the Arctic Ocean in Radarsat synthetic aperture radar data. J Geophys Res 108(C11):3363. doi:101029/2002JC001363

Kwok R, Cunningham GF, Pang SS (2004a) Fram Strait sea ice outflow. J Geophys Res 109:C01009. doi:101029/2003JC001785

Kwok R, Zwally J, Yi D (2004b) ICESat observations of Arctic sea ice: a first look. Geophys Res Lett 31:L16401. doi:101029/2004GL020309

Lange MA (1991) Antarctic sea ice: its development and basic properties. In: Proceedings of international conference on role of polar regions in global change, 11–15 June 1990. University of Alaska, Fairbanks, pp 275–283

Lange MA, Eicken H (1990) The sea ice thickness distribution in the north-western Weddell Sea. J Geophys Res 95:4821–4837

Lange MA, Ackley SF, Dieckmann GS, Eicken H, Wadhams P (1989) Development of sea ice in the Weddell Sea Antarctica. Ann Glaciol 12:92–96

Laxon S (1994) Sea ice altimeter processing scheme at the EODC. Int J Remote Sens 15(4):112–116

Laxon S, Peacock N, Smith D (2003) High inter-annual variability of sea ice thickness in the Arctic region. Nature 425:947–950. doi:101038/Nat02050

LeDrew EF, Barber DG (1994) The SIMMS Programme: a study of change and variability within the marine cryosphere. Arctic 47:256–264

Lemke P (1987) A coupled one-dimensional sea ice-ocean model. J Geophys Res 92:13164–13172

Lemke P, Owens WB, Hibler WD III (1990) A coupled sea ice-mixed layer-pycnocline model for the Weddell Sea. J Geophys Res 95:9513–9525

Leppäranta M, Hakala R (1992) The structure and strength of first-year ice ridges in the Baltic Sea. Cold Reg Sci Technol 20:295–311

Leppäranta M, Lensu M, Kosloff P, Veitch B (1995) The life story of a first-year sea ice ridge. Cold Reg Sci Technol 23:279–290

Lewis EL (1967) Heat flow through winter ice. In: Proceedings of international conference on low temperature science. Hokkaido University, Sapporo, 1966, pp 611–631

Lewis JK, Richter-Menge JA (1998) Motion-induced stresses in pack ice. J Geophys Res 103(C10):21831–21843

Lindsay RW (1998) Temporal variability of the energy balance of thick Arctic pack ice. J Clim 11:313–333

Lindsay RW (2002) Ice deformation near SHEBA. J Geophys Res 107(C10):8042. doi:10.1029/2000JC000445

Lindsay RW (2003) Changes in the modelled ice thickness distribution near the Surface Heat Budget of the Arctic Ocean (SHEBA) drifting ice camp. J Geophys Res 108(C6):3194. doi:101029/2001JC000805

Lindsay RW, Rothrock DA (1994) Arctic sea ice albedo from AVHRR. J Clim 7:1737–1749

Lindsay RW, Rothrock DA (1995) Arctic sea ice leads from advanced very high resolution radiometer images. J Geophys Res 100(C3):4533–4544

Lindsay RW, Stern HL (2003) Radarsat geophysical processor system products: quality of sea-ice trajectory and deformation estimates. J Atmos Ocean Technol 20:1333–1347

Liu AK, Martin S, Kwok R (1997) Tracking ice edges and ice floes by wavelet analysis of SAR images. J Atmos Ocean Technol 14:1187–1198

Liu AK, Zhao Y, Wu SY (1999) Arctic sea ice drift from wavelet analysis of NSCAT and special sensor microwave imager data. J Geophys Res 104(C5):11529–11538

Loder JW, Petrie B, Gawarkiewicz G (1998) The coastal ocean off northeastern North America: a large-scale view. In: Robinson AR, Brink KH (eds) The sea volume 11. The global coastal ocean. Wiley, New York, pp 105–133

Lyon WK (1984) The navigation of Arctic polar submarines. J Navig 37:155–179

Lytle VI, Ackley SF (1991) Sea ice ridging in the eastern Weddell Sea. J Geophys Res 96(C10):18411–18416

Lytle VI, Ackley SF (1996) Heat flux through sea ice in the western Weddell Sea convective and conductive transfer processes. J Geophys Res 101:8853–8868

Lytle VI, Worby AP, Massom RA (1998) Sea-ice pressure ridges in East Antarctica. Ann Glaciol 27:449–454

Lytle VI, Massom R, Bindoff N, Worby A, Allison I (2000) Wintertime heat flux to the underside of East Antarctic pack ice. J Geophys Res 105(C12):28759–28769

Maksym T, Jeffries MO (2000) A one-dimensional percolation model of flooding and snow ice formation on Antarctic sea ice. J Geophys Res 105(C11):26313–26331

Marko JR, Thomson RE (1977) Rectilinear leads and internal motions in the ice pack of the western Arctic Ocean. J Geophys Res 82:979–987

Martin T, Augstein E (2000) Large-scale drift of Arctic sea ice retrieved from passive microwave satellite data. J Geophys Res 105(C4):8775–8788

Martin S, Cavalieri DJ (1989) Contribution of the Siberian Sea polynyas to the Arctic Ocean intermediate and deep water. J Geophys Res 94:12725–12738

Martin T, Wadhams P (1999) Sea-ice flux in the East Greenland current. Deep Sea Res II 46(6–7):1063–1082

Martinson DG, Steele M (2001) Future of Arctic sea ice cover: implications of an Antarctic analog. Geophys Res Lett 28(2):307–310

Maslanik JA, Serreze MC, Barry RG (1996) Recent decreases in Arctic summer ice cover and linkages to atmospheric circulation anomalies. Geophys Res Lett 23(13):1677–1680

Maslanik JA, Fowler C, Key J, Scambos T, Hutchinson T, Emery W (1998) AVHRR-based polar pathfinder products for modelling applications. Ann Glaciol 25:388–392

Maslanik JA, Serreze MC, Agnew T (1999) On the record reduction in 1998 western Arctic sea-ice cover. Geophys Res Lett 26(13):1905–1908

Massom RA (1992) Observing the advection of sea ice in the Weddell Sea using buoy and satellite passive microwave data. J Geophys Res 97:15559–15572

Massom RA, Drinkwater MR, Haas C (1997) Winter snow cover on sea ice in the Weddell Sea. J Geophys Res 102(1):1101–1117

Maykut GA (1978) Energy exchange over young sea ice in the central Arctic. J Geophys Res 83:3646–3658

Maykut GA (1982) Large-scale heat exchange and ice production in the central Arctic. J Geophys Res 87:7971–7984

Maykut GA (1986) Surface heat and mass balance. In: Untersteiner N (ed) The geophysics of sea ice. Plenum, New York

Maykut GA, McPhee MG (1995) Solar heating of the Arctic mixed layer. J Geophys Res 100(C12):24691–24704

Maykut GA, Perovich DK (1987) The role of shortwave radiation in the summer decay of a sea ice cover. J Geophys Res 92:7032–7044

Maykut GA, Untersteiner N (1971) Some results from a time-dependent thermodynamic model of sea ice. J Geophys Res 76:1550–1575

McLaren AS, Wadhams P, Weintraub R (1984) The sea ice topography of M'Clure Strait in winter and summer of 1960 from submarine profiles. Arctic 37:110–120

McLaren AS, Serreze MC, Barry RG (1987) Seasonal variations of sea ice motion in the Canada Basin and their implications. Geophys Res Lett 14(11):1123–1126

McLaren AS, Barry RG, Bourke RH (1990) Could Arctic ice be thinning? Nature 345:762

McLaren AS, Walsh JE, Bourke RH, Weaver RL, Wittmann W (1992) Variability in sea-ice thickness over the North Pole from 1977 to 1990. Nature 358:224–226

McLaren AS, Bourke RH, Walsh JE, Weaver RL (1994) Variability in sea-ice thickness over the North Pole from 1958–1992. In: The polar oceans and their role in shaping the global environment. Geophysical monograph 85. American Geophysical Union, Washington, DC, pp 363–371

McPhee MG (1978) A simulation of inertial oscillations in drifting pack ice. Dyn Atmos Ocean 2:107–122

McPhee MG (1980) Heat transfer across the salinity-stabilized pycnocline of the Arctic Ocean. In: Carstens T, McClimans T (eds) Proceedings of the 2nd international symposium on stratified flows, Trondheim, Norway, pp 527–537

McPhee MG (1983) Turbulent heat and momentum transfer in the oceanic boundary layer under melting pack ice. J Geophys Res 88:2827–2835

McPhee MG (1987) A time-dependent model for turbulent transfer in a stratified-oceanic boundary layer. J Geophys Res 92:6977–6986

McPhee MG (1992) Turbulent heat flux in the upper ocean under sea ice. J Geophys Res 97:5365–5379

McPhee MG, Smith JD (1976) Measurements of the turbulent boundary layer under pack ice. J Phys Oceanogr 6:696–711

McPhee MG, Maykut GA, Morison JH (1987) Dynamics and thermodynamics of the ice/upper ocean system in the marginal ice zone of the Greenland Sea. J Geophys Res 92:7017–7031

Meier WN, Maslanik JA, Fowler CW (2000) Error analysis and assimilation of remotely sensed ice motion within an Arctic sea ice model. J Geophys Res 105(C2):3339–3356

Melling H (1983) In situ determination of the thermal diffusivity of sea ice. Canadian Technical Reports of Hydrography and Ocean Science 27. Institute of Ocean Sciences, Sidney, 24 pp

Melling H (1993) The formation of a haline shelf front in wintertime in an ice-covered arctic sea. Cont Shelf Res 13(10):1123–1147

Melling H (1998a) Detection of features in first-year pack ice by synthetic aperture radar (SAR). Int J Remote Sens 19(6):1223–1249

Melling H (1998b) Sound scattering from sea ice: Aspects relevant to ice-draft profiling by sonar. J Atmos Ocean Technol 15(6):274–297

Melling H (2000a) Exchanges of freshwater through the shallow straits of the North American Arctic. In: The freshwater budget of the Arctic Ocean. NATO/WCRP/AOSB. Kluwer Academic, Amsterdam. Proceedings of a WCRP/AOSB/NATO advanced research workshop, Tallinn, Estonia, Apr 1998, pp 479–502

Melling H (2000b) Scientific interpretation of measurements of sea-ice draft measurements by moored sonar. In: Lemke P, Colony R (eds) Joint report 4th session of the ACSYS Sea Ice/Ocean Modeling (SIOM) panel and the ACSYS workshop on sea-ice thickness measurements and data analysis, Monterey, CA, 7–11 Apr 1997. WMO/TD No 991. World Meteorological Organization, Geneva A510, 5pp

Melling H (2002) Sea ice of the northern Canadian Arctic Archipelago. J Geophys Res 107(C11):3181. doi:101029/2001JC001102

Melling H, Lewis EL (1982) Shelf drainage flows in the Beaufort Sea and their effect on the Arctic Ocean pycnocline. Deep Sea Res 29:967–985

Melling H, Moore RM (1995) Modification of halocline source waters during freezing on the Beaufort Sea shelf: evidence from oxygen isotopes and dissolved nutrients. Cont Shelf Res 15(1):89–113

Melling H, Riedel DA (1993) Draft and movement of pack ice in the Beaufort Sea, Apr 1990–Mar 1991. Canadian Technical reports of Hydrography and Ocean Science 151, Institute of Ocean Sciences, Sidney, 79 pp

Melling H, Riedel DA (1994) Draft and movement of pack ice in the Beaufort Sea, Apr 1991–Apr 1992. Canadian Technical Reports of Hydrography and Ocean Science 162, Institute of Ocean Sciences, Sidney, 109 pp

Melling H, Riedel DA (1995) The underside topography of sea ice over the continental shelf of the Beaufort Sea in the winter of 1990. J Geophys Res 100(C7):13641–13653

Melling H, Riedel DA (1996a) Development of seasonal pack ice in the Beaufort Sea during the winter of 1991–92: a view from below. J Geophys Res 101(C5):11975–11992

Melling H, Riedel DA (1996b) The thickness and ridging of pack ice causing difficult shipping conditions in the Beaufort Sea summer 1991. Atmos Ocean 31(3):457–487

Melling H, Riedel DA (2004) Draft and movement of pack ice in the Beaufort Sea: a time-series presentation, Apr 1990–Aug 1999. Canadian Technical reports of Hydrography and Ocean Science, vol 238, 24 pp

Melling H, Lake RA, Topham DR, Fissel DB (1984) Oceanic thermal structure in the western Canadian Arctic. Cont Shelf Res 3:233–258

Melling H, Topham DR, Riedel DR (1993) Topography of the upper and lower surfaces of 10 hectares of deformed sea ice. Cold Reg Sci Technol 21:349–369

Melling H, Johnston PH, Riedel DA (1995) Measurement of the topography of sea ice by moored sub-sea sonar. J Atmos Ocean Technol 12(3):589–602

Melling H, Gratton Y, Ingram RG (2001) Ocean circulation within the North Water Polynya of Baffin Bay. Atmos Ocean 39(3):301–325

Mellor GL, McPhee MG, Steele M (1986) Ice-saltwater turbulent boundary layer interaction with melting or freezing. J Phys Oceanogr 16:1829–1846

Midttun L (1985) Formation of dense bottom water in the Barents Sea. Deep Sea Res 32: 1233–1241

Miles MW, Barry RG (1998) A 5-year satellite climatology of winter sea ice leads in the western Arctic. J Geophys Res 103(C10):21723–21734

Morison JH, McPhee MG, Maykut GA (1987) Boundary layer upper ocean and ice observations in the Greenland Sea marginal ice zone. J Geophys Res 92:6987–7011

Morison JH, McPhee MG, Curtin TB, Paulson CA (1992) The oceanography of winter leads. J Geophys Res 97(C7):11199–11218

Morison JH, Aagaard K, Falkner KK, Hatakeyama K, Moritz R, Overland JE, Perovich D, Shimada K, Steele M, Takizawa T, Woodgate R (2002) North Pole environmental observatory delivers early results. EOS Trans Am Geophys Union 83(33):357, 360–361

Moritz RE (1992) Seasonal and regional variability of sea ice thickness distribution. In: Proceedings of 3rd conference on polar meteorology and oceanography, Portland, OR, 29 Sept–2 Oct 1992. American Meteorological Society, Boston, pp 68–71

Mullison J, Melling H, Johns W, Freitag P (2004) The role of moored current profilers in climate variability research. Sea Technol 45(2):17–28

Multala J, Hautaniemi H, Oksama M, Leppäranta M, Haapala J, Herlevi A, Riska K, Lensu M (1996) An airborne electromagnetic system on a fixed wing aircraft for sea ice thickness mapping. Cold Reg Sci Technol 24:355–373

Muramoto K, Matsuura K, Endoh T (1994) A technique of continuous analysis of sea ice distribution using video images taken from a ship. 16th Symposium on polar meteorology and glaciology, Japan national institute for polar research, Tokyo, 4-5 August 1993. In: Proceedings of NIPR Symposium on Polar Meteorology and Glaciology, vol 8, pp 161–168

Ninnis RM, Emery WJ, Collins MJ (1986) Automated extraction of pack ice motion from advanced very high resolution radiometer imagery. J Geophys Res 91:10725–10734

Nutt DC (1966) The drift of ice island WH-5. Arctic 19(3):244–262

Oelke C (1997) Atmospheric signatures in sea-ice concentration estimates from passive microwaves: modelled and observed. Int J Remote Sens 18(5):1113–1136

Ohshima KI, Yoshida K, Shimoda H, Wakatsuchi M, Endoh T, Fukuchi M (1998) Relationship between the upper ocean and sea ice during the Antarctic melting season. J Geophys Res 103(C4):7601–7615

Onstott RG (1992) SAR and scatterometer signatures of sea ice. In: Carsey FD (ed) Microwave remote sensing of sea ice. Geophysical monograph 68. American Geophysical Union 2000, Washington, DC, pp 73–104

Ostlund HG, Hut G (1984) Arctic Ocean water mass balance from isotope data. J Geophys Res 89:6373–6381

Overland JE, Walter BA, Curtin TB, Turet P (1995) Hierarchy and sea ice mechanics. J Geophys Res 100(C3):4559–4572

Overland JE, McNutt SL, Salo S, Groves J, Li S (1998) Arctic sea ice as a granular plastic. J Geophys Res 103(C10):21845–21868

Overland JE, McNutt SL, Groves J, Salo S, Andreas EL, Persson POG (2000) Regional sensible and radiative heat flux estimates for the winter Arctic during the Surface Heat Budget of the Arctic Ocean (SHEBA) experiment. J Geophys Res 105(C6):14093–14102

Parkinson CL (1991) Inter-annual variability of the spatial distribution of sea ice in the north polar region. J Geophys Res 96(C3):4791–4801

Parkinson CL (1992a) Inter-annual variability of monthly Southern Ocean sea ice distributions. J Geophys Res 97(C4):5349–5363

Parkinson CL (1992b) Spatial patterns of increases and decreases in the length of the sea ice season in the north polar region 1979–1985. J Geophys Res 97(C9):14377–14388

Parkinson CL, Cavalieri D (1989) Arctic sea ice 1973–1987: seasonal regional and inter-annual variability. J Geophys Res 94(C10):14499–14523

Parkinson CL, Cavalieri DJ, Gloersen P, Zwally HJ, Comiso JC (1999a) Variability of the Arctic sea ice cover 1978–1996. J Geophys Res 104(C9):20837–20856

Parkinson CL, Cavalieri DJ, Gloersen P, Zwally HJ, Comiso JC (1999b) Arctic sea ice extents areas and trends 1978–1996. J Geophys Res 104(C9):20837–20856

Partington K, Flynn T, Lamb D, Bertoia C, Dedrick K (2003) Late twentieth century Northern Hemisphere sea-ice record from the US National Ice Center ice charts. J Geophys Res 108(C11):3343. doi:101029/2002JC001623

Peacock N, Laxon S (2004) Sea surface height determination in the Arctic Ocean from ERS altimetry. J Geophys Res 109:C07001. doi:101029/2001JC001026

Pease CH (1980) Eastern Bering Sea ice processes. Mon Weather Rev 108:2015–2023

Perovich DK, Elder BC (2001) Temporal evolution of Arctic sea ice temperature. Ann Glaciol 33:207–211

Perovich DK, Elder BC (2002) Estimates of ocean heat flux during SHEBA. Geophys Res Lett 29(9):doi:101029/2001GL014171

Perovich DK, Maykut GA (1990) Solar heating of a stratified ocean in the presence of a static ice cover. J Geophys Res 95(C10):18233–18245

Perovich DK, Richter-Menge JA (2000) Ice growth and solar heating in springtime leads. J Geophys Res 105(C3):6541–6548

Perovich DK, Tucker WB III (1989) Oceanic heat flux in the Fram Strait measured by a drifting buoy. Geophys Res Lett 16(9):995–998

Perovich DK, Elder BC, Richter-Menge JA (1997) Observations of the annual cycle of sea ice temperature and mass balance. Geophys Res Lett 5:555–558

Perovich DK, Richter-Menge JA, Tucker WB III (2001) Seasonal changes in Arctic sea ice morphology. Ann Glaciol 33:171–176

Perovich DK, Tucker WB III, Ligett KA (2002a) Aerial observations of the evolution of ice surface conditions during summer. J Geophys Res 107(C10):doi:101029/2000JC000449

Perovich DK, Grenfell TC, Light B, Hobbs PV (2002b) Seasonal evolution of the albedo of multi-year Arctic sea ice. J Geophys Res 107(C10):doi:101029/2000JC000438

Perovich DK, Grenfell TC, Richter-Menge JA, Light B, Tucker WB III, Eicken H (2003) Thin and thinner: ice mass balance measurements during SHEBA. J Geophys Res 108(C3): doi:101029/2001JC001079

Pfirman SL, Colony R, Nürnberg D, Eicken H, Rigor IG (1997) Reconstructing the origin and trajectory of drifting Arctic sea ice. J Geophys Res 102(6):12575–12586

Pilkington GR, Wright BD (1991) Beaufort Sea ice thickness measurements from an acoustic under-ice upward-looking ice keel profiler. In: Proceedings of the 1st international offshore and polar engineering conference, Edinburgh, UK, 11–16 Aug 1991. International Offshore and Polar Engineering Conference, pp 456–461

Podgorny IA, Grenfell TC (1996) Partitioning of solar energy in melt ponds from measurements of pond albedo and depth. J Geophys Res 101(C10):22737–22748

Polyakov IV, Proshutinsky AY, Johnson MA (1999) Seasonal cycles in two circulation of Arctic climate. J Geophys Res 104(C11):25761–25788

Polyakov IV, Alekseev GV, Berkryaev RV, Bhatt US, Colony R, Johnson MA, Karklin VP, Walsh D, Yulin AV (2003) Long-term ice variability in Arctic marginal seas. J Clim 16(12):2078–2085

Prinsenberg SJ, Peterson IK, Holladay S (1996) Comparison of airborne electromagnetic ice thickness data with NOAA/AVHRR and ERS-1/SAR images. Atmos Ocean 34(1):185–205

Pritchard RS (1975) An elastic-plastic constitutive law for sea ice. J Appl Mech 42:379–384

Pritchard RS (1998) Ice conditions in an anisotropic sea-ice dynamics model. Int J Offshore Pol Eng 8:9–15

Pritchard RS, Coon MD, McPhee MG (1977) Simulation of sea ice dynamics during AIDJEX. J Press Vessel Technol 99:491–497

Proshutinsky AY, Johnson MA (1997) Two circulation regimes of the wind-driven Arctic Ocean. J Geophys Res 102(C6):12493–12514

Remund QP, Long DG (1999) Sea-ice extent mapping using Ku-band scatterometer data. J Geophys Res 104(C5):11515–11528

Richter-Menge JA, Elder BC (1998) Characteristics of pack ice stress in the Alaskan Beaufort Sea. J Geophys Res 103(C10):21817–21829

Richter-Menge JA, McNutt SL, Overland JE, Kwok R (2002) Relating Arctic pack ice stress and deformation under winter conditions. J Geophys Res 107(C10):8040. doi:101029/2000JC000477

Rigor I (2002) IABP drifting buoy pressure temperature position and interpolated ice velocity. Compiled by the Polar Science Center, Applied Physics Laboratory, University of Washington Seattle, in association with NSIDC. National Snow and Ice Data Center. Digital media, Boulder

Rigor IG, Wallace JM (2004) Variations in the age of Arctic sea ice and summer sea-ice extent. Geophys Res Lett 31:L09401. doi:101029/2004GL019492

Rigor IG, Wallace JM, Colony RL (2002) Response of sea ice to the Arctic oscillation. J Clim 15(18):2648–2668

Romanov IP (1995) Atlas of ice and snow of the Arctic Basin and Siberian Shelf Seas, 2nd edn (trans: Tunik A). Backbone Publishing, Washington, DC, 277 pp

Rothrock DA, Thomas DR (1992) Ice modeling and data assimilation with the Kalman smoother. In: Carsey FD (ed) Microwave remote sensing of sea ice, AGU Geophysical monograph no 68. American Geophysical Union, Washington, DC, pp 405–418

Rothrock DA, Thorndike AS (1980) Geometric properties of the underside of sea ice. J Geophys Res 85:3955–3963

Rothrock DA, Thomas DR, Thorndike AS (1988) Principal component analysis of satellite passive microwave data over sea ice. J Geophys Res 93:2321–2332

Rothrock DA, Yu Y, Maykut GA (1999) Thinning of Arctic sea-ice cover. Geophys Res Lett 26(23):3469–3472

Rothrock DA, Kwok R, Groves D (2000) Satellite views of the Arctic Ocean freshwater balance. In: The freshwater budget of the Arctic Ocean. NATO/WCRP/AOSB. Kluwer Academic, Amsterdam. Proceedings of WCRP/AOSB/NATO advanced research workshop, Tallinn, Estonia, Apr 1998, pp 409–452

Rothrock DA, Zhang J, Yu Y (2003) The Arctic ice thickness anomaly of the 1990s: a consistent view from observations and models. J Geophys Res 108(C3):3083. doi:101029/2001JC001208

Rudels B, Anderson LG, Jones EP (1996) Formation and evolution of the surface mixed layer and halocline of the Arctic Ocean. J Geophys Res 101(C4):8807–8822

Ruffieux D, Oersson POG, Fairall CW, Wolfe DE (1995) Open ocean and lead surface energy budgets during LEADEX. J Geophys Res 100(C3):4593–4612

Sater JE (1964) Arctic drifting stations. Arctic Institute of North America, Washington, DC, 475 pp

Schramm J, Flato GM, Curry JA (2000) Toward modelling of enhanced basal melting in ridge keels. J Geophys Res 105(C6):14081–14092

Schulson EM, Hibler WD III (1991) The fracture of ice on scales large and small: Arctic leads and wing cracks. J Glaciol 37:319–322

Sear CB, Wadhams P (1992) Statistical properties of arctic sea ice morphology derived from side-scan sonar images. Prog in Oceanogr 29:133–160

Serreze MC, Maslanik JA, Scambos TA, Fetterer F, Stroeve J, Knowles K, Fowler C, Drobot S, Barry RG, Haran TM (2003) A record minimum Arctic sea ice extent and area in 2002. Geophys Res Lett 30(3):1110. doi:101029/2002GL016406

Shine KP, Crane RG (1984) The sensitivity of a one-dimensional thermodynamic sea ice model to changes in cloudiness. J Geophys Res 89:615–622

Shy TL, Walsh JE (1996) North Pole ice thickness and association with ice motion history 1977–1992. Geophys Res Lett 23(21):2975–2978

Simmonds I, Jacka TH (1995) Relationships between the inter-annual variability of Antarctic sea ice and the Southern Oscillation. J Clim 8(3):637–647

Sinha AK (1976) A field study for sea ice thickness determination by electromagnetic means. Geologi Surv Can Pap 76(1C):255–228. Avail Natural Resources Canada, Booth Street Ottawa Canada

Skyllingstad ED, Paulson CA, Pegau WS, McPhee MG, Stanton T (2003) Effect of keels on ice bottom turbulence exchange. J Geophys Res 108(C12):3372

Smith DM (1998a) Recent increase in the length of the melt season of perennial Arctic sea ice. Geophys Res Lett 25(5):655–658

Smith DM (1998b) Observations of perennial Arctic sea ice melt and freeze-up using passive microwave data. J Geophys Res 103:27753–27769

Steele M, Boyd T (1998) Retreat of the cold halocline layer in the Arctic Ocean. J Geophys Res 103(C5):10419–10436

Steele M, Flato GM (2000) Sea ice growth melt and modelling: a survey. In: The freshwater budget of the Arctic Ocean. NATO/WCRP/AOSB. Kluwer Academic, Amsterdam. Proceedings of WCRP/AOSB/NATO advanced research workshop, Tallinn, Estonia. Apr 1998, pp 549–587

Steele M, Morison JH, Curtin TB (1995) Halocline water formation in the Barents Sea. J Geophys Res 100(C1):881–894

Steffen K, Schweiger A (1991) NASA team algorithm for sea ice concentration retrieval from DMSP special sensor microwave imager: comparison with landsat satellite imagery. J Geophys Res 96(C12):21971–21987. doi:10.1029/91JC02334

Stern HL, Moritz RE (2002) Sea ice kinematics and surface properties from Radarsat synthetic aperture radar during the SHEBA drift. J Geophys Res 107(C10):8028. doi:101029/2000JC000472

Stern HL, Rothrock DA, Kwok R (1995) Open water production in Arctic sea ice: Satellite measurements and model parameterizations. J Geophys Res 100(C10):20601–20512

Strass VH (1998) Measuring sea ice draft and coverage with moored upward looking sonars. Deep Sea Res 45:795–818

Strass VH, Fahrbach E (1998) Temporal and regional variation of sea ice draft and coverage in the Weddell Sea obtained from upward looking sonars. Antarct Res Ser 74:123–140. American Geophysical Union

Sturm M, Morris K, Massom RA (1996) A description of the snow cover on the winter sea ice of the Amundsen and Ross Seas. Antarct J USA 30:1–4

Sturm M, Morris K, Massom RA (1998) The winter snow cover of the West Antarctic pack ice its spatial and temporal variability. In: Jeffries MO (ed) Antarctic sea ice physical processes interactions and variability. Antarctic Research Service 74. American Geophysical Union, Washington, DC, pp 19–40

Sturm M, Holmgren J, Perovich D (2001) Spatial variations in the winter heat flux at SHEBA: estimates from snow-ice interface temperatures. Ann Glaciol 33:213–220

Sturm M, Perovich D, Holmgren J (2002) Thermal conductivity and heat transfer through the snow on the ice of the Beaufort Sea. J Geophys Res 107(C10):doi: 101029/2000JC000409

Thomas D, Martin S, Rothrock D, Steele M (1996) Assimilating satellite concentration data into an Arctic sea ice mass balance model 1979–1985. J Geophys Res 101(C9):20849–20868

Thorndike AS, Colony R (1982) Sea ice motion in response to geostrophic winds. J Geophys Res 87:5845–5852

Thorndike AS, Rothrock DA, Maykut GA, Colony R (1975) The thickness distribution of sea ice. J Geophys Res 80:4501–4513

Topham DR, Bowen RG (1996) An evaluation of the performance of the airborne remote ice thickness sensor 'Ice Probe' over a first-year pressure ridge in 1991. Canadian Contractor Report of Hydrography and Ocean Sciences 47. Institute of Ocean Sciences, Sidney, 75 pp

Toyota T, Kawamura T, Ohshima KI, Shimoda H, Wakatsuchi M (2004) Thickness distribution texture and stratigraphy and a simple probabilistic model for dynamical thickening of sea ice in the southern Sea of Okhotsk. J Geophys Res 109:C06001. doi:101029/2003JC002090

Tucker WB III, Westhall VH (1973) Arctic sea ice ridge frequency distribution derived from laser profiles. AIDJEX Bull 21. Division of Marine Resources. University of Washington, Seattle, pp 171–180

Tucker WB III, Weeks WF, Frank M (1979) Sea ice ridging over the Alaskan continental shelf. J Geophys Res 84:4885–4897

Tucker WB III, Perovich DK, Hopkins MA, Hibler WD III (1991) On the relationship between local stresses and strains in Arctic pack ice. Ann Glaciol 15:265–270

Tucker WB III, Gow A, Meese D, Bosworth H (1999) Physical characteristics of summer sea ice across the Arctic Ocean. J Geophys Res 104(C1):1489–1504

Tucker WB III, Weatherly JW, Eppler DT, Farmer LD, Bentley DL (2001) Evidence for rapid thinning of sea ice in the western Arctic Ocean at the end of the 1980s. Geophys Res Lett 28(14):2851–2854

Tuhkuri J, Lensu M (2002) Laboratory tests on ridging and rafting of ice sheets. J Geophys Res 107(C9):3125. doi:101029/2001JC000848

Tunik AL (1994) Route-specific ice thickness distribution in the Arctic Ocean during a North Pole crossing in August 1990. Cold Reg Sci Technol 22:205–217

Ukita J, Moritz RE (1995) Yield curves and flow rules of pack ice. J Geophys Res 100(C3):4545–4558

Untersteiner N (1961) On the mass and heat budget of Arctic sea ice. Arch Meteorol Geophys Bioklim A12:151–182

Untersteiner N (1964) Calculations of temperature regime and heat budget of sea ice in the central Arctic. J Geophys Res 69:4755–4765

Vihma T, Launiainen J (1993) Ice drift in the Weddell Sea in 1990–91 as tracked by a satellite buoy. J Geophys Res 98(C8):14471–14485

Vihma T, Launiainen J, Uotila J (1996) Weddell Sea ice drift: kinematics and wind forcing. J Geophys Res 101(C8):18279–18296

Vinje TK (2001a) Anomalies and trends of sea-ice extent and atmospheric circulation in the Nordic Seas during the period 1864–1998. J Clim 14(3):255–267

Vinje TK (2001b) Fram Strait ice fluxes and atmospheric circulation 1950–2000. J Clim 14(16):3508–3517

Vinje TK, Berge T (1989) Upward looking sonar recordings at 75N 12W from 22 June 1987 to 20 June 1988. Norsk Polarinstitutt Rapportserie NR Norsk Polarinstitutt 1330, Oslo, nr-51

Vinje TK, Finnekåsa O (1986) The ice transport through the Fram Strait. Norsk Polarinstitutt Skrifter 186, Oslo

Vinje TK, Kvambekk AS (1991) Barents Sea drift ice characteristics. Polar Rec 10(1):59–68

Vinje TK, Nordlund N, Kvambekk A (1998) Monitoring ice thickness in Fram Strait. J Geophys Res 103(C5):10437–10450

Vinnikov KY, Robock A, Cavalieri DJ, Parkinson CL (2002) Analysis of seasonal cycles in climatic trends with application to satellite observations of sea ice extent. Geophys Res Lett 29(9):24–28

Visbeck M, Fischer J (1995) Sea surface conditions remotely sensed by upward-looking ADCP's. J Atmos Ocean Technol 12(2):141–149

Vowinckel E, Orvig S (1962) Water balance and heat flow of the Arctic Ocean. Arctic 15:205–223

Wadhams P (1977) Characteristics of deep pressure ridges in the Arctic Ocean. In: Proceedings of POAC'77, 26–30 Sept 1977. St John's, pp 544–555

Wadhams P (1978) Side-scan sonar imagery of sea ice in the Arctic Ocean. Can J Remote Sens 4:161–173

Wadhams P (1981) Sea ice topography of the Arctic Ocean in the region 70W to 25E. Phiosl Trans R Soc Lond A302:45–85

Wadhams P (1983a) Arctic sea ice morphology and its measurement. J Soc Underw Technol 9(2):1–12

Wadhams P (1983b) Sea-ice thickness distribution in Fram Strait. Nature 305:108–111

Wadhams P (1988) The underside of sea ice imaged by side-scan sonar. Nature 333:161–164

Wadhams P (1989) Sea-ice thickness distribution in the trans-polar drift stream. Rapp P V Reun Cons Int Explor Mer 188:59–65

Wadhams P (1990) Evidence for thinning of the Arctic ice cover north of Greenland. Nature 345:795–797

Wadhams P (1991) Variations in sea ice thickness in the polar regions. In: Proceedings of international conference on the role of the polar regions in global change, 11–15 June 1990. University of Alaska Fairbanks, Fairbanks, pp 4–13

Wadhams P (1992) Sea ice thickness distribution in the Greenland Sea and Eurasian Basin May 1987. J Geophys Res 97:5331–5348

Wadhams P (1994) Sea ice thickness changes and their relation to climate. In: The polar oceans and their role in shaping the global environment. Geophysical monograph 85. American Geophysical Union, Washington, DC, pp 337–361

Wadhams P (1997) Ice thickness in the Arctic Ocean: the statistical reliability of experimental data. J Geophys Res 102(C2):27951–27959

Wadhams P, Davis NR (2000) Further evidence of ice thinning in the Arctic Ocean. Geophys Res Lett 27(24):3973–3976

Wadhams P, Horne RJ (1980) An analysis of ice profiles obtained by submarine sonar in the Beaufort Sea. J Glaciol 25(93):401–424

Wadhams P, Lowry RT (1977) A joint topside-bottomside remote sensing experiment on Arctic sea ice. In: Proceedings of the 4th canadian symposium on remote sensing, Quebec City, May 1977, pp 407–423

Wadhams P, Martin S (1990) Processes determining the bottom topography of multiyear Arctic sea ice. In: Ackley SF, Weeks WF (eds) Sea ice properties and processes. Proceedings of WF weeks sea ice symposium. CRREL monograph no 90–1. CRREL, Hanover, pp 136–141

Wadhams P, McLaren AS, Weintraub R (1985) Ice thickness distribution in Davis Strait in February from submarine sonar profiles. J Geophys Res 90(C1):1069–1077

Wadhams P, Lange MA, Ackley SF (1987) The ice thickness distribution across the Atlantic Sector of the Antarctic Ocean in mid-winter. J Geophys Res 92:14535–14552

Wadhams P, Davis NR, Comiso JC, Kutz R, Crawford J, Jackson G, Krabill W, Sear CB, Swift R, Tucker WB III (1991) Concurrent remote sensing of arctic sea ice from submarine and aircraft. Int J Remote Sens 12:1829–1840

Wadhams P, Tucker WB III, Krabill WB, Swift RN, Comiso JC, Davis NR (1992) The relationship between sea ice freeboard and draft in the Arctic Basin and implications for ice thickness monitoring. J Geophys Res 97(C12):20325–20334

Walker ER, Wadhams P (1979) Thick sea-ice floes. Arctic 32:140–147

Warren SG, Rigor IG, Untersteiner N, Radionov VF, Bryazgin NN, Aleksandrov YI, Colony R (1999) Snow depth on Arctic sea ice. J Clim 12(6):1814–1829

WCRP-41 (1990) Sea-ice and climate report of the fourth session of the working group no 41. Rome, 20–23 Nov 1989. WMO/TD-No 377

Weaver RLS, Steffen K, Heinrichs J, Maslanik JA, Flato GM (2000) Data assimilation in sea-ice monitoring. Ann Glaciol 31:327–332

Weeks WF, Ackley SF, Govoni J (1989) Sea ice ridging in the Ross Sea Antarctica as compared with sites in the Arctic. J Geophys Res 94:4984–4988

Weingartner TJ, Cavalieri DJ, Aagaard K, Sasaki Y (1998) Circulation dense water formation and outflow on the northeast Chukchi shelf. J Geophys Res 103(C4):7547–7661

Williams E, Swithinbank C, de Robin GdeQ (1975) A submarine sonar study of Arctic pack ice. J Glaciol 15:349–362

Winebrenner DP, Nelson ED, Colony R, West RD (1994) Observation of melt onset on multi-year Arctic sea ice using the ERS-1 synthetic aperture radar. J Geophys Res 99:22425–22441

Winebrenner DP, Long DG, Holt B (1998) Mapping the progression of melt onset and freeze-up using SAR and scatterometry. In: Tsatsoulis C, Kwok R (eds) Recent advances in the analysis of SAR data of the polar oceans. Springer, New York, pp 29–144

Winsor P (2001) Arctic sea ice thickness remained constant during the 1990's. Geophys Res Lett 28(6):1039–1041

Winsor P, Bjork G (2000) Polynya activity in the Arctic Ocean from 1958–1997. J Geophys Res 105(C4):8789–8803

Winsor P, Chapman DC (2002) Distribution and inter-annual variability of dense water production from coastal polynyas on the Chukchi shelf. J Geophys Res 107(C7):doi:101029/2001JC000984

WMO (1999) Sea-ice charts of the Arctic. In: Proceedings of a workshop, Seattle, 5–7 Aug 1998. WMO/TD No 949, p 38, Append

WMO (2000) In: Lemke P, Colony R (eds) Joint report on the 4th session of the ACSYS Sea Ice/ Ocean Modelling (SIOM) panel and the ACSYS workshop on sea-ice thickness measurements and data analysis, Monterey, 7–11 Apr 1997. WMO/TD No 991, World Meteorological Organization, Geneva, p A510

WMO (2003) Sea-ice extent and the global climate system and long-term variability of the Barents Sea region. Summary report of a joint meeting, Toulouse, 15–19 Apr 2002. . WCRP Informal Report No 9/2003, p 22, Append

Worby AP, Jeffries MO, Weeks WF, Morris K, Jana R (1996) The thickness distribution of sea ice and snow cover during late winter in the Bellingshausen and Amundsen Seas Antarctica. J Geophys Res 101(C12):28441–28456

Worby AP, Massom RA, Allison I, Lytle VI, Heil P (1998) East Antarctic sea ice: a review of its structure properties and drift. In: Jeffries MO (ed) Antarctic Sea ice physical processes interactions and variability. American Geophysical Union. Antarctic Research Service 74. AGU, Washington, DC, pp 41–68

Worby AP, Griffin PW, Lytle VI, Massom RA (1999) On the use of electromagnetic induction sounding to determine winter and spring sea ice thickness in the Antarctic. Cold Reg Sci Technol 29:49–58

Worby AP, Bush GM, Allison I (2001) Seasonal development of the sea-ice thickness distribution in East Antarctica: measurement from upward-looking sonar. Ann Glaciol 33:177–180

Yackel JJ, Barber DG (2000) Melt ponds on sea ice in the Canadian Archipelago: 2 on the use of radarsat-1 synthetic aperture radar for geophysical inversion. J Geophys Res 105(C9):22061–22070

Yackel JJ, Barber DG, Hanesiak JM (2000) Melt ponds on sea ice in the Canadian Archipelago: 1 variability in morphological and radiative properties. J Geophys Res 105(C9):22049–22060

Yackel J, Barber DG, Papakyriakou TN (2001) On the examination of spring melt in the North Water polynya using Radarsat-1. Atmos Ocean 39(3):195–208

Yu Y, Lindsay RW (2003) Comparison of thin ice thickness distributions derived from Radarsat geophysical processor system and advanced very high resolution radiometer data sets. J Geophys Res 108(C12):3387. doi:1029/2002JC001319

Yu Y, Rothrock DA (1996) Thin ice thickness from satellite thermal imagery. J Geophys Res 101(C11):25753–25766

Yu Y, Rothrock DA, Lindsay RW (1995) Accuracy of sea ice temperature derived from advanced very high resolution radiometer. J Geophys Res 100(C3):4525–4532

Yu Y, Rothrock DA, Zhang J (2001) Thin ice impacts on surface salt flux and ice strength: Inferences from advanced very high resolution radiometer. J Geophys Res 106(C7):13975–13988

Zedel L, Crawford GB, Gordon L (1996) On the determination of wind direction using an upward looking acoustic Doppler current profiler. J Geophys Res 101(C5):12163–12176

Zhang J, Thomas RR, Rothrock DA, Lindsay RW, Yu Y, Kwok R (2003) Assimilation of ice motion observations and comparisons with submarine ice thickness data. J Geophys Res 108(C6):3170. doi:101029/2001JC001041

Zhao Y, Liu AK, Long DG (2002) Validation of sea ice motion from QuikSCAT with those from SSM/I and buoy. IEEE Trans Geosci Remote Sens 40(6):1241–1246

Chapter 4
Observations in the Ocean

Bert Rudels, Leif Anderson, Patrick Eriksson, Eberhard Fahrbach, Martin Jakobsson, E. Peter Jones, Humfrey Melling, Simon Prinsenberg, Ursula Schauer, and Tom Yao

Abstract The chapter begins with an overview of the exploratory work done in the Arctic Ocean from the mid nineteenth century to 1980, when its main features became known and a systematic study of the Arctic Ocean evolved. The following section concentrates on the decade between 1980 and 1990, when the first scientific icebreaker expeditions penetrated into the Arctic Ocean, when large international programme were launched, and the understanding of the circulation and of the processes active in the Arctic Ocean deepened. The main third section deals with the studies and the advances made during the ACSYS decade. The section has three headings: the circulation and the transformation of water masses; the changes that have been observed in the Arctic Ocean, especially during the last decades; and the transports between the Arctic Ocean and the surrounding world ocean through the different passages, Fram Strait, Barents Sea, Bering Strait and the Canadian Arctic Archipelago. In section

B. Rudels (✉)
Department of Physics, University of Helsinki, P.O. Box 64, FI-00014 Helsinki, Finland

Finnish Meteorological Institute, P.O. Box 503, FI-00101 Helsinki, Finland
e-mail: bert.rudels@helsinki.fi; bert.rudels@fmi.fi

L. Anderson
Department of Chemistry, University of Gothenburg, SE-41296 Göteborg, Sweden
e-mail: leifand@chalmers.se

P. Eriksson
Finnish Meteorological Institute, P.O. Box 503, FI-00101 Helsinki, Finland
e-mail: Patrick.Eriksson@fmi.fi

E. Fahrbach • U. Schauer
Alfred-Wegener-Institute for Polar and Marine Research, P.O. Box 120161, D-27515 Bremerhaven, Germany
e-mail: Eberhard.Fahrbach@awi.de; Ursula.schauer@awi.de

M. Jakobsson
Department of Geological Sciences, Stockholm University, SE-10691 Stockholm, Sweden
e-mail: martin.jakobsson@geo.su.se

P. Lemke and H.-W. Jacobi (eds.), *Arctic Climate Change: The ACSYS Decade and Beyond,* Atmospheric and Oceanographic Sciences Library 43,
DOI 10.1007/978-94-007-2027-5_4, © Springer Science+Business Media B.V. 2012

four, the Arctic Ocean is considered as a part of the Arctic Mediterranean Sea, and the impacts of possible climatic changes on the circulation in the Arctic Mediterranean and on the exchanges with the world ocean are discussed.

Keywords Arctic Ocean • Arctic Mediterranean Sea • Ocean Circulation • Water mass formation • Water mass transformation • Mixing • Open ocean convection Thermohaline circulation • Boundary convection • Intrusions • Double-diffusive convection

4.1   Mid 1800–1980: Exploration

Because of the severe high-latitude climate and the perennial ice cover, little was known about the Arctic Ocean 100 years before the beginning of the ACSYS decade. Was it an ocean, or did large, yet undetected, landmasses exist? Was open water to be found behind the forbidding ice fields? Large polynyas had been observed beyond the coastal ice cover, and the northernmost extension of the Gulf Stream was believed to transport warm water into the Arctic Ocean west of Svalbard. Because sea ice contains so little salt, it was also speculated that only freshwater freezes. Once a ship got beyond the pack ice formed in the low-salinity water influenced by runoff from the continents, it would encounter an ice-free ocean with saline water supplied by the Gulf Stream (Petermann 1865).

Driftwood, originating from Siberia, had been found on the Greenland coasts, indicating a flow of ice, or water, across the Polar Sea, exiting between Greenland and Svalbard. Finally, remnants from the wreck of the steamship *Jeanette*, crushed by the ice in the East Siberian Sea in 1881 in its attempt to reach the North Pole from Bering Strait, were found at Julianehåb in southwestern Greenland in 1884. This confirmed the existence of a rapid stream carrying ice across the Arctic Ocean from the East Siberian Sea to the East Greenland Current, flowing southward along the east coast of Greenland. This discovery was, perhaps, decisive for Nansen's plan to reach the North Pole by allowing a ship to be frozen into the ice – no open water was expected by Nansen – upstream of this transpolar stream and then drift across the Polar Sea, passing the pole.

E.P. Jones • S. Prinsenberg
Department of Fisheries and Oceans, Bedford Institute of Oceanography, P.O. Box 1006, Dartmouth NS B2Y 4A2, Canada
e-mail: Peter.Jones@dfo-mpo.gc.ca; Simon.Prinsenberg@dfo-mpo.gc.ca

H. Melling
Institute of Ocean Sciences, Fisheries and Oceans Canada, Sidney, BC, Canada, V8L 4B2
e-mail: Humfrey.Melling@dfo-mpo.gc.ca

T. Yao
3989 W 18th Ave, Vancouver, BC, Canada V6S 1B6
e-mail: tomyao@telus.net

The drift of *Fram* 1893–1896, although it never reached the pole, provided the first information of the Arctic Ocean beneath the ice (Nansen 1902). It was deep, >3,000 m, and below the ice and the cold, low-salinity surface water, the salinity was observed to increase towards the bottom, and a layer with temperatures above 0°C was found between 200 and 500 m depth. Nansen concluded that the warm layer was Atlantic water advected into the Arctic Ocean from the Norwegian Sea through the passage west of Svalbard–Petermann's Gulf Stream. Nansen also suggested that the high salinity of the deep water was caused by freezing and brine rejection on the Arctic Ocean shelves, where the brine-enriched water sinks to and accumulates at the bottom. The salinity of the shelf bottom water increases during winter, and as it eventually crosses the shelf break, it sinks into the deep Arctic Ocean basins. In a later work, largely based on Amundsen's oceanographic observations on *Gjöa* in 1901, Nansen examined the possibility to form dense, saline water in the eastern Barents Sea (Nansen 1906). In that work, he also discussed open ocean convection and deep-water formation in the Greenland Sea. The salinities determined on *Fram* were later found erroneous, and Nansen adopted the view that the deep waters of the Arctic Ocean were advected from deep-water formation area in the Greenland Sea (Nansen 1915). Nansen also noted that sea ice drifted to the right of the wind and assumed that this was an effect of the earth's rotation. This observation eventually led to the formulation of the theory of wind-driven ocean currents and to the discovery of the Ekman spiral (Ekman 1905).

During the following 80 years, up to 1980, the study of the Arctic Ocean retained much of its early exploratory character. Amundsen attempted to reach the North Pole from Bering Strait with *Maud* 1919–1925, but the vessel got trapped in the ice in the East Siberian Sea and never crossed the shelf break. An effort to enter the Arctic Ocean through Fram Strait, the passage between Greenland and Svalbard, using a discarded submarine, *Nautilus*, was attempted in 1931, but due to technical problems, *Nautilus* only reached slightly north of Svalbard. The icebreaker *Sedov*, involuntarily, repeated the drift of *Fram* 1937–1940, and the first Soviet ice drift station, North-pole 1, lead by Papanin, was established by aircraft at the North Pole in May 1937, and the research group was picked up by the icebreaker *Taymyr* in the East Greenland Current in February 1938 (Libin 1946; Buinitsky 1951). This drift station was followed by many others, most of them launched by the Soviet Union, but also other countries contributed as with e.g. T3 ice island, and the AIDJEX and LOREX ice camps. A comprehensive oceanographic survey of the entire Arctic Basin was made using aviation during two spring seasons of 1955 and 1956. In 1973–1979, seven such surveys were made following each other with a yearly interval and a total number of stations of 1,229. These data were used for preparation of charts in the Atlas of the Arctic Ocean (Gorshov 1980) and in the Atlas of the Arctic (Treshnikov 1985).

The knowledge of the bathymetry improved steadily, and the large extension of the shelves and the existence and location of the major ridges and deep basins became gradually known. More than half (53%) of the 9.5×10^{12} m^2 large Arctic Ocean is now known to consist of shelf areas (Jakobsson et al. 2004a). The deep ocean is divided into two main basins, the Eurasian Basin and the Canadian Basin,

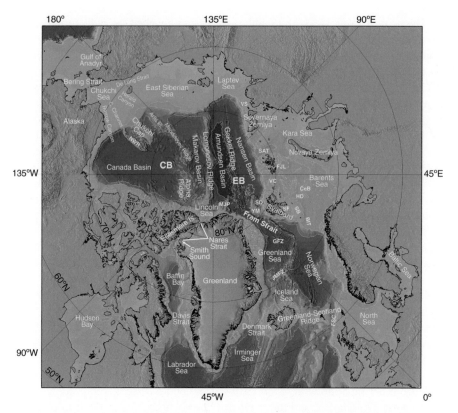

Fig. 4.1 Map of the Arctic Mediterranean Sea showing geographical and bathymetric features. The bathymetry is from IBCAO updated data base (Jakobsson et al. 2004b; Jakobson et al. 2008), and the projection is Lambert Equal Area. The 200, 500, 2,000 and 4,000 m isobaths are shown. *BIT* Bear Island Trough, *CB* Canadian Basin, *CeB* Central Bank, *EB* Eurasian Basin, *FJL* Franz Josef Land, *GFZ* Greenland Fracture Zone, *FSC* Faroe Shetland Channel, *HD* Hopen Deep, *JMFZ* Jan Mayen Fracture Zone, *MJP* Morris Jessup Plateau, *NwR* Northwind Ridge, *SAT* St. Anna Trough, *SB* Svalbrad Bank, *SD* Sofia Deep, *SF* Storfjorden, *YP* Yermak Plateau, *VC* Victoria Channel, *VS* Vilkiltskij Strait (Rudels 2009)

by the Lomonosov Ridge, sill depth ca 1,600 m. The Eurasian Basin is further separated into the deeper (4,500 m) Amundsen Basin and the somewhat shallower Nansen Basin (4,000 m) by the Gakkel Ridge. The Canadian Basin is separated by the Alpha and Mendeleyev ridges into the 4,000 m deep Makarov Basin and the shallower (3,800 m) and larger Canada Basin (Fig. 4.1). Airborne expeditions during spring allowed for hydrographic observations from the ice, covering extensive regions. Especially in the 1950s and in the mid 1970s, the Soviet airborne expeditions extended almost over the entire Arctic Ocean.

The picture of the deep Arctic Ocean, the northernmost part of the North Atlantic, that evolved was one of an ice-covered, strongly stratified ocean, whose water column was dominated by advection from the neighbouring seas and oceans. The large river runoff and the inflow of low-salinity Pacific water through Bering Strait form the low-salinity polar surface water, with an upper, winter homogenised Polar Mixed Layer (PML) closest to the ice. In summer, seasonal ice melt creates a 10–20-m surface layer of still lower salinity (Coachman and Barnes 1961; Coachman and Aagaard 1974). The salinity starts to increase below the PML, at about 50 m, but the temperature remains close to freezing until it suddenly increases as the Atlantic layer is encountered between 100 and 250 m depth. This cold layer, between the PML and the Atlantic water, was named the Arctic Ocean halocline (Coachman and Aagaard 1974). In the Canadian Basin the PML is less saline than in the Eurasian Basin due to the inflow of low-salinity Pacific water. The Pacific water entering during winter, the Bering Sea Winter Water (BSWW), is colder and more saline and also contributes to the halocline (Coachman and Barnes 1961). In the Eurasian Basin the PML is more saline and Pacific water is practically absent, and additional sources for the halocline water had to be considered (Fig. 4.2). It was realised that it could not be created by direct mixing between the PML and the Atlantic water, and Coachman and Barnes (1962) suggested that Atlantic water is brought onto the shelves through deeper canyons, becomes cooled and diluted by less-saline surface water and eventually returns to the deep basins, intruding between the PML and the Atlantic layer.

The Atlantic layer was observed throughout the Arctic Ocean, being colder and less saline in the Canada Basin than at its entrance through Fram Strait as a part of the West Spitsbergen Current (Timofeyev 1960; Coachman and Barnes 1963). The deep water in the Arctic Ocean was assumed supplied from the Greenland Sea through Fram Strait (Nansen 1915; Wüst 1941). It is warmer than in the Greenland Sea, and the deep Canadian Basin is warmer than the deep Eurasian Basin. The latter was explained by the presence of the Lomonosov Ridge, which would prevent the densest water from the Greenland Sea to penetrate further into the Arctic Ocean. This temperature difference was in fact used as evidence for the existence of a ridge dividing the Arctic Ocean in two major basins before the Lomonosov Ridge was properly mapped (Worthington 1953). Another indication of the presence of a submarine ridge was the retarding of the tidal wave entering through Fram Strait (Harris 1911; Fjeldstad 1936).

The circulation of the ice and the surface water was found to be anti-cyclonic, centred around the Beaufort Gyre north of the North American continent, with the transpolar drift (TPD) moving across the Arctic Ocean in two branches, the Siberian branch from the Siberian shelves and the transpolar branch detaching from the western side of the Beaufort Gyre (Fig. 4.3). The movements of the Atlantic and deeper water masses, by contrast, were deduced from water mass properties, mainly temperature, to be cyclonic around the Arctic Ocean (e.g. Timofeyev 1960; Coachman and Barnes 1963; Coachman and Aagaard 1974).

The ice camps offered opportunities to measure the velocity profiles underneath the ice and Hunkins (1966) documented the existence of the Ekman spiral, the velocity at the surface was 45° to the right of the wind and decreased and turned further to the right with increasing depth. Hunkins separated the velocity into three components: a geostrophic velocity caused by the sea surface slope, a turbulent, 1–2 m thick, boundary layer just below the ice and moving in the same direction as the ice and the Ekman spiral caused by the wind as predicted by Ekman (1905).

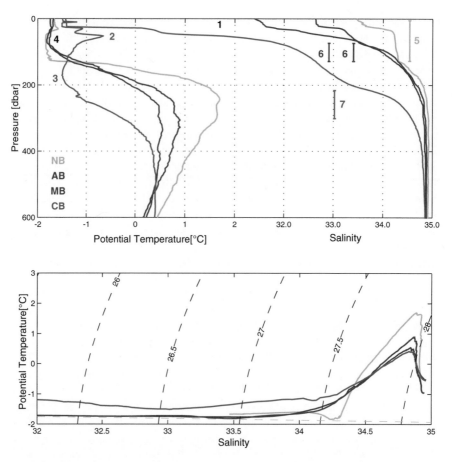

Fig. 4.2 The characteristics of the upper layers in the different basins of the Arctic Ocean. (**a**) potential temperature and salinity profiles, and (**b**) ΘS curves. *Yellow* Nansen Basin, *Green* Amundsen Basin, *Magenta* Makarov Basin, *Blue* Canada Basin. *1* Summer melt water layers, *2* Temperature maximum related to the Bering Strait summer inflow, *3* upper halocline, *4* Temperature minimum created by winter convection, marking the lower limit of the Polar Mixed Layer, *5* Winter-mixed layer in the Nansen Basin, *6* lower halocline in the Amundsen (*green*) and Makarov (*magenta*) basins, *7* lower halocline in the Canada Basin (Adapted from Rudels et al. 2004a). The positions of the stations are shown in Fig. 4.19a

High velocity events in the upper part of the water column were noticed already during the drift of the first ice station, NP-1, and were suggested to be connected with high-energy eddies (Shirshov 1944). Similar events were encountered on the later ice stations (Belyakov and Volkov 1980) and during the AIDJEX experiment (Hunkins 1974; Newton et al. 1974). The eddies were 10 to 20 km in diameter and highly energetic with the maximum azimuthal velocities located around 150 m depth and reaching 40–60 cm s⁻¹. These eddies, mostly anti-cyclonic, were mainly found in the Canada Basin and were commonly associated with anomalous Θ-S characteristics suggesting the presence of a different water mass. Hunkins (1974)

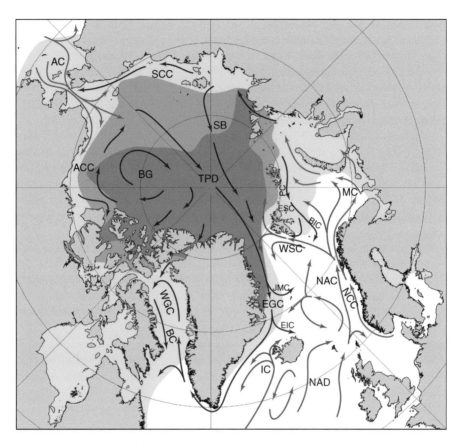

Fig. 4.3 The circulation of the upper layers of the Arctic Mediterranean Sea. Warm Atlantic currents are indicated by red arrows and cold, less-saline polar and arctic currents by *blue arrows*. Low-salinity transformed currents are shown by *green arrows*. The maximum ice extent is shown in *blue* and the minimum ice extent in *red*. The absolute minimum, to date, of 2007 is shown in *dark red*. *AC* Anadyr Current, *ACC* Alaskan Coastal Current, *BC* Baffin Current, *BIC* Bear Island Current, *BG* Beaufort Gyre, *EGS* East Greenland Current, *EIC* East Iceland Current, *ESC* East Spitsbergen Current, *IC* Irminger Current, *JMC* Jan Mayen Current, *MC* Murman Current, *NAD* North Atlantic Drift, *NAC* Norwegian Atlantic Current, *NCC* Norwegian Coastal Current, *PS* Persey Current, *SB* Siberian branch (of the Transpolar Drift), *SCC* Siberian Coastal Current, *TPD* Transpolar Drift, *WGC* West Greenland Current, *WSC* West Spitsbergen Current (Adapted from Rudels 2001)

assumed that the eddies were generated by baroclinic instability as the inflowing Pacific water entered the Canada Basin in the Barrow Canyon, north of Alaska. This idea was later supported by theoretical work by Hart and Killworth (1976).

The exchanges between the Arctic Ocean and the world ocean through the different passages were studied by hydrographic observations and by direct current measurements, and several budgets for the Arctic Ocean and for the Nordic Seas (the Greenland, Iceland and Norwegian seas) were presented (e.g. Mosby 1962; Vowinckel and Orvig 1970; Aagaard and Greisman 1975; Nikiforov and Shpaiker 1980).

Fram Strait, connecting the Arctic Ocean with the Nordic Seas, is the only deep (2,600 m) passage and the only one where steady in- and outflows occur. A second connection between the Arctic Ocean and the Nordic Seas is over the broad and fairly shallow (200–300 m) Barents Sea, where mainly an inflow to the Arctic Ocean takes place. An inflow from the Pacific Ocean occurs through the narrow and shallow (50 m) Bering Strait, while the restricted, shallow (125–230 m) passages in the Canadian Arctic Archipelago are dominated by outflow of low-salinity polar surface water.

The most cited transport estimates through the different passages around 1980 were (+ in, − out); Bering Strait; +1.5 Sv, the Canadian Arctic Archipelago; −2.1 Sv, the Barents Sea; +0.6 Sv, Fram Strait; Atlantic water +7.1 Sv, polar surface water −1.8 Sv, modified Atlantic water −5.3 Sv provided by Aagaard and Greisman (1975). In this budget, the transports through Fram Strait were assumed to balance. Such large inflow of warm Atlantic water raised the question: What will happen if the freshwater input to the Arctic Ocean diminishes? Will the heat stored in the Atlantic layer then reach the surface and melt the ice? The importance of the halocline, acting as a barrier to such vertical heat flux, was then also appreciated (Aagaard and Coachman 1975).

The knowledge of the Arctic Ocean oceanography gained during the first eight decades of the twentieth century is admirably summarised and presented in the Atlas of the Arctic Ocean (Gorshov 1980). The oceanographic observations from the Russian drifting stations continued until 1991, and during the period 1937–1991, there were 1,800 soundings. Winter oceanographic observations were conducted at 7,200 stations using airplanes. During the summers of 1950–1993, oceanographic surveys of the Siberian Arctic Sea conducted 31,000 soundings (McClimans personal communication 2010).

4.2   1980–1990: Interpretation

The period from 1980 until the early 1990s saw the, hesitant, beginnings of the scientific icebreaker expeditions to the Arctic Ocean: *Ymer* in 1980, *Polarstern* in 1984 and 1987 and *Oden* and *Polarstern* together reaching the North Pole in 1991. It also saw the first large multi-national research programmes, the Marginal Ice Zone Experiment (MIZEX) in 1983 and 1984 and a winter MIZEX in 1987; a multi-national programme to study the convection in the Greenland Sea, the Greenland Sea Project (GSP); and the first systematic effort to determine the exchanges through Fram Strait with hydrographic sections and current metre arrays extending across the entire strait, the Fram Strait Project. In spite of the wealth of observations gathered by these programmes, the perhaps most important progress was in understanding the circulation and the processes active in the Arctic Mediterranean (see Chap. 6).

The old idea, advanced by Nansen, about ice formation, brine rejection and accumulation of saline water on the shelves was revived. Aagaard et al. (1981) proposed this as a mechanism for producing the Arctic Ocean halocline. They estimated that about 2 Sv ought to be produced on the shelves and examined the potential of the different shelf areas to supply water to the halocline. The main production areas were found to be the Barents and the Kara Seas and the Chukchi Sea. Melling and

Lewis (1982) studied plumes entering the halocline in the Canada Basin, and Jones and Anderson (1986) suggested that the nutrient maximum at a salinity of 33.1 in the Canadian Basin originated from the Chukchi shelf, where dense brine-enriched bottom water becomes rich in nutrients from the re-mineralization of organic matter at the bottom before it crosses the shelf break into the deep basin and enters the halocline.

These findings had some bearing on models describing the exchanges between the Arctic Ocean and the surrounding seas as a two-layer fjord circulation, where Atlantic water enters through Fram Strait, becomes entrained into the PML and then exported, together with the Pacific water and the runoff, as low salinity Polar surface water through Fram Strait and the Canadian Arctic Archipelago in geostrophically balanced boundary flows (Stigebrandt 1981). The thickness of the ice cover and the freshwater content in the PML then depend upon the heat loss to the atmosphere and on the amount of heat entrained into the mixed layer from below, and the outflow is controlled by the depth of the mixed layer and the density difference between the PML and Atlantic water. Assuming that the heat loss to the atmosphere, the river runoff and the Pacific inflow are known, it is possible to close the system and compute the exchanges between the Arctic Ocean and the North Atlantic. If the straits are wider than the internal Rossby radius, the transports are only determined by the density difference between the layers and the depth of the upper layer, and since the Canadian Arctic Archipelago has more openings than Fram Strait, it can sustain a larger transport (Stigebrandt 1981). However, if the entrainment of Atlantic water in the interior is prevented by the halocline, this coupling disappears and the shelf processes have to be taken explicitly into account.

Björk (1989) assumed that the outflow from the Arctic Ocean was geostrophic and determined by the stratification. He computed the production of saline shelf water necessary to reproduce the stratification and maintain the exchange. He found that about 1.2 Sv had to be supplied from the shelves. Rudels (1989) took the inflow from the Bering Strait to be short-circuited to the Canadian Arctic Archipelago and only considered the exchanges with the Nordic Seas. Assuming that the river runoff that enters the shelves also enters the basins, it was possible to compute, using salinities observed on the shelves and an estimate of the ice production related to the observed heat loss, how much of the runoff is exported as ice; how much as less saline, compared to the PML, surface water and how much as brine-enriched denser water. The salt balance was closed by requiring a compensating flow from the basin onto the shelves, at the bottom in summer and at the surface in winter. Assuming geostrophic flow in Fram Strait, an Arctic Ocean water column could be constructed that was capable of exporting the created volumes, allowing for a compensating inflow of Atlantic water, divided between Fram Strait and the Barents Sea. It was found that the export from the shelves mainly freshened the PML, and additional ice formation had to occur in the basins to reach the observed salinity of the PML. Only the Barents and Kara Seas produced water dense enough to supply the halocline and only a small amount, 0.2 Sv.

The deep and bottom waters in the Arctic Ocean are warmer and more saline than the Greenland Sea deep water, and Aagaard (1980) noticed that the Canadian Basin

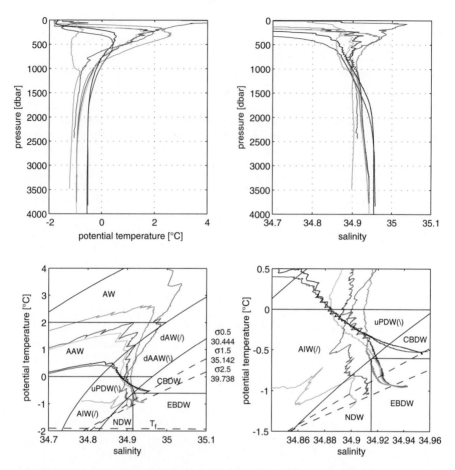

Fig. 4.4 Characteristics of the water columns in different parts of the Arctic Mediterranean. (*upper left*) potential temperature profiles, (*upper right*) salinity profiles, (*lower panel*) ΘS curves. *Green* Greenland Sea, *Red* Fram Strait (West Spitsbergen Current); *Dark yellow* Nansen Basin (Fram Strait branch), *Magenta* Nansen Basin (interior), *Cyan* Amundsen Basin, *Black* Makarov Basin, *Blue* Canada Basin (From Rudels 2009). The positions of the stations are shown in Fig. 4.19b

deep water was not only warmer but also more saline than the Eurasian Basin deep water (Fig. 4.4). This implies that some deep-water formation and convection must take place in the Arctic Ocean. The stratification is too strong for this to occur in the interior of the basins, and shelf-slope convection, as Nansen originally suggested, is the only possibility. Either entraining (Rudels 1986a) or shaving (Aagaard et al. 1985) boundary plumes was suggested as the mode of convection. Dense shelf bottom water had been observed in the eastern Barents Sea (e.g. Nansen 1906; Midttun 1985) and in Storfjorden in southern Svalbard (Midttun 1985; Anderson et al. 1988), and in 1986, a warmer, saline bottom layer was observed on the continental slope west of Svalbard, indicating a plume sinking from Storfjorden into Fram Strait entraining warm Atlantic water on its way down the slope (Quadfasel et al. 1988).

Circulation schemes for the deep waters in the Arctic Mediterranean Sea connecting the two deep water sources, the Arctic Ocean and the Greenland Sea, through Fram Strait were proposed to explain the differences in deep water characteristics in the different basins (Aagaard et al. 1985; Rudels 1986a). The more saline deep waters of the Arctic Ocean, the warmer, less-dense Canadian Basin Deep Water (CBDW) and the colder, denser Eurasian Basin Deep Water (EBDW) pass south through Fram Strait into the Nordic Seas. Part of the outflow remains at the rim of the Greenland Sea, interacting and mixing with the waters of the Greenland Sea. This mixing product then follows the Jan Mayen Fracture Zone into the Norwegian Sea to renew the Norwegian Sea Deep Water (NSDW). The NSDW then partly returns to the Arctic Ocean via Fram Strait. The other fraction of the Arctic Ocean deep waters enters the Greenland Sea, where it, together with the locally convecting waters, creates the Greenland Sea Deep Water (GSDW), the presence of the EBDW being revealed by the deep salinity maximum in the central Greenland Sea (Aagaard et al. 1985; Rudels 1986a) (Fig. 4.5).

Some efforts to quantify the circulation were made, either just from the Θ-S structures (Rudels 1986a) or using additional information from tracer observations (e.g. Smethie et al. 1988; Heinze et al. 1990; Schlosser et al. 1990). To match the transports with the observed Θ-S characteristics, Rudels (1986a, 1987) found that more deep water was exchanged through Fram Strait than was formed in the Arctic Ocean and in the Greenland Sea. This suggests that the deep exchanges between Arctic Ocean and the Nordic Seas are not driven by the deep-water formation but forced by other processes, e.g. the wind fields. Changes in the strength of the deep-water formation in one source area, e.g. the Greenland Sea, could, however, change the pathways of the circulation. A situation with a strong convection and deep-water formation in the Greenland Sea could force the Arctic Ocean deep waters to bypass the Greenland Sea and cross the Jan Mayen Fracture Zone into the Iceland Sea and perhaps also allow them to exit through Denmark Strait. Such variations in the circulation could also explain the fairly rapid changes in deep-water temperatures reported from the Greenland Sea (Aagaard 1968). Cooling is easily explained by an increased convection in the Greenland Sea, while it is more difficult to account for a rapid warming of the GSDW by local processes. However, if more Arctic Ocean deep water penetrates into the Greenland Sea in periods when the convection is weak, the heating can be explained (Rudels 1986a).

A situation with only deep-water sources is not possible, and implicit in all efforts to quantify the deep-water production is the assumption that the corresponding volume either upwells into the upper layer or is exported elsewhere. However, the circulation connecting the two deep-water sources, the Greenland Sea and the Arctic Ocean shelves, was assumed to be largely internal to the Arctic Mediterranean, and the overflow of dense water from the Arctic Mediterranean to the North Atlantic was believed to involve Arctic intermediate water formed mainly in the Iceland Sea (Swift et al. 1980; Swift and Aagaard 1981) and perhaps in the Greenland Sea (Smethie and Swift 1989).

Fram Strait was studied intensely. A large cyclonic recirculation of Atlantic water, as had been suggested already in the end of the nineteenth century (Ryder 1891),

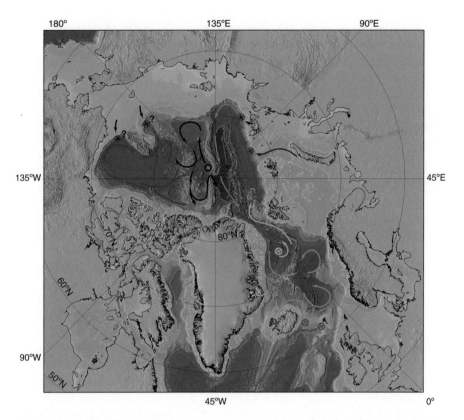

Fig. 4.5 Circulation of the deep waters. Advective exchanges through Fram Strait and advective input down the St. Anna Trough. Possible sources for slope convection are indicated by *crosses*. Crosses in the interior of the basins indicates convection or sinking from intermediate to deeper levels as for Amundsen Basin intermediate water sinking into the deep Makarov Basin. Uncertain sources and pathways are indicated by '?' (Adapted from Jones et al. 1995)

was found in the strait (Quadfasel et al. 1987; Rudels 1987; Bourke et al. 1988), and the amount of Atlantic water that really entered the Arctic Ocean was estimated, by geostrophy, to be closer to 1 or 2 Sv than to 7 Sv (e.g. Rudels 1987; Bourke et al. 1988). By contrast it was realised that the volume entering the Arctic Ocean over the Barents Sea via St. Anna Trough was as large as that passing through Fram Strait. However, the heat transport was considerably smaller because of the large heat loss taking place in the Barents Sea (Rudels 1987; Blindheim 1989).

The understanding of the role and the importance of the processes and the circulation in the Arctic Mediterranean Sea (Krümmel 1907; Wüst 1941; Sverdrup et al. 1942) for the global climate 100 years after the drift of *Fram* were summarised by Aagaard at the Nansen Centennial Meeting in Bergen 1993 (Aagaard and Carmack 1994). In 1997–1998, information collected by that time in the AARI database was used for generating climatic atlases of the Arctic Ocean for the winter (EWG 1997) and summer (EWG 1998) seasons.

4.3   The ACSYS Decade: New Insights, Variability and Change

4.3.1   Circulation and Transformation of Water Masses

4.3.1.1   Atlantic and Intermediate Water Circulation

The movement of the Atlantic water in the Arctic Ocean was known to be cyclonic but mainly thought of as a broad stream slowly moving around the central basin (e.g. Timofeyev 1960; Coachman and Barnes 1963), with perhaps a smaller anti-cyclonic gyre in the Beaufort Sea. The icebreaker expeditions to the Arctic Ocean in the late 1980s and early 1990s provided several new observational clues, and a more detailed picture of the circulation could be constructed. On hydrographic sections across the Eurasian Basin, signs of intrusive mixing and of a freshening of the Atlantic layer from the Nansen Basin to the Amundsen Basin were observed. Based upon XBT observations from the icebreaker *Rossia*, Quadfasel et al. (1993) proposed that the intrusions were created north of the Laptev Sea, as cold, brine-enriched water sinks off the shelf into the Atlantic layer. The intrusions were then assumed advected with the mean flow along the Gakkel Ridge toward Fram Strait. This interpretation is contrary to the one suggested by Perkin and Lewis (1984), who observed similar intrusions north of the Yermak Plateau during the EUBEX experiment in 1981 and assumed that these were created as the inflowing warm Atlantic water of the West Spitsbergen Current encountered and interacted with the 'old' Atlantic water of the Arctic Ocean water column north of Svalbard.

On the *Oden* 1991 expedition, intrusions and interleaving structures were observed in the northern Nansen Basin, in the Amundsen Basin and at the Lomonosov Ridge. The structures extended over too large depth interval and were too regular to be the result of occasional plumes leaving the Laptev shelf, and a more likely source would be the encounter between the inflow of Atlantic water through Fram Strait and the inflow of Atlantic water from the Norwegian Sea across the Barents Sea, which has been cooled and freshened in the Barents Sea before it enters the Arctic Ocean via the St. Anna Trough. Not having any observations from the continental slope, Rudels et al. (1994) postulated that the two inflow branches, the warmer Fram Strait branch and the colder, less-saline Barents Sea branch, meet at the continental slope east of the St. Anna Trough and that intense mixing takes place creating inter-leaving structures, which, once formed, would penetrate further across the front between the two branches, probably driven by the potential energy released by dou-ble-diffusive convection. When the driving property contrasts have been removed, the intrusions would remain as fossil structures and become advected with the mean circulation. These intrusive layers could then be used as tracers, marking the flow of the now combined branches, especially their return flow in the Eurasian Basin towards Fram Strait. As the two branches return towards Fram Strait, the Barents Sea branch water dominates in the Amundsen Basin and over the Lomonosov Ridge, while the Fram Strait branch water makes a tighter turn and is more prominent over the Gakkel Ridge and in the northern Nansen Basin.

The *Oden* 1991 observations also showed abrupt, and large, changes in water mass properties across the Lomonosov Ridge (Anderson et al. 1994; Rudels et al. 1994). The temperature in the Atlantic layer was much lower, while the intermediate water below the Atlantic layer was warmer and more saline in the Makarov Basin than in the Amundsen Basin. These differences were found at levels shallower than the sill depth of the Lomonosov Ridge and were interpreted as the result of shelf-slope convection acting on the boundary current flowing along the continental slope (Rudels et al. 1994).

Cold, saline and dense shelf water plumes sink down the slope, and the plumes merge with their surroundings when they reach their appropriate density level, entering, or bypassing, the Atlantic layer. The less-dense plumes thus cool and freshen the Atlantic layer, while denser, deeper sinking plumes entrain warm Atlantic water during their descent, and their temperature increases and, as they merge into the water column, they make the deeper layers warmer and more saline (Fig. 4.4; Rudels 1986a; Rudels et al. 1994). To attain the characteristic Θ-S properties of the upper Polar Deep Water (uPDW), with temperature decreasing and salinity increasing with depth, forming a straight line in the Θ-S diagram (Rudels et al. 1994), which are observed below the Atlantic layer on the Makarov Basin side of the Lomonosov Ridge, the water column must have moved a considerable distance along the continental slope, after it has crossed the Lomonosov Ridge (Fig. 4.4 and Table 4.1). This implies that the water on the Makarov Basin side of the Lomonosov Ridge flows along the ridge from the American side of the basin, which suggested that the boundary current splits at the Mendeleyev Ridge and makes a loop in the Makarov Basin and then flows along the Lomonosov Ridge in opposite direction as the flow on the Amundsen Basin side (Rudels et al. 1994).

The intermediate water in the Amundsen Basin is colder and less saline than that in both the Nansen and the Makarov basins, indicating that the Amundsen Basin intermediate layers are dominated by water from the Barents Sea branch, which has been less affected by slope convection. In the Atlantic layer, the temperature maximum found over the Lomonosov Ridge on the Amundsen Basin side of the front was warmer than in both the Amundsen Basin and the Makarov Basin, which could be due to a more rapid return flow along the ridge than in the interior Amundsen Basin (Rudels et al. 1994), taking into account the higher temperatures of the Fram Strait Atlantic water observed in the early 1990s (Quadfasel et al. 1991) (see Sect. 4.3.2.1).

Silicate and CFC concentrations on the Oden stations in the outflow area between the Morris Jesup Plateau and the Yermak Plateau showed differences between the intermediate waters closer to Greenland and those farther offshore. This implied that the boundary current here consists of streams from the different basins that still can be distinguished as the boundary current approaches Fram Strait. A higher silicate concentration closer to Greenland indicates the presence of water that has circulated around the Canada Basin, receiving its higher silicate content by incorporating Pacific water. Farther from Greenland, the water column became more similar to that of the Makarov Basin, while the waters from the loops in the Amundsen and Nansen basins were found closer to the Yermak Plateau (Rudels et al. 1994) (Fig. 4.6).

Table 4.1 Water mass definitions originally derived for Fram Strait (Friedrich et al. 1995; Rudels et al. 1999c) modified to include the less-dense Polar waters. The separation of water masses in the Nordic Seas is rudimentary

Water mass boundaries	Water masses	Origins, remarks
$\sigma_\theta \leq 27.70$ Upper waters		
$\sigma_\theta \leq 27.20$, $\Theta \leq 0$	Polar Water I (PWI)	Includes the Pacific inflow, the PML, shelf water and the upper halocline
$27.20 < \sigma_\theta \leq 27.70$, $\Theta \leq 0$	Polar Water II (PWII)	Includes the Atlantic-derived lower halocline and the winter-mixed layer in the Nansen Basin and the Barents Sea
$27.70 < \sigma_\theta \leq 27.97$ Atlantic waters		
(a) $27.70 < \sigma_\theta \leq 27.97$, $2 < \Theta$;	Atlantic Water (AW), and Re-circulating Atlantic Water	Norwegian Sea and West Spitsbergen Current
(a) $27.70 < \sigma_\theta \leq 29.97$, $0 < \Theta \leq 2$	Arctic Atlantic Water (AAW)	Arctic Ocean: (b) includes the Arctic Ocean thermocline
(b) $27.70 < \sigma_\theta \leq 27.97$, $\Theta \leq 0$, $S \leq 34.676 + 0.232\Theta$		
$27.97 < \sigma_\theta$, $\sigma_{0.5} \leq 30.444$ Intermediate waters		
$27.97 < \sigma_\theta$, $\sigma_{0.5} \leq 30.444$, $0 < \Theta$	Dense Atlantic (DAW) (/) and Re-circulating dense Atlantic Water	Θ decreasing, S decreasing with depth
$27.97 < \sigma_\theta$, $\sigma_{0.5} \leq 30.444$, $0 < \Theta$;	Dense Arctic Atlantic Water (DAAW) (\)	S increasing, Θ decreasing with depth
$27.97 < \sigma_\theta$, $\sigma_{0.5} \leq 30.444$, $\Theta \leq 0$;	Arctic Intermediate Water AIW (/)	Greenland Sea: Includes a salinity minimum, in the Greenland Sea also a temperature minimum
$27.97 < \sigma_\theta$, $\sigma_{0.5} \leq 30.444$, $\Theta \leq 0$	Upper Polar Deep Water (uPDW) (\)	Arctic Ocean: S increasing, Θ decreasing with depth
$30.444 < \sigma_{0.5}$ Deep Waters		
$30.444 < \sigma_{0.5}$, $S \leq 34.915$	Nordic seas Deep Water (NDW)	Greenland Sea: Includes GSDW, ISDW, NSDW
$30.444 < \sigma_{0.5}$, $-0.6 < \Theta$, $34.915 < S$	Canadian Basin Deep Water (CBDW)	Canadian Basin
$30.444 < \sigma_{0.5}$, $\Theta \leq -0.6$, $34.915 < S$	Eurasian Basin Deep Water (EBDW)	Eurasian Basin

The $\sigma_{1.5}$ and $\sigma_{2.5}$ isopycnals shown in the ΘS diagrams correspond to the densities at the sill depth of the Lomonosov Ridges and Fram Strait, respectively. For an alternative water mass classification see Carmack (1990)

Several of these conjectures regarding the circulation were vindicated in the following years. In 1993, *Polarstern* studied the Laptev Sea shelf and slope. The less-saline Barents Sea branch water column was observed: the maximum temperature and salinity in the Atlantic layer were distinctly lower north of the Laptev Sea than at stations taken west of Franz Josef Land, and at some stations, strong interleaving between the two branches was observed (Schauer et al. 1997). At the slope

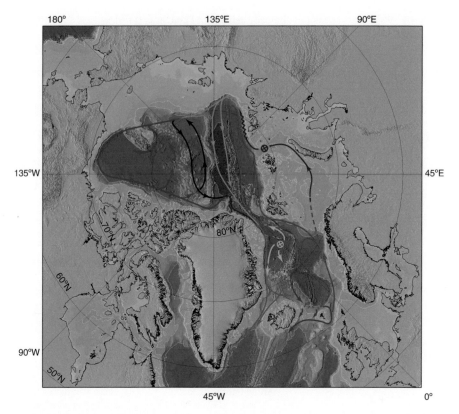

Fig. 4.6 Schematics showing the circulation in the subsurface Atlantic and intermediate layers of the Arctic Ocean and the Nordic Seas. The interactions between the Barents Sea and Fram Strait inflow branches north of the Kara Sea as well as the recirculation and different inflow streams in Fram Strait and the overflows across the Greenland–Scotland Ridge are indicated (Adapted from Rudels et al. 1994)

north of Severnaya Zemlya, the expected strong interleaving between the two inflow branches was observed in 1995 (Rudels et al. 2000a). The interleaving extended over several hundred metres depth and was present in all kinds of background stratifications: saltfinger unstable, diffusively unstable and also when the water column was stable in both properties. The thickness of the individual layers ranged from 5 to 50 m (Fig. 4.7).

In 1996, *Polarstern* occupied a section across the St. Anna Trough. It showed, as had the *Northwind* observations in 1965 (Hanzlick and Aagaard 1980), the colder, less-saline Barents Sea branch water entering the Nansen Basin in the deepest part and along the eastern flank of the trough (Schauer et al. 2002a, b). The 1996 *Polarstern* cruise then continued with a section from the eastern Kara Sea, across the Nansen and Amundsen basins, over the Lomonosov Ridge and into to the Makarov Basin (Schauer et al. 2002a). The Barents Sea branch was observed at the slope, but the mixing between the branches appeared somewhat weaker than in 1995. Farther from

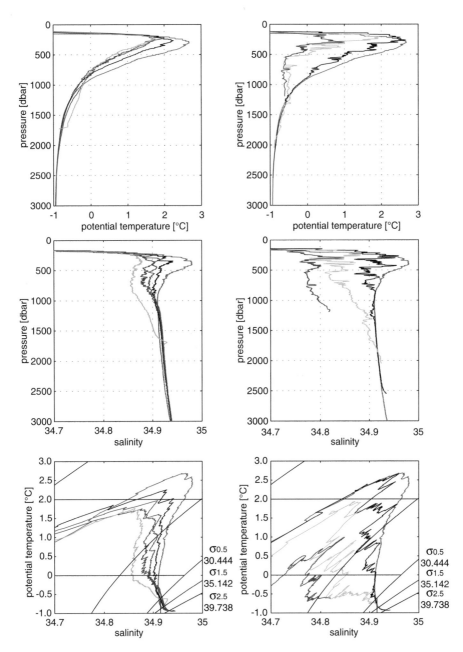

Fig. 4.7 Potential temperature and salinity profiles and ΘS curves showing the interaction and interleaving between the Fram Strait branch and the Barents Sea branch north of Severnaya Zemlya (*right column*) and the water mass properties in the Nansen and Amundsen basins seaward of the Fram Strait branch (*left column*). *Red station* Fram Strait branch, *blue station* Barents Sea branch, *black* and *cyan stations* show mixing between the branches. Note that the interleaving is present not only in the Atlantic layer but also in the intermediate levels (~1,000 m). Seaward of the Fram Strait branch (*left column, red station*), the warm, saline interleaving in the Nansen Basin (*black* and *magenta stations*) suggests a close recirculation of the Fram Strait branch in the Nansen Basin while the colder interleaving in the Amundsen Basin (*green* and *blue stations*) indicates that the intermediate part of the water column here is dominated by Barents Sea branch water (From Rudels 2009). Station positions are shown in Fig. 4.19c

the slope, an almost undisturbed, warm Fram Strait branch Atlantic core was observed, while on the basin side of this core, regular inversions and extensive layer structures were encountered in the Atlantic layer as well as in the intermediate water below. These inversions were similar to those observed on the *Oden* sections but with larger layer thickness and with overall higher temperature and salinity. Also these layers were interpreted as created by interactions between Barents Sea branch and Fram Strait branch waters (Rudels et al. 1999a) (Fig. 4.7).

The Barents Sea branch water was prominent in the Amundsen Basin and at the Lomonosov Ridge, consistent with a flow of Barents Sea branch, and to a lesser degree Farm Strait branch, waters towards Fram Strait in the Amundsen Basin and at the Lomonosov Ridge. However, the interleaving structures present on the off-shore side of the warmer Fram Strait branch water may not only be an indication of intruding Barents Sea branch water but could also, as suggested by Rudels et al. (1994), be a sign that a major fraction of the Fram Strait branch is returning towards Fram Strait in the northern Nansen Basin and over the Gakkel Ridge, advecting intrusions already formed as the two branches flowed eastward along the continental slope. However, the hypotheses of the formation, dynamics and movements of the intrusions in the Arctic Ocean diverge and we shall return to this below.

In 1994, on the transpolar section taken by the icebreaker *Louis S. St-Laurent*, the stations in the Makarov Basin close to the Mendeleyev Ridge showed warm cores in the Atlantic layer and colder, less-saline water at 1,200–1,600 m depth. This is consistent with the picture of the boundary current crossing the Lomonosov Ridge and then being partly deflected into the Makarov Basin at the Mendeleyev Ridge. No stations were obtained at the continental slope in the Makarov Basin during the AO94 expedition, and it was not possible to determine, with certainty, the origin of the intermediate depth, cold, low-salinity lenses. Were they outflows from a nearby shelf, or did they derive from the Barents Sea branch (Aagaard et al. 1996; Carmack et al. 1997; Swift et al. 1997)? In 1995, *Polarstern* worked on the slope from the eastern Kara Sea to the western East Siberian Sea. Less-saline intermediate water could then be followed along the slope from the Kara Sea to the East Siberian Sea, where it extended down to 1,600 m, indicating the sill depth at the Lomonosov Ridge encountered by the boundary current. This is in agreement with less-saline intermediate Barents Sea branch water entering the Canadian Basin and the interior of the Makarov Basin (Rudels et al. 2000a).

No large property contrasts between the waters above sill level on each side of the Lomonosov Ridge were seen close to the continental slope as were found at the *Oden* section further into the basins. This suggests that the boundary current broadens, or makes a loop along the ridge, before it splits into one part flowing northward along the ridge and the rest entering the Canadian Basin (Rudels et al. 2000a; Schauer et al. 2002a).

During the *Polarstern* cruises in 1995 and 1996, three moorings were deployed in 1995 and recovered the following year. Two of the moorings were located at the 1,700 m isobath on the continental slope, north of the Laptev Sea and north of the East Siberian Sea and close to the Lomonosov Ridge, while the third was deployed at the same bottom depth on the Amundsen Basin side of the ridge. The flow in the boundary current was found to be largely barotropic, and the calculated transports indicated 5 Sv flowing eastward above the Laptev Sea slope. The current then bifurcated with about 2.5 Sv moving north along the Lomonosov Ridge and 2.5 Sv crossing the ridge into the Makarov Basin (Woodgate et al. 2001).

Further to the east, in the Canada Basin, analysis of hydrographic and tracer data from the SCICEX-96 expedition together with data obtained from the AO94 and the *Polarstern* expeditions indicated that a further bifurcation of the boundary current occurred in the Canada Basin west of the Chukchi Cap (Smith et al. 1999; Smethie et al. 2000). The boundary current in the Canadian Basin thus separates into 3 loops, a splitting occurring at each topographic feature present at the slope. Similar bifurcations are then to be expected on the American side of the basin, at the Alpha Ridge and at the Lomonosov Ridge, bringing older and more transformed Atlantic and intermediate layer waters into the interior Canada and Makarov basins forming, at least rudimentary, gyres in these basins.

4.3.1.2   Formation of the 'Lower Halocline'

The origin of the more saline, nutrient-poor lower halocline has been found more elusive than that of the upper halocline (Jones and Anderson 1986; Jones et al. 1991; Anderson and Jones 1992). The low nutrient content indicates that the contact with bottom sediments is brief or non-existent, and an accumulation of brine-enriched water at the bottom of the shelves appears less likely. Rudels et al. (1991) proposed that the haline convection in the northern Barents Sea and north of Svalbard was limited, not by the shelf bottom but by the pycnocline above the Atlantic water and that it was strong enough to homogenise the less-saline upper layer, which would continue into the Arctic Ocean, eventually supplying water to the lower halocline. Steele et al. (1995) suggested that the melting of sea ice, as the Atlantic water meets the ice in the marginal ice zone in the northern Barents Sea, would lead to the formation of water with the characteristics of the lower halocline. Rudels et al. (1996) proposed that as the West Spitsbergen Current enters the Arctic Ocean through Fram Strait, the Atlantic water encounters, and melts, sea ice. The melt water is mixed into the upper part of the Atlantic water, cooling it and reducing its salinity. A less-saline upper layer is created that is homogenised in winter by convection. Close to Fram Strait, this convection is initially thermal, but when the upper layer reaches freezing temperature, the convection becomes haline, driven by freezing and brine rejection.

This upper layer extends down to the thermocline above the Atlantic layer and is advected eastward together with, and above, the main part of the Fram Strait branch. In summer, a low-salinity melt water layer develops at the surface, which again is removed by ice formation the following winter. This situation extends over the entire Nansen Basin, and the cold halocline, present elsewhere in the Arctic Ocean, is missing. Rudels et al. (1996) proposed that this 'winter-mixed layer' reaching down to a pycnocline with coinciding thermocline and halocline is the embryo of the 'lower' halocline water found in the other Arctic Ocean basins.

Steele et al. (1995), following Moore and Wallace (1988), assumed that the salinity, S_1, of the upper 'melt water' layer should be close to $S_1 = S_A(1 - c\Delta TL^{-1})$, where S_A is the salinity of the Atlantic water and ΔT the temperature difference between the upper layer and the Atlantic water. L is the latent heat of melting and c the heat capacity of sea water. The minimum salinity is reached when the surface water is at the freezing point, and all available heat of the Atlantic water has been used to melt sea ice. This results in lower salinities than those observed in the lower halocline (34.2–34.4), unless the Atlantic water has been cooled considerably before it encounters the ice.

Rudels et al. (1999b) proposed that cooling in the presence of sea ice always leads to a loss of oceanic sensible heat both to the atmosphere and to ice melt and proposed that the fraction, f, going to ice melt is such that the ice melt rate is a minimum. A larger fraction evidently leads to more ice melt, but so does also a smaller fraction, since the stability at the base of the mixed layer becomes weaker, allowing for stronger entrainment and larger transport of heat into the mixed layer from below. The fraction was found to be $f \approx 2\alpha L(c\beta S_A)^{-1}$ where α is the coefficient of heat expansion and β the coefficient of salinity contraction. The salinity of the upper layer becomes $S_1 = S_A(1 + 2\alpha\Delta T(\beta S_A)^{-1})^{-1}$, and with the temperatures and salinities observed in the Atlantic water north of Svalbard, this expression gives a salinity range of 34.2–34.4, close to that observed in the upper layer north of Svalbard and in the lower halocline. The salinity minimum is attained when the surface layer reaches freezing temperature. Once ice starts to form, the salinity again rises above this value.

The winter-mixed layer remains in contact with the sea surface and is ventilated down to the thermocline each winter, until it is overrun by shelf water of still lower salinity that crosses the shelf break. A massive outflow of low-salinity shelf water commonly occurs from the Laptev Sea and is, in reality, a further interaction between the two inflow branches. The Barents Sea branch does not just supply water to the Atlantic, intermediate and deeper layers of the Arctic Ocean via the St. Anna Trough. One part, comprising much of the water of the Norwegian Coastal Current, enters the Kara Sea through the Kara Gate south of Novaya Zemlya. In the Kara Sea, it receives the runoff from Ob and Yenisey and its salinity decreases. Most of the shelf water appears to continue eastward through the Vilkiltskij Strait into the Laptev Sea, and only a smaller fraction crosses the Kara Sea shelf break into the Nansen Basin. Once in the Laptev Sea, the runoff from the Lena, Yana and Khatanga rivers is added to the shelf water before it crosses the shelf break and enters the deep Arctic Ocean basins.

The inflow of less-saline shelf water on top of the winter-mixed layer in the Nansen Basin creates a 'cold' salinity gradient between the shelf water and the winter-mixed layer. Winter convection becomes limited to the 'shelf water', which evolves into the PML, while the now isolated 'winter-mixed layer' becomes the halocline water mass commonly referred to as the lower halocline (Rudels et al. 1996). This implies that the lower halocline initially is a mode water (McCartney 1977), a halostad with vertically almost constant temperature and salinity, which, after it has been isolated from the surface convection, becomes stratified by internal turbulent mixing. The fact that the term halocline in the Arctic Ocean is used to signify both a distinct water mass between the Polar Mixed Layer and the Atlantic layer and a sharp salinity gradient may cause some confusion (see below Sect. 4.3.2.2).

This mechanism for halocline formation differs from the shelf scenarios previously proposed (Coachman and Barnes 1962; Aagaard et al. 1981). The mechanisms do not exclude each other. In the Eurasian Basin, the waters crossing the shelf break are either dense enough to enter, or sink below, the Atlantic layer (see e.g. Schauer et al. 1997; Rudels and Friedrich 2000) or are less dense and enter at the surface. There, they either become stirred into the mixed layer in winter or, if the volumes are large, evolve into the PML. As the PML and the halocline both enter the Canadian Basin, the density difference between them is large enough for water of intermediate density to sink off the shelves and penetrate between the PML and the lower halocline, as e.g. the upper halocline water discussed by Jones and Anderson (1986).

A similar homogenisation of melt water above the Atlantic layer also takes place in the northern and eastern Barents Sea, and Rudels et al. (2001, 2004a) traced the part of the Barents Sea winter-mixed layer, which enters the Arctic Ocean with the Barents Sea branch in the St. Anna Trough, along the Eurasian continental slope into the Canada Basin. The halocline water deriving from the Barents Sea branch is more saline and becomes warmer than that of the Fram Strait branch due to stronger mixing with underlying Atlantic water at the continental slope (Rudels et al. 2004a) (Fig. 4.8).

At the mooring deployed on 1995/1996 north of the Laptev Sea, halocline water from either the Fram Strait branch or from the Barents Sea branch was observed at any particular time (Woodgate et al. 2001). This suggests that the two branches move side by side at least as far as the Lomonosov Ridge. The Fram Strait branch supplies the colder, and slightly less-saline, halocline observed in the Amundsen Basin but also contributes to the lower halocline in Makarov basins and in the Canada Basin west of the Chukchi Cap. The Barents Sea branch halocline remains at the slope until it crosses the Chukchi Cap. Beyond the Chukchi Cap, it appears to leave the slope to flow into the basin, supplying the warmer, more saline lower halocline water in the remotest (seen from the Atlantic) part of the Canada Basin. Some of the Barents Sea branch halocline water passes through Nares strait into Baffin Bay, but most of it exits the Arctic Ocean together with the Fram Strait branch halocline water in the East Greenland Current, where strong isopycnal mixing between the two halocline water masses has been observed in the temperature profiles (Rudels et al. 2005).

4.3.1.3   Pacific Water

Pacific water in the Arctic Ocean is distinguished from the Atlantic-derived waters by its lower salinity and density. This implies that it mostly contributes to the upper layers of the Arctic Ocean water column – the PML and the halocline. The Pacific water enters through the shallow Bering Strait and starts as three separate water

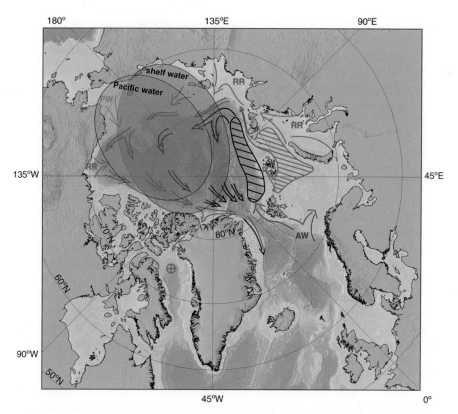

Fig. 4.8 Circulation of the Atlantic-derived halocline waters. The proposed source areas for the Fram Strait branch lower halocline water (*black*) and the Barents Sea branch lower halocline water (*blue*) are shown by diagonals and the circulation of these halocline waters in the Arctic Ocean is indicated. *RR* river runoff, *PW* Pacific water, *AW* Atlantic water. The *green* and *yellow* transparent ovals show the distributions of the Eurasian Basin shelf input and the Pacific water, respectively. The cross indicates possible contribution of Barents Sea branch lower halocline water to the Baffin Bay bottom water (Adapted from Rudels et al. 2004)

masses; farthest to the west is the more saline Gulf of Anadyr water, $32.8 < S < 33.2$, then the Bering Shelf water, $32.5 < S < 32.8$ and Alaskan Coastal Water $S \approx 32$ to the east (Coachman et al. 1975). Once these waters enter the Chukchi Sea, the two westernmost waters merge in the central part to form the Bering Strait Water (BSW), while the Alaskan Coastal Water (ACW) remains close to the Alaskan coast in the Alaskan Coastal Current (ACC). It becomes further diluted by runoff and enters the southern Canada Basin along the Barrow Canyon. The Bering Strait Water mainly flows northward and enters the Arctic Ocean primarily down the Herald Canyon (Coachman et al. 1975) (see also Fig. 4.3 above).

The characteristics of the Pacific water depend upon season, and Coachman and Barnes (1961) distinguished between the Bering Strait Summer Water (BSSW) with temperature clearly above the freezing point and salinity around 32.5, which shows up as a temperature maximum around 75 m in the Canada Basin, and Bering Strait Winter Water (BSWW) with temperature at the freezing point and salinity around 33.1 and located between 150 and 200 m in the Canada Basin water column.

Steele et al. (2004) identified two temperature maxima, one less dense deriving from the Alaskan Coastal Water (ACW) and one denser originating from the BSSW. Since these two water masses enter the Arctic Ocean by different paths, down the Barrow Canyon and down the Herald Canyon respectively, the salinity of the upper temperature maximum can be used to determine which water mass is present and thus trace the circulation of the less-dense part of the Pacific water in the interior of the Arctic Ocean (see Sect. 4.3.2 below).

The denser and colder BSWW mainly enters the Canada Basin west of the Chukchi Cap (Coachman et al. 1975), but recent studies have suggested that a less-dense fraction of BSWW also reaches the Canada Basin east of the Chukchi Cap (McLaughlin et al. 2004). The temperature minimum is also associated with an oxygen minimum and a nutrient, especially a silicate, maximum. Jones and Anderson (1986) proposed that this nutrient maximum was the result of remineralisation of organic matter from the shelf bottom, which implies convection of brine-enriched water to the shelf bottom that subsequently leaves the shelf, sinking into the deep Canada Basin forming the upper halocline.

The Pacific water differs from the Atlantic water also in other parameters. The nutrient concentrations are generally higher, and the ratio of nitrate and phosphate concentrations is different from that in the Atlantic water. This is because the oxygen concentration in the North Pacific, and especially in the Bering Sea, is low, and nitrate is used to oxidise organic matter, which results in a deviation from the Redfield ratio (Redfield et al. 1963) and changes the nitrate–phosphate ratio. The nitrate–phosphate ratio has therefore been used, often together with the silicate concentration, to determine the Pacific fraction in the water masses both within the Arctic Ocean and in the outflows through Fram Strait and the Canadian Arctic Archipelago (Jones et al. 1998; Jones et al. 2003).

That the Pacific water dominates the upper layers in the Canada Basin is rather obvious, considering its proximity to Bering Strait, but also the upper layers of the entire Canadian Basin during the early ice camp expeditions were found to exhibit Pacific water characteristics (Kinney et al. 1970; Gorshov 1980; Moore et al. 1983). This situation has presently changed (see Sect. 4.3.3 below). The Pacific water dominates in the western channels of the Canadian Arctic Archipelago (Jones and Coote 1980), and only in the deeper (230 m) Nares Strait is Atlantic water found in the deeper layers (Jones et al. 2003). Pacific water has also been observed leaving the Arctic Ocean through Fram Strait (e.g. Anderson and Dyrssen 1981; Jones et al. 1998, 2003).

Elevated silicate concentrations have been found in the lower halocline for salinities at least as high as 33.6 (Salmon and McRoy 1994), and the nitrate and phosphate concentrations and also the vertical gradient in I^{129} concentration (Swift et al. 1997; McLaughlin et al. 2002) show that Pacific water penetrates deeper than the 33.1 of the upper halocline (Jones and Anderson 1986). This could be due to excessive brine release, especially in polynyas on the Chukchi shelf close to the

Alaskan coast, leading to salinities above 34 (Weingartner et al. 1998) occasionally above 36 (Aagaard et al. 1985), high enough for the Pacific water to sink into the lower halocline. The difference between the deeper parts of the water columns in the Eurasian and Canadian basins, from the Atlantic layer to the bottom, has also been interpreted as sign of deep-reaching shelf-slope convection occurring in the Canadian Basin, which, by necessity, would involve Pacific water (Aagaard et al. 1985; Rudels 1986a). These volumes of Pacific water would, however, be small < 0.1 Sv (Rudels et al. 1994; Jones et al. 1995).

Another possibility could be that large amplitude vertical displacements observed at upwelling events (Carmack and Kulikov 1998; Kulikov et al. 1998) bring lower halocline and Atlantic water onto the shelf. Atlantic water has been observed moving along the bottom of the Barrow Canyon towards the Chukchi shelf (Weingartner et al. 1998), and it could occur also elsewhere. This situation, similar to the mechanism suggested by Coachman and Barnes (1962) for the formation of the lower halocline in the Eurasian Basin (see Sect. 4.2), would enhance the mixing between the Atlantic and Pacific water and stir Pacific waters into the lower halocline.

4.3.1.4 Shelf Processes, River Runoff, Ice Melt, Freezing and Brine Rejection

The atmospheric meridional transport of water vapour leads to net precipitation in the Arctic. The low air temperatures limit the vapour content, and the net precipitation over the Arctic Ocean is small. But because of its large catchment area, the Arctic Ocean, comprising 2.5% of the area of the world ocean, receives 7% of the runoff of the earth's rivers (Baumgartner and Reichel 1975). The runoff is strongly seasonal with the largest discharge in spring and early summer. The river runoff forms low-salinity surface plumes that commonly flow eastward along the coast. The runoff from Ob and Yenisey in the Kara Sea mostly passes through the Vilkiltskij Strait into the Laptev Sea before it, together with the runoff from the Lena and Yana rivers, crosses the shelf break into the Amundsen Basin. The rivers discharging into the East Siberian Sea contribute to the Siberian Coastal Current, which may reach as far east as Bering Strait, before it becomes mixed into the northward-flowing Pacific water; occasionally, it even passes through the strait into the Bering Sea (Roach et al. 1995). The runoff from the Yukon enters the Arctic Ocean in the Alaskan Coastal Current (ACC) via Bering Strait, and more runoff is added as the ACC moves towards the Mackenzie delta. The river plumes may easily become disrupted by offshore winds, which spread the water from the coast, thinning the low-salinity layer and thus preventing e.g. the shallow Siberian Coastal Current from developing (Münchow et al. 1999; Weingartner et al. 1999). Changes in the large-scale atmospheric circulation pattern such as the Arctic Oscillation (AO) (or the Northern Annular Mode (NAM)) (Thompson and Wallace 1998) and the North Atlantic Oscillation (NAO) (Hurrell 1995) may also affect the paths of the larger runoff volumes from the Ob, Yenisey and Lena rivers, making them enter either the

Amundsen Basin or flow eastward into the East Siberian Sea (Proshutinsky and Johnson 1997) (see Sect. 4.3.2 below).

The river water enters the deep Arctic Ocean basins as the low-salinity shelf waters cross the shelf break. River water can be distinguished from sea ice melt water by its higher total alkalinity and different ^{18}O concentrations, which can be used to track the river water across the Arctic Ocean (Anderson et al. 1994; Anderson et al. 2004). Tracers like barium may be used to identify water from different rivers (Guay et al. 2001). ^{18}O makes it possible to separate river runoff, ice melt and ice formation (Östlund and Hut 1984) (see below, Sects. 4.3.1.5 and 4.3.2).

The shore-fast ice and the grounded ridges 'stamukhies' outside the coast prevent the river water from easily escaping to the outer shelf areas, and large, under ice, freshwater lakes may form seasonally as in the Mackenzie delta (Carmack and MacDonald 2002). The ice formation and the ice melt, since the ice moves differently than the water, add a further path for the freshwater to circulate and become redistributed in the Arctic Ocean surface layer (Macdonald 2000). On occasion the melting and the northward retreat of the drift ice may act as a buoyancy source that creates a low-salinity surface flow. This has been observed in the East Siberian and eastern Chukchi Sea where, in 1995, instead of the expected eastward-flowing Siberian Coastal Current, a westward surface flow was observed (Münchow et al. 1999; Weingartner et al. 1999).

The interaction between sea ice and the underlying ocean is simple in most parts of the deep Arctic Ocean. Ice melts in summer, mostly from above, due to the strong solar radiation. In autumn and winter, the freshwater is removed by freezing, and released brine homogenises the PML. Since the PML is located above the cold halocline, no heat is added by entrainment from below during winter, and the ice grows as a response to the heat loss to the atmosphere and to space. Only in the marginal ice zone does the ice come in direct contact with warmer water and melts from below. The formation of a melt water layer below the ice will, however, reduce the heat transfer and the melting unless the ice is drifting very rapidly, keeping the melt water layer thin, and continuously forcing contact with new, warm water (McPhee 1990; Boyd and D'Asaro 1994).

A somewhat similar situation occurs in the Nansen Basin, where the cold halocline is absent and, during winter, warm Atlantic water may be entrained into the winter-mixed layer, influencing the ice formation rate as well as supplying heat to the atmosphere. In the Nansen Basin, the stratification and the deep winter-mixed layer will limit the entrainment, and the only effect is a slightly reduced ice formation. However, in less-stratified waters like the Greenland Sea and the Weddell Sea in Antarctica, this vertical heat flux from below could lead to the disappearance of the ice already in midwinter, when the cooling is the strongest (Martinson 1990; Lemke 1993; Walin 1993; Rudels et al. 1999b).

As it became clear that sinking of cold, saline water from the shelves down the slope was the one process that could explain not only the formation of water in the density range of the halocline but also the transformations of the deeper layers of the Arctic Ocean basins (Nansen 1906; Aagaard et al. 1985; Rudels 1986a) and after the observation of dense bottom water sinking from Storfjorden into Fram

Strait (Quadfasel et al. 1988), the studies of freezing, brine rejection and the accumulation of cold saline water on the shelves and of the sinking of dense water down the continental slopes were intensified. The ice production and the corresponding formation of saline water in polynyas were estimated from satellite observations of polynya extent and the wind and temperature fields (Martin and Cavalieri 1989; Cavalieri and Martin 1994; Winsor and Björk 2000). Winsor and Björk examined 28 different polynya areas forced by 39 years of reanalysed meteorological data. Using a polynya model originally developed by Pease (1987), they obtained a much lower (0.2 Sv) estimate of halocline water formation than Björk (1989), Martin and Cavalieri (1989) and Cavalieri and Martin (1994) (around 1.2 Sv), but similar to the estimate by Rudels (1989) (0.2 Sv). They concluded that the areas most likely to form dense water, capable of renewing the deeper layers of the Eurasian Basin and the Canadian Basin, were the Barents Sea and the Chukchi Sea, respectively.

In numerical studies of dense water production in polynyas, Gawarkiewicz and Chapman (1995) found that water was transported across the shelf in small-scale eddies formed by baroclinic instabilities around the polynya. The production of dense water and large density anomalies were thus limited by the eddy flux (Chapman and Gawarkiewicz 1997; Gawarkiewicz et al. 1998), and Gawarkiewicz (2000) found that eddies could transport dense water across the shelf break and down the slope, giving another mechanism in addition to baroclinic instability (Hart and Killworth 1976) and frictional spin-up (D'Asaro 1988a, b) for forming and injecting eddies into the upper Arctic Ocean. Winsor and Chapman (2002) combined the polynya model used by Winsor and Björk (2000) with a primitive equation ocean model to estimate the production and export of dense water from the Chukchi shelf. They found that the amount of dense water production, halocline water and denser, was small and that the density anomalies were as much due to the initial conditions, set by the inflowing Pacific water, as to the ice formation in the polynya. This is in agreement with the observations by Roach et al. (1995) and Weingartner at al. (1998).

Convection, thermal and haline, and the formation and spreading of dense water over the shelf and down the slope were modelled by Backhaus et al. (1997), using different process models. Especially the outflow from Storfjorden was examined and compared with observations (Jungclaus et al. 1995; Backhaus et al. 1997). The path of the plume was well simulated as was the change of properties caused by the entrainment. The Storfjorden polynya and the outflow of dense water from Storfjorden into Fram Strait have been thoroughly investigated during the ACSYS period (e.g. Schauer 1995; Haarpainter 1999; Haarpainter et al. 2001; Schauer and Fahrbach 1999; Fer et al. 2003a; Fer et al. 2003b; Skogseth et al. 2005). The 1990s was a period when little highly dense water was observed in Storfjorden, but in 2002, salinities above 35.8 were observed in Storfjorden from *Oden* during the Arctic Ocean 2002 expedition. This was perhaps the highest salinity reported from Storfjorden to that date (Rudels et al. 2005).

Sinking of dense water from the shelves into the deep Arctic basins was not noticed until 1995 (Rudels et al. 2000a), when thin layers of saline, denser but warmer bottom water was found deeper than 2,000 m on the continental slope north of Severnaya Zemlya. Similar bottom layers were also seen north of the eastern Kara Sea in 1996 and were actually present also on some *Polarstern* stations taken north of the Laptev Sea in 1993 (Fig. 4.9). The origin of these dense bottom layers has, so far, not been determined, and the paths of the water have not been reconstructed. Detailed plume models, taking into account the bathymetry and the origins and paths of the plumes, have therefore not yet been used to describe the deepest reaching boundary convection within the Arctic Ocean.

Some efforts have been made to use simple balance models to estimate the properties the shelf water initially must have and the rate of entrainment into the plume that is necessary to reproduce the characteristics observed in the deep Arctic Ocean basins. In the Eurasian Basin, the intermediate waters are dominated by the two inflow branches, and the effects of shelf–slope convection are mainly seen in the deepest part of the water column, where the temperature is almost constant with depth and the salinity increases towards the bottom. In the Canadian Basin, however, the changes in the Atlantic and intermediate waters that cross the Lomonosov Ridge are due to slope convection. Rudels et al. (1994) assumed that the entrainment rate of ambient water into the plumes was constant with depth and added for every 150 m of descent a volume of entrained water corresponding to the initial volume passing the 200 m isobath. Since the dense shelf water is formed by brine rejection, the initial temperature of all plumes should be the same, and the depths the plumes reach would only depend upon their salinity. The assumptions made by Rudels et al. (1994) did not lead to contradiction with the observed water mass properties in the interior of the Canadian (Makarov) Basin. The entrained water quickly dominated the characteristics of the plumes, making them warmer than their surroundings once they had passed the Atlantic layer. To reach the deeper parts of the water column, the initial salinity had to be high, close to or above 36, but perhaps not inconceivable. The volume of shelf water contributing to the ventilation of the deeper layers was very small, around 0.02 Sv (Rudels et al. 1994).

Goldner (1999a, 1999b) examined slope convection using an inverse box model to determine the transport from the boxes and the renewal from the shelves necessary to maintain the mass, heat and salt balances in the boxes. He separated the Canadian Basin and the Eurasian Basin and found that the Canadian Basin shelves could supply about 0.2 Sv and the Eurasian Basin shelves 1.4 Sv, including the Barents Sea inflow, to the deep basins. In these models, there is no balance of the transport across the shelf break as in Björk (1989) and Rudels (1989), and the mass and salt budgets were assumed closed by inflows from behind, through Bering Strait, over the Barents Sea and by river runoff. For the Bering Strait inflow, this is obvious, but considering how especially the Norwegian Coastal Water of the Barents Sea branch spreads from one shelf sea to the next (see e.g. Rudels et al. 1999c), this is probably realistic also for the Eurasian Basin. In Goldner (1999a), the shelves were found to supply mainly dense water to the deep basins, while in Goldner (1999b), the input of freshwater from the shelves was larger. The first result is likely due to the inclusion of the Barents Sea branch as a shelf outflow rather than a direct inflow from the Norwegian Sea to the Nansen Basin. The second result, which implies a net ice formation in the deep basins, appears more realistic.

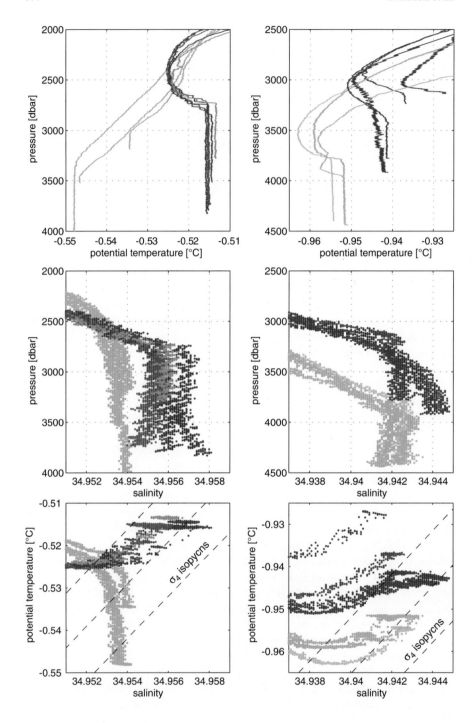

4.3.1.5 Tracking the Waters: Insights from Tracers

The chemical signature is useful to trace the source, distribution and circulation of most Arctic Ocean water masses. Chemical tracers can be separated into the three groups, conservative (e.g. ^{18}O), bioactive (e.g. oxygen and nutrients) and transient (e.g. CFCs, ^{3}H, ^{14}C), contributing different information.

Two freshwater sources dominate in the PML, sea ice melt and river runoff, and they have very distinct chemical signatures. Sea ice melt has a signature that is a function of salinity and the composition of the seawater from which it was formed. Runoff, on the other hand, has low fraction of $^{18}O/^{16}O$ ($\delta^{18}O$) and high barium (Ba) and total alkalinity (A_T) concentrations. The low $\delta^{18}O$ signal is a signature of meteoric water, while the high Ba and A_T concentrations are caused by weathering of minerals in the drainage basins. The latter often results in high, but variable, concentrations of Ba and A_T in the waters of individual rivers.

When applying $\delta^{18}O$ or A_T to evaluate the freshwater sources, one also gets negative sea ice melt water, and this is an indication of brine being added to the water during sea ice production. Already in the beginning of the 1980s, $\delta^{18}O$ was used to evaluate the distribution of sea ice melt water (Tan and Strain 1980) and to make balances of mass (volume) and freshwaters for the Arctic (Östlund and Hut 1984). Since then, $\delta^{18}O$ has been a commonly used tracer, often in combination with others, to evaluate the water mass composition of the Arctic Ocean (e.g. Schlosser et al. 1994a; Bauch et al. 1995; Melling and Moore 1995; Macdonald et al. 1995; Macdonald et al. 1999; Ekwurzel et al. 2001).

By combining data from five cruises performed in 1987, 1991, 1993, 1994 and 1995, Ekwurzel et al. (2001) show that the runoff is fairly evenly distributed, adding ~10 m of freshwater over the whole central Arctic Ocean except for the Nansen Basin, where much less is found. This estimate includes freshwater removed as sea ice.

Fig. 4.9 *Left column*, deep-water characteristics in the Makarov Basin (*green*) and the Canada Basin (*red* and *violet*). *Right column* deep-water characteristics in the Amundsen Basin (*cyan* (1991), *green* and *yellow* (1996)) and Nansen Basin (*blue* (1996)). *Violet* (1995) and *magenta* (1993) indicate stations taken at the slope of the Nansen and Amundsen basins. The deepest of these stations is at the base of the slope north of the Laptev Sea. The higher temperatures and salinities at the bottom suggest that the bottom water is created by slope convection and entraining plumes. The variability in salinity and temperature in the Amundsen Basin bottom water could either be due to geothermal heating or to varying characteristics of the waters derived from the slope convection, as is seen by the high temperature and salinity found at the base of the Laptev Sea slope (deepest *violet station*). The temperature minimum in the Nansen and Amundsen basins could be due to less-saline inflow from the St. Anna Trough. In the Canada Basin, the temperature minimum could be explained by the advection of deep waters from the Makarov Basin across the sill between the Alpha and the Mendeleyev ridges. The Makarov Basin is the only basin with no deep temperature minimum. The salinity first reaches a maximum value in a 1,200 m deep layer while the temperature decreases for another 400 m before the isothermal and isohaline bottom water is reached. This is in agreement with the densest water mostly deriving from a spillover of intermediate water from the Amundsen Basin across the Lomonosov Ridge (Jones et al. 1995), a process yet to be confirmed. Stations positions are shown in Fig. 4.19d

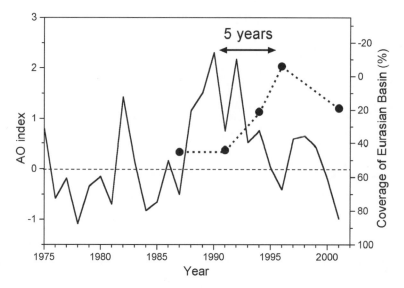

Fig. 4.10 Changes in the shelf (river) water front. The *solid line* represents the winter (November–February) values of the Arctic Oscillation, and the *dots* show the percentage of the deep Eurasian Basin covered by low-salinity shelf water (river runoff) (Adapted from Anderson et al. 2004)

The sea ice melt is mainly found north of the Barents Sea, where the warm inflowing Atlantic Water meets and melts the exiting sea ice, while brine equivalent to ~6 m sea ice production is found over the Canadian and western Amundsen Basins, with little north of the Laptev Sea, where much of the ice in the Arctic initially is formed. The distribution of sea ice melt and brine reflects a combination of ice dynamics and circulation of the upper waters.

Total alkalinity is an almost conservative tracer in the Arctic Ocean, and high concentrations are found in river runoff as a result of a combination of decay of organic matter and dissolution of metal carbonates. It has been used to trace the river runoff front within the Eurasian Basin (e.g. Anderson and Jones 1992), which reflects the extent of the 'cold halocline' as further discussed under 4.3.2.3. By applying the variability of the front during five cruises between 1987 and 2001, it was shown that the runoff has a minimum coverage over the Eurasian Basin during years of high AO indexes, but with a lag of ~5 years (Anderson et al. 2004) (Fig. 4.10).

Dissolved barium concentrations in the upper Arctic Ocean (<200 m) range between 19 and 168 nmol L^{-1} in a manner geographically consistent with sources and sinks. The sources are runoff, with the highest concentrations from the American continent, and sinks are biological removals (Guay and Falkner 1998). As a result of these variable sources and sinks, barium has to be used more as a qualitative than as a quantitative tracer.

One of the most useful bioactive chemical tracers of Arctic Ocean water masses is silicate, which is high in the Pacific water entering over the Bering and Chukchi shelves. Through fixation by marine plankton with subsequent sedimentation and decay at the sediment surface, very high concentrations are found in waters of salinities around 33.1 flowing off the Chukchi shelf (Jones and Anderson 1986). Furthermore, elevated silicate concentrations are found at all depths below the 33.1 isohaline at the Chukchi shelf slope, suggesting the penetration of high-density plumes of high salinity from the Chukchi Sea.

The distribution of the silicate concentration in the intermediate depth waters (from the upper halocline all the way to below the Atlantic layer) collected during the *Oden*-91 expedition was one of the key signatures behind the circulation pattern suggested by Rudels et al. (1994). This circulation pattern was supported by the silicate concentration distribution as observed from ice camps during the 1980s (Jones and Anderson 1986; Jones et al. 1991; Anderson and Jones 1992), suggesting a steady-state situation with a consistent circulation pattern. However, during the ACSYS, decade changes have been observed in the silicate concentration distribution, especially within the Canadian Basin, with decreasing concentrations in the Makarov Basin and central Canada Basin (McLaughlin et al. 1996). These changes have been accompanied by a penetration of warmer Atlantic Layer water into the Makarov Basin (e.g. Morison et al. 1998) (see Sect. 4.3.2.1 below).

The time history of water masses can be deduced from transient tracer distribution and an appropriate model approach. One of the first estimates of Arctic Ocean water mass residence times was made for the freshwater component using observed seawater tritium concentrations compared to tritium in precipitation (Östlund 1982). At most observational sites, a linear relationship between salinity and tritium concentration was found, indicating mixtures of Atlantic layer water and freshwater. The resulting freshwater tritium concentration points to an average residence time of 11 ± 1 years since the freshwater left the river mouth for the data collected in the Nansen Basin and somewhat higher for those collected in the Canada Basin (Östlund 1982). This approach was extended by Schlosser et al. (1994a) by adding ^{3}He and ^{18}O tracer data to those of tritium. With this extra information, it is possible to estimate the residence time of the river runoff on the shelf as well as the time since the halocline water left the shelf. This is possible since the tritium/^{3}He age is set to zero as long as ^{3}He can be lost to the atmosphere by gas exchange. When correcting for the limited exchange of ^{3}He on the shelf, the mean residence time of the runoff component on the shelf becomes 3.5 ± 2 years (Schlosser et al. 1994a). The tritium/^{3}He age along a section crossing the Nansen Basin at about 30°E shows low surface water age (~1 year) at the Barents Sea shelf break, increasing to about 5 years over the Gakkel Ridge. Over the Gakkel Ridge, the tritium/^{3}He age increases with depth to a maximum of about 15 years in the lower halocline (Schlosser et al. 1994a, b). A similar age distribution was also found in the upper and intermediate waters using CFCs (Wallace et al. 1992).

The transient tracer distribution was used by Frank et al. (1998) to investigate the subsurface flow along the continental slope of the Eurasian Basin. The observed tritium/^{3}He age distribution in the Atlantic Layer was found to be 2 years along the continental slope north of the Barents Sea, while it increased to 6 years at the Laptev Sea continental margin for both the Fram Strait and Barents Sea branches. The transit time of the Fram Strait branch of 4 years from the Barents Sea continental margin to that of the Laptev Sea estimated from the tritium/^{3}He age was consistent with the

CFC ages of 4–5 years (Frank et al. 1998). The comparably high tritium/^{3}He age of the Barents Sea branch at the Laptev Sea continental margin was explained by a combination of a 'non-zero' age when the water left the Barents Sea and mixed with older water along the Kara Sea continental slope. Smethie et al. (2000) expanded the transient tracer study of the intermediate waters into the Canadian Basin using samples collected from the Lomonosov Ridge (88°N, 44°W) to the Alaskan North slope (71°N, 145°W) on the SCICEX 96 cruise. Their results show tritium/^{3}He ages in the Barents Sea branch of the Atlantic Layer of 14.5 years over the Lomonosov Ridge, 18.5 years at 78°N off the Chukchi Cap and at the Alaskan North slope, and older water between 78°N and 88°N. They suggest that a fraction of the Barents Sea branch follows the Lomonosov Ridge from the Laptev Sea, while some follows the Siberian continental margin, where a fraction flows off the Chukchi Cap into the Canada Basin with some flow continuing along the boundary to the southern Canada Basin (Smethie et al. 2000). This flow path is consistent with that described in Sect. 4.3.1.1 but adds a time perspective to the flow.

The most common use of transient tracers in Arctic Ocean research has been to deduce the deep-water ventilation (Smethie et al. 1988; Heinze et al. 1990; Schlosser et al. 1990; Schlosser et al. 1994b; Jones et al. 1995; Schlosser et al. 1995; Schlosser et al. 1997; Anderson et al. 1999). Different approaches (model formulations) are used to evaluate the 'ages' of the deep and bottom waters of the different Arctic basins, but with consistent results considering the different approaches. In general, the results give mean residence times and ventilation times in the order of 100–250 years for the Eurasian Basin Deep Water, 200–350 years for the Eurasian Basin Bottom Water and 300–360 years for the Canadian Basin Deep and Bottom Waters. One investigation of special interest is that of Livingston et al. (1984), which used the artificial radionuclide signal from the Sellafield (at that time Windscale) reprocessing plant on the Irish Sea to find a transit time of 8–10 years from the source to 1,500 m depth near the North Pole (LOREX site). This is a remarkable short transit time but cannot be ruled out in light of the flow path of Atlantic water over the Lomonosov Ridge and the very high current speeds observed over the ridge close to the Laptev Sea (Woodgate et al. 2001).

4.3.1.6 Mixing in the Interior

The deep, interior Arctic Ocean is a low-energy environment, where processes normally overwhelmed by mechanical stirring and turbulence may be of importance (Carmack 1986; Padman 1995). The thermocline, where the temperature is unstably stratified, is a region where molecular processes such as double-diffusive convection through diffusive interfaces could occur. In the Canadian Basin, regular staircases with homogenous layers, 5–10 m thick, separated by thin interfaces with steps in temperature and salinity, have been observed (Neal et al. 1969; Neshyba et al. 1971; Padman and Dillon 1987, 1988, 1989). These staircases may extend over depth intervals of 100–200 m and could be instrumental in cooling the Atlantic layer in the deep Canada Basin. In the Eurasian Basin, the thermocline is stronger but has smaller vertical extent, and the observed steps are larger than those in the Canadian Basin and often appear to be connected with weak inversions. Merryfield (2000) has suggested that horizontal advection, similar but weaker than the advection that creates interleaving and inversions, might operate in the formation of step structures. Extensive layers and step structures have also been observed closer to the bottom in the deep Canada Basin (Timmermans et al. 2003). These layers will be discussed further in connection with the deep-water formation and transformation (Sect. 4.3.1.7).

The intermediate layers of the Arctic Ocean are often characterised by large, regular intrusions, mainly in the Atlantic layer but also deeper in the water column (see Sect. 4.3.1.1 and Fig. 4.4). These layers pose several questions: How do they form? Do they have dynamics of their own? How important are they for the transfer of heat, for example, from the warm Atlantic water in the boundary current into the interior of the basins and for the mixing and transformation of the Arctic Ocean water masses?

Carmack et al. (1995) noticed that the anomalously high temperatures (see discussion in 4.3.2.1) observed in the Atlantic layer in the Makarov Basin in 1993 were associated with prominent intrusions and postulated that the intrusions were formed as the warmer Atlantic water of the boundary current interacted with the colder, and older, Atlantic layer in the Makarov Basin. This explanation is akin to the one offered by Perkin and Lewis (1984), when they observed inversions in the Atlantic layer north of Svalbard during the EUBEX expedition. Similar interleaving structures were found in the Makarov, Amundsen and Nansen basins on the AO94 expedition (Carmack et al. 1997) and had been seen in the Nansen and Amundsen basins on the *Oden*-91 expedition and on the *Polarstern* expeditions to the Eurasian Basin in the 1990s (e.g. Rudels et al. 1994; Rudels et al. 1999a). The layers appeared to reach across the basins, implying a horizontal extension of 100–1,000 km, and Carmack et al. (1997) suggested that the layers were self-propelled, formed at frontal zones, especially as the Atlantic water enters through Fram Strait, and then expanded, driven by double-diffusive processes, into the interior of the basins. The formation and expansion of interleaving layers could then be a major process in bringing heat from the boundary current into the interior of the basins (Carmack et al. 1997).

The observed interleaving structures often have alternating diffusive and salt-finger interfaces, and double-diffusive transports are likely to be present. Layers are observed in all kinds of background stratifications, diffusively unstable (cold and fresh on top), saltfinger unstable (warm and saline on top) and stable-stable (temperature decreasing, salinity increasing with depth). Stability analyses show that layers may form at thermohaline fronts, when one of the components is unstably stratified (Stern 1967; Toole and Georgi 1981; Ruddick 1992; Ruddick and Walsh 1995), but with both component stably stratified, finite disturbances are needed to carry water parcels across, or at least sufficiently far into, the front to form inversions and allow double-diffusive convection to commence.

Walsh and Carmack (2002) explored the spreading of the intrusions encountered in the upper part of the water column, above the Atlantic layer, where the temperature stratification is unstable, and double-diffusive transports due through diffusive interfaces could occur. They examined the possibility that the non-linear equation of state, which would lead to contraction of mixing, or cabbeling, could reduce the density changes of the warmer intrusions penetrating into colder surroundings. The reduced density changes would then slow down the spreading of the intrusions and limit their extent. Considering the suggested basin-wide coherence of the individual layers, this effect, should it be present, does not seem to prevent the layer structures of the Arctic Ocean to reach the, perhaps largest, horizontal extent and the largest ratio of horizontal to vertical scale found in the world ocean.

May and Kelley (2001) examined the older data from the EUBEX expedition north of Svalbard in 1981 (Perkin and Lewis 1984) and proposed that the baroclinic-ity of the flow field is important, and perhaps necessary, for intrusions to develop. In the inflow region close to Svalbard, such dynamic situation could exist. However, inversions and layers are also observed in the stable-stable stratified upper Polar Deep Water (uPDW) around 1,000 m depth. These layers are in the Nansen, Amundsen and Makarov basins associated with the presence of less-saline Barents Sea branch water. Merryfield (2002) suggested that the deep layer structures could be caused by differential diffusion due to the faster diffusivity of heat. This effect could lead to the formation of layers even when both components are stably strati-fied. The time for the layers to form by this process is of the order of years, which appears long, since similar layers are found further to the east in the Nansen and Amundsen Basin, closer to the input of the Barents Sea branch (Rudels et al. 1999a) and at the Mendeleyev Ridge as the deeper part of the boundary current penetrates into the Makarov Basin (Swift et al. 1997). It is, however, shorter than the residence time of the uPDW.

These interpretations are all different from the one offered by Rudels et al. (1994) and discussed earlier in connection with the circulation of the Atlantic water. Rudels et al. (1994) also assumed that the interleaving structures were formed in frontal zones, in the Eurasian Basin primarily in the confluence area of the Barents Sea and in the Fram Strait branches, but expected the intrusions to evolve and expand only during a short period, running down the potential energy stored in the unstably dis-tributed component. Afterward, fossil intrusions would remain and be passively advected by the mean circulation around the basins. Interleaving structures could also form, when there are changes in the characteristics of the boundary current, which create larger property contrasts and intrusions within the boundary current. These intrusions are then carried along by the boundary current and become reinforced as the boundary current enters the different basins and encounters differ-ent water properties. Such interpretation would be consistent with the similar shape of the intrusions seen in the boundary current as the warm pulse of Atlantic water penetrates further into the Canada Basin (see e.g. McLaughlin et al. 2004 and also Sect. 4.3.2.1 below).

The formation and the dynamics of the interleaving layers, and their importance for the mixing, heat balance and water mass transformations in the Arctic Ocean, are thus questions far from solved. Are they as extensive as they appear, or do they belong to several distinct water bodies? If interleaving structures are created in dif-ferent areas, they will, because of the small layer thickness, nevertheless appear to be connected when looked upon in a Θ-S diagram. These layer structures may, in the end, be found less important than prominent, but they do pose intriguing theoretical and observational challenges.

4.3.1.7   The Bottom Water: Stagnant or Ventilated?

Because of the higher salinities and temperatures found in the deep Arctic Ocean basins, as compared to the Greenland Sea, the renewal of the deep and bottom waters in the different basins has been assumed due to slope convection that brings cold, brine-enriched shelf water into the deep (e.g. Nansen 1906). In the Nansen Basin, this process is modified by the inflow of Nordic Sea deep water through Fram Strait and the input of denser water of the Barents Sea branch, sinking down the St. Anna Trough. In the Canadian Basin, no such deep inflows occur, and differences between the water characteristics in the Amundsen Basin at the sill depth of the Lomonosov Ridge and in the deep Makarov Basin, the Makarov Basin being warmer and more saline must be caused by slope convection. Jones et al. (1995) estimated the ventilation time of the deeper (>1,700 m) layers of the Makarov Basin using potential temperature and salinity profiles and C^{14} concentrations and the plume model applied by Rudels et al. (1994) for the intermediate layers. The deeper layers in the Makarov Basin do not exhibit an increasing salinity and constant temperature with depth, as do the other Arctic Ocean basins. Instead, the salinity reaches a constant value, while the temperature continues to decrease with depth until an 800 m deep isothermal and isohaline bottom layer is encountered. No temperature minimum is present above this layer (Fig. 4.9).

This is not what to expect from slope convection, where the temperature would attain a constant value as the plumes eventually reach the same temperature as their surroundings, and the plumes with the highest salinity (density) will reach the bottom and fill up the deepest layers, leading to increasing salinity with depth. The explanation offered by Jones et al. (1995) was that a spillover of colder water from the Amundsen Basin could take place through deeper, undetected gaps in the central Lomonosov Ridge. Once colder Amundsen Basin water enters the warmer Makarov Basin, it would sink due to its higher compressibility (the thermobaric effect). Although newer bathymetry (Jakobsson et al. 2004b) indicated a narrow, deep (2,500 m) passage through the central part of the Lomonosov Ridge, no oceanographic observations had so far been 'knowingly' made in this passage. The ventilation times determined by Jones et al. (1995) for the deep Makarov Basin were of the same order, 200–300 years, as was found independently by Schlosser et al. (1994b).

However, on the *Healy–Oden* crossing of the Arctic Ocean in 2005, the HOTRAX expedition, this area was closely examined. A 2,700-m deep depression referred to as the *intra-basin* was found in the Lomonosov Ridge with a sill depth of 2,400 m to the Amundsen Basin but a sill depth of only 1870 m to the Makarov Basin (Björk et al. 2007). The layers in the intra-basin between 1,700 and 2,300 m were dominated by deep water from the Makarov Basin overlaying colder, denser Amundsen Basin deep water. No continuous flow from the Amundsen Basin to the Makarov

Basin was observed. Should such flow exist, it must be intermittent and perhaps depend upon external forcing. The Canadian Basin Deep Water (CBDW) was present not only in the intra-basin but also beyond the sill in the Amundsen Basin, implying that CBDW passes from the Makarov Basin through the intra-basin into the Amundsen Basin. The CBDW did not spread into the Amundsen Basin but appeared to follow the Lomonosov Ridge towards Greenland. The passage through the intra-basin might thus be an important, perhaps the most important, passage for CBDW to the Eurasian Basin and to Fram Strait (Björk et al. 2007).

In the Canada, Amundsen and Nansen basins, the salinity increases, and the temperature in the deep waters decreases with depth until a weak, deep temperature minimum is encountered. Below the minimum, both the temperature and the salinity increase with depth until a thick (800 m in the Amundsen Basin, >1,000 m in the Canada Basin) isothermal and isohaline bottom layer is reached (Fig. 4.9). The salinity increase towards the bottom suggests bottom water renewal by slope convection. However, the temperature minimum above the homogenous bottom layer cannot be explained by slope convection.

Extrema often indicate advective features. The minimum present in the Canada Basin can be explained by an inflow of colder, less-saline and less-dense deep water from the Makarov Basin, passing the sill between the Mendeleyev Ridge and the Alpha Ridge (Fig. 4.9). In contrast to the situation in the Makarov Basin, the slope convection in the Canada Basin then creates water denser than that crossing the sill, and the inflow from the Makarov Basin stays above the bottom layer. The similar structure in the Amundsen and Nansen basins is more difficult to understand. One possibility would be that very dense Barents Sea branch water is formed intermittently and sinks along the bottom of the St. Anna Trough, reaching a sufficiently large 'initial' depth at the slope that it avoids entraining warmer, less-dense intermediate water. A smaller salinity excess would then be sufficient for dense plumes originating from the Barents Sea branch to sink deep into the basin, and the temperature increase would be less if they start to entrain higher up on the slope, as would be the case for the high-salinity water formed around Severnaya Zemlya (Rudels et al. 2000a).

However, another interpretation has been proposed. Timmermans et al. (2003) assume that the deepest layer in the Canada Basin is a remnant from a convection event occurring several hundred years ago (MacDonald and Carmack 1991). The temperature in the bottom layer is slowly increasing due to geothermal heating, which stirs and homogenises the bottom layer. The heat is transferred out of the bottom layer, perhaps by double-diffusive convection through the diffusive interfaces, which form an extensive step structure extending over most of the Canada Basin (see Fig. 4.9), or by increased diapycnal mixing at the lateral boundaries at the continental slopes. This transfer prevents the temperature of the bottom layer from increasing, and the density from decreasing, too rapidly. However, in the next decade or so, the density of the bottom layer might have decreased sufficiently for the bottom layer to overturn and mix with the overlying water (Timmermans et al. 2003).

Björk and Winsor (2006) noticed a temperature change in the bottom layer in the Amundsen Basin between 1991, 1996 and 2001, which they interpreted as an effect of geothermal heating. However, the temperature and salinity in the bottom layers in the Amundsen and Nansen basins also vary along the basins, and higher temperatures and salinities were observed closer to the Laptev Sea in 1995 than in the central part of the basin in 1996. This is where dense water, formed by slope convection, is expected to enter the deep Amundsen Basin. It is thus at least conceivable that the observed changes are due to advection of water with higher temperatures and salinities recently created by slope convection (Fig. 4.9). However, the bottom layers are likely to be influenced by geothermal heating. The earth is losing heat, and it is to be expected that this heat is stored in a stagnant layer. The deep and homogenous layer also suggests that a mixing agent such as a weak thermal convection stirs the bottom water. Geothermal heating and boundary convection are two completely independent processes and can well operate simultaneously, together forming the characteristics and the controlling evolution of the bottom waters in the deep basins. The slope convection would supply the high-salinity water, causing part of the temperature increase in the deep and bottom waters, while the geothermal heating adds further heat and also stirs the bottom layers.

The Canadian and Eurasian basins are different below mid-depth because of the presence of the Lomonosov Ridge, but within the Eurasian Basin, the Nansen and Amundsen basins also exhibit differences below the sill depth of the Lomonosov Ridge. The Nansen Basin has a lower salinity between 1,500 and 2,500 m due to the inflow of less-saline Arctic Intermediate Water (AIW) and Nordic Seas Deep Water (NDW) through Fram Strait. This is the inflow of deep water from the Greenland Sea postulated by Nansen (1915). The lower salinity extends upward almost to the Atlantic layer in the Nansen Basin, but the Amundsen Basin becomes less saline than the Nansen Basin above 1,500 m because of the presence of the even less-saline Barents Sea inflow branch at these levels in the Amundsen Basin. Below the Barents Sea branch water, the Amundsen Basin has a salinity maximum around 1,500 m (Fig. 4.4) due to returning, saline deep water from the Canadian Basin, which, after crossing the Lomonosov Ridge, perhaps mainly through the intra-basin (Björk et al. 2007), bifurcates at the Morris Jesup Plateau with one part penetrating into the Amundsen and along the Gakkel Ridge but not entering the Nansen Basin. The rest joins the East Greenland Current and exit the Arctic Ocean through Fram Strait (Jones et al. 1995; Fig. 5). Warm saline Canadian Basin Deep Water (CBDW) has also been observed on the Amundsen Basin side of the Lomonosov Ridge close to the Laptev Sea. These volumes appear to be much smaller than those involved in the exchanges through the intra-basin (Rudels et al. 2000a; Woodgate et al. 2001).

4.3.1.8   Eddies

The description of the circulation presented above gives the impression that the water movements below the halocline all occur as streams, or rivers, moving around the basins. However, just as eddies have been observed in the upper layers of the Canadian Basin (e.g. Hunkins 1974, Newton et al. 1974; see Sect. 4.2 above), eddies have also been seen in the deeper layers. The low-salinity lens observed in the

Makarov Basin from *Louis S. St-Laurent* in 1994 (Swift et al. 1997) might well have been an eddy, and on the *Polarstern* 96 section across the Eurasian Basin, 4 or 5 eddies were observed. They were found at different depths and displayed different Θ-S characteristics with respect to the neighbouring stations (Fig. 4.11). The origin of the different eddies could therefore be determined (Schauer et al. 2002a).

The eddy with the largest vertical extent was observed over the Gakkel Ridge and was centred at 1,700 m depth, and its presence was detected vertically over more than 2,000 m. It was a warm and saline anomaly implying the presence of Canadian Basin Deep Water (CBDW) (Fig. 4.11). In the Amundsen Basin, a cold, less-saline eddy was found at 1,000 m depth, showing the presence of almost undiluted Barents Sea branch water. The deeper interleaving thermohaline layers present in the Nansen Basin did not show a continuous change across the front but had a station with extreme properties in the centre, suggesting layers evolving out of (or into) a lens of Barents Sea branch water, just as layers are forming around 'meddies' in the North Atlantic. Two eddies were also found in the Atlantic layer. One was identified in the interleaving structure on the basin side of the Fram Strait branch by having much weaker intrusions, suggesting inflowing rather than returning Fram Strait branch water. The second was a warm, saline fairly deep eddy of Fram Strait branch water in the Amundsen Basin. The CBDW eddy made an especially large impact on the water column, displacing the isolines downward, giving the impression that the homogenous bottom layer thickness had been reduced compared with the neighbouring stations (Fig. 4.9). The eddies, except the one surrounded by interleaving layers, were only observed at single stations and thus had a horizontal extent of less than 30 nautical miles.

How numerous are these eddies and what is their impact on the observed property distributions? How much of the deep salinity maximum in the Amundsen Basin is due to eddies moving from Greenland towards the Laptev Sea, slowly losing their properties, compared to a continuous, direct flow? The eddies appear to move differently than the mean circulation, as is suggested by their passing through ambient water masses without affecting the interleaving thermohaline layers, creating unexpected water mass characteristics in far-off areas (Schauer et al. 2002a).

4.3.2   Change

An appreciation of the variability of Arctic Ocean water masses was implicit in the rhetoric question asked by Quadfasel et al. (1991) 'Is the Arctic warming?' This question was based upon comparison between the climatological temperature field found in the Atlas of the Arctic Ocean (Gorshov 1980) and XBT observations made from the icebreaker *Rossia* in 1990. These indications have been followed by many others during the 'ACSYS' decade. First, the changes were mainly considered as possible evidence that the Arctic now was showing signs of what climate models long have been implying – the global climate is getting warmer and the largest changes are expected at high northern latitudes. While this may well be the case, the

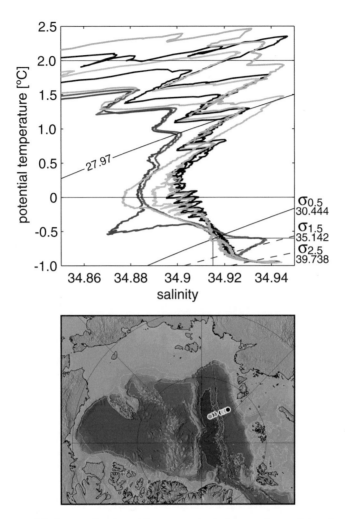

Fig. 4.11 (*Top*) ΘS diagram showing eddies present in the Intermediate and deep layers on the *Polarstern* 96 section across the Eurasian Basin. The *red station* shows the warm and saline Canadian Basin Deep Water; the *blue station* indicates an isolated lens of cold, low-salinity Barents Sea branch water. The *cyan* and *black stations* show an eddy of Barents Sea branch water with warmer and more saline water in both the slope and the basin directions and surrounded by inter-leaving structures like a 'meddy'. This is in contrast to the slope to basin decrease in salinity and temperature seen in the interleaving in the Atlantic layer, although also here an eddy of warm Atlantic water was detected (*green station* located in space between the *black* and *cyan stations*). Finally, an eddy of warm, saline Fram Strait branch water, *yellow station*, was present in the colder, Barents Sea branch-dominated Atlantic layer in the Amundsen Basin (Adapted from Rudels 2009). (*Bottom*) Blow-up map showing the positions of the stations

awareness that conditions in the deep Arctic Ocean could change initiated a search for variability in the Arctic Ocean and lead to efforts to relate this variability to changes in the forcing. These investigations have also gone back in time, examining older data. Did similar changes occur earlier but pass unnoticed? Intense study of the different meteorological conditions has revealed changing, and alternating, dominant climatic regimes in the Arctic, e.g. the Arctic Oscillation (AO) (Thompson and Wallace 1998) and the North Atlantic Oscillation (NAO) (Hurrell 1995), and imposed the question how the different regimes in these oscillations influence and interact with the oceanic circulation regimes in the Arctic Mediterranean (Proshutinsky and Johnson 1997). The strength and importance of the recent changes have also lead to reanalyses of earlier, historic observations to find if similar rapid changes in the water mass characteristics had occurred (e.g. Swift et al. 2005; Polyakov et al. 2004).

4.3.2.1   The Atlantic Layer

An increase in temperature in the Atlantic layer as the one observed by Quadfasel et al. (1991) could be due to a stronger and/or warmer inflow of Atlantic water through Fram Strait. Probably, it is a consequence of both. The higher temperatures were first observed at the continental slope in the Nansen Basin in 1990 (Quadfasel et al. 1991), north of the Laptev Sea in 1993 (Schauer et al. 1997), at the Arlis Plateau in 1993 (Carmack et al. 1995), in the interior of the Makarov Basin and at the Lomonosov Ridge close to the pole in 1993 during the SCICEX-93 expedition with *USS Pago* (Morison et al. 1998), in 1994 from *CCGS Louis S. St Laurent* (Aagaard et al. 1996; Swift et al. 1997) and in the eastern Amundsen Basin from *RV Polarstern* in 1995 (Rudels et al. 2000a) and 1996 (Schauer et al. 2002a). In 1997, a warmer Atlantic layer with intrusions of less-saline and colder Barents Sea branch water was seen north of Fram Strait, returning from, presumably, the Nansen Basin (Rudels et al. 2000b).

The eastward spreading in the boundary current in the Canada Basin has taken longer time, and only in 1998 was warmer Atlantic water observed north of the Chukchi Cap and the Northwind Ridge (McLaughlin et al. 2004). The 'pulse' of warm Atlantic water thus works as a Lagrangian tracer delineating the paths of the Atlantic water in the Arctic Ocean. The warm Atlantic core was here, as almost everywhere else, characterised by inversions in temperature and salinity in the vertical profiles. This indicates mixing with ambient waters, or since the intrusions are very similar to those observed upstream in the boundary current, they could also have formed at the first encounter between the different waters, and these initial, advected, intrusions might precondition the further evolution of the interleaving (see Sect. 4.3.1.6 above).

After a short period of colder Atlantic water entering the Arctic Ocean in the late 1990s, the temperature of the inflowing Atlantic water has again increased. The new inflow pulses have been traced both upstream through the Norwegian Sea and downstream along the continental slope and ridges in the Arctic Ocean (Polyakov et al. 2005; Walczowski and Piechura 2006). Long-term changes in the Atlantic water are examined further in Chap. 6.

4.3.2.2 Redistribution of Water Masses in the Upper Layers

In connection with the presence of warmer Atlantic water in the Eurasian Basin, it was also found that the thickness of the upper, low-salinity surface layer had been reduced, and the Atlantic layer was closer to the sea surface and the ice (Morison et al. 1998). This vertical shift of the temperature and salinity profiles accounts for a large part of the change relative to the climatology. A similar change was also seen in the Makarov Basin (Carmack et al. 1995; Morison et al. 1998), where both the salinity and the temperature were higher than the climatological values. Changes were also seen in the nutrient and oxygen distributions. The high nutrient and low oxygen concentrations observed from the ice camps in the 1960s and 1970s (e.g. Kinney et al. 1970; Jones and Anderson 1986) were largely absent, when the icebreakers in the 1990s started to penetrate into the Makarov Basin. Excess of low-salinity surface water, by contrast, was found in the Canada Basin (McPhee et al. 1998; MacDonald et al. 1999), where the salinity of the upper layer was reduced. Carmack et al. (1995), McLaughlin et al. (1996) and Morison et al. (1998) explained this as a shift of the front between upper waters deriving from the Atlantic and the Pacific Oceans. The Pacific upper water lens no longer extended to the Lomonosov Ridge but was confined to the Canada Basin, and the Atlantic derived upper layer was in contact with the surface up to the Mendeleyev Ridge. The Pacific water, being forced into the Beaufort Sea and towards the North American slope and shelf, was becoming more prominent in the Lincoln Sea north of Ellesmere Island (Newton and Sotirin 1997). This could indicate an increased eastward boundary flow along the continental slope towards Nares Strait and Fram Strait, compensating the shift in Pacific water on the Siberian side of the Arctic Ocean. The draining of the Pacific water through the western channels of the Canadian Arctic Archipelago might also have increased.

The frontal shift is a manifestation of the contraction of the anti-cyclonic Beaufort Gyre, which is maintained by the high-pressure cell that develops during winter over the Beaufort Sea. When the advection of warmer air from lower latitudes is strong, especially through the Atlantic sector, the high-pressure cell weakens, and the oceanic anticyclone diminishes and is forced closer to the American continent, while the cyclonic circulation is strengthened. This leads to changes in the circulation of the low-salinity shelf waters on the Eurasian shelves (Proshutinsky and Johnson 1997; Polyakov et al. 1999; Polyakov and Johnson 2000; Maslowski et al. 2000).

In the anti-cyclonic regimes, the shelf water crosses the shelf break in the eastern Laptev Sea and enters the Amundsen Basin, but in the cyclonic regime (Proshutinsky and Johnson 1997) that has dominated during the 1990s, the shelf water has instead moved into the East Siberian Sea and then entered either the Makarov Basin or the Canada Basin (Maslowski et al. 2000). Since the Siberian shelves receive the runoff from the large rivers, this has led to increased salinities in the upper layers of the Amundsen and partly in the Makarov Basin and, likely, to lower salinities in the Canada Basin. The latter effect is enhanced by the closer confinement of the low-salinity Pacific water to the Canada Basin.

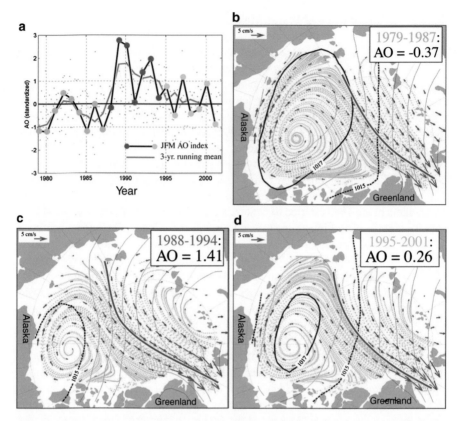

Fig. 4.12 Different circulation modes in the Pacific water. (**a**) Time series of the Arctic Oscillation (*AO*) index for January–March and its running means. Monthly values are shown as *grey dots*. Three periods are distinguished, 1979–1987 (*green*, AO negative), 1988–1994 (*red*, AO positive) and 1995–2001 (*blue*, AO weakly positive). (**b**, **c**, and **d**) show the mean sea-ice motion (vectors) and sea level pressure (contours). The *blue streaks* are the Lagrangian paths forced by the mean vector fields. The magenta line indicates the zero vorticity contour (From Steele et al. 2004)

Steele et al. (2004) traced the distribution of the upper temperature maxima of the Pacific water, deriving from the Alaskan Coastal Water (ACW) and Bering Sea Summer Water (BSSW), respectively (see above Sect. 4.3.1.3). They concluded that during a negative or weakly positive AO state with a well-developed Beaufort Gyre, the BSSW becomes trapped in the Beaufort Gyre and the upper waters north of Ellesmere Island and Greenland derive from the Siberian branch of the Transpolar Drift. In the strong positive AO state with a diminished Beaufort Gyre, the Siberian branch starts farther to the east, and the BSSW becomes incorporated into the Transpolar Drift and passes directly to the Lincoln Sea. It is also possible that the shifting of the low-salinity Beaufort Gyre surface water lens towards the North American continent strengthens the boundary flow of the Alaskan Coastal Current (ACC) to the Lincoln Sea (Steele et al. 2004) (Fig. 4.12).

4.3.2.3   Changes in the Halocline

The deflection of the low-salinity shelf water occurring in the cyclonic regime has caused a weakening and even occasional disappearance of the two-layer Polar Mixed Layer – halocline structure in parts of the Amundsen Basin. This was first noted by Steele and Boyd (1998), who compared observations close to the North Pole from SCICEX cruises in 1993 and 1995 and from *Louis S. St-Laurent* 1994 with the measurements from *Oden* in 1991. This change implies that warm water of the Atlantic layer is in direct contact with and could be entrained into the winter-mixed layer in the Amundsen Basin as well as in the Nansen Basin, which might lead to melting of ice, or rather, could reduce the ice formation rate (Martinson and Steele 2001). Whether this important implication takes place or not, depends upon the mixing mechanisms. If the main stirring is due to mechanically generated turbulence, heat can continuously be brought to the surface. Some heat would be used to melt ice; the rest would be lost to the atmosphere (Walin 1993; Rudels et al. 1999b). If the mixing is due to haline convection, as a consequence of ice formation, the heat brought into the winter-mixed layer causes ice to melt or at least lowers the ice formation rate. This reduces the convection (Martinson 1990; Martinson and Steele 2001). This in turn weakens the entrainment and the vertical heat transfer from the Atlantic layer to the mixed layer and to the ice. In either case, mechanical mixing or convection, the vertical heat flux to the ice only occurs in winter, when the strong stratification induced by summer ice melt has been removed by freezing.

The presence of the ice cover and the heat advected to the Arctic Ocean by the Atlantic water are both important components in the heat balance of the Arctic, and the formation and maintenance of the halocline as an insulating layer has therefore drawn some attention. The change in the upper layer Θ-S distribution has been described as 'the retreat of the cold halocline *layer*' (Steele and Boyd 1998) This, perhaps, to take into consideration that the halocline in the Arctic Ocean is a water mass in its own right, having a vertical extension of 50–200 m, not being just a salinity gradient (Coachman and Aagaard 1974) (see Sect. 4.2 above). To speak of the retreat of the halocline layer might then suggest that the formation of halocline water has diminished. This is not the case. In the 1990s, a winter-mixed layer has been present not only in the Nansen Basin but also in the central Amundsen Basin up to the Lomonosov Ridge, and perhaps also in the Makarov Basin. What was missing was the upper, cold low-salinity shelf water because it had been deflected farther to the east, and no cold salinity gradient had formed above the winter-mixed layer. This prevents the winter-mixed layer from becoming a halocline water mass, but it does not mean that the production of potential halocline water has diminished. The 'retreat of the cold halocline layer' had thus perhaps better been rendered 'the diversion of the low-salinity surface water'.

Recent observations have shown the return of the low-salinity shelf water to the central Amundsen Basin, and a cold halocline has again been established between the shelf water layer (the Polar Mixed Layer) and the winter-mixed layer (Björk et al. 2002; Boyd et al. 2002). Anderson et al. (2004) determined from total alkalinity the presence of river water and sea ice melt water in the upper layers of the

Eurasian Basin between 1987 and 2001 using data from 5 icebreaker cruises (see also Sect. 4.3.1.5 above). They found that in 1987 (Anderson et al. 1989) the river water front, defined as the line along which 5% of the surface water was river water, reached to the Gakkel Ridge. The extent of the river water layer then diminished from 1991 and 1994 to a minimum in 1996, when the front was located over the Lomonosov Ridge and river water was practically absent in the Amundsen Basin. In 2001, the front again had moved into the Amundsen Basin but not as far as the Gakkel Ridge (Fig. 4.10).

Another feature, which may be connected with the changed distribution of the river water, was observed in 1998 in the Canada Basin. At this time, a cold, close to freezing, highly oxygenised temperature minimum with salinity 33.9 was found close to the northern part of the Chukchi Cap (McLaughlin et al. 2004). McLaughlin et al. (2004) suggest that this halocline water mass derives from a new source of lower halocline water, the East Siberian Sea. Before 1990, the East Siberian Sea had been ignored as a possible source of halocline water because of its low salinity (Aagaard et al. 1981). During the 1990s, when most of the runoff from the western Siberian rivers entered the East Siberian Sea and not the Amundsen Basin, the salinity could be even less, and the East Siberian Sea as a source for lower halocline water appears unlikely. However, the change in the river water distribution might have caused the low-salinity surface water to be absent also from parts of the Makarov Basin. This would extend the area without a cold halocline and with convection down to the thermocline from the Nansen Basin and parts of the Amundsen Basin into the Makarov Basin, and it would maintain the cold winter-mixed layer that is the embryo of the Fram Strait branch halocline in these basins (Rudels et al. 1996; Rudels et al. 2004a). The location of the saline temperature minimum north of the Chukchi Cap rather than between the Chukchi Cap and the continental slope is in agreement with the different paths of the Barents Sea branch and the Fram Strait branch halocline waters as suggested by Rudels et al. (2001, 2004a).

In the Nansen Basin, the Polar Mixed Layer – halocline structure is absent during both the anti-cyclonic and the cyclonic regimes. Consequently, the area where the Atlantic layer and the ice cover were in contact did not increase dramatically during the 1990s and therefore probably did not contribute much to the reported reduction of the ice cover extent and thickness in the Arctic Ocean (McPhee et al. 1998; Rothrock et al. 1999; Winsor 2001; Holloway 2001; Holloway and Sou 2002).

4.3.2.4 The Intermediate Layers

The variability observed in the 1990s was not confined to the upper and Atlantic layers. Changes were also seen at intermediate depths. The less-saline Barents Sea inflow branch became more prominent in the Eurasian Basin between 1991 and 1995/1996, and a distinct salinity minimum was present below the Atlantic layer in 1995 (see e.g. Fig. 4.7). This change could have several causes. A larger precipitation in the Norwegian Sea and the Barents Sea, as a result of the strong positive NAO state dominating in the early 1990s, would lower the salinity of the Barents

Sea branch, and the warmer Atlantic water of the Fram Strait branch could be less dense than before, leading to a, relatively, deeper input from the St. Anna Trough that penetrated below rather than into the Atlantic layer, creating the mid-depth salinity minimum (Rudels and Friedrich 2000). The low-salinity Barents Sea branch water mass could be traced in the interior Nansen Basin and Amundsen Basin and along the Lomonosov Ridge. It was also seen entering the Makarov Basin in the boundary current (Rudels et al. 2000a).

The cold, less-saline intermediate water found in the Makarov Basin during AO94 expedition (Swift et al. 1997) and the cooling and freshening of the intermediate waters observed in the Canada Basin between 1989 and 1995 (McLaughlin et al. 2002, 2004) indicate an inflow of Barents Sea branch intermediate water into the interior of the Makarov and Canada basins. However, this colder, less-saline water must have left the Barents Sea earlier than the low-salinity water observed at the Lomonosov Ridge and north of the East Siberian Sea in 1995. Are we then observing a trend in decreasing salinities in the intermediate waters supplied by the Barents Sea, or are we just beginning to observe the variability of the Barents Sea branch inflow characteristics?

A lower salinity at the Lomonosov Ridge and in the boundary current changes the characteristics of the intermediate water of the Makarov and Canada basins by lateral advection and mixing. It will also affect the slope convection, since the entrained water will be colder and less saline. This gradually leads to different characteristics of the intermediate and deep waters in the Canadian Basin, even if the formation of dense shelf water and the entrainment rate do not change (Rudels et al. 2000a). Most likely they do, and the deep Makarov and Canada Basin water columns will never attain a state, where the advection in the boundary current, the slope convection and the exchanges between slope and basin interior are in equilibrium.

4.3.3 Exchanges

The exchanges, not just of volume but also of heat and freshwater (salt), have a significance extending beyond pure bookkeeping. The oceanic transports of heat have impact on the Arctic climate, and the freshwater balance of the Arctic Ocean is important for the existence of the ice cover as well as for the dense water formation, in the Arctic Ocean and in the downstream convection sites: the Greenland Sea, the Iceland Sea and the Labrador Sea. Through its influence on the deep-water formation, the freshwater transport affects the Meridional Overturning Circulation and thus the global climate. Much of the recent work on the exchanges between the Arctic Ocean and the world ocean has been carried out within the international Arctic–Subarctic Ocean Fluxes (ASOF) programme, which has acted as an umbrella for several international and national studies. Results of these efforts have now been published in the volume 'Arctic–Subarctic Ocean Fluxes' edited by Dickson, Meincke and Rhines (Dickson et al. 2008a). The summary of the transport estimates through the different passages given below represents the state of knowledge at the early stage of ASOF. Rather than trying to update the present section, we refer the reader to the ASOF volume for more recent results. The North Atlantic Meridional Overturning Circulation (MOC) is discussed in Chap. 6 with reference to some diagnostic model calculations.

4.3.3.1   Fram Strait Transports

As the deepest connection between the Arctic Ocean and its surrounding seas, the passage between Greenland and Svalbard – Fram Strait –, has long been recognised as being of prime importance not only for the hydrography of the Arctic Ocean but also for the climate of the Arctic. It is the main (>90%) exit for the sea ice formed in the Arctic Ocean. It is also the passage where most of the oceanic sensible heat is advected into the Arctic Ocean. Although this transport is not large enough to keep the Arctic Ocean free of ice, as was speculated in the mid nineteenth century, the heat transport in the narrow meridional band occupied by the West Spitsbergen Current is comparable to the atmospheric heat transport across 80°N. In addition, the export of ice, or latent 'cold', further emphasises the importance of Fram Strait for the heat balance of the Arctic.

Fram Strait is also the only deep passage through which the Arctic Ocean deep waters may be advectively renewed, and for a long time, it was assumed that the Arctic Ocean deep waters derived from the Greenland Sea. When it was found that the Arctic Ocean deep waters to a large extent are locally produced, or at least locally modified, Fram Strait nevertheless remained the only passage, where an exchange of deep water could occur. A study of the water masses present in Fram Strait thus provides information of the integral effects of the water mass (trans) formation processes in the Arctic Ocean.

The main focus, however, has always been the transports. The inflow of Atlantic water was first estimated by geostrophic computations on hydrographic sections extending from Svalbard to the ice edge (e.g. Timofeyev 1962). Some hydrographic sections extended across the strait to the Greenland side, e.g. Ob in 1956 and Edisto in 1965 (Palfrey 1967). The northward transport was estimated to be 3–4 Sv (Timofeyev 1962). The outflow of polar surface water in the East Greenland Current has been more difficult to determine but has commonly been estimated to be 1–2 Sv (e.g. Mosby 1962; Vowinckel and Orvig 1970).

In 1971–1972, the first yearlong current measurements in the West Spitsbergen Current were obtained (Aagaard et al. 1973). They showed a much larger northward transport, 8 Sv, than previously assumed. The larger transport was due to a strong barotropic component of the current, which could not be assessed by the geostrophic computations. These observations formed the backbone of the mass, salt and heat budget for the Arctic Ocean made by Aagaard and Greisman (1975), where the inflow of 7.1 Sv of Atlantic water was assumed balanced by an outflow of 1.8 Sv of polar water and 5.3 Sv of returning, colder Atlantic water. The current measurements were continued for several years, and in 1984–1985 and 1985–1986, current metre arrays extending across the entire strait were deployed. The observations showed a westward turning of the current in the central part of the strait, and the southward flow in the East Greenland Current above 600 m was estimated to be 3 Sv, 1 Sv polar water and 2 Sv Atlantic water (Foldvik et al. 1988).

In the 1980s, several complete sections were obtained across Fram Strait, and geostrophic transports were computed. The obtained transports of Atlantic water (1–2 Sv) were generally much lower than those estimated from direct current measurements, and a large recirculation of the West Spitsbergen Current was found in the strait (Rudels 1987; Bourke et al. 1988). That warmer Atlantic water is moving south beneath the polar water in the East Greenland Current has been known for a century, and its source had been deduced as the recirculation of the Atlantic water of the West Spitsbergen Current (Ryder 1891).

To allow for the unknown reference velocities, Rudels (1987) used a variational approach with constraints on the mass and salt transports imposed by requirement of mass and salt balance, estimates of the transports through the other passages and constraints on the deep circulation based on considerations of the Θ-S structure of the deep waters. The reference velocities were then calculated by minimising the total kinetic energy of the flow field. The obtained transports were about half of those estimated by Aagaard and Greisman (1975): The northward transport of Atlantic water was close to 2 Sv with 1 Sv recirculating in the strait. The applied mass balance also required a net southward flow of 1 Sv through the strait. Schlichtholz and Houssais (1999a, b) used the entire hydrographic data set from the MIZEX 1984 experiment but no other data and determined the circulation in the strait using a 3-dimensional variational model. They obtained a still smaller northward transport; less than 1 Sv of Atlantic water was found to enter the Arctic Ocean. By the end of the 1980s, the general opinion was that the transport estimates based on the direct current measurements had been too large and that the recirculation in the strait had been underestimated. Aagaard and Carmack (1989) in their new mass and freshwater balance of the Arctic Ocean gave 1 Sv as the inflow of Atlantic water through Fram Strait.

After a few years with less activity, the study of the Fram Strait exchanges took new life by the end of the 1990s. In connection with the VEINS (Variability of Exchanges in the Northern Seas) programme, a current metre array comprising 14 moorings was deployed across the strait along the sill at 79°N. This mooring array has, with small changes, additions and reductions, been in position ever since, first as a part of the ASOF programme and now within the DAMOCLES (Developing Arctic Modelling and Observing Capabilities for Long-term Environmental Studies). Hydrographic sections have been taken at least once every year in connection with the replacement of the moorings, but often more than one section have been taken in a year.

The results from the current measurements confirm the large northward transport and the barotropic flow obtained by the earlier moorings in the West Spitsbergen Current (Aagaard et al. 1973). The total, northward and southward, monthly mean transports have been around 10 Sv, with a net southward flow ranging from 1 to 3 Sv in the different years (Fahrbach et al. 2001; Schauer et al. 2004; Schauer et al. 2008). The estimated northward transport of Atlantic water has been around 4 Sv.

The transports have shown variations not only between the years but also seasonally with the strongest flow in winter. The northward heat flux has also varied between the years, but in 1999, the amount of heat entering the Arctic Ocean through Fram Strait was estimated to be sufficient to generate a temperature increase in the Atlantic layer comparable to that observed in the early 1990s (Schauer et al. 2004; Schauer et al. 2008).

The different streams of the West Spitsbergen Current have also become better known. In addition to the inflow stream over the shelf north of Svalbard, a second stream follows the western and northern slope around the Yermak Plateau. This stream transports Atlantic water and also intermediate and deep waters from the Nordic Seas into the Arctic Ocean (Aagaard et al. 1987; Rudels et al. 2000b; Rudels et al. 2005). The flow around the Yermak Plateau may be as strong as that closer to Svalbard (Schauer et al. 2004). Gascard et al. (1995) found from sub-surface drifters that a part of the West Spitsbergen Current may also cross the Yermak Plateau into the Sofia Deep along a smaller trough in the plateau. The recirculation in Fram Strait is clearly seen in the water mass properties, but it has been difficult to chart the recirculating branch(es), and it may be that the exchange between the West Spitsbergen Current and the East Greenland Current mainly takes place by the transfer of eddies (Gascard et al. 1988; Gascard et al. 1995; Rudels et al. 2005).

During VEINS and ASOF, the West Spitsbergen Current has been extensively explored during yearly expeditions with the sailing research vessel *Oceania*. Recently, the hydrography work has been augmented by current observations using ship-mounted ADCP as well as lowered (L)ADCP. These snapshots also show the large difference between geostrophically computed velocity and direct current measurements. Especially, the LADCP observations indicate transports in the West Spitsbergen Current that are even larger than those estimated from the mooring array (Walczowski et al. 2005). One complete CTD section made by *Polarstern* was complemented by current measurements from a ship mounted ADCP. Also, this estimate showed high transport and a net southward flow, 11.5 Sv to the north and 13.1 Sv to the south (Cisewski et al. 2003).

In the context of the ASOF programme, geostrophic transport computations were made also on historic sections. These, surprisingly, also show an increase in transports between the early 1980s and the late 1990s. A part, but not all, of this increase could be due to the closer station spacing on the more recent sections, which allows for the resolution of smaller eddies, which will add to the total northward and southward flows. The net southward flow ranged from 1.5 to above 4 Sv (Rudels et al. 2004b; Rudels et al. 2008). A net outflow is expected because of the Barents Sea inflow, but more than 3 Sv appears unrealistic. However, geostrophic computations, here combined with constraints on the deep salt and mass fluxes, are always susceptible to incomplete station spacing, not resolving the eddies, the effects of the sections not being synoptic, and to the unknown reference velocity. It is also likely that the geostrophic computations more severely underestimate the transports in the West Spitsbergen Current than in the more baroclinic East Greenland Current, resulting in a too large net outflow.

Rudels et al. (2000b) compared the water mass distributions on hydrographic sections taken in 1984 and in 1997 and found that the area occupied by recirculating

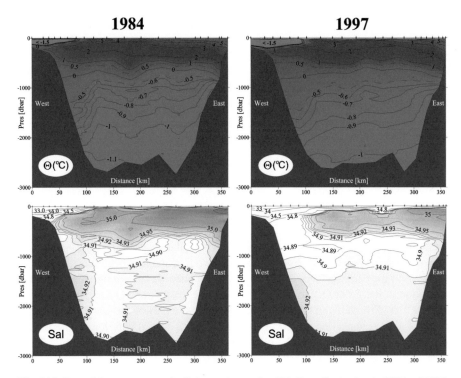

Fig. 4.13 Potential temperature and salinity sections at the sill in Fram Strait taken in 1984 and 1997. The recirculation, indicated by the lateral extent of the high salinity Atlantic water, was more confined to the central part of the strait in 1997 than in 1984, allowing a broader passage for the Arctic Ocean water masses to exit the Arctic Ocean and suggesting a closer coupling between the Arctic Ocean and the Nordic Sea in the 1990s. The outflow of Arctic Ocean intermediate and deep waters are seen by the larger separation of the +0.5°C and −0.5°C isotherms and the higher salinity at the Greenland slope. Recent years have shown a return to the 1980s situation with larger recirculation in the strait and the saline Atlantic water reaching to the Greenland slope (Adapted from Rudels 2001)

Atlantic water in 1984 extended up to the Greenland continental slope, while in 1997, it was more confined to the central part of the strait, opening a free passage for the Atlantic, intermediate and deep water of the Arctic Ocean (Fig. 4.13). This was interpreted as a change from a state with strong recirculation in the strait to one where more Atlantic water enters the Arctic Ocean and takes part in the circulation of the boundary current around the Arctic Ocean basins. A similar interpretation was made by Blindheim et al. (2000) in a discussion of the recent freshening trend occurring in the Nordic Seas, which could partly result from a larger presence of Arctic Ocean waters at the expense of the Recirculating Atlantic Water (RAW) in the Nordic Seas.

The ice export, although temporally varying, appears to have a fairly steady mean transport around 0.09 Sv (e.g. Dickson et al. 2008a). The estimates of the Fram Strait export of liquid freshwater, by contrast, seems to vary between less than half of the ice export (Aagaard and Carmack 1989) and as large, and perhaps larger, than the ice export (Meredith et al. 2001). The presence in the East Greenland Current and the export of Pacific water through Fram Strait are also changing with time. After being present in the 1980s and 1990s up to the early 2000 (Anderson and Dyrssen 1981; Jones et al. 2003), Pacific water was not observed in 2004 (Falck et al. 2005). Falck et al. (2005) suggest that this change could be caused by changes in the state of the Arctic Oscillation, a more negative AO index would indicate a larger Beaufort Gyre and a westward shift of the Transpolar Drift, confining the Pacific water to north of the Canadian Arctic Archipelago (Steele et al. 2004, see also Fig. 4.12).

4.3.3.2 The Barents Sea

The widest entrance to the Arctic Ocean is over the Barents Sea. It is one of the Arctic shelf seas and thus considered a part of the Arctic Ocean. Its southern and western half is ice free and dominated by inflowing Atlantic water from the Norwegian Sea, while the northern and eastern half is more polar, and ice covered a large part of the year. In the early Russian literature, this difference was taken into account in the definition of the Polar Ocean, which included the deep Arctic Ocean and the Kara, Laptev, East Siberian and Chukchi Seas but only the part of the Barents Sea lying north of a line connecting the north-eastern part of Svalbard with the northern cape of Novaya Zemlya, excluding the mainly ice-free southern part of the Barents Sea.

The Norwegian Atlantic Current bifurcates when it reaches the latitudes of northern Norway, and the eastern stream enters the Barents Sea in the Bear Island Trough together with the Norwegian Coastal Current. The low-salinity ($S \approx 34.4$) coastal current flows eastward close to the coast, while the Atlantic water again splits with one branch continuing eastward south of the Central Bank in the middle of the Barents Sea and the second moving north into the Hopen Deep west of the bank. As it encounters the sill between Edgeøya and the Great Bank, the stream separates into 3 parts, one part turns eastward north of the Central Bank, a smaller part crosses the sill into the northern Barents Sea and the third part recirculates in the Hopen Deep towards the Norwegian Sea (Loeng 1991; Pfirman et al. 1994).

The Atlantic water is cooled and freshened by net precipitation and by ice melt. Few rivers enter the Barents Sea, and the main freshwater source is the Norwegian Coastal Current, carrying low-salinity water from the Baltic Sea augmented by run-off from the Norwegian coast. The cooling increases the density of the Atlantic water, and it occupies the main part of the water column above the deeper depressions. However, the densest water is formed over the shallow shelf areas in Storfjorden (see Sects. 4.2 and 4.3.1.5 above), close to Novaya Zemlya, and over the Central Bank. Here, less-saline surface water deriving, west of Novaya Zemlya from the Norwegian Coastal Current and from ice melt over the Central Bank, is cooled to freezing temperature. Ice is formed and the released brine convects into the water column. The convection eventually reaches the bottom and creates an almost homogenous water column at freezing temperature. Over the Central Bank, the water becomes less saline, but denser, than the Atlantic water in the Hopen Deep (Quadfasel et al. 1992), but on the shallow area west of Novaya Zemlya, the water may occasionally become more saline than the Atlantic water in the eastern depression. This higher salinity is mainly due to the presence of a lee polynya close to Novaya Zemlya, while the ice remains above the Central Bank and the convection just homogenises the underlying water column and creates a dense anti-cyclonic vortex above the bank, trapping the ice. The water eventually drains from the shallow areas into the deeper depressions, entraining some Atlantic water forming a cold, dense bottom layer (Quadfasel et al. 1992).

Sea ice is advected with the wind and the surface currents and will ultimately melt, often at a position different from its formation. In the northern Barents Sea, this creates a less-saline upper layer on top of the Atlantic water. The stratification due to the separator effect caused by freezing and melting sea ice is augmented by the advection of sea ice and low-salinity surface water from the Arctic Ocean and from the Kara Sea into the northern Barents Sea. Aagaard and Woodgate (2001) proposed that ice melt was fundamental in lowering the salinity of the Atlantic water creating the less-saline deep-water masses of the Arctic Ocean that eventually also supply water to the Greenland–Scotland overflow. However, extremely strong wind mixing and/or a partial refreezing and convection are needed to mix the melt water against the stratification into the underlying Atlantic water.

The Atlantic water entering the Barents Sea from the west thus becomes transformed into three different waters: (1) less-saline and less-dense upper layer, (2) cooled and denser Atlantic water and (3) cold, dense and brine-enriched bottom water. The main outflow from the Barents Sea occurs between Franz Josef Land and Novaya Zemlya and then continues into the Arctic Ocean via the St. Anna Trough. A considerable mixing takes place between the denser water of the two inflow branches, but the less-dense upper layer of the Barents Sea branch remains distinct and eventually evolves into Barents Sea branch lower halocline water (Rudels et al. 2001, 2004a). Some less-saline water, mainly Norwegian Coastal Current water, enters the Kara Sea south of Novaya Zemlya, and a weaker outflow of dense water to the Arctic Ocean takes place in the Victoria Channel to the west of Franz Josef Land (Rudels 1986a; Schauer et al. 1997). The water that recirculates in the Hopen Deep returns as denser and colder water to the Norwegian Sea via the northern part of the Bear Island Trough (Blindheim 1989; Quadfasel et al. 1992). Here, an outflow of cold, less-saline and less-dense Arctic water is also present along the Svalbard Bank (Loeng 1991). This is partly a continuation of the Persey Current to the south of Franz Josef Land, which flows from the Kara Sea westward along the polar front and partly a southward transport of polar surface water from the Arctic Ocean. The waters entering the Norwegian Sea from the Barents Sea are carried northward into the Arctic Ocean by the West Spitsbergen Current.

In the early 1900s it was assumed that the main inflow to the Arctic Ocean occurred through the Barents Sea (Nansen 1906), and in Russian literature an inflow of 1.5 Sv was estimated (Nikiforov and Shpaiker 1980). Later, the strength of the Barents Sea inflow was assumed to be less, and Aagaard and Greisman (1975) gave a total flow of 0.7 Sv into the Kara Sea between Franz Josef Land and the Eurasian continent, only 10% of their estimated inflow through Fram Strait. In the 1980s, the transport estimates for the Barents Sea again began to increase.

Rudels (1987) derived 1.2 Sv from heat budget considerations, about the same as his estimate of the inflow of Atlantic water through Fram Strait, and Blindheim (1989) obtained a net transport of 1.9 Sv into the Barents Sea from a 3-week record from a current metre array deployed in the Bear Island Channel.

Loeng et al. (1993, 1997) and Schauer et al. (2002b) analysed the measurements from a yearlong deployment of a mooring array between Franz Josef Land and Novaya Zemlya in 1991. They found a net inflow to the Kara Sea of about 2 Sv. The largest transport occurred in the deeper southern part of the passage, and to the north, a weak westward flow of Atlantic water deriving from the Fram Strait inflow branch was observed, suggesting an anti-cyclonic circulation around Franz Josef Land. Schauer et al. (2002a, 2002b) also examined the hydrography in the St. Anna Trough and in the passage between Franz Josef Land and Novaya Zemlya and concluded that the high-density cold water present in the deepest part and at the eastern flank of the St. Anna Trough was a continuation of the flow passing eastward between Franz Josef Land and Novaya Zemlya.

The direct current measurements between Franz Josef Land and Novaya Zemlya have not been repeated yet, but since 1997, a current metre array has been deployed within the VEINS and ASOF programme in the western Barents Sea opening, being replaced every year. It has revealed a strong barotropic component and large short-time variability in the transport of Atlantic water. The flow is mainly eastward with a net transport around 1.5 Sv. However, not only days but also periods extended over almost a month (mostly April) with westward net transport through the Barents Sea opening have been recorded (Ingvaldsen et al. 2004a; Ingvaldsen et al. 2004b).

To get the total inflow over the Barents Sea, the transport of 0.7 Sv of the Norwegian Coastal Current has to be added (Blindheim 1989). A total transport of 2.2 Sv is in agreement with Blindheim (1989) and Loeng et al. (1993) and Schauer et al. (2002b). The Norwegian Coastal Current, as it enters the Barents Sea, still carries about half of the freshwater exported from the Baltic Sea and a large fraction of the runoff from the Norwegian coast. Its low salinity (34.4) and comparably large transport make it the largest freshwater source in the Barents Sea, and it contributes significantly to the freshwater balance of the other Siberian shelf seas and of the Arctic Ocean.

4.3.3.3 Bering Strait

Bering Strait connects the two opposite poles of the world ocean, the deeply venti-lated North Atlantic and the almost stagnant North Pacific, and allows for a direct communication of surface waters between the two oceans. The flow is northward, into the Arctic Ocean, although southward flow may take place in the presence of strong northerly winds. In 'The Oceans', Sverdrup estimated the transport to be 0.3 Sv (Sverdrup et al. 1942). Russian work during the mid century indicated a larger transport and a significant seasonal variability, 1.2 Sv in summer and 0.4 Sv in winter (Maksimov 1945; Fedorova and Yankina 1964). The early direct measurements of the currents in and north of the strait suggested a somewhat larger flow 1.0–1.5 Sv, and no seasonal variation was reported (see e.g. Coachman et al. 1975).

The driving force for the flow was early realised to be the higher sea level in the North Pacific as compared to the Arctic Ocean, giving a pressure head that forces an essentially barotropic flow through the strait. The force balance becomes one-dimensional between the downstream pressure gradient, the field acceleration and the retarding friction (Coachman and Aagaard 1966). Coachman and Aagaard (1966) related the sea level slope to the local wind field and found a relation between the wind measured in Nome, Alaska and the sea level slope and thus the transport.

Stigebrandt (1984) took a more global view and showed that the cause for the higher sea level in the Pacific was the less-saline and less-dense water column in the North Pacific as compared to the water column in the North Atlantic. Assuming a level of compensating pressure at 1,600 m, the upper boundary of the North Atlantic Deep Water, which moves from the North Atlantic towards the Pacific, he obtained a value of the sea level slope and estimated the transport through Bering Strait to be 1.5 Sv, consistent with the current measurements. The higher sea level in the Pacific has partly been explained by the westward atmospheric transport of water vapour from the North Atlantic across the Isthmus of Panama to the North Pacific, making the Pacific water column less and the Atlantic water column more saline (Weyl 1968). The transport of low-salinity Pacific water through Bering Strait is thus a shortcut to correct this freshwater imbalance.

The observational work in the Bering Strait area has continued and has been especially intense in the last years. The more recent transport estimates vindicate the early Russian results (Coachman and Aagaard 1981; Coachman and Aagaard 1988; Roach et al. 1995). The transport varies seasonally with the strongest flow in summer. The yearly mean transport through the strait is about 0.8 Sv, less than what was believed in the 1970s and 1980s. The annual variability of the transport appears to be small, but the salinity as well as the temperature of the inflow varies both seasonally and annually (Roach et al. 1995). Woodgate et al. (2005) deduced from yearlong current measurements in 1991–1992 that the Pacific water leaves the Chukchi Sea at four main locations: (1) through De Long Strait into the East Siberian Sea, (2) along Herald Canyon, (3) through the Central Channel and (4) along the Barrow Canyon. These flows appeared to be equal in strength but with different properties. The outflow to the East Siberian Sea is likely compensated, in volume if not in salt, by an inflow of the Siberian Coastal Current. From current observations between 1990 and 2004 Woodgate and Aagaard (2005) also reported on the transport of the Alaskan Coastal Current, a strong, maximum velocity 1.7 m/s, and narrow low-salinity surface current present along the Alaskan coast from early summer to late fall. Woodgate and Aagaard (2005) estimated that this current, previously ignored, could add about 0.01–0.02 Sv or 20% to the freshwater transport through Bering Strait.

4.3.3.4 The Canadian Arctic Archipelago

The narrow channels of the Canadian Arctic Archipelago connect the Arctic Ocean with the Baffin Bay and the Labrador Sea and provide a second outlet for the less-dense Arctic Ocean waters to the North Atlantic, which does not pass through the

Nordic Seas. In the first half of the 1900s, it was assumed that the outflow through the archipelago was small, and in his mass balance for the Arctic Mediterranean, Sverdrup ignored the flow through the Canadian Arctic Archipelago and assumed that the water masses in Baffin Bay derived from the Labrador Sea with a deep water renewal by local convection (Sverdrup et al. 1942).

Baffin Bay is a deep (>2,300 m) narrow bay connected to the Labrador Sea and the North Atlantic by the 600 m deep Davis Strait. Baffin Bay is ice covered most of the year, and its upper layer has a temperature minimum close to freezing around 100 m with a salinity of ~33.6, indicating the depth of local winter convection. Below this low-salinity layer, a temperature maximum with temperatures of 2–3°C and a salinity of 34.5 is encountered, showing Atlantic water entering Baffin Bay from the south in the West Greenland Current. Below sill depth, the temperature decreases to –0.45°C at the bottom with a salinity of 34.45 (Riis-Carstensen 1936; Kiilerich 1939).

The channels in the Canadian Arctic Archipelago are shallow, 100 m to 230 m, and covered by land-fast ice most of the year, but the low-salinity surface water present in Baffin Bay suggests an inflow from the Arctic Ocean, and Sverdrup's view was later challenged. Several estimates, from geostrophic computations and occasionally from current observations (Day 1968; Sadler 1976), were made during the 1950s and 1960s, either in the individual straits or in Baffin Bay and Davis Strait, and have been summarised by Bailey (1957), Collin (1962), Collin and Dunbar (1963), Muench (1971) and Coachman and Aagaard (1974). A net southward flow of 1–2 Sv was obtained but with large temporal variations and with indications that changes in the transports through the different passages may compensate each other.

Bailey (1956) noticed that the characteristics of the Baffin Bay deep and bottom water were similar to those observed at 250 m in the Beaufort Sea and suggested that the deeper layers were supplied from the Arctic Ocean. This view has been contested by Bourke et al. (1990) and Bourke and Paquette (1991) who, during an expedition in 1986, did not observe any connection between the water found in the northern Nares Strait with the bottom water in Baffin Bay even though the characteristics were similar. They proposed, as did Sverdrup, that the Baffin Bay deep and bottom water is formed by local convection. Since neither an advective renewal of the Baffin Bay bottom water nor local convection has been observed in Baffin Bay, the question is still open. Should the Arctic Ocean supply the deep and bottom water of Baffin Bay, it would primarily be a contribution from the Barents Sea branch halocline water (Rudels et al. 2004a).

Simple model estimates of the outflow through the Archipelago have been provided by Stigebrandt (1981), who considered the entire Arctic Ocean and assumed a two layer geostrophically controlled outflow through Fram Strait and the Canadian Arctic Archipelago. He found the outflow through the archipelago to be larger (2 Sv) than that through Fram Strait (1.5 Sv). Rudels (1986b) assumed geostrophic two-layer flow through the channels in the archipelago as well as through Davis Strait and combined this with a heat, mass and salt balance for Baffin Bay, finding an outflow in the upper layers of 0.7 Sv and postulating an additional deeper outflow of 0.3 Sv. Steele et al. (1996) estimated the freshwater balance in different parts of the Arctic Ocean, assuming geostrophic transports across the boundaries between the different areas. They obtained an outflow of low-salinity water through the archipelago of 0.54 Sv.

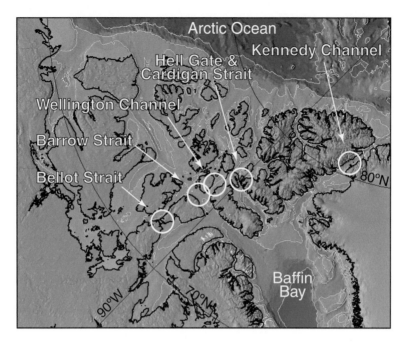

Fig. 4.14 Map showing the Canadian Arctic Archipelago and the main choke points

All transports through the Canadian Arctic Archipelago must cross the sills at one (or two) of six constrictions – Bellot Strait, Barrow Strait (Lancaster Sound), Wellington Channel, Cardigan Strait and Hell Gate (Jones Sound) and Kennedy Channel (Nares Strait). As Bellot Strait cross-section is small compared to the other straits and Wellington Channel enters into eastern Barrow Strait, monitoring surveys of the fluxes through the archipelago have been concentrated on the three main pathways: Kennedy Channel, Hell Gate and Cardigan Strait and eastern Barrow Strait (Fig. 4.14).

In the early 1980s, efforts were made to reference geostrophic calculations to measure currents in a number of channels (Prinsenberg and Bennett 1987; Fissel et al. 1988). Four factors posed significant challenges to the technology available at that time – unsatisfactory direction reference, small (less than 5 km) coherence scales for the flows, large seasonal variability and the logistics. However, a general pattern emerged from this study; currents near the surface flow towards Baffin Bay on the right-hand side of channels (looking south-east), and counter flow on the left-hand side creates a generally cyclonic circulation within the channels of the archipelago. In the late 1980s and early 1990s, the Bedford Institute of Oceanography ran a large hydrographic and mooring programme in Baffin Bay. The current metre moorings were mostly concentrated to Davis Strait, but deployments were also made in the central and northern Baffin Bay. The net volume transport through Davis Strait, based upon these current measurements, was 2.4 Sv (Tang et al. 2004).

Successful year-round use of acoustic Doppler profiling sonar in the Arctic (Melling et al. 1995) has stimulated new initiatives in measuring flow within the Archipelago. An RDI Work Horse ADCP was deployed on one of two moorings in

Fig. 4.15 Sketch of torsionally rigid mooring (Adapted from Melling 2004)

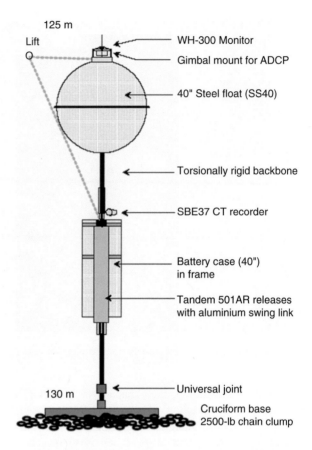

Smith Sound during the North Water Project, and in 1998, Fisheries and Oceans, Canada began a mooring study in Cardigan Strait and in eastern Barrow Strait. Along with the re-deployments in Cardigan Strait, the combined work established for the first time simultaneous observations in two of the three pathways through the archipelago with two goals: (1) to develop a reliable and cost-effective method of measuring current direction near the geomagnetic pole and (2) to acquire a better knowledge of the structure of Arctic channel flows.

For shallow straits, torsionally rigid moorings for ADCPs, holding the instrument at a fixed geographic heading so that a geomagnetic direction reference is not needed, have been successfully used. The unique mooring was designed to meet the special challenges of the environment. It is compact, rising less than 4 m from the seafloor to minimise vulnerability to icebergs and to reduce sensitivity to the strong, 2–3 ms^{-1}, currents. The Work Horse Monitor and in later years a 75-kHz Work Horse Long Ranger both of RD Instruments are mounted in gimbals to remain zenith-pointing despite lay-down of the mooring in strong current. A universal joint in the backbone permits the mooring to stand upright regardless of seabed roughness and slope (Melling 2004) (Fig. 4.15).

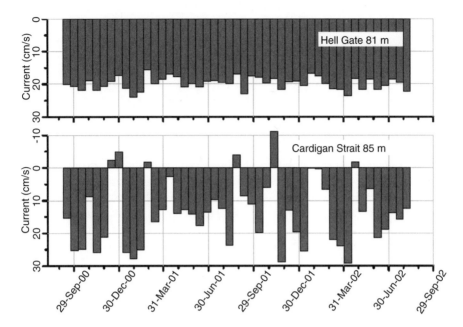

Fig. 4.16 15-day average of the current at 85 m depth in the Cardigan Strait and in the Hell Gate (Adapted from Melling 2004)

A second method to measure current direction near the magnetic pole especially for mid-water flow has been developed for the eastern Barrow Strait moorings (Hamilton 2001). The system uses precision heading reference systems to measure the orientation of ADCPs mounted in streamlined buoyancy packages. The measured magnetic heading is then adjusted for the varying declination with measurements from the Natural Resources Canada (NRCAN) geomagnetic observatory in Resolute to obtain direction relative to true north which, at the mooring site, varies by as much as ±30° over the course of a day. Thus, three commercially available products have been combined into a single package to provide reliable current direction and speeds in areas with any depths where the horizontal component of the earth's magnetic field is too low for conventional instrumentation to be useful.

Yearly mean flows through Cardigan Strait and Hell Gate during 1998–2002 show little year-to-year variations (Melling 2004). Shear is concentrated near the seabed and surface and suggested to be generated near the bottom by the benthic drag and by hydraulics at the sill. The surface shear is in part a consequence of wind, which accelerates between the high lands bordering the straits. The surface velocity is derived from tracking ice, which is sometimes land-fast, and the effect of annual averaging is not the same as for the current. Figure 4.16 displays the variation of a 15-day average current at 85 m depth for Hell Gate and Cardigan Strait. This measure of flow is clearly covariant in Cardigan Strait and Hell Gate, but the amplitude of the variation in Hell Gate is less than one seventh of that in Cardigan Strait, only 17 km away.

With support from the US National Science Foundation and from Fisheries and Oceans Canada, insights and technology from the study in Cardigan Strait and Hell Gate are being applied to the measurement of the fluxes of volume, freshwater and heat through Nares Strait, the third major pathway through the archipelago (Melling 2004). Twenty-three moorings were deployed during a 4-week expedition using the USCGC *Healy* in the summer of 2003. Eight torsionally rigid moorings carrying 75 kHz Long Ranger ADCPs were positioned at 5 km spacing across southern Kennedy Channel. These moorings were interleaved with eight taut-line moorings rising to 30 m depth and carrying SBE37 temperature-salinity-pressure recorders. In addition, two moorings carried ice-profiling sonar to measure the flux of freshwater by ice (Melling and Riedel 1996). Nestled into sheltered coastal locations were 5 very stable moorings carrying instruments to measure the total hydrostatic pressure (plus temperature and salinity). The deployment of the mooring array went according to plan, but the intended recovery from the ice in spring had to be cancelled, and the camp abandoned due to high-speed winds. This postponed the recovery by more than a year. However, the instruments were eventually recovered, and the measurements are being analysed (Müchow et al. 2006; Müchow et al. 2007; Müchow and Melling 2008; Melling et al. 2008).

At the mooring array in eastern Barrow Strait, the channel is 65 km wide and reaches a maximum depth of 285 m. Mobile and land-fast pack ice conditions are normally found for 10 months of the year. Salinity and temperature profiles collected during the August surveys showed that the coldest water ($-1.7°C$) was located at mid-depth, had a salinity of 32.8–33.0 and represented remnants of the winter-surface-mixed layer. Above this water mass, a very stable surface layer, diluted by ice melt and local runoff and warmed by summer heating, was found. The warmest and least saline water was observed at the surface along the two coasts, indicating the existence of coastal currents along both shores. Below the cold mid-depth water mass, the water became warmer and saltier with depth. This warmer deep water had entered the area from northern Baffin Bay. Geostrophic current field derived from the August 1998 density distribution shows an eastward-flowing current decreasing with depth that extends from the southern shore of Lancaster Sound to two-thirds of the way across the sound. A depth-varying current along the northern shore appears to be restricted to one-third of the northern part of the sound (Prinsenberg and Hamilton 2004).

Seasonal mean velocity profiles, derived from the daily mean along-strait data, show that different flow regimes exist at the two sides of the sound and that the currents exhibit seasonal and inter-annual variability. Along the southern shore, the yearly mean and seasonal flows are eastward, very homogeneous and mainly barotropic. Along the northern shore, the seasonal mean flow is weak, variable and baroclinic and generally westward in the surface and bottom layers. As was seen earlier in the geostrophic flow field, the strongest eastward flow along the southern shore occurs during the late summer, while along the northern shore, a definite three-layer flow regime exists.

To estimate fluxes from site-specific time series, it is assumed that the site-specific mooring data can represent, through weighting, the cross-sectional fluxes. Preliminary analysis of data from August 2001 to August 2002 of a modified array provided

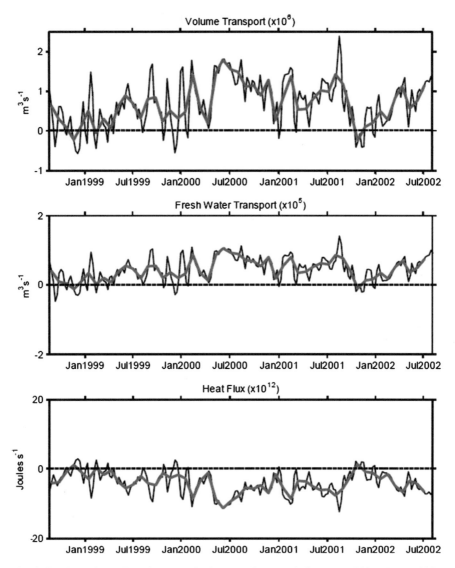

Fig. 4.17 Flux estimate through Barrow Strait over a 4-year period, August 1998 to August 2002 (Adapted from Prinsenberg and Hamilton 2004)

surface layer (0–60 m) current measurements from ¼ to ½ way across the strait as well as measurements at the southern and northern sites. A preliminary look at this data indicates that in winter and spring the eastward currents are similar across the southern two-thirds of the strait, while in summer and fall, the southern array should be reduced to 55% to represent the mean currents observed over two-thirds of the strait. The estimated transports of volume, heat and freshwater, calculated relative to 34.8 and −0.1°C, through Lancaster Sound during 3 years, are shown in Fig. 4.17.

A regional numerical model was developed for the Canadian Arctic Archipelago to provide a basis for evaluating the mooring programmes and to extrapolate ocean-ographic properties through the entire domain, as not all areas can be instrumented. The intent is to calibrate the model with the mooring observations to present day conditions and then to simulate changes in sea-ice regime and oceanography within the archipelago due to global climate change. The modelling work has successfully simulated the barotropic tidal heights and currents as well as the mean circulation (and fluxes) in response to sea level difference between the Arctic Ocean and Baffin Bay (Kliem and Greenberg 2003). The model is diagnostic, and the temperature and salinity fields are thus treated as fixed in time and used as forcing for the model. The open boundaries in the Arctic Ocean and Baffin Bay have both inflow and outflow regions. The inflow is given by prescribing the elevation along the boundary as specified for summer conditions by a large-scale ocean model (Holloway and Sou 2002). The simulations highlight the huge impact of the density field on the flow, emphasising the importance of the baroclinic forces. Secondly, the fluxes depend on the Arctic Ocean–Baffin Bay elevation difference, and a 5-cm Baffin Bay set-down can double the fluxes (Kliem and Greenberg 2003). The actual value is unknown but may be accessible once GPS referenced tide gauge data become available.

Fluxes derived from these observations in eastern Barrow Strait, and to a lesser extent in Cardigan Strait and Hell Gate, show large seasonal and inter-annual vari-ability. Model simulations indicate that summer fluxes through the eastern Barrow Strait make up 35% of the flux through the Canadian Arctic Archipelago (Kliem and Greenberg 2003). Other models with different grid sizes and differently simu-lated physical processes indicate that fluxes through eastern Barrow Strait can be up to 50% of the total fluxes through the archipelago (Maslowski 2003). If these frac-tions also apply to the yearly mean fluxes (0.7 Sv), this indicates that the yearly mean volume flux through the archipelago would range from 1.4 to 2.0 Sv, similar to present literature values (Melling 2000; Prinsenberg and Bennett 1987). The archipelago model indicates 0.3 Sv through Jones Sound (sum of Hell Gate and Cardigan Strait). A 10-cm set-down is required to match the estimated summer flux of 1.0 Sv through eastern Barrow Strait (Kliem and Greenberg 2003). Both data sets thus indicate a Baffin Bay set-down in sea surface relative to the Arctic Ocean that is not simulated by existing coarse global models.

4.4   The Arctic Ocean as Part of the Arctic Mediterranean Sea

The larger Arctic Mediterranean Sea comprises not only the Arctic Ocean but also the Nordic Seas; the Greenland, Iceland and Norwegian Seas. In the first part of the twentieth century, the part played by the Arctic Mediterranean in the global circula-tion was assumed small, as was reflected in the mass balance for the Arctic Mediterranean compiled by Sverdrup in 'The Oceans' (Sverdrup et al. 1942). He estimated that 3 Sv entered the Arctic Mediterranean from the North Atlantic and 0.3 Sv from the North Pacific through Bering Strait. Net precipitation and runoff were estimated to be 0.09 Sv and 0.16 Sv, respectively. The outflow, 3.55 Sv, was assumed to occur almost exclusively through Denmark Strait and to consist mainly of less-saline surface water. The amount of dense overflow water from the Arctic Mediterranean to the North Atlantic was considered negligible, and the North Atlantic Deep Water was believed to form in the North Atlantic southeast of Greenland.

This view has been completely revised. The inflow from the North Atlantic is now estimated to be 8 Sv, and most of it, 75%, returns as dense overflow water, 3 Sv between Scotland and Iceland and 3 Sv through Denmark Strait. The rest is transformed into less-dense water masses (Hansen and Østerhus 2000). The Bering Strait inflow is estimated to be 0.8 Sv, implying an outflow of 3 Sv of less-dense polar water, which likely is divided equally between Denmark Strait and the Arctic Archipelago (Hansen and Østerhus 2000). These findings indicate that the Arctic Mediterranean, together with the Southern Ocean, ventilates the deep world ocean. The Southern Ocean supplies the bottom waters and the Antarctic Intermediate Waters (AAIW), while the deep waters between 1,500 and 4,000 m derives from the Arctic Mediterranean, with less-dense additions from the Mediterranean outflow and from the Labrador Sea (Worthington 1976; Talley and McCartney 1982; McCartney and Talley 1984). In recent years, the possibility that the deep convection taking place in the Irminger Sea, in the northern limp of the subpolar gyre, also contributes to the North Atlantic Deep Water has again been revived (Pickart et al. 2003). However, the major contribution to the North Atlantic Deep Water are the overflows crossing the Greenland–Scotland Ridge, mainly in Denmark Strait and through the Faroe–Shetland Channel, and the ambient water entrained into the descending overflow plumes (Dickson and Brown 1994).

In the early 1980s, it was widely assumed that the intermediate convection in the Nordic Seas, especially in the Iceland Sea, supplied the Denmark Strait overflow water (Swift et al. 1980; Swift and Aagaard 1981), while the deep water formed in the Greenland Sea contributed to the deep waters of the Nordic Seas and the Arctic Ocean (Nansen 1906; Wüst 1941; Kiilerich 1945; Mosby 1959). The insights gained in the 1980s have shown that the Arctic Ocean is a source of deep water and that the Arctic Ocean and the Greenland Sea both contribute to the formation of the Norwegian Sea Deep Water (NSDW) (Aagaard et al. 1985; Rudels 1986a; Swift and Koltermann 1988). The deep flow through Fram Strait goes in both directions, exchanging deep waters between the two source regions. Since the depths of the basins are larger than the sill depth in Fram Strait, the bottom waters in the Arctic Ocean basins and in the Greenland Sea are locally produced and very distinct, the Arctic Ocean deep and bottom waters being warmer and more saline than the Greenland Sea bottom water. This deep circulation was, however, thought mainly confined to the Arctic Mediterranean.

During the Greenland Sea project in the late 1980s and the early 1990s, saline deep water from the Arctic Ocean was observed in Denmark Strait (Buch et al. 1996), and the possibility that the Arctic Ocean could contribute directly to the overflow was then considered seriously. This had been suggested earlier by Rudels (1986a) based on continuity arguments, and it had been implicit in all work on the deep circulation in the Arctic Mediterranean from the late 1980s (Aagaard et al.

1985; Smethie et al. 1988; Heinze et al. 1990). It is not possible to continuously produce deep and bottom water without allowing it to exit somewhere.

Mauritzen (1996a, 1996b), using an inverse box model, concluded that the Atlantic layer in the Arctic Ocean and the Atlantic water recirculating in Fram Strait were the main suppliers of the Denmark Strait overflow water. In this model, the contributions from the central gyres in the Greenland and Iceland seas were found almost negligible. Anderson et al. (1999) estimated the Arctic Ocean contribution to the overflow water using a plume entrainment model constraint with CFCs and concluded that the Arctic Ocean likely supplied about 40% of the overflow water.

The question of the importance of the different source areas is, however, far from settled. In addition to the time variability in the source waters, the overflow waters may also be drawn from different sources at different periods. This is especially the case for the Denmark Strait overflow water. Rudels et al. (2002) examined the different waters in the East Greenland Current and found that, in 1998, the Arctic and Recirculating Atlantic waters (AAW & RAW) and the upper Polar Deep Water (uPDW) were light enough to cross the Jan Mayen Fracture Zone and the sill in Denmark Strait. The Canadian Basin Deep Water was mostly too dense to cross the Jan Mayen Fracture Zone and remained in the Greenland Sea, while Arctic Intermediate Water from the Greenland Sea entered the East Greenland Current and crossed the sill into the Iceland Sea. Water masses too dense to cross the sill in Denmark Strait then flow eastward north of Iceland into the Norwegian Sea (Rudels et al. 1999c). The densest water crossing the sill appears to be transported by a narrow barotropic jet formed north of Iceland (Jónsson and Waldimarsson 2004), but whether this means that the jet carries waters formed in the Iceland Sea or the deeper part of the East Greenland Current is shifted towards Iceland north of the strait before it crosses the sill was not clarified by the end of the ACSYS decade.

The intensified observational activity during the decades at the turn of the millennium has revealed changes also in the characteristics of the deep waters in the Arctic Mediterranean, especially in the smaller basins in the Nordic Seas. Rudels et al. (1999c) examined data from the Greenland Sea taken in the late 1980s and concluded that, at least in 1988, a ventilation deeper than 2,000 m had occurred (see Fig. 4.18). The deep salinity maximum, deriving from the Eurasian Basin Deep Water (EBDW), had been reduced, and the Arctic Intermediate Water formed this year was denser than the Canadian Basin Deep Water (CBDW), which did not penetrate into the basin but remained at the slope, staying high enough to partly cross the Jan Mayen Fracture Zone and enter the Iceland Sea and then continue towards Denmark Strait (Rudels et al. 1999c). In 1993, the convection was limited to the upper 1,000 m, and the deep salinity maximum became more saline and shifted downward. Rudels (1995) and Meincke et al. (1997) suggested that the absence of deep convection relaxes the high-density dome in the central Greenland Sea. The isopycnals, and the salinity maximum, then move downward, and the EBDW at the rim can penetrate, isopycnally, into the central gyre without its salinity being reduced by the input of convecting, less-saline and colder water from above. The volume of the Greenland Sea Bottom Water (GSBW) below the salinity maximum has decreased, and its temperature and salinity have increased.

This agrees with the observations made on transient tracers, which suggest a reduction of the Greenland Sea deep water formation (Schlosser et al. 1991; Bönisch and Schlosser 1995; Bönisch et al. 1997). Meincke et al. (1997) used the changes in the salinity maximum to estimate the strength of the mechanically driven diapycnal mixing in the deep Greenland Sea and obtained a surprisingly large vertical diffusion coefficient, $K_v \approx 3.7 \times 10^{-3}$ m^2 s^{-1}. It should be emphasised that the EBDW and CBDW signals encountered in the Greenland Sea are much diluted by mixing in Fram Strait and along the Greenland slope from Fram Strait to the Greenland Sea, and probably less than 10% of the water in the temperature and salinity maxima comes from the Canadian and Eurasian basins.

After 1993, an intermediate temperature maximum started to develop in the Greenland Sea, indicating that Canadian Basin Deep Water (CBDW) was penetrating from the rim into the central gyre and that the convection in the Greenland Sea was producing intermediate water, less dense than the CBDW (Rudels 1995). This was a change from the 1980s, when the Arctic Intermediate Water (AIW) was denser than the CBDW and less CBDW entered the Greenland Sea Gyre, and the part that did enter the central basin was redistributed vertically by the deeper reaching local convection (Rudels et al. 1999c). The temperature and salinity of the CBDW-derived temperature maximum have gradually increased, and the maximum has been shifted towards deeper levels (Budéus et al. 1998). A possible cause for this downward displacement could be that the recent Greenland Sea convection has been confined to levels above the temperature maximum and that it is easier to bring water at the surface into the central Greenland Sea than remove it as AIW at intermediate levels, causing the storage and the thickness of the AIW layer in the Greenland Sea to increase with time (Fig. 4.18).

In 1996, low-salinity surface water was present in the Greenland Sea due to a large ice export through Fram Strait during the winter of 1994/1995, and ice formation occurred in the winter of 1996/1997. This led to a deepening of the convection from 800 to 1,200 m, and a less-saline, thick intermediate layer of Arctic Intermediate water was created. The winter of 1996/1997 was also the first winter after the Tracer Release Experiment in the Greenland Sea, when a cloud of SF$_6$ was injected at the 28.049 σ_θ surface. In the centre of the Greenland Sea Gyre, the convection reached deeper than this density surface, and the maximum tracer concentration was shifted to deeper (denser) levels, while outside of the gyre the SF$_6$, except for turbulent diffusion, remained on the initial density surface (Watson et al. 1999). This suggests that the convection in 1996/1997 was mainly haline, penetrating through the SF$_6$ cloud, entraining intermediate, SF$_6$ rich, water and bringing it to denser levels. The convection in the following years has been thermal and reached 1,600 m to 1,800 m, except locally within deeper vortices to be discussed below. The convecting water, the AIW, has been less dense than the CBDW, and the ventilated layer has remained above the temperature maximum.

A question, which has been ignored lately because of the presently almost ice-free conditions in the Greenland Sea, is the importance of ice formation for deep convection in the Greenland Sea. Most of the deep waters in the world ocean derive from high latitudes, where the stratification in the upper layers is in salinity, not in temperature, and ice formation is necessary to attain sufficiently high surface densities to induce convection, on the shelves and slopes as well as in the deep basins

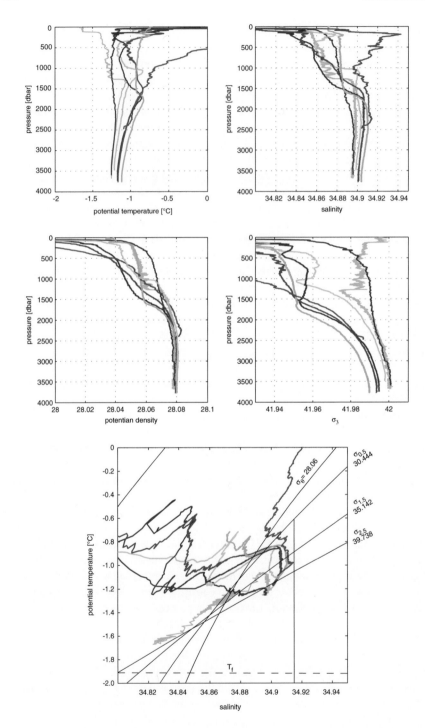

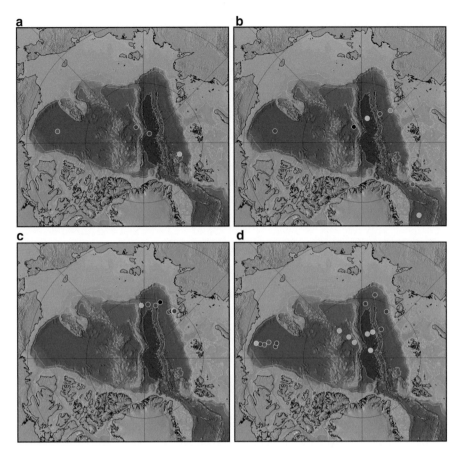

Fig. 4.19 (**a**) Positions of stations shown in Fig. 4.2. (**b**) Positions of stations shown in Fig. 4.4. (**c**) Positions of stations shown in Fig. 4.7. (**d**) Positions of stations shown in Fig. 4.9

Fig. 4.18 Changes in the ΘS properties in the water column in the central Greenland Sea between 1988 and 2002. (*upper left*) potential temperature profiles, (*upper right*) salinity profiles, (*centre left*) potential density profiles referenced to the sea surface, (*centre right*) potential density profiles referenced to 3,000db (*lower panel*) ΘS curves. The cyan station is from a winter station in 1988 with active convection down to 1,250 m (see also Rudels et al. 1989), while the *blue station* is from June the same year, indicating that the convection eventually reached deeper than 2,000 m. The *green station* is from 1993, the *violet* from 1997, the *grey-blue* from 1998 and the *yellow* from 2002, showing the advent and the deepening of the mid-depth temperature maximum derived from the Canadian Basin Deep Water, as well as the increase in temperature and salinity of the deeper layers due to inflow of EBDW. The red station is from the continental slope in 2002 showing the CBDW characteristics as it enters the Greenland Sea Gyre. The colder, less-saline layer found close to the bottom at the slope has the same characteristics as water at 3,300 m in the central Greenland Sea Gyre. This indicates that the previous doming of the isopycnals towards the surface in the central Greenland Sea, prominent in previous years, has now in the deeper layers been replaced by a depression of the isopycnals as a response to the inflow of Arctic Ocean deep waters and the accumulation of AIW in the upper part of the water column. The salinity and the temperature of the Arctic Intermediate Water have increased after the haline convection event in 1997. The upper, convectively ventilated part of the water column has become more homogenous and less stable during the thermal convection operating in recent years compared to the haline convection that occurred e.g. in 1988 (Adapted from Rudels 2010)

(Rudels 1993). Carmack (2000) separated the world ocean in a mid-latitude α-ocean and a high-latitude β-ocean, α and β being the coefficients of heat expansion and salt contraction, respectively, where the stability was determined by either heat or salt, and pointed out that the open ocean deep convection sites were mostly located at the boundary between these two oceans. Is the deepest convection expected to occur on the α-side or on the β-side of the boundary?

In the Greenland Sea, the common assumption has been that ice-free conditions lead to larger heat loss and to deep thermal convection into a weakly stratified water column, where the density surfaces dome upwards, preconditioned by the wind (Nansen 1906; Helland-Hansen and Nansen 1909; Mosby 1959; Killworth 1979). Rudels (1986a) and Aagaard and Carmack (1989) pointed out that for the same heat loss, the density increase of a water parcel due to ice formation and brine release is substantially larger than that due to cooling. If the ice cover never grows thick enough to significantly lower the heat flux to the atmosphere, ice formation would be more efficient in creating dense water.

Rudels (1986a) noted the deep salinity maximum always present in the Greenland Sea water column and suggested that the density anomalies created by brine rejection lead to a large-enough density increase in the upper layer that it could convect through and bypass the intermediate layers below, in the case of bottom water formation also through the salinity maximum, instead of gradually deepening into the underlying water. As the upper layer is emptied into the deep, warmer intermediate water is brought to the surface and starts to melt ice, re-forming the less-saline surface layer that again becomes cooled to freezing temperature. Ice formation recommences and a new deep convection event occurs. This would continue until so much freshwater is convected into the deeper parts of the water column that the surface layer does not reach the freezing point before the water column becomes unstable and convects (Rudels et al. 1999b). A stage of thermal convection with a slow deepening of the mixed layer then begins, which does not allow for any further deep convection events and for renewal of the deep and bottom waters, unless the thermal convection manages to homogenise the entire water column before the end of winter. The mainly ice-free conditions during the last 10 years have prevented any rigorous test of these ideas and they remain speculations.

There could be several causes behind the recent reduced convective deep and bottom water renewal in the Greenland Sea. The most obvious cause would be that the climate has become warmer, leading to less cooling and less dense water formation. The high NAO + state during the 1990s with warmer, windier winters and more precipitation could also reduce the salinity, and density, of the surface water in the Greenland Sea. Disregarding variability and changes in the atmospheric forcing, the largest difference between the 1980s and the present is the thick >1,000 m upper layer with higher temperature and initially a fairly low salinity, which likely was established after the large freshwater import from the Arctic Ocean in 1994/1995. The density difference between this layer and the temperature maximum below is so great that cooling of the entire layer to freezing temperature is required to reach the density of the temperature maximum. The heat content in the upper layer is too large for this to occur during one winter, and in the last years the convection in the Greenland Sea has been thermal and confined to above the intermediate temperature maximum.

During the last few years, several, deep homogenous vortices have been observed in the Greenland Sea. These vortices are 10–20 km in diameter, and some have penetrated down to 2,400 m. They are denser than their surroundings in their upper part and less dense in their lower part, and the isopycnals below the vortices are depressed (Gascard et al. 2002; Wadhams et al. 2002; Budéus et al. 2004). Apart from a much larger vertical extent, they are similar to the mesoscale eddies observed in the Arctic Ocean (see above, Sect. 4.2). The water mass characteristics of the eddies can be reproduced by a mixing of about 1/3 surface water with 2/3 intermediate water (Gascard et al. 2002). This suggests a homogenisation of the water column by convection. However, the scales of the meteorological forcing are larger than 20 km, and to create such a localised convection, the water column must be preconditioned on that scale. Wadhams et al. (2002) proposed that the passage of intermediate water between the Boreas Basin and the Greenland Basin through a narrow gap in the Greenland Fracture Zone might spin up the intermediate water and bring it closer to the surface. This is somewhat similar to the suggestion by D'Asaro (1988) for formation of the Arctic eddies by a frictional spin-up of dense Bering Strait inflow water in the Barrow Canyon. The possibility for the convection to locally break through the surface layer and homogenise a narrow water column thus increases. Once the convection has commenced, the surface water from a larger area may convect down the funnel, increase the depth of the convection and depress the isopycnals below the maximum convection depth. In summer, the eddy will be covered by a less-dense surface layer, which will be rapidly removed the next fall, and the eddy will be further cooled and 'recharged' thus being able to survive several seasons. Because their density is lower than that of the mid-depth temperature maximum, these vortices do not appear to be the mode by which the deep and bottom waters of the Greenland Sea might locally be renewed. It is also not known how much they contribute to the formation of Arctic Intermediate Water that presently takes place in the Greenland Sea. The nature of these eddies and of the convection is presently subject to intense research.

Contrary to the earlier situation, when the isopycnals domed upwards toward the sea surface, the density surfaces in the Greenland Sea below the AIW now are depressed downwards (Budéus and Ronski 2009). The wind stress curl, which has commonly been assumed to cause the doming of the isopycnals, has not, except during some years in the mid 1980s, shown any sign of decreasing (Jakobsen et al. 2003) and should be as effective in creating a doming of the density surfaces in the Greenland Sea as in the 1970s. This suggests that the dome of dense deep water observed earlier would rather be due to the active formation and convection of denser water than an effect of wind forcing.

The formation of Arctic Intermediate Water (AIW) has also led to a freshening of the Norwegian Sea (Blindheim et al. 2000; Mork and Blindheim 2000). The intermediate and deep waters have become less dense, and the isopycnals crossing the Iceland–Scotland Ridge at the Faroe–Shetland Channel have been displaced downwards. The pressure head driving dense water through the Faroe–Shetland Channel then becomes reduced, which could lead to a decrease in the overflow. A possible slowing down of the Faroe–Shetland overflow has also been reported (Hansen et al. 2001). However, this trend appears to have been reversed in recent years (Hansen and Østerhus 2007). The strength of the Denmark Strait overflow has remained unaffected, and a slowdown of the Meridional Overturning Circulation has not been documented (Dickson et al. 2008b).

The importance of the Arctic Ocean freshwater balance for the Atlantic Overturning Circulation has been the focus of much recent research (Peterson et al. 2006; Serreze et al. 2006; Dickson et al. 2007). The freshwater discharged into the Arctic Ocean is exported as ice and as liquid water within the water column. Will a reduced ice cover affect the distribution of the fluxes between these two routes and will this lead to a change in the residence time of the freshwater in the Arctic Ocean, and in what direction? Will such a change influence the convection in the downstream ventilation sites, the Greenland Sea and the Labrador Sea? Will the freshwater be only temporarily stored in the Nordic Seas and then continue with the overflow water into the North Atlantic, causing changes like the recently documented freshening of the North Atlantic Deep Water (Curry et al. 2003)?

Many sources, in the Arctic Ocean as well as in the Nordic Seas, contribute to the overflow (Rudels et al. 2002; Rudels et al. 2003), and also the Arctic Intermediate Water presently created in the Greenland Sea, in spite of its reduced density, is still dense enough to barely pass the sill in the Denmark Strait. SF_6, which has become an identifier of the AIW and its spreading into the Iceland and Norwegian seas and into the Arctic Ocean through Fram Strait, has been followed. The main export appears to be towards the Arctic Ocean, not towards the Norwegian and Iceland seas and the Greenland–Scotland Ridge (Messias et al. 2008; but see also Marnela et al. 2008). This suggests that the main part of the AIW makes a loop into the Arctic Ocean, or at least into Fram Strait, before it joins the East Greenland Current and becomes transported southward. The direct southward transport of SF_6 reached the Greenland–Scotland Ridge first in Denmark Strait and later in the Faroe–Shetland Channel. It appears, however, that the SF_6 first crossed the ridge into the North Atlantic through the Faroe–Shetland Channel and only later through Denmark Strait (Messias et al. 2008). In the Arctic Mediterranean Sea, the overall production and storage of water dense enough to supply the Greenland–Scotland overflow thus do not appear to have diminished, and the renewal of the North Atlantic Deep Water and the forcing of the Meridional Overturning Circulation are not presently threatened.

Acknowledgement The writing and revision of this study have extended over a considerable period of time, and economic support has been obtained from various sources during various phases of the writing process. The more important of these sources are: ASOF-N (contract No EVK2-CT-2002–00139) (BR, PE, EF, US), DAMOCLES (Contract No 018509) (BR, LA, PE, EF, US,) The Academy of Finland (No. 210551) (BR), the Canadian Panel of Energy Research and Development (EPJ).

References

Aagaard K (1968) Temperature variations in the Greenland Sea deep water. Deep Sea Res 15:281–296

Aagaard K (1980) On the deep circulation in the Arctic Ocean. Deep Sea Res 27:251–268

Aagaard K, Carmack EC (1989) The role of sea ice and other freshwater in the Arctic circulation. J Geophys Res 94:14485–14498

Aagaard K, Carmack EC (1994) The Arctic Ocean and climate: a perspective. In: Johannessen OM, Muench RD, Overland JE (eds) The polar oceans and their role in shaping the global climate. American Geophysical Union, Washington, DC, pp 5–20

Aagaard K, Coachman LK (1975) Toward an ice-free Arctic Ocean. Eos 56:484–486

Aagaard K, Greisman P (1975) Towards a new mass and heat budget for the Arctic Ocean. J Geophys Res 80:3821–3827

Aagaard K, Woodgate RA (2001) Some thoughts on the freezing and melting of sea ice and their effects on the ocean. Ocean Model 3:127–135

Aagaard K, Darnall C, Greisman P (1973) Year-long measurements in the Greenland-Spitsbergen passage. Deep Sea Res 20:743–746

Aagaard K, Coachman LK, Carmack EC (1981) On the halocline of the Arctic Ocean. Deep Sea Res 28:529–545

Aagaard K, Swift JH, Carmack EC (1985) Thermohaline circulation in the Arctic Mediterranean Seas. J Geophys Res 90:4833–4846

Aagaard K, Foldvik A, Hillman SR (1987) The West Spitsbergen Current: disposition and water mass transformation. J Geophys Res 92:3778–3784

Aagaard K, Barrie LA, Carmack EC, Garrity C, Jones EP, Lubin D, Macdonald RW, Swift JH, Tucker WB, Wheeler PA, Whritner RH (1996) U.S., Canadian researchers explore the Arctic Ocean. Eos 177:209–213

Anderson LG, Dyrssen D (1981) Chemical constituents of the Arctic Ocean in the Svalbard area. Oceanol Acta 4:305–311

Anderson LG, Jones EP (1992) Tracing upper waters of the Nansen Basin in the Arctic Ocean. Deep Sea Res 39:425–443

Anderson LG, Jones EP, Lindegren R, Rudels B, Sehlstedt P-I (1988) Nutrient regeneration in cold, high salinity bottom water of the Arctic shelves. Cont Shelf Res 8:1345–1355

Anderson LG, Jones EP, Koltermann K-P, Schlosser P, Swift JH, Wallace DWR (1989) The first oceanographic section across the Nansen Basin in the Arctic Ocean. Deep Sea Res 36:475–482

Anderson LG, Björk G, Holby O, Jones EP, Kattner G, Koltermann K-P, Liljeblad B, Lindegren R, Rudels B, Swift JH (1994) Water masses and circulation in the Eurasian Basin: results from the Oden 91 expedition. J Geophys Res 99:3273–3283

Anderson LG, Jones EP, Rudels B (1999) Ventilation of the Arctic Ocean estimated from a plume entrainment model constrained by CFCs. J Geophys Res 104:13423–13429

Anderson LG, Jutterström S, Kaltin S, Jones EP, Björk G (2004) Variability in river runoff distribution in the Eurasian Basin of the Arctic Ocean. J Geophys Res 109:C01016. doi:10.1029/2003JC00173

Backhaus JO, Fohrmann H, Kämpf J, Rubino A (1997) Formation and export of water masses produced in Arctic shelf polynyas – process studies of oceanic convection. ICES J Mar Sci 54:366–382

Bailey WB (1956) On the origin of Baffin Bay deep water. J Fish Res Board Can 13(3):303–308

Bailey WD (1957) Oceanographic features of the Canadian archipelago. J Fish Res Board Can 14:731–769

Bauch D, Schlosser P, Fairbanks RG (1995) Freshwater balance and the source of deep and bottom waters in the Arctic Ocean inferred from the distribution of $H_2{}^{18}O$. Prog Oceanogr 35:53–80

Baumgartner A, Reichel E (1975) The world water balance. Elsevier, Amsterdam, 179 pp

Belyakov LN, Volkov VA (1980) Doklady Akad. Nauk SSR 254(3):752–754

Björk G (1989) A one-dimensional time-dependent model for the vertical stratification of the upper Arctic Ocean. J Phys Oceanogr 19:52–67

Björk G, Winsor P (2006) The deep waters of the Eurasian Basin, Arctic Ocean: geothermal heat flow, mixing and renewal. Deep Sea Res I 53:1253–1271

Björk G, Söderqvist J, Winsor P, Nikolopoulos A, Steele M (2002) Return of the cold halocline to the Amundsen Basin of the Arctic Ocean: implication for the sea ice mass balance. Geophys Res Lett 29(11):1513. doi:10.1029/2001GL014157

Björk G, Jakobsson M, Rudels B, Swift JH, Anderson L, Darby DA, Backman J, Coakley B, Winsor P, Polyak L, Edwards M (2007) Bathymetry and deep-water exchange across the central Lomonosov Ridge at 88–89°N. Deep Sea Res I 54:1197–1208. doi:10.1016/j.dsr.2007.05.010

Blindheim J (1989) Cascading of Barents Sea bottom water into the Norwegian Sea. Rapp P-v Réun Cons Int Explor Mer 188:49–58

Blindheim J, Borovkov V, Hansen B, S-Aa M, Turrell WR, Østerhus S (2000) Upper layer cooling and freshening in the Norwegian Sea in relation to atmospheric forcing. Deep Sea Res I 47:655–680

Bönisch G, Schlosser P (1995) Deep water formation and exchange rates in the Greenland/Norwegian Seas and the Eurasian Basin of the Arctic Ocean. Prog Oceanogr 35:29–52

Bönisch G, Blindheim J, Bullister JL, Schlosser P, Wallace DWR (1997) Long-term co-ordinated trends of temperature, density and transient tracers in the central Greenland Sea. J Geophys Res 102:18553–18577

Bourke RH, Paquette RG (1991) Formation of Baffin Bay bottom and deep waters. In: Chu PC, Gascard J-C (eds) Deep convection and deep water formation in the oceans. Elsevier, Amsterdam, pp 135–155

Bourke RH, Weigel AM, Paquette RG (1988) The westward turning branch of the West Spitsbergen Current. J Geophys Res 93:14065–14077

Bourke RH, Addison VG, Paquette RG (1989) Oceanography of Nares Strait and Northern Baffin Bay in 1986 with emphasis on deep and bottom water formation. J Geophys Res 94:8289–8302

Boyd TJ, D'Asaro EA (1994) Cooling of the West Spitsbergen Current: wintertime observations west of Svalbard. J Geophys Res 99:22597–22618

Boyd TJ, Steele M, Muench RD, Gunn JT (2002) Partial recovery of the Arctic Ocean halocline. Geophys Res Lett 29(14). doi:10.1029/2001GL014047

Buch E, S-Aa M, Kristmannsson SS (1996) Arctic Ocean deep water masses in the western Iceland Sea. J Geophys Res 101:11965–11973

Budéus G, Ronski S (2009) An integral view of the hydrographic development in the Greenland Sea over a decade. The Open Oceanogr J 3:8–39

Budéus G, Schneider W, Krause G (1998) Winter convection events and bottom water warming in the Greenland Sea. J Geophys Res 103:18513–18527

Budéus G, Cisewski B, Ronski S, Dietrich D, Weitere M (2004) Structure and effects of a long lived vortex in the Greenland Sea. Geophys Res Lett 31:L05304. doi:10.1029/2003GL017983

Buinitsky VKH (1951) Ice formation and drift in the Arctic Basin. In: Proceedings of the drifting expedition of Glavsevmorput onboard the Icebreaker "G. Sedov" 1937–149, V.4M-L, pp 74–179

Carmack EC (1986) Circulation and mixing in ice covered waters. In: Untersteiner N (ed) The geophysics of sea ice. Plenum, New York, pp 641–712

Carmack EC (1990) Large-scale physical oceanography of polar oceans. In: Smith WO Jr (ed) Polar oceanography, Part A. California Academic Press, San Diego, pp 171–212

Carmack EC (2000) The Arctic Ocean's freshwater budget: sources, storage and export. In: Lewis EL et al (eds) The freshwater budget of the Arctic Ocean, vol 70, NATO science series 2 environmental security. Kluwer, Dordrecht, pp 91–126

Carmack EC, Kulikov YA (1998) Wind-forced upwelling and internal wave generation in Mackenzie Canyon, Beaufort Sea. J Geophys Res 103:18447–18458

Carmack EC, MacDonald RW (2002) Oceanography of the Canadian shelf of the Beaufort Sea: a setting for marine life. Arctic 55(Supp 1):29–45

Carmack EC, Macdonald RW, Perkin RG, McLaughlin FA (1995) Evidence for warming of Atlantic water in the southern Canadian Basin. Geophys Res Lett 22:1961–1964

Carmack EC, Aagaard K, Swift JH, Macdonald RW, McLaughlin FA, Jones EP, Perkin RG, Smith JN, Ellis KM, Killius LR (1997) Changes in temperature and tracer distributions within the Arctic Ocean: results from the 1994 Arctic Ocean section. Deep Sea Res II 44:1487–1502

Cavalieri DJ, Martin S (1994) The contribution of Alaskan, Siberian and Canadian coastal polynyas to the cold halocline layer of the Arctic Ocean. J Geophys Res 99:18343–18362

Chapman DC, Gawarkiewicz G (1997) Shallow convection and buoyancy equilibrium in an idealized coastal polynya. J Phys Oceanogr 27:555–566

Cisewski B, Budéus G, Krause G (2003) Absolute transport estimates of total and individual water masses in the northern Greenland Sea derived from hydrographic and acoustic Doppler current profiler measurements. J Geophys Res 108(C9):3298. doi:10.1029/2002JCC001530

Coachman LK, Aagaard K (1966) On water exchange through Bering Strait. Limnol Oceanogr 44–59 pp

Coachman LK, Aagaard K (1974) Physical oceanography of the arctic and subarctic seas. In: Herman Y (ed) Marine geology and oceanography of the arctic seas. Springer, New York, pp 1–72

Coachman LK, Aagaard K (1981) Reevaluation of water transports in the vicinity of Bering Strait. In: Wood DW, Calder JA (eds) The eastern Bering Sea shelf: Oceanography and resources, vol 1. University of Washington Press, Seattle, pp 95–110

Coachman LK, Aagaard K (1988) Transports through Bering Strait: annual and interannual variability. J Geophys Res 93:15535–15539

Coachman LK, Barnes CA (1961) The contribution of Bering Sea water to the Arctic Ocean. Arctic 14:147–161

Coachman LK, Barnes CA (1962) Surface waters in the Eurasian Basin of the Arctic Ocean. Arctic 15:251–277

Coachman LK, Barnes CA (1963) The movement of Atlantic water in the Arctic Ocean. Arctic 16:8–16

Coachman LK, Aagaard K, Tripp R (1975) Bering Strait: the regional physical oceanography. University of Washington Press, Seattle, 172 pp

Collin AE (1962) Oceanographic observations in the Canadian Arctic and the adjacent Arctic Ocean. Arctic 15:194–201

Collin AE, Dunbar MJ (1963) Physical oceanography in Arctic Canada. In: Barnes HE (ed) Oceanography and biology: an annual review. G. Allen & Unwin, London, pp 45–77

Curry R, Dickson RR, Yashayaev I (2003) A change in the freshwater balance of the Atlantic over the past four decades. Nature 426:826–829

D'Asaro EA (1988a) Observations of small eddies in the Beaufort Sea. J Geophys Res 93:6669–6684

D'Asaro EA (1988b) Generation of submesoscale vortices: a new mechanism. J Geophys Res 93:6685–6693

Day CG (1968) Current measurements in Smith Sound, summer 1963. USCG Oceanogr Rep 16:75–84

Dickson RR, Brown J (1994) The production of North Atlantic Deep Water. Sources, rates and pathways. J Geophys Res 99:12319–12341

Dickson R, Rudels B, Dye S, Karcher M, Meincke J, Yashayaev I (2007) Current estimates of freshwater flux through the Arctic and Subarctic seas. Prog Oceanogr 73:210–230. doi:10.1016/j.pocean.2006.12.003

Dickson RR, Meincke J, Rhines P (2008a) Arctic-Subarctic Ocean fluxes. Springer, Dordrecht, X+736 pp

Dickson RR, Dye S, Jónsson S, Köhl A, Macrander A, Marnela M, Meincke J, Olsen S, Rudels B, Valdimarsson H, Voet G (2008b) The overflow flux west of Iceland: variability, origins and forcing. In: Dickson RR, Meincke J, Rhines P (eds) Arctic-Subarctic Ocean fluxes. Springer, Dordrecht, pp 443–474

Ekman VW (1905) On the influence of earth's rotation on ocean currents. Arch Math Astron Phys 2(11)

Ekwurzel B, Schlosser P, Mortlock RA, Fairbanks RG, Swift JH (2001) River runoff, sea ice melt water, and Pacific water distribution and mean residence times in the Arctic Ocean. J Geophys Res 106:9075–9092

Environmental Worging Group (EWG) (1997) Joint U.S. Russian atlas of the Arctic Ocean: ocean-ography atlas for the Winter period. National Snow and Ice Data Center, Boulder

Environmental Worging Group (EWG) (1998) Joint U.S. Russian atlas ofthe Arctic Ocean: ocean-ography atlas for the Summer period. National Snow and Ice Data Center, Boulder

Fahrbach E, Meincke J, Østerhus S, Rohardt G, Schauer U, Tverberg V, Verduin J (2001) Direct measurements of volume transports through Fram Strait. Polar Res 20:217–224

Falck E, Kattner G, Budéus G (2005) Disappearance of Pacific water in the northwestern Fram Strait. Geophys Res Lett 32:L14169. doi:10.1029/2005GL023400

Fedorova AP, Yankina AS (1964) The passage of Pacific Ocean water through the Bering Strait into the Chukchi Sea. Deep Sea Res 11:427–434

Fer I, Skogseth R, Haugan PM (2003a) Mixing of the Storfjorden (Svalbard Archipelago) overflow inferred from density overturns. J Geophys Res 109:C01005. doi:10.1029/2003JC001968

Fer I, Skogseth R, Haugan PM, Jaccard P (2003b) Observations of the Storfjorden overflow. Deep Sea Res I 50:1283–1303

Fissel DB, Birch JR, Melling H, Lake RA (1988) Non-tidal flows in the Northwest Passage. Canadian technical report of hydrography and ocean sciences 8, Institute of Ocean Sciences, Sidney, Canada, V8L 4B2, 143 pp

Fjeldstad JE (1936) Results of tidal observations. Norwegian North Polar Expedition "Maud" 1918–1925. Sci Results 4(4):88pp

Foldvik A, Aagaard K, Törresen T (1988) On the velocity field of the East Greenland Current. Deep Sea Res 35:1335–1354

Frank M, Smethie WM, Bayer R (1998) Investigation of subsurface water flow along the continen-tal margin of the Eurasian Basin using transient tracers tritium, ^{3}He, and CFCs. J Geophys Res 103:30773–30792

Friedrich H, Houssais M-N, Quadfasel D, Rudels B (1995) On Fram Strait water masses. In: Extended abstract, Nordic seas symposium, Hamburg, 7/3–9/3 1995, pp 69–72

Gascard J-C, Kergomard C, Jeannin PF, Fily M (1988) Diagnostic study of the Fram Strait mar-ginal ice zone during summer from Marginal Ice Zone Experiment 84 and 84 Lagrangian observations. J Geophys Res 93:3613–3641

Gascard J-C, Richez C, Roaualt C (1995) New insights on large-scale oceanography in Fram Strait: the West Spitsbergen Current. In: Smith WO Jr, Grebmeier JM (eds) Arctic oceanogra-phy, marginal ice zones and continental shelves, vol 49. American Geophysical Union, Washington, DC, pp 131–182

Gascard J-C, Watson AJ, Messias M-J, Olsson KA, Johannessen T, Simonsen K (2002) Long-lived vortices as a mode of deep ventilation in the Greenland Sea. Nature 416:525–527

Gawarkiewicz G (2000) Effects of ambient stratification and shelf-break topography buoyancy equilibrium in an idealised coastal polynya. J Geophys Res 105:3307–3324

Gawarkiewicz G, Chapman DC (1995) O numerical study of dense water formation and transport on a shallow, sloping continental shelf. J Geophys Res 100:4489–4508

Gawarkiewicz G, Weingartner T, Chapman DC (1998) Sea-ice processes and water mass modifica-tion and transports over Arctic shelves. In: Brink KH, Robinson AR (eds) The Sea, vol 10. Wiley, New York, pp 171–190

Goldner DR (1999a) Steady models of Arctic shelf-basin exchange. J Geophys Res 104:29733–29755

Goldner DR (1999b) On the uncertainty of the mass, heat and salt budgets of the Arctic Ocean. J Geophys Res 104:29757–29770

Gorshov SG (1980) World Ocean Atlas 3. The Arctic Ocean. Leningrad USSR Ministry of Defence XIV, 180 pp

Guay CK, Falkner KK (1998) A survey of dissolved barium in the estuaries of major Arctic rivers and adjacent seas. Cont Shelf Res 18:859–882

Guay CK, Falkner KK, Muench RD, Mensch M, Frank M, Bayer R (2001) Wind-driven transport pathways for Eurasian Arctic river discharge. J Geophys Res 106:11469–11480

Haarpainter J (1999) The Storfjorden polynya: ERS-2 SAR observations and overview. Polar Res 18:175–182

Haarpainter J, O'Dwyer J, Gascard J-C, Haugan PM, Schauer U, Østerhus S (2001) Seasonal transformations of water masses, circulation and brine formation observed in Storfjorden, Svalbard. Ann Glaciol 33:437–443

Hamilton JM (2001) Accurate ocean current direction measurements near magnetic poles. The 11th international offshore and polar engineering conference proceedings. ISOPE, Stavanger, vol I:815–846

Hansen B, Østerhus S (2000) North Atlantic – Nordic Seas exchanges. Prog Oceanogr 45:109–208

Hansen B, Østerhus S (2007) Faroe bank Channel overflow 1995–2005. Prog Oceanogr 75:817–856. doi:10.1016/j.pocean.2007.09.004

Hansen B, Turrell WT, Østerhus S (2001) Decreasing overflow from the Nordic seas into the Atlantic Ocean through the Faroe Bank Channel since 1950. Nature 411:927–930

Hanzlick D, Aagaard K (1980) Freshwater and Atlantic Water in the Kara Sea. J Geophys Res 85:4937–4942

Harris RA (1911) Arctic tides. US Printing Office, Washington, DC, 103 pp

Hart JE, Killworth PD (1976) On open ocean baroclinic instability in the Arctic. Deep Sea Res 23:637–645

Heinze C, Schlosser P, Koltermann K-P, Meincke J (1990) A tracer study of the deep water renewal in the European polar seas. Deep Sea Res 37:1425–1453

Helland-Hansen B, Nansen F (1909) The Norwegian Sea. Its physical oceanography based upon the Norwegian researches 1900–1904. Kristiania, Report on Norway fishery and marine investigations II(1):390 pp

Holloway G (2001) Is the arctic sea ice rapidly thinning? Ice and climate news, the Arctic Climate System Study/Climate and Cryosphere project newsletter 1:2–5

Holloway G, Sou T (2002) Has the arctic sea-ice rapidly thinned? J Clim 15:1691–1701

Hunkins K (1966) Ekman drift currents in the Arctic Ocean. Deep-Sea Res 13:607–620

Hunkins K (1974) Subsurface eddies in the Arctic Ocean. Deep-Sea Res 21:1017–1033

Hurrell JW (1995) Decadal trends in the North Atlantic Oscillation: regional temperatures and precipitation. Science 269:677–679

Ingvaldsen RB, Asplin L, Loeng H (2004a) Velocity field of the western entrance to the Barents Sea. J Geophys Res 109:C03021. doi:101029/2003JC001811

Ingvaldsen RB, Asplin L, Loeng H (2004b) The seasonal cycle in the Atlantic transport to the Barents Sea during the years 1997–2001. Cont Shelf Res 24:1015–1032

Jakobsen PK, Nielsen MH, Quadfasel D, Schmith T (2003) Variability of the surface circulation of the Nordic Seas during the 1990s. In: Hydrobiological variability in the ICES area 1990–1999. ICES Marine Science Symposia, vol 219, pp 367–370

Jakobson M, Macnab R, Mayer L, Anderson R, Edwards M, Hatzky J, Schenke HW, Johnson P (2008) An improved bathymetric portrayal of the Arctic Ocean: implications for ocean modelling and geological, geophysical and oceanographic analyses. Geophys Res Lett 35:L07602. doi:10.1029/2008GL033520

Jakobsson M, Grantz A, Kristoffersen Y, MacNab R (2004a) Bathymetry and physiography of the Arctic Ocean and its constituent seas. In: Stein R, Macdonald RW (eds) The organic carbon cycle in the Arctic Ocean. Springer, Berlin, pp 1–6

Jakobsson M, MacNab R, Cherkis N, Schenke H-W (2004b) The international bathymetric chart of the Arctic Ocean (IBCAO). Research Publication RP-2, National Geophysical Data Center, Boulder

Jones EP, Anderson LG (1986) On the origin of the chemical properties of the Arctic Ocean halocline. J Geophys Res 91:10759–10767

Jones EP, Coote AR (1980) Nutrient distributions in the Canadian Archipelago: indictors of summer water mass and flow characteristics. Can J Fish Aquat Sci 37(4):589–599

Jones EP, Anderson LG, Wallace DWR (1991) Tracers of near-surface, halocline and deep waters in the Arctic Ocean: implications for circulation. J Mar Syst 2:241–255

Jones EP, Rudels B, Anderson LG (1995) Deep waters of the Arctic Ocean: origins and circulation. Deep-Sea Res 42:737–760

Jones EP, Anderson LG, Swift JH (1998) Distribution of Atlantic and Pacific waters in the upper Arctic Ocean: implications for circulation. Geophys Res Lett 25:765–768

Jones EP, Swift JH, Anderson LG, Lipizer M, Civitarese G, Falkner KK, Kattner G, McLaughlin FA (2003) Tracing Pacific water in the North Atlantic Ocean. J Geophys Res 108(C4):3116. doi:10.1029/2001JC001141

Jónsson S, Waldimarsson H (2004) A new path for the Denmark Strait overflow water from the Iceland Sea to Denmark Strait. Geophys Res Lett 31:L03305. doi:10.1029/2003GL019214

Jungclaus JH, Backhaus JO, Fohrmann H (1995) Outflow of dense water from the Storfjord in Svalbard: a numerical model study. J Geophys Res 100:24719–24728

Kiilerich A (1939) The Godthaab Expedition 1928. A theoretical treatment of the hydrographical observational material. Medd. om Grönland 78(5):149 pp

Kiilerich A (1945) On the hydrography of the Greenland Sea. Medd om Grönland 144(2):63 pp

Killworth PD (1979) On chimney formations in the oceans. J Phys Oceanogr 9:531–554

Kinney P, Arhelger ME, Burrell DC (1970) Chemical characteristics of water masses in the Amerasian basin of the Arctic Ocean. J Geophys Res 75:4097–4104

Kliem N, Greenberg DA (2003) Diagnostic simulations of the summer circulation in the Canadian Arctic Archipelago. Atmosphere-Ocean 41(4):273–289. doi:10.3137/ao.410402

Krümmel O (1907) Handbuch der Ozeanographie. Band I, Stuttgart, 526 pp

Kulikov EA, Carmack EC, MacDonald RW (1998) Flow variability at the continental shelf break of the Mackenzie shelf in the Beaufort Sea. J Geophys Res 104:12725–12741

Lemke P (1993) Modelling sea ice – mixed layer interaction. In: J Willebrandt, DLT Anderson (eds) Modelling oceanic climate interactions. Proceedings of the NATO advanced study institute on modelling of oceanic climate interactions, Les Houches, 17–28 Feb 1992. NATO ASI series I: global environmental change, vol 11. Springer, Berlin, pp 243–269

Libin YaS (1946) Hydrological observations/expedition onboard "H-169" air plane to the "pole of inaccessibility" scientific results.-M-L, Izdatelstvo Glavsevmorputi, pp 74–123

Livingston HD, Kufferman SL, Bowen VT, Moore RM (1984) Vertical profile of artificial radionuclide concentration in the central Arctic Ocean. Geochim Cosmochim Acta 48:2195–2203

Loeng H (1991) Features of the physical oceanographic conditions in the central parts of the Barents Sea. Polar Res 10:5–18

Loeng H, Ozhigin V, Ådlandsvik B, Sagen H (1993) Current measurements in the northeastern Barents Sea. ICES C.M. 1993/C:41 Hydrographic Committee, 22 pp

Loeng H, Ozhigin V, Ådlandsvik B (1997) Water fluxes through the Barents Sea. ICES J Mar Sci 54:310–317

MacDonald RW (2000) Arctic estuaries and ice: a positive-negative estuarine couple. In: Lewis EL et al (eds) The freshwater budget of the Arctic Ocean. NATO science series 2: environmental security – vol 70. Kluwer, Dordrecht, pp 383–407

Macdonald RW, Carmack EC (1991) Age of Canada Basin deep waters: a way to estimate primary production for the Arctic Ocean. Science 254:1348–1350

Macdonald RW, Paton DW, Carmack EC, Omstedt A (1995) The freshwater budget and under-ice spreading of the Mackenzie River water in the Canadian Beaufort Sea based on salinity and $^{18}O/^{16}O$ measurements in water and ice. J Geophys Res 100:895–919

Macdonald RW, Carmack EC, McLaughlin FA, Falkner KK, Swift JH (1999) Connections among ice, runoff and atmospheric forcing in the Beaufort Gyre. Geophys Res Lett 26: 2223–2226

Maksimov IV (1945) Determining the relative volume of the annual flow of Pacific water into the Arctic Ocean through Bering Strait. Probl Arktiki 2:51–58

Marnela M, Rudels B, Olsson KA, Anderson LG, Jeansson E, Torres DJ, Messias MJ, Swift JH, Watson AJ (2008) Transport of Nordic Sea water masses and excess SF_6 through Fram Strait to the Arctic Ocean. Prog Oceanogr 78:1–11

Martin S, Cavalieri DJ (1989) Contributions of the Siberian shelf polynyas to the Arctic Ocean intermediate and deep water. J Geophys Res 94:12725–12738

Martinson DG (1990) Evolution of the southern Ocean winter mixed layer and sea ice: open ocean deep water formation and ventilation. J Geophys Res 95:11641–11654

Martinson DG, Steele M (2001) Future of the Arctic sea ice cover: implications of an Antarctic analog. Geophys Res Lett 28:307–310

Maslowski W (2003) High resolution modelling of the Arctic Ocean: a decade of progress. In: ACSYS final science conference, St. Petersburg, Russia, 11–14 Nov 2003: WCRP-118(CD) WMO/TD(1232)

Maslowski W, Newton B, Schlosser P, Semtner A, Martinson D (2000) Modelling recent climate variability in the Arctic. Geophys Res Lett 27:3743–3746

Mauritzen C (1996a) Production of dense overflow waters feeding the North Atlantic across the Greenland Sea-Scotland Ridge. Part 1: evidence for a revised circulation scheme. Deep Sea Res 43:769–806

Mauritzen C (1996b) Production of dense overflow waters feeding the North Atlantic across the Greenland Sea-Scotland Ridge. Part 2: an inverse model. Deep Sea Res 43:807–835

May BD, Kelley DE (2001) Growth and steady state stages of thermohaline intrusions in the Arctic Ocean. J Geophys Res 106:783–794

McCartney MS (1977) Subantarctic mode water. In: Angel M (ed) A voyage of discovery. Pergamon Press, Oxford, pp 103–119

McCartney MS, Talley LD (1984) Warm-to-cold water conversions in the Northern North Atlantic Ocean. J Phys Oceanogr 14:922–935

McLaughlin FA, Carmack EC, Macdonald RW, Bishop JKB (1996) Physical and geochemical properties across the Atlantic/Pacific water mass boundary in the southern Canada Basin. J Geophys Res 101:1183–1197

McLaughlin FA, Carmack EC, MacDonald RW, Weaver AJ, Smith J (2002) The Canada Basin 1989–1995: upstream events and far-field effects of the Barents Sea branch. J Geophys Res 107. doi:1029/2001JC000904

McLaughlin FA, Carmack EC, Macdonald RW, Melling H, Swift JH, Wheeler PA, Sherr BF, Sherr EB (2004) The joint roles of Pacific and Atlantic-origin waters in the Canada Basin, 1997–1998. Deep Sea Res I 51:107–128

McPhee MG (1990) Small-scale processes. In: Smith WO Jr (ed) Polar oceanography, Part A: physical science. Academic, San Diego, pp 287–334

McPhee MG, Stanton TP, Morison JH, Martinson DG (1998) Freshening of the upper ocean in the Arctic: is the perennial sea ice disappearing? Geophys Res Lett 25:1729–1732

Meincke J, Rudels B, Friedrich H (1997) The Arctic Ocean-Nordic Seas thermohaline system. ICES J Mar Sci 54:283–299

Melling H (2000) Exchanges of freshwater through the shallow straits of the North American Arctic. In: EL Lewis et al (eds) The freshwater budget of the Arctic Ocean. Proceedings of a NATO advanced research workshop, Tallinn, Estonia, 27 Apr–1 May 1998. Dordrecht, Kluwer Academic, Dordrecht, pp 479–502

Melling H (2004) Fluxes through the northern Canadian Arctic Archipelago. ASOF Newsl 2:3–7

Melling H, Lewis EL (1982) Shelf drainage flows in the Beaufort Sea and their effect on the Arctic Ocean pycnocline. Deep Sea Res 29:967–985

Melling H, Moore RM (1995) Modification of halocline source waters during freezing on the Beaufort Sea shelf: evidence from oxygen isotopes and dissolved nutrients. Cont Shelf Res 15:89–113

Melling H, Riedel DA (1996) Development of seasonal pack ice in the Beaufort Sea during the winter of 1991–1992: a view from below. J Geophys Res 101(C5):11975–11991

Melling H, Johnston PH, Riedel DA (1995) Measurement of the underside topography of sea ice by moored subsea sonar. J Atmos Ocean Technol 12(3):589–602

Melling H, Agnew TA, Falkner KK, Greenberg DA, Craig ML, Münchow A, Petrie B, Prinsenberg SJ, Samelson RM, Woodgate RA (2008) In: Dickson RR, Meincke J, Rhines P (eds) Arctic-Subarctic ocean fluxes. Springer, Dordrecht, pp 193–247

Meredith MP, Heywood KJ, Dennis PF, Goldson LE, White R, Fahrbach E, Østerhus S (2001) Freshwater fluxes through the western Fram Strait. Geophys Res Lett 28:615–618

Merryfield WJ (2000) Origin of thermohaline staircases. J Phys Oceanogr 30:1046–1068

Merryfield WJ (2002) Intrusions in double-diffusively stable Arctic waters: evidence for differential mixing? J Phys Oceanogr 32:1452–1459

Messias M-J, Watson AJ, Johannessen T, Oliver KIC, Olsson KA, Fogelqvist E, Olafsson J, Bacon S, Balle J, Bergman N, Budéus G, Danielsen M, Gascard J-C, Jeansson E, Olafsdottir SR, Simonsen K, Tanhua T, Van Scoy K, Ledwell JR (2008) The Greenland Sea tracer experiment 1996–2002: horizontal mixing and transport of Greenland Sea intermediate water. Prog Oceanogr 78:85–105. doi:1016/jpocean.2007.06.005

Midttun L (1985) Formation of dense bottom water in the Barents Sea. Deep Sea Res 32:1233–1241

Melling H (2004) Fluxes through the northern Canadian Arctic Archipelago. ASOF Newsl 2:3–7

Moore RM, Wallace DWR (1988) A relationship between heat transfer to sea ice and temperature-salinity properties of the Arctic Ocean waters. J Geophys Res 93:565–571

Moore RM, Lowings MG, Tan FC (1983) Geochemical profiles in the central Arctic Ocean. Their relation to freezing and shallow circulation. J Geophys Res 88:2667–2674

Morison JH, Steele M, Anderson R (1998) Hydrography of the upper Arctic Ocean measured form the nuclear submarine USS Pargo. Deep Sea Res I 45:15–38

Mork K-A, Blindheim J (2000) Variations in the Atlantic inflow to the Nordic Seas, 1955–1996. Deep Sea Res I 47:1035–1057

Mosby H (1959) Deep water in the Norwegian Sea. Geofys Publ 21, 62 pp

Mosby H (1962) Water, mass and heat balance of the North Polar Sea and of the Norwegian Sea. Geofys Publ 24(II):289–313

Müchow A, Melling H (2008) Ocean current observations from Nares Strait to the west of Greenland: interannual to tidal variability and forcing. J Mar Res 66:801–833

Müchow A, Melling H, Falkner KK (2006) An observational estimate of volume and freshwater flux leaving the Arctic Ocean through Nares Strait. J Phys Oceanogr 36:2025–2041

Müchow A, Falkner KK, Melling H (2007) Spatial continuity of measured seawater and tracer fluxes through Nares Strait, a dynamically wide channel bordering the Canadian Archipelago. J Mar Res 65:759–788

Muench RD (1971) The Physical oceanography of the northern Baffin Bay region. The Baffin Bay – North Water Project. Scientific report no 1, Arctic Institute of North America, Washington, DC, 150 pp

Münchow A, Weingartner TJ, Cooper LW (1999) The summer hydrography and surface circulation of the East Siberian Shelf Sea. J Phys Oceanogr 29:2167–2182

Nansen F (1902) Oceanography of the North Polar Basin. The Norwegian North polar expedition 1893–1896. Sci Res (9):427 pp

Nansen F (1906) Northern waters, Captain Roald Amundsen's oceanographic observations in the Arctic Seas in 1901. Videnskab-Selskabets Skrifter 1. Matematisk-Naturvidenskabelig Klasse 1, Kristiania, pp 1–145

Nansen F (1915) Spitsbergen waters. Videnskabs-selskabets skrifter I. Matematisk-Naturvidenskabelig klasse I(3):145 pp

Neal VT, Neshyba S, Denner W (1969) Thermal stratification in the Arctic Ocean. Science 166:373–374

Neshyba S, Neal VT, Denner W (1971) Temperature and conductivity measurements under Ice Island T-3. J Geophys Res 76:8107–8120

Newton JL, Sotirin BJ (1997) Boundary undercurrent and water mass exchanges in the Lincoln Sea. J Geophys Res 102:3393–3403

Newton JL, Aagaard K, Coachman LK (1974) Baroclinic eddies in the Arctic Ocean. Deep Sea Res 21:707–719

Nikiforov EG, Shpaiker AO (1980) Principles of large-scale variations of the Arctic Ocean hydrography. Hydrometeizdat, Leningrad, 269 pp

Östlund HG (1982) The residence time of the freshwater component in the Arctic Ocean. J Geophys Res 87:2035–2043

Östlund HG, Hut G (1984) Arctic Ocean water mass balance from isotope data. J Geophys Res 89:6373–6381

Padman L (1995) Small-scale physical processes in the Arctic Ocean. In: Smith WO Jr, Grebmeier JM (eds) Arctic oceanography, marginal ice zones and continental shelves, vol 49. American Geophysical Union, Washington, DC, pp 131–182

Padman L, Dillon TM (1987) Vertical heat fluxes through the Beaufort Sea thermohaline staircases. J Geophys Res 92:10799–10806

Padman L, Dillon TM (1988) On the horizontal extent of the Canada Basin thermohaline steps. J Phys Oceanogr 18:1458–1462

Padman L, Dillon TM (1989) Thermal microstructure and internal waves in an oceanic diffusive staircase. Deep Sea Res 36:531–542

Palfrey KM (1967) Physical oceanography in the northern part of the Greenland Sea in summer of 1964. MS thesis, University of Washington, 101 pp

Pease CH (1987) The size of wind-driven polynyas. J Geophys Res 92:7049–7059

Perkin RG, Lewis EL (1984) Mixing in the West Spitsbergen Current. J Phys Oceanogr 14:1315–1325

Petermann A (1865) Der Nordpol und Südpol, die Wichtigkeit ihrer Erforschung in geographischer und kulturhistorischer Beziehung. Mit Bemerkungen über die Strömungen der Polar-Meere. Pet Mitt Gotha, pp 146–160

Peterson BJ, McClelland J, Curry R, Holmes RM, Walsh JE, Aagaard K (2006) Trajectory shifts in the Arctic and Subarctic freshwater cycle. Science 313:1061–1066

Pfirman SL, Bauch D, Gammelsröd T (1994) The northern Barents Sea: water mass distribution and modification. In: Johannessen OM, Muench RD, Overland JE (eds) The polar oceans and their role in shaping the global environment. American Geophysical Union, Washington, DC, pp 77–94

Pickart RS, Spall MA, Ribergaard MH, Moore GWK, Milliff RF (2003) Deep convection in the Irminger Sea forced by the Greenland tip jet. Nature 424:152–156

Polyakov IV, Proshutinsky AY, Johnson MA (1999) Seasonal in the two regimes of Arctic Climate. J Geophys Res 104:25761–25788

Polyakov IV, Johnson MA (2000) Arctic decadal and interdecadal variability. Geophys Res Lett 27:4097–4100

Polyakov IV, Alekseev GV, Timokhov LA, Bhatt US, Colony RL, Simmons HL, Walsh D, Walsh JE, Zakharov VF (2004) Variability of the Intermediate Atlantic water of the Arctic Ocean over the last 100 years. J Clim 17:4484–4494

Polyakov IV, Beszczynska A, Carmack EC, Dmitrenko IA, Fahrbach E, Frolov IF, Gerdes R, Hansen E, Holfort J, Ivanov VV, Johnson MA, Karcher M, Kauker F, Morison J, Orvik KA, Schauer U, Simmons HL, Skagseth Ø, Sokolov VT, Steele M, Timokhov LA (2005) One more step toward a warmer Arctic. Geophys Res Lett 32(17):1–4. doi:101029/2005GL023740

Prinsenberg SJ, Bennett EB (1987) Mixing and transports in Barrow Strait, the central part of the Northwest passage. Cont Shelf Res 7:913–935

Prinsenberg SJ, Hamilton J (2004) The oceanic fluxes through Lancaster Sound of the Canadian Arctic Archipelago. ASOF Newsl 2:8–11

Proshutinsky AY, Johnson MA (1997) Two circulation regimes of the wind-driven Arctic Ocean. J Geophys Res 102:12493–12514

Quadfasel D, Gascard J-C, Koltermann K-P (1987) Large-scale oceanography in Fram Strait during the 1984 marginal ice zone experiment. J Geophys Res 92:6719–6728

Quadfasel D, Rudels B, Kurz K (1988) Outflow of dense water from a Svalbard fjord into the Fram Strait. Deep Sea Res 35:1143–1150

Quadfasel D, Sy A, Wells D, Tunik A (1991) Warming in the Arctic. Nature 350:385

Quadfasel D, Rudels B, Selchow S (1992) The Central Bank vortex in the Barents Sea: water mass transformation and circulation. ICES Mar Sci Symp 195:40–51

Quadfasel D, Sy A, Rudels B (1993) A ship of opportunity section to the North Pole: upper ocean temperature observations. Deep Sea Res 40:777–789

Redfield AC, Ketchum BH, Richard FA (1963) The influence of organisms on the composition of sea water. In: Hill MN (ed) The Sea, vol 2. Interscience, New York, pp 26–77

Riis-Carstensen E (1936) The "Godthaab" expedition 1928. The hydrographic work and material. Meddeleser om Grönland 78(3):101 pp

Roach AT, Aagaard K, Pease CH, Salo A, Weingartner T, Pavlov V, Kulakov M (1995) Direct measurements of transport and water properties through the Bering Strait. J Geophys Res 100:18443–18457

Rothrock DA, Yu Y, Maykut GA (1999) Thinning of the Arctic sea-ice cover. Geophys Res Lett 26:3469–3472

Ruddick B (1992) Intrusive mixing in a Mediterranean salt lens: intrusion slopes and dynamical mechanisms. J Phys Oceanogr 22:1274–1285

Ruddick B, Walsh D (1995) Observation of the density perturbations which drive thermohaline intrusions. In: Brandt A, Fernando HJS (eds) Double-diffusive convection. American Geophysical Union, Washington, DC, pp 329–334

Rudels B (1986a) The θ-S relations in the northern seas: implications for the deep circulation. Polar Res 4 ns:133–159

Rudels B (1986b) The outflow of polar water through the Arctic Archipelago and the oceanographic conditions in Baffin Bay. Polar Res 4 ns:161–180

Rudels B (1987) On the mass balance of the Polar Ocean, with special emphasis on the Fram Strait. Norsk Polarinstitutt Skrifter 188, 53 pp

Rudels B (1989) The formation of polar surface water, the ice export and the exchanges through the Fram Strait. Prog Oceanogr 22:205–248

Rudels B (1993) High latitude ocean convection. In: Stone DB, Runcorn SK (eds) Flow and creep in the solar system: observations, modeling and theory. Kluwer, Dordrecht, pp 323–356

Rudels B (1995) The thermohaline circulation of the Arctic Ocean and the Greenland Sea. Philos Trans R Soc Lond A 352:287–299

Rudels B (2001) Arctic basin circulation. In: Steele JH, Turekian KK, Thorpe SA (eds) Encyclopedia of ocean sciences. Academic, San Diego, pp 177–187. doi:10.1016/rwos.2001.0372

Rudels B (2009) Arctic Ocean circulation. In: Steele JH, Turekian KK, Thorpe SA (eds) Encyclopedia of ocean sciences 2nd ed. Academic, San Diego, pp 211–225. doi:016/B978–012374473–9.00601–9

Rudels B (2010) Constraints on the exchanges in the Arctic Mediterranean – do they exist and can they be of use? Tellus 62A:109–122

Rudels B, Friedrich HJ (2000) The transformations of Atlantic water in the Arctic Ocean and their significance for the freshwater budget. In: Lewis EL et al (eds) The freshwater budget of the Arctic Ocean. NATO science series 2: environmental security – vol 70. Kluwer, Dordrecht, pp 503–532

Rudels B, Quadfasel D, Friedrich H, Houssais M-N (1989) Greenland Sea convection in the winter of 1987–1988. J Geophys Res 94:3223–3227

Rudels B, Larsson A-M, Sehlstedt P-I (1991) Stratification and water mass formation in the Arctic Ocean: some implications for the nutrient distribution. Polar Res 10:19–31

Rudels B, Jones EP, Anderson LG, Kattner G (1994) On the intermediate depth waters of the Arctic Ocean. In: Johannessen OM, Muench RD, Overland JE (eds) The role of the polar oceans in shaping the global climate. American Geophys Union, Washington, DC, pp 33–46

Rudels B, Anderson LG, Jones EP (1996) Formation and evolution of the surface mixed layer and the halocline of the Arctic Ocean. J Geophys Res 101:8807–8821

Rudels B, Björk G, Muench RD, Schauer U (1999a) Double-diffusive layering in the Eurasian Basin of the Arctic Ocean. J Mar Syst 21:3–27

Rudels B, Friedrich HJ, Hainbucher D, Lohmann G (1999b) On the parameterisation of oceanic sensible heat loss to the atmosphere and to ice in an ice-covered mixed layer in winter. Deep Sea Res II 46:1385–1425

Rudels B, Friedrich HJ, Quadfasel D (1999c) The Arctic circumpolar boundary current. Deep Sea Res II 46:1023–1062

Rudels B, Muench RD, Gunn J, Schauer U, Friedrich HJ (2000a) Evolution of the Arctic Ocean boundary current north of the Siberian shelves. J Mar Syst 25:77–99

Rudels B, Meyer R, Fahrbach E, Ivanov VV, Østerhus S, Quadfasel D, Schauer U, Tverberg V, Woodgate RA (2000b) Water mass distribution in Fram Strait and over the Yermak Plateau in summer 1997. Ann Geophys 18:687–705

Rudels B, Jones EP, Schauer U, Eriksson P (2001) Two sources for the lower halocline in the Arctic Ocean. ICES CM 2001/W:15, 18 pp

Rudels B, Fahrbach E, Meincke J, Budéus G, Eriksson P (2002) The East Greenland Current and its contribution to the Denmark Strait overflow. ICES J Mar Sci 59:1133–1154

Rudels B, Eriksson P, Buch E, Budéus G, Fahrbach E, Malmberg S-Aa, Meincke J, Mälkki P (2003) Temporal switching between sources of the Denmark Strait overflow water. ICES marine science symposia: hydrobiological variability in the ICES area, 1990–1999, Edinburgh, 8–10 Aug 2001: Actes du symposium: posters presented at the symposium, vol 219, pp 319–325

Rudels B, Jones EP, Schauer U, Eriksson P (2004a) Atlantic sources of the Arctic Ocean surface and halocline waters. Polar Res 23:181–208

Rudels B, Marnela M, Eriksson P, Schauer U (2004b) Variability of volume, heat and freshwater transports through Fram Strait. In: ACSYS final science conference, St. Petersburg, Russia, 11–14 Nov 2003, extended abstracts. WCRP-118(CD)WMO/TD(1232)

Rudels B, Björk G, Nilsson J, Winsor P, Lake I, Nohr C (2005) The interactions between waters from the Arctic Ocean and the Nordic Seas north of Fram Strait and along the East Greenland Current: results from the Arctic Ocean-02 Oden expedition. J Mar Syst 55:1–30. doi:10.1016/j. jmarsys.2004.06.008

Rudels B, Marnela M, Eriksson P (2008) Constraints on estimating mass, heat and freshwater transport in the Arctic Ocean: an exercise. In: Dickson RR, Meincke J, Rhines P (eds) Arctic-Subarctic ocean fluxes. Springer, Dordrecht, pp 315–341

Ryder C (1891–1892) Tidligere Ekspeditioner til Grønlands Østkyst nordfor 66° Nr Br Geogr Tids Board 11, København, pp 62–107

Sadler HE (1976) Water, heat and salt transports through Nares Strait, Ellesmere Island. J Fish Res Board Can 33(10):2286–2295

Salmon DK, McRoy CP (1994) Nutrient-based tracers in the Western Arctic: a new lower halocline defined. In: Johannessen OM, Muench RD, Overland JE (eds) The polar oceans and their role in shaping the global environment, AGU geophysical monographs 85. American Geophysical Union, Washington, DC, pp 47–61

Schauer U (1995) The release of brine-enriched shelf water from Storfjord into the Norwegian Sea. J Geophys Res 100:16015–16028

Schauer U, Fahrbach E (1999) A dense bottom water plume in the western Barents Sea: downstream modification and interannual variability. Deep Sea Res I 46:2095–2108

Schauer U, Muench RD, Rudels B, Timokhov L (1997) Impact of eastern Arctic shelf water on the Nansen Basin intermediate layers. J Geophys Res 102:3371–3382

Schauer U, Loeng H, Rudels B, Ozhigin VK, Dieck W (2002a) Atlantic water flow through the Barents and Kara Seas. Deep Sea Res I 49:2281–2298

Schauer U, Rudels B, Jones EP, Anderson LG, Muench RD, BjörkG SJH, Ivanov V, Larsson A-M (2002b) Confluence and redistribution of Atlantic water in the Nansen, Amundsen and Makarov basins. Ann Geophys 20:257–273

Schauer U, Fahrbach E, Østerhus S, Rohardt G (2004) Arctic Warming through the Fram Strait: oceanic heat transports from 3 years of measurements. J Geophys Res 109:C06026. doi:10.1029/2003JC001823

Schauer U, Beszczynska-Möller A, Walczowski W, Fahrbahr E, Piechura J, Hansen E (2008) Variation of measured heat flow through the Farm Strait between 1997 and 2006. In: Dickson RR, Meincke J, Rhines P (eds) Arctic-Subarctic ocean fluxes. Springer, Dordrecht, pp 65–85

Schlichtholz P, Houssais M-N (1999a) An inverse modeling study in Fram Strait. Part I: dynamics and circulation. Deep Sea Res II 46:1083–1135

Schlichtholz P, Houssais M-N (1999b) An inverse modeling study in Fram Strait. Part II: water mass distribution and transport. Deep Sea Res II 46:1137–1168

Schlosser P, Bönisch G, Kromer B, Münnich KO, Koltermann K-P (1990) Ventilation rates of the waters in the Nansen Basin of the Arctic Ocean derived from a multi-tracer approach. J Geophys Res 95:3265–3272

Schlosser P, Böhnisch G, Rhein M, Bayer R (1991) Reduction of deep water formation in the Greenland Sea during the 1980s: evidence from tracer data. Science 251:1054–1056

Schlosser P, Bauch D, Fairbanks R, Bönisch G (1994a) Arctic river-runoff: mean residence time on the shelves and in the halocline. Deep Sea Res Part A 41:1053–1068

Schlosser P, Kromer B, Östlund G, Ekwurzel B, Bönisch G, Loosli HH, Purtschert R (1994b) On the ^{14}C and ^{39}Ar distribution in the central Arctic Ocean: implications for deep water formation. Radiocarbon 36:327–343

Schlosser P, Bönisch G, Kromer B, Loosli HH, Büler B, Bayer R, Bonan G, Koltermann K-P (1995) Mid 1980s distribution of tritium, ^{3}He, ^{14}C, ^{39}Ar, in the Greenland/Norwegian seas and in the Nansen Basin of the Arctic Ocean. Prog Oceanogr 35:1–28

Schlosser P, Kromer B, Ekwurzel B, Bönisch G, McNichol A, Schneider R, von Reden K, Östlund HG, Swift JH (1997) The first trans-Arctic ^{14}C section: comparison of the mean ages of the deep waters in the Eurasian and Canadian basins of the Arctic Ocean. Nucl Instrum Methods Phys Res Sect B 123:431–437

Serreze MC, Barrett A, Slater AJ, Woodgate RA, Aagaard K, Steele M, Moritz R, Meredith M, Lee C (2006) The large-scale freshwater cycle in the Arctic. J Geophys Res 111:C11010. doi:10.1029/2005JC003424

Shirshov PP (1944) Scientific results of the drift of station "North Pole". Doklad na obshchem sobr. AN SSSR, Moscow, 14–17, fevralya

Skogseth R, Haugan PM, Jakobsson M (2005) Water mass transformations in Storfjorden. Cont Shelf Res 25:667–695. doi:10.1016/j.csr.2004.10.005

Smethie WM, Swift JH (1989) The tritium:krypton −85 age of the Denmark Strait overflow water and Gibbs Fracture zone water just south of Denmark Strait. J Geophys Res 94:8265–8275

Smethie WM, Chipman DW, Swift JH, Koltermann K-P (1988) Chlorofluoromethanes in the Arctic Mediterranean seas: evidence for formation of bottom water in the Eurasian Basin and the deep-water exchange through Fram Strait. Deep Sea Res 35:347–369

Smethie WM, Schlosser P, Bönisch G (2000) Renewal and circulation of intermediate waters in the Canadian Basin observed on the SCICEX 96 cruise. J Geophys Res 105:1105–1121

Smith JN, Ellis KM, Boyd T (1999) Circulation features in the central Arctic Ocean revealed by nuclear fuel reprocessing tracers from Scientific Ice Expeditions 1995 and 1996. J Geophys Res 104:29663–29667

Steele M, Boyd T (1998) Retreat of the cold halocline layer in the Arctic Ocean. J Geophys Res 103:10419–10435

Steele M, Morison JH, Curtin TB (1995) Halocline water formation in the Barents Sea. J Geophys Res 100:881–894

Steele M, Thomas D, Rothrock D, Martin S (1996) A simple model study of the Arctic Ocean freshwater balance, 1979–1985. J Geophys Res 101:20833–20848

Steele M, Morison JH, Ermold W, Rigor I, Ortmeyer M (2004) Circulation of summer Pacific water in the Arctic Ocean. J Geophys Res 109:C02027. doi:10.1029/2003JC002009

Stern ME (1967) Lateral mixing of water masses. Deep Sea Res 14:747–753

Stigebrandt A (1981) A model for the thickness and salinity of the upper layers of the Arctic Ocean and the relation between the ice thickness and some external parameters. J Phys Oceanogr 11:1407–1422

Stigebrandt A (1984) The North Pacific: a global-scale estuary. J Phys Oceanogr 14:464–470

Sverdrup HU, Johnson MW, Fleming RH (1942) The oceans: their physics, chemistry and general biology. Prentice-Hall, New York, 1042 pp

Swift JH, Aagaard K (1981) Seasonal transitions and water mass formation in the Icelandic and Greenland Seas. Deep Sea Res 28:1107–1129

Swift JH, Koltermann K-P (1988) The origin of the Norwegian Sea deep water. J Geophys Res 93:3563–3569

Swift JH, Aagaard K, S-Aa M (1980) The contribution of the Denmark Strait overflow to the deep North Atlantic. Deep Sea Res 27:29–42

Swift JH, Jones EP, Carmack EC, Hingston M, Macdonald RW, McLaughlin FA, Perkin RG (1997) Waters of the Makarov and Canada Basins. Deep Sea Res II 44:1503–1529

Swift JH, Aagaard K, Timokhov L, Nikiforov EG (2005) Long-term variability of Arctic Ocean water: evidence from a reanalysis of the EWG data set. J Geophys Res 110:C03012. doi:10.1029/2004JC002312

Talley LD, McCartney MS (1982) Distribution and circulation of Labrador sea water. J Phys Oceanogr 12:1189–1205

Tan FC, Strain PM (1980) The distribution of sea ice melt water in the eastern Canadian Arctic. J Geophys Res 85:1925–1932

Tang CCL, Ross C-K, Yao T, Petrie B, DeTracey BM, Dunlap E (2004) The circulation, water masses and sea-ice of Baffin Bay. Prog Oceanogr 63:183–228

Thompson DWJ, Wallace JM (1998) The Arctic oscillation signature in the wintertime geopotential height and temperature fields. Geophys Res Lett 25:1297–1300

Timmermans M-L, Garrett C, Carmack E (2003) The thermohaline structure and evolution of the deep waters in the Canada Basin, Arctic Ocean. Deep Sea Res I 50:1305–1321. doi:10.1016/S0967(03), 00125-0

Timofeyev VT (1960) Water masses of the Arctic basin. Gidromet. Izdat, Leningrad

Timofeyev VT (1962) The movement of Atlantic water and heat into the Arctic sea basin. Deep Sea Res 9:358–361

Toole JM, Georgi DT (1981) On the dynamics and effects of double-diffusively driven intrusions. Prog Oceanogr 10:121–145

Treshnikov AF (ed) (1985) Atlas of the Arctic. GUGiK, Moscow, 204 pp

Vowinckel E, Orvig S (1970) The climate of the North Polar Basin. In: Orvig S (ed) World climate survey, vol 14. Climates of the polar regions, vol 14. Elsevier, Amsterdam, 370 pp

Wadhams P, Holfort J, Hansen E, Wilkinson JP (2002) A deep convective chimney in the winter Greenland Sea. Geophys Res Lett 29. doi:0.1029/2001GL014306

Walczowski W, Piechura J (2006) New evidence of warming propagating toward the Arctic Ocean. Geophys Res Lett 33:L12601. doi:10129/2006GL025872

Walczowski W, Piechura J, Osinski R (2005) The West Spitsbergen Current volume and heat transport from synoptic observations in summer. Deep Sea Res I 52:1374–1391. doi:1016./j.dsr.2005.03.009

Walin G (1993) On the formation of ice on deep weakly stratified water. Tellus 45A:143–157

Wallace DWR, Schlosser P, Krysell M, Bönisch G (1992) Halocarbon ratio and tritium/^{3}He dating of water masses in the Nansen Basin, Arctic Ocean. Deep Sea Res A 39:443–458

Walsh D, Carmack E (2002) A note on the evanescent behavior of Arctic thermohaline intrusions. J Mar Res 60:281–310

Watson AJ, Messias M-J, Fogelqvist E, Van Scoy KA, Johannessen T, Oliver KIC, Stevens DP, Tanhua T, Olsson KA, Carse F, Simonsen K, Ledwell JR, Jansen J, Cooper DJ, Kruepke JA, Guilyardi E (1999) Mixing and convection in the Greenland Sea from a tracer release experiment. Nature 401:902–904

Weingartner TJ, Cavalieri DJ, Aagaard K, Sasaki Y (1998) Circulation, dense water formation and outflow on the northeast Chukchi Sea shelf. J Geophys Res 103:7647–7662

Weingartner TJ, Danielson S, Sasai Y, Pavlov V, Kulakov M (1999) The Siberian Coastal Current: a wind and buoyancy-forced arctic coastal current. J Geophys Res 104:29697–29713

Weyl PK (1968) The role of the oceans in climate change. Meteorol Monogr 8:37–62

Winsor P (2001) Arctic sea ice thickness remained constant during the 1990s. Geophys Res Lett 28:1039–1041

Winsor P, Björk G (2000) Polynya activity in the Arctic Ocean from 1958 to 1997. J Geophys Res 105:8789–8803

Winsor P, Chapman DC (2002) Distribution and interannual variability of dense water production from coastal polynyas on the Chukchi shelf. J Geophys Res 107(C7):3079. doi:10.1029/2001JC000984

Woodgate RA, Aagaard K (2005) Revising the Bering Strait freshwater flux into the Arctic Ocean. Geophys Res Lett 32:L02602. doi:1029/204GL021747

Woodgate RA, Aagaard K, Muench RD, Gunn J, Björk G, Rudels B, Roach AT, Schauer U (2001) The Arctic Ocean boundary current along the Eurasian slope and the adjacent Lomonosov Ridge: water mass properties, transports and transformations from moored instruments. Deep Sea Res I 48:1757–1792

Woodgate RA, Aagaard K, Weingartner T (2005) A year in the physical oceanography of the Chukchi Sea: Moored measurements from autumn 1990–1991. Deep Sea Res II 52:3116–3149. doi:10.1016/j.dsr2.2005.10.016

Worthington LV (1953) Oceanographic results of Project Skijump I and II. Trans Am Geophys Union 34(4):543–551

Worthington LV (1976) On the North Atlantic circulation, Johns Hopkins oceanographic studies no 6. The Johns Hopkins University Press, Baltimore, 110 pp

Wüst G (1941) Relief und Bodenwasser in Nordpolarbecken. Zeitschrift der Gesellschaft für Erdkunde zu Berlin 5(6):163–180

Chapter 5
Observed Hydrological Cycle

Hermann Mächel, Bruno Rudolf, Thomas Maurer, Stefan Hagemann, Reinhard Hagenbrock, Lev Kitaev, Eirik J. Førland, Vjacheslav Rasuvaev, and Ole Einar Tveito

Abstract The transition between the liquid and solid phase affects all processes in the Arctic. The solid phase is a special challenge to the instruments and the scientists who develop new instruments or analyze, correct, and interpret the observed data.

H. Mächel (✉) • B. Rudolf
German Weatherservice, Frankfurter Str. 135, D-63067 Offenbach, Germany
e-mail: hermann.maechel@dwd.de; bruno.rudolf@dwd.de

T. Maurer
Federal Institute of Hydrology (BfG), Am Mainzer Tor 1, D-56002 Koblenz, Germany
e-mail: Thomas.Maurer@bafg.de

S. Hagemann
Max Planck Institute for Meteorology, Bundesstraße 53, D-20146 Hamburg, Germany
e-mail: Stefan.Hagemann@zmaw.de

R. Hagenbrock
WetterOnline Meteorologische Dienstleistungen GmbH, Am Rheindorfer Ufer 2,
D-53117 Bonn, Germany
e-mail: Reinhard.Hagenbrock@wetteronline.de

L. Kitaev
Institute of Geography RAS, Staromonetniy 29, 109017 Moscow, Russia
e-mail: lkitaev@mail.ru

E.J. Førland
The Norwegian Meteorological Institute, P.O. Box 43, Blindern 0313 Oslo, Norway
e-mail: eirikjf@met.no

V. Rasuvaev
All-Russian Scientific Research Institute of the Hydrological and Meteorological
Information – World Data Center, Koroleva 6, 249020 Obninsk, Russia
e-mail: razuvaev@meteo.ru

O.E. Tveito
The Norwegian Meteorological Institute, P.O. Box 43, Blindern 0313 Oslo, Norway
e-mail: Ole.Einar.Tveito@met.no

P. Lemke and H.-W. Jacobi (eds.), *Arctic Climate Change: The ACSYS Decade and Beyond,* Atmospheric and Oceanographic Sciences Library 43, DOI 10.1007/978-94-007-2027-5_5, © Springer Science+Business Media B.V. 2012

This chapter shows some results from observed parameters of the hydrological cycle, that is, precipitation, snow, runoff, and the atmospheric moisture flux into the polar cap. The precipitation in the Arctic catchments shows only for central Siberia a slight decreasing trend in summer and in North Europe an increasing trend in winter. In Northern Eurasia in winter, the snow depth and snow cover duration increase, as well as the temperature. The runoff and the atmospheric moisture flux derived from radiosonde data show no temporal changes. The uncertainties are still high due to the sparse measuring network, and satellite data are not yet usable for climatological purposes.

Abbreviations

APDA	Arctic Precipitation Data Archive
ARDB	Arctic Runoff Data Base
CIS	Commonwealth of Independent States (FSU Former Soviet Union)
CMAP	CPC Merged Analysis of Precipitation
cog	center of gravity
CPC	Climate Prediction Center (NOAA)
CRU	Climate Research Unit
DARE	WMO Data Rescue Program
ECMWF	European Centre for Medium-Range Weather Forecasts
ERA15	ECMWF 15 Years Re-Analysis
ERA40	ECMWF 40 Years Re-Analysis
FAO	UN Food and Agricultural Organization
FSU	Former Soviet Union
GDCN	Global Daily Climatology Network
GHCN	Global Historical Climate Network
GIS	Geographic Information system
GPCC	Global Precipitation Climatology Centre
GPCP	Global Precipitation Climatology Project
GPM	Global Precipitation Mission
GRDC	Global Runoff Data Centre
GTS	Global Telecommunication System
HARA	Historical Arctic Rawinsonde Archive
HD	hydrological discharge model
NCAR	National Center for Atmospheric Research
NCEP	National Centers for Environmental Prediction
NESDIS	National Environmental Satellite Data, and Information Service (NOAA)
NHS	National Hydrological Services
NOAA	National Oceanic and Atmospheric Administration
NRA	NCEP Reanalysis
NSIDC	National Snow and Ice Data Center

NWP	Numerical Weather Prediction models
R-ArcticNet	Regional Electronic, Hydrographic Data Network For the Arctic Region
RC	runoff coefficient
SL	land surface scheme
TOVS	TIROS Operational Vertical Sounder
TRMM	Tropical Rainfall Measuring Mission
USSR	Union of Soviet Socialist Republics
WMO	World Meteorological Organization

5.1   Introduction

The existence of water in its three phases on the Earth is the basis of life in general. The transition between these phases, including transport processes, is known as the global hydrological cycle.

The hydrological cycle plays an important role in the Earth's energy balance. It is a link between radiation and atmospheric thermodynamics via latent heat release, which involves complex interactions and feedbacks of individual components (evaporation, precipitation, runoff, ice, albedo, clouds, ocean circulation, etc.).

In the *global* view, the hydrological cycle or the water balance is a closed system in which the global evaporation is equal to the global precipitation of about 505,000 km^3/year liquid water.

If the hydrological cycle is separated into an ocean and a land part (in the pre-satellite era no precipitation observations for the oceans exist), over land the difference between evaporation (E) and precipitation (P) is balanced by the total runoff (R).

$$(P-E)_{Land} = (E-P)_{Ocean} = R \qquad (5.1)$$

Until today, the calculations of the individual components of the hydrological cycle from observational data contain many uncertainties of an unknown range, which are the results of sparse measurement networks, different measurement techniques, and different computational algorithms. This is particularly valid for evaporation over land areas (E_{Land}), but also for evaporation over the oceans, which is not operationally observed. Moreover, the operational precipitation measurements over the land are affected by many biases, and over the oceans, the precipitation (P_{Ocean}) is derived from satellite observations only for the last 25 years. Therefore, the calculations of the global hydrological cycle components vary in the range of about $\pm 20\%$ by several authors.

In the *Arctic*, the individual elements of the hydrological cycle differ in intensity, magnitude, and significance, in comparison with those in mid-latitudes or in the Tropics. These differences are largely determined by variations in the regional heat

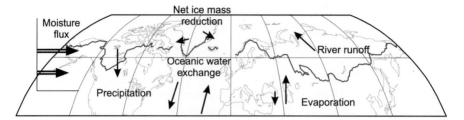

Fig. 5.1 Scheme of the Arctic hydrological cycle

Table 5.1 Components of the Arctic hydrological cycle (in mm/year) for the north polar cap north of 70°N prior to ACSYS (estimates from the ACSYS projects are shown in Table 5.11)

Source	Period	P	P–E	R
Baumgartner and Reichel (1975)	Multiyear	170	61	61
Korzoun et al. (1977a, b)	Multiyear	–	204	–
Masuda (1990)	1979	–	155	–
Overland and Turet (1994)	1965–1990	–	214	–
Peixoto and Oort (1983)	1963–1973	–	122	–
Sellers (1965)	Multiyear	175	50	50
Serreze et al. (1995)	1974–1991	–	163	–

balance, the atmospheric circulation, the presence of water in all its three phases, and the presence of permafrost. By the continuous occurrence of permafrost, the soil infiltration is considerably reduced, and consequently, the surface and river run-off is enhanced.

According to MacKay and Løken (1974), in tundra regions, the runoff is on average 60–70% and evaporation in the order of 30–40% of the precipitation, assuming precipitation measurements are correct. The groundwater contribution is probably less than 10% of the total runoff, but this can increase in the discontinuous permafrost zone to as much as 20–40% of the total, with consequent changes in other components.

The Arctic hydrological processes are controlled by the extraordinary different energy regimes in summer and winter. The surface runoff and the river discharge are greatly influenced by snow and/or glacier melt processes during the short summer season, for example.

In contrast to the global hydrological cycle, the Arctic water cycle is not a closed system. It is influenced by different factors, such as horizontal atmospheric moisture advection, oceanic salinity, and water mass exchanges, as well as ice and glacier changes (Fig. 5.1).

Quantitative characteristics of these factors are, however, unknown or contain large errors and/or are calculated for different areas and periods. Thus, an intercomparison of the estimates given by different authors is difficult (cf. Table 5.1).

In view of climate change, the Arctic hydrological cycle plays an important role. According to the global circulation models, the strongest warming is expected in the Polar Regions, which directly affects the hydrological cycle by increasing precipitation, ice melting, and runoff. Finally, the added fresh water input to the oceans modifies the thermohaline circulation. In this context, an accurate quantification of the Arctic hydrological cycle components is an essential task.

To determine the Arctic hydrological cycle, we can quantify at least the following observed components with more or less errors:

- River runoff (discharge)
- Precipitation (solid, liquid)
- Moisture flux convergence
- Evolution of the ice mass, permafrost
- Oceanic transports

To compute the total water budget for the Arctic, we also need model data because for some subregions, no observational data exist and some components of the hydrological cycle are not directly measured.

Output from operational global weather forecast models have been used to estimate unmeasured variables, but these models generally showed extreme large spin-up effects in precipitation forecasts. Available reanalysis data potentially offer an opportunity to assess the global hydrological cycle more accurately since they already include a large number of observations. The model results cover all parameters of Eq. 5.1, and the difference between the left and right side is a measure of the bias. However, these hydrological parameters are direct model output variables (cf. Sect. 5.3).

The following sections will give an overview of currently available observational data (precipitation, snow depth, runoff, and radiosonde data), their pilot examination, and their comparison with reanalysis data (NCEP, ERA15, and ERA40).

5.2 Observational Data

In situ observations are very important not only for climate monitoring but also for the detection of physical processes and teleconnection patterns as well as to verify forecast and other models.

In the high latitudes, the development of the rain gauge network is characterized by two major steps: a first phase in 1935 (Alaska and USSR or Former Soviet Union, FSU) and the second one at the end of the 1950s (Fig. 5.2).

Special observations in the inner Arctic start with the Russian drifting ice stations during 1937 and from 1950 onward.

To archive and analyze all these data, NOAA designates in 1982 the National Snow and Ice Data Center (NSIDC). In 1988 the Global Runoff Data Centre (GRDC) and the Global Precipitation Climatology Centre (GPCC) were established including the Arctic region.

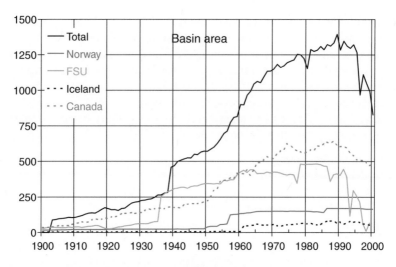

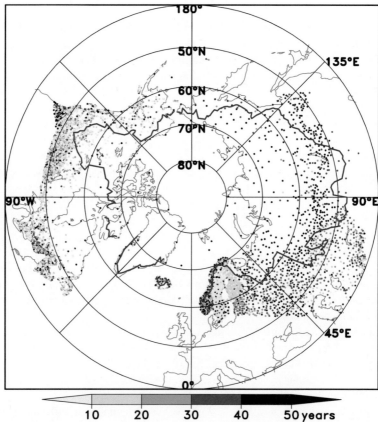

Fig. 5.2 Time evolution of the number of stations reporting precipitation and their spatial distribution for the period 1950–2000 (*purple line*: Arctic basin; source: APDA database 2003)

Table 5.2 Status of precipitation observations available at GPCC/APDA for some countries prior to and at the end of the ACSYS decade

Status 1994			Status 2003	
Period	No. of stations	Country	No. of stations	Period
1986–1992	622	FSU	2,004	1891–1999
1986–1992	320	Canada	7,281	1840–2000
1986–1992	77	Norway	649	1950–2000
1986–1992	153	Alaska	514	1891–2000
1986–1992	1,172	Sum	10,448	1840–2000

5.2.1   Precipitation

Precipitation is the most variable parameter of the hydrological cycle. Therefore many observations are needed to get a reliable area average.

Because of different observing instruments, the WMO initiated some International Instrument Comparison Studies. Sevruk and Klemm (1989) compiled a catalog of worldwide standard gauges, and Sevruk (1990), a summary of instrument comparison studies.

The progress during the ACSYS decade is visible in the increase of archived precipitation data from gauge observations at the Arctic Precipitation Data Archive (APDA) (Table 5.2), the start of snow depth data collection in GPCC, new full global satellite-based precipitation estimates, and in a new Solid Precipitation Measurement Comparison Study initiated by the WMO (Goodison et al. 1998). The US Weather Service and the Meteorological Service of Canada made great efforts to digitize their own historical climatic data and made them available for the public on CD-ROMs.

5.2.1.1   Systematic Gauge Measuring Error and Its Correction

Precipitation measured by rain gauges is systematically underestimated due to evaporation, wetting losses, and the drift of snow and droplets across the gauge funnel. This undercatch can amount up to 200–300% of the measured precipitation (Fig. 5.3). For the estimation of reliable global or regional precipitation amounts, an adequate correction of the data used or of the derived product is required (Legates 1987).

The correction model developed by Rubel and Hantel (1999) is based on individual, daily meteorological information and specific instrument characteristics: correction of daily precipitation data from GTS synoptic reports according to Sevruk (1986) and Førland et al. (1996) reports with:

$$Z_{cor} = k(Z_m + \Delta Z_w + \Delta Z_e) \tag{5.2}$$

with Z_m = measured precipitation depth, ΔZ_w = losses by wetting (climatological estimate), ΔZ_e = losses by evaporation (climatological estimate), and k = instrument-specific factor for wind dependence for *liquid* precipitation as a function of wind

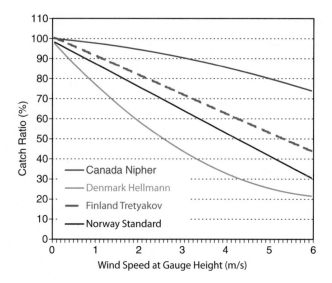

Fig. 5.3 Catch ratio in % of reference as a function of wind speed for dry snow. Function parameters strongly depend on the instrument type (Figure after Goodison et al. 1998)

velocity and precipitation intensity, and for *solid* precipitation as a function of wind velocity and air temperature.

Such "on-event" corrections are applicable only to similar conditions from which they are derived, e.g., the corrections of Fig. 5.3 are valid only for individual events or daily data. However, most precipitation data are only available as monthly sums and without any other meteorological information. One way to apply these corrections to monthly data is to calculate monthly correction factors between the observed and corrected daily data at the synoptic stations. These correction factors can then be interpolated to the uncorrected stations in their vicinity.

The new "on-event" correction method of GPCC is applicable because it is determined by information being actually available in data from synoptic stations. The new "on-event" correction method supplies much more reliable precipitation data than the use of climatological mean correction factors. This climatological mean correction factors were estimated in a time when it was colder and the solid precipitation had a greater portion than today.

5.2.1.2 Comparison of Gridded Precipitation Data Based on Different Data Sets

This part deals with some aspects concerning gridded precipitation that were used to calculate the area mean of the river basins. An average from regular grid points represents better the area average than the mean of irregularly distributed stations. The gridded values primarily depend on the number of available stations and secondarily on the interpolation method. The Global Precipitation Climatology

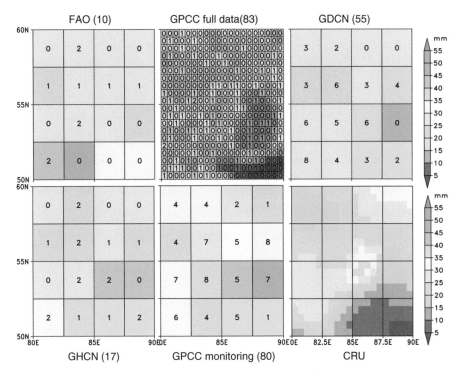

Fig. 5.4 Comparison of different gridded monthly precipitation data sets for January 1987 for the area 80–90°E, 50–60°N (*numbers* indicate the available observations per grid box). Abbreviations: GPCC Monitoring (real-time data from GTS, 2003); GPCC Full Data (including additional data; Version 2, 2003); CRU (Climate Research Unit, Norwich/UK (New et al. 1999, 2000); for the CRU data set, the number of observations is not available); FAO (UN Food and Agricultural Organization, FAOCLIM version 1.2, 1999); GHCN (Global Historical Climatology Network version 2B and CAMS (Chen et al. 2002)); GDCN (Global Daily Climatology Network, V1.0 (Gleason 2002))

Centre (GPCC) maintains the worldwide largest data collection of monthly precipitation data (including data from National Weather Services of about 170 countries and also CRU, FAO, and GHCN data; for details see Fig. 5.4). For this comparison we choose a month with comparatively high station density (January 1987). Figure 5.4 shows the different spatial resolution of some available gridded data sets with the number of observations in each grid box. The differences in the spatial patterns are obviously the result of the different number of stations included in each data set.

The time series of the yearly sums of the area average (80–90°E, 50–60°N) show little differences (Fig. 5.5). Only at the end of the 1990s, the series of the FAO and GDCN show higher discrepancies due to the lack of observations (cf. Fig. 5.6). The decreased availability of data is mainly caused by the delay of 1–5 years between the observation and receiving of the data at the data centers. An additional reason is the closure of stations worldwide due to budget reductions. A special situation

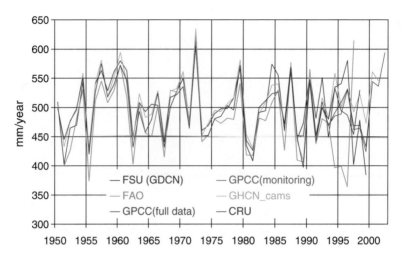

Fig. 5.5 Yearly precipitation sums in mm/year for the area 80–90°E, 50–60°N from different data sets (abbr. see Fig. 5.4)

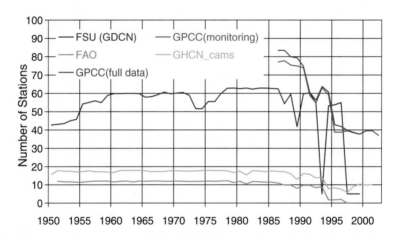

Fig. 5.6 Time evolution of the annual number of stations reporting precipitation for the area 80–90°E, 50–60°N (abbr. see Fig. 5.4; GPCC data starts in 1986)

prevails in the Former Soviet Union after the separation of some republics where nowadays no superior authority exists, which collects and distributes the data.

5.2.1.3 Comparison of Annual Precipitation and Runoff

The comparison of precipitation with independent measurements could verify the station data and also the interpolation method. Such an independent measurement is the runoff for the same catchment area. As Fig. 5.7 shows, there are remarkable – varying

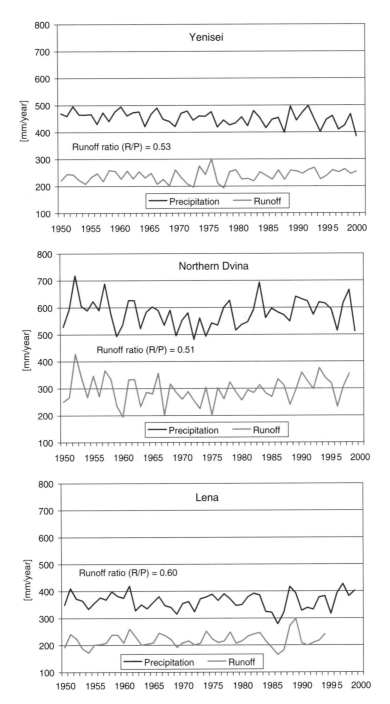

Fig. 5.7 Comparison of runoff and area mean precipitation for the drainage system of river Yenisei, Northern Dvina, and Lena in mm/year

from catchment to catchment – differences between precipitation and runoff. This is the so-called runoff ratio (R/P) or runoff coefficient (RC) and is basically a measure of the evaporation within the basin area, which differs between the rivers.

From the time series of the annual precipitation sums and runoff for three rivers in Russia for the period 1950–2000, it can be derived that significant long-term trends in both parameters are not detectable. Nevertheless, in the Yenisei and Lena catchment area, slightly opposite tendencies in precipitation (decrease) and runoff (increase) are visible. The correlation of the yearly values of precipitation and run-off yields $r=0.10$ for Yenisei, $r=0.69$ for Northern Dvina, and $r=0.73$ for Lena. The low correlation for Yenisei may be due to the relatively high autumnal snow fall from the preceding year.

5.2.2   Runoff

On behalf of ACSYS the Global Runoff Data Centre (GRDC) in Koblenz, Germany, maintains the Arctic Runoff Data Base (ARDB). This collection of discharge data originates from the networks of the National Hydrological Services (NHS) of the countries within the Arctic Ocean drainage basin. Accordingly, the observational methods for gauging river discharge vary from country to country and within a country also from region to region.

In general, river discharge is not monitored directly, as measuring discharge is a costly business, especially on large streams. Rather, continuous measurements of the water levels are taken in the simplest case by staff gauges, typically by float-operated shaft encoders but also by bubble gauges, pressure probes or radar sensors. The resulting water level measurements are transformed in river discharge values by applying the so-called stage-discharge curve. This curve is based on sporadic measurements of the velocity profile at the gauging station at different water levels. More-advanced methods use an electromagnetic flow sensor instead of a mechanical propeller or apply a stationary ultrasonic metering system (Grabs et al. 2000).

Typically, measurements at high floods are rare. Moreover, the relation between stage and discharge is not constant in time; seasonal effects due to vegetation as well as to morphological changes over time require a continuous recalibration of the relationship. Gauging stations in high latitudes impose additional problems related to freeze and thaw in a cold climate, e.g., ice jams.

5.2.2.1   Database

As summarized in Table 5.3, the ARDB currently holds river discharge time series data from 8 countries, featuring 2,404 runoff gauging stations in the arctic hydro-logical region (including stations in the Yukon River basin and in the catchment of Hudson Bay). 1,023 of these stations feature daily discharge data. Figure 5.8 gives a geographical overview of the locations of all ARDB stations.

Table 5.3 Summary statistics of river discharge data kept in the Arctic Runoff Data Base (ARDB) by country (Status April 2004)

Country	Data from 8 countries	Number of stations	Station years	Average time series length (years)	Period (years)
Arctic	Monthly (total)	2,404	78,036	32.5	1877–2002
	original daily	1,023	34,392	33.6	1883–2002
Canada	Monthly (total)	820	27,825	33.9	1892–2001
	original daily	817	27,653	33.8	1892–2001
Finland	Monthly (total)	3	96	32.0	1949–2001
	original daily	3	96	32.0	1949–2001
Iceland	Monthly (total)	11	546	49.6	1932–2002
	original daily	10	494	49.4	1932–2002
Kazakhstan	Monthly (total)	97	2,810	29.0	1936–1989
	original daily	6	126	21.0	1938–1987
Mongolia	Monthly (total)	5	152	30.4	1945–1984
	original daily	1	5	5.0	1978–1982
Norway	Monthly (total)	24	1,985	82.7	1892–2001
	original daily	22	1,866	84.8	1892–2001
Russia	Monthly (total)	1407	43,810	31.1	1877–2000
	original daily	128	3,363	26.3	1883–1999
United States	Monthly (total)	37	812	21.9	1932–2002
	original daily	36	789	21.9	1932–2002

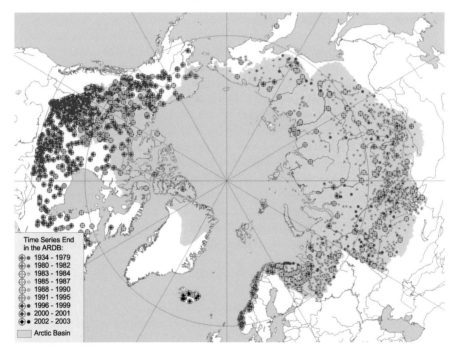

Time Series End
in the ARDB:
- 1934 - 1979
- 1980 - 1982
- 1983 - 1984
- 1985 - 1987
- 1988 - 1990
- 1991 - 1995
- 1996 - 1999
- 2000 - 2001
- 2002 - 2003

Arctic Basin

Fig. 5.8 Map of river discharge stations kept in the Arctic Runoff Data Base (ARDB) color-coded by latest record in the database. (Status April 2004, daily discharge values are available for stations with a cross in the background, monthly discharge values otherwise)

It should be mentioned that there is another database providing data from the Arctic region called R-ArcticNet (A Regional, Electronic, Hydrographic Data Network For the Arctic Region, http://www.r-arcticnet.sr.unh.edu/v3.0) and available at the University of New Hampshire, USA. This database is freely available on the internet. It holds no daily data, but only monthly data. Some of the data originates from a research project cooperation with Russian scientists. As GRDC acts as the global discharge inventory with an official UN mandate, it is committed to this task on a long-term basis rather than project oriented. Consequently, GRDC makes an effort to incorporate R-ArcticNet data into the ARDB which were not available to GRDC from original sources.

5.2.2.2 Analysis and Results

Freshwater discharge from continents into the oceans is of major interest for monitoring the global freshwater resources. Changes in the freshwater influx from the large northbound streams in Siberia and North America affect the salinity of the Arctic shelf seas and thus are an important component in the sensitive feedbacks controlling the thermohaline circulation. In the framework of the GRDC product "Long term mean annual freshwater surface water fluxes into the world oceans," average fresh water fluxes have been estimated, including fluxes into the Arctic Ocean (de Couet and Maurer 2009).

Discharge from land areas not integrally captured by GRDC stations has been determined via estimating mean annual runoff coefficients (*RC*, defined as quotient of runoff and precipitation) by means of regionalization from nearby monitored areas and applying precipitation data from the Global Precipitation Climatology Centre (GPCC).

Application of GIS analysis on a 0.5° elevation grid optimized for flow path detection allowed determining the catchments of all the individual grid cells that form the fringe of the continents, i.e., all continental grid cells were co-registered with their respective fringe grid cell through which they drain to the oceans. Furthermore, each grid cell was assigned either a calculated or estimated RC. Thus, it was possible to calculate for each fringe grid cell the integral flux from its adjacent catchment as the spatially weighted product of RC and precipitation over all co-registered grid cells. Summarizing the fluxes of subsets of continental fringe cells allows estimating fluxes for arbitrary coastline sections, including the Arctic Ocean.

Table 5.4a displays the estimated runoff volumes across the coastlines between two latitudes and by continent. Table 5.4b provides the corresponding coverage by GRDC stations. Table 5.4c gives the aggregated precipitation integrated over the associated spatial areas as estimated by GPCC for the period 1961–1990. This corresponds to estimated runoff coefficients as indicated in Table 5.4d.

Shiklomanov et al. (2000) provide a time series of estimated annual total inflow to the Arctic Ocean between 1921 and 1996 based on the R-ArcticNet database. Their mean value of total inflow amounts to 5,249 km³/a or 4,301 km³/a excluding Hudson Bay. While their time series of total inflow shows no trend, the time series excluding Hudson Bay features a slight upward trend of 0.88 km³/a or 0.2‰/a.

Table 5.4 GRDC estimated average quantities on hinterland catchment areas of Arctic coastal- and 5°-latitude zones (Status August 2003)

Latitude	NAM	EUR	ASI	AO
(a) GRDC estimated runoff volume in km³/a				
N 85–80	8	53	4	65
80–75	18	36	34	88
75–70	78	102	1,551	1,731
70–65	460	314	687	1,461
65–60	284	154	4	442
90–60	847	659	2,281	3,788
(b) Ratio GRDC observed/estimated runoff volume in %				
N 85–80	0	0	0	0
80–75	0	0	0	0
75–70	16	0	82	75
70–65	76	51	77	71
65–60	84	89	0	85
90–60	71	45	79	71
(c) GRDC estimated GPCC precipitation volume in km³/a				
N 85–80	12	82	6	99
80–75	25	56	41	122
75–70	104	152	2,518	2,773
70–65	890	427	1,611	2,928
65–60	351	294	5	650
90–60	1,382	1,011	4,180	6,573
(d) GRDC estimated runoff coefficient in %				
N 85–80	72	65	61	66
80–75	70	64	85	72
75–70	75	67	62	62
70–65	52	73	43	50
65–60	81	52	85	68
90–60	61	65	55	58

NAM North America, *EUR* Europe, *ASI* Asia, *AO* Arctic Ocean

Table 5.5 lists some aggregations of freshwater fluxes into the Arctic Ocean and relates them to a reduced value of the long-term mean annual flux as estimated in Table 5.4a. (The value was reduced, as the Table 5.4a estimate considers a slightly larger area than the Arctic drainage basin depicted in Fig. 5.8, i.e., includes fluxes from Yukon River basin (201 km³/a), Island (36 km³/a), northern parts of Hudson Bay, southern Baffin Island, and southern Greenland.) It is interesting to note that the long-term average flux from the six largest rivers is exactly the same as for the reference period 1979–1993.

Figure 5.9 displays the time series of average annual discharges in the seven largest rivers draining to the Arctic Ocean, while Fig. 5.10 shows their average monthly hydrographs over the 15-year period 1979–1993 including the hull of maximum and minimum monthly flows during these 15 years.

Figure 5.9 reveals practically no trend in the summary annual discharge time series of all 7 rivers during the 15-year period considered. However, Peterson et al. (2002)

Table 5.5 Freshwater flux into the Arctic Ocean

#	Freshwater flux into the Arctic Ocean from	Annual average km³/a	m³/s	% of line #2	Reference Period
1	Coastlines between latitude 60–90°N	3,788	120,117	106.7	1961–1990
2	Line #1 reduced by Yukon and Island flux	3,551	112,601	100.0	1961–1990
3	34 rivers draining into the Arctic Ocean	2,609	82,731	73.5	Irregular
4	7 largest rivers as given in Fig. 5.10	2,149	68,144	60.5	Complete data
5	6 largest rivers in Russia	1,863	59,075	52.5	Complete data
6	7 largest rivers as given in Fig. 5.10	2,142	67,922	60.3	1979–1993
7	6 largest rivers in Russia	1,863	59,075	52.5	1979–1993

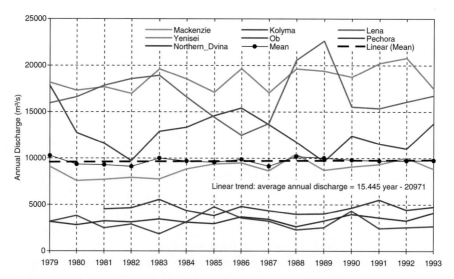

Fig. 5.9 Annual discharge of the seven largest rivers draining to the Arctic Ocean and trend line of the mean flow (multiply by 7 to get the total flow of 67,922 m³/s or 2,142 km³/a). Stations as in Fig. 5.10

analyzed a 7% discharge increase for the 6 Russian rivers between 1936 and 1999 corresponding to 2.0±0.7 km³/a or 1.1‰/a. They related the increase in discharge from the Arctic basin to the increase in global mean annual temperature and thus extrapolated discharge based on IPCC projections of 1.4–5.8°C for the year 2100. The straightforward linear extrapolation – explicitly not termed "prediction" – by Peterson et al. (2002), which yields a corresponding estimate of 18% to 70% increase in Eurasian Arctic river discharge over present conditions, has attracted great attention

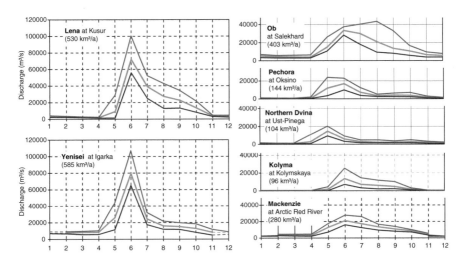

Fig. 5.10 Average monthly flow hydrographs (*green line*) for the period 1979–1993 of the 7 largest rivers draining to the Arctic Ocean (capturing approximately 60% of the total freshwater flux) and maximum (*red line*) and minimum (*blue line*) monthly flows during this period

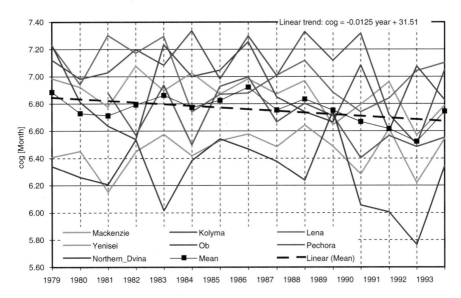

Fig. 5.11 Time series of center of gravity of all one-year, monthly flow hydrographs of the seven largest rivers draining to the Arctic Ocean (5 indicating May, 6: June, 7: July) and trend line of the mean center of gravity. Stations as in Fig. 5.10

in the climate research community and has been the subject of some debate (see, e.g., Sherwood and Idso 2003).

To further investigate changes in the runoff regime in the Arctic basin, the shape of the discharge hydrographs was examined. Figure 5.11 shows time series of the

Table 5.6 Trends for the period 1979–1993 of the center of gravity (cog) of all one-year, monthly flow hydrographs of the seven largest rivers draining to the Arctic Ocean (stations as in Fig. 5.10)

River	Average cog (month)	Slope (days/year)	Shift over 15 years (days)
Mackenzie	6.88	−0.492*	−7.37
Kolyma	7.07	−0.466	−6.99
Lena	7.01	−0.498	−7.47
Yenisei	6.46	0.145	2.18
Ob	6.91	−0.180	−2.70
Pechora	6.69	−0.792*	−11.8
N. Dvina	6.29	−0.444	−6.67
Average	6.76	−0.374*	−5.61

*Significant at the 95% level

center of gravity (*cog*) of all one-year, monthly flow hydrographs of the 7 largest rivers for the considered 15-year period calculated as:

$$cog_j = \frac{\sum\limits_{i=1}^{12} i \cdot Q_{ij}}{\sum\limits_{i=1}^{12} Q_{ij}} \tag{5.3}$$

where i is a counter over the 12 months of the year j, and Q_{ij} is the discharge of month i in year j. As the linear trend line of mean cog shows, there is an overall slight, statistically not significant shift to earlier discharge of about 0.0125 month/a which corresponds to 5.6 days in the 15-year period.

However, the trends of cog for longer time series show considerably lower, though also decreasing rates. In Table 5.6 river-wise trends of cog for the 15-year period 1979–1993 are listed. The average shift of the hydrographs to earlier discharge amounts here to about 5.6 days in the 15-year period. Overall, these findings accord with the behavior that one would expect under warming conditions (earlier snowmelt, earlier runoff).

5.2.2.3 Problems in Runoff

Water balance: River discharge is highly variable, both spatially and temporally. It is a very noisy signal, being the residual of other – usually much larger – components of the water balance equation, i.e., precipitation, evapotranspiration, and soil moisture and groundwater storage change, all of which are more difficult to measure than discharge, which is the only variable that constitutes an integral over a large area (river basins). This is especially true in the almost unpopulated and inhospitable far North. Moreover, discharge is often severely affected by anthropogenic flow regulation activities, such as dams or irrigation. To analyze changes in the Arctic water cycle and its linkage to the energy cycle, it is thus necessary to have a good coverage of measurements for all these variables.

Accuracy of measurements is another problem, especially in high latitudes. Extreme climatic conditions make it difficult, dangerous, and costly to take reliable measurements (e.g., ice jams).

Table 5.7 Characteristics of snow surveys in Russia and CIS-countries

Type of landscape	Length of route (m)	Sampling Distance (m) Snow depth	Snow density
Open sites	2,000	20	200
Forest with great number of glades	1,000	20	100
Continuous forest	500	10	100

Data availability   continues to be a problem. Besides the general difficulty of getting access to measured data, monitoring rivers in the far North is especially costly. For quite a number of gauging stations in the United States, Canada, and Russia, operation was thus discontinued in recent years. This is an unbearable situation and – given the potential impacts on, e.g., the thermohaline circulation – is a dangerous neglect.

5.2.3   Snow Cover, Snow Depth

5.2.3.1   Observational Methods and Databases

The first systematic measurements of snow on the ground in Russia started in the nineteenth century. It consisted of daily measurements of snow depth using fixed sticks, and visual estimations of snow cover. In the first quarter of the twentieth century, snow surveys were initiated in order to estimate the snow water equivalent (Manual… 1985; Kopanev 1971). In this section the spatial and temporal variability of snow depth, snow water equivalent, and number of days with snow cover is studied for Russia, CIS countries (Commonwealth of Independent States), Fennoscandia, and Canada.

Russia and many CIS countries provide 10-day (5-day) snow surveys for different landscapes (Manual… 1985), where snow depth and snow density are measured (Table 5.7). Snow depth (accuracy of 0.5 cm) is determined by means of metal and wooden scales, and snow density is measured (in kg/m³) by gravimetric method. Snow water equivalent (in mm) is calculated by multiplying snow depth with snow density. In addition, daily observations of snow depth on open and sheltered sites are performed by the use of fixed measuring sticks. Snow cover is a visual estimation of the fraction of the ground which is covered by snow.

In Canada, field surveys of snow water equivalent, which have been performed since the 1930s, are carried out by various institutions. The number of measurements on a route differs from 4 to 50. Synchronism of observations is lacking. However, it should be mentioned that the Weather Service has unified data and monitored the information quality, when preparing collection of observational data since 1955 (McKay and Gray 1981; Schmidlin 1995).

A total number of 1,400 weather stations in the north of Eurasia (Russia and CIS countries) were used to analyze the snow water equivalent for the period 1966–1990 (number varies from 500 to 1,300 in various years). For the period 1991–2000 the analysis comprised 1,000 stations (number varies from 500 to 800 in various years). The analysis of snow depth, snow coverage, and winter air temperature was based on 223 stations in the CIS countries and 100 stations in Fennoscandia. The maximum number of weather stations presented in the snow database developed by the Weather Service of Canada sums up to 2,000.

5.2.3.2   Results

Mean snow depth of February and the number of days with snow cover above 50% for November–May (1936–2000) were investigated for an estimation of regional patterns and for long-term variability over the Northern Eurasia (CIS countries, Fennoscandia).

The spatial variability of snow depth over Northern Eurasia corresponds to features of atmospheric circulation and orography (Fig. 5.12a; Kitaev et al. 2002). The maximum values of snow depth are found in the Ural region and the western part of East Siberian low mountains. High values are also found in a zone influenced by the Aleutian low pressure system at the Far East. Small snow depths are found in a zone influenced by the Siberian high pressure system, in the south of East European plain. The long-term mean values of snow depth have regional differences and vary from 11 cm in the Kazakhstan region to 39 cm in Western Siberia (Table 5.8).

Number of days with snow cover >50% is smoothly increasing from the south to the north according to the spatial variability of mean winter air temperature (November–May, Fig. 5.12b). The regional mean values of number of days with snow cover >50% varies from 86 in Kazakhstan to 220 in Eastern Siberia (Table 5.8). These features correspond to the regional variations of air temperatures (November–May, Fig. 5.12c). Typical regional mean winter temperatures vary from 1.2°C in Kazakhstan to −22.8°C in Eastern Siberia (Table 5.8).

A long-term increase of snow depth and number of days with snow cover is typical for most parts of Northern Eurasia (Fig. 5.13, Table 5.9). A long-term decrease of snow depth takes place in some southern regions. These features occur on a background prevailing positive long-term trends of winter temperature in most parts of the region, with the largest increase in Kazakhstan. Over some separate northern regions, there are negative temperature trends (Fig. 5.13c). The highest positive linear trend of snow depth is found in the Far East and the East European Plain. In the latter region also the largest increase in number of days with snow cover is found. Early dates of stable snow cover formation are predetermined by the tendency to air temperature decrease in autumn, during formation of snow water equivalents. This fact is true for both the southern and northern regions of the investigated territory. Long-term trend of increasing winter precipitation amount also contributes to increase the period with stable snow cover in the north as well as in the south of the East European Plain.

Increase of snow cover duration in Northern Siberia and in the Amur Basin occurs against a background of long-term decrease in winter precipitation amount. This decreasing winter precipitation is compensated by an increased portion of solid precipitation in spring. At the Eurasian Arctic coast, snow cover duration increases in connection with decreasing autumn temperatures and winter precipitation.

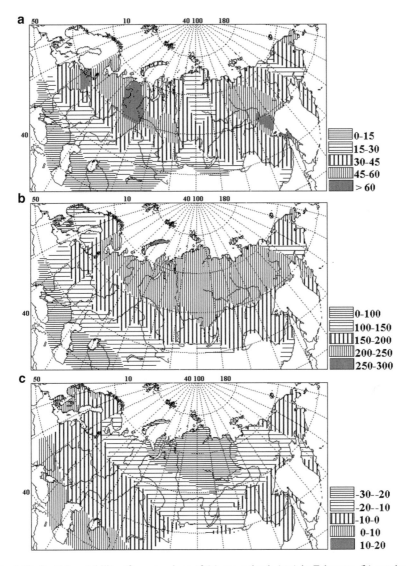

Fig. 5.12 Spatial variability of mean values of (**a**) snow depth (cm) in February, (**b**) number of days with snow cover >50% (days), and (**c**) winter air temperature (November–May; °C) for 1936–2000

For the total region of Northern Eurasia, the analysis indicates positive trends (Fig. 5.14) of snow depth (+0.091 cm/year), number of days with snow cover >50% (+0.119 days/year), and winter air temperature (+0.015°C/year). The reason for increasing snow indices in regions with increasing winter temperatures is probably that higher temperatures are associated with higher amounts of perceptible water. Because of the low winter temperature (−11.5°C) in this region, most of this increased precipitation is falling as snow.

Table 5.8 Mean values for the period 1936–2000 of snow depth, number of days with snow cover, and winter (November–May) temperature for different regions in Northern Eurasia

Region	Snow depth (cm)	Snow cover (days)	Winter temperature (°C)
Fennoscandia	–	119	0.1
East European Plain	37	169	−8.0
Western Siberia	39	195	−14.6
Eastern Siberia	34	220	−22.8
Far East	38	192	−9.9
Kazakhstan	11	86	1.2
Northern Eurasia (total)	34	179	−11.5

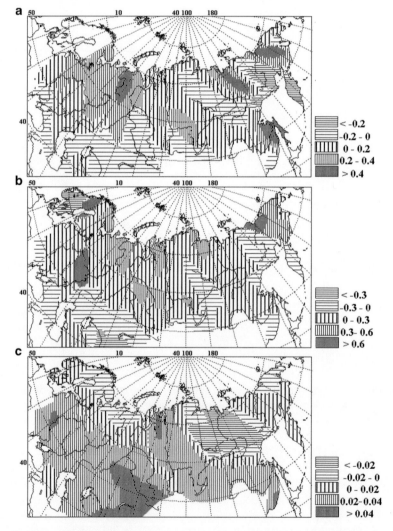

Fig. 5.13 Long-term trends for 1936–2000 of (**a**) snow depth (cm/year), (**b**) number of days with snow cover >50% (days/year), and (**c**) winter air temperature (November–May; °C/year)

Table 5.9 Linear trends (1936–2000) of snow depth, number of days with snow cover (>50%), and winter air temperature (November–May) for different regions in Northern Eurasia

Region	Snow depth (cm/year)	Snow cover (day/year)	Temperature (°C/year)
Fennoscandia	–	0.145	0.005
East European Plain	0.114	0.245	0.016
Western Siberia	0.082	0.211	0.029
Eastern Siberia	0.047	0.143	0.000
Far East	0.113	0.094	0.005
Kazakhstan	−0.077	−0.191	0.044
Northern Eurasia (total)	0.091	0.119	0.015

The comparative characteristic of changes of snow water equivalent for Northern Eurasia (CIS countries) and northern part of North America (Canada) was carried out for the period 1966–1996. The North America (within the confines of Canada) was subdivided into three regions, i.e., the western, the central, and the eastern part. The western region includes middle-high mountains, the height of which is not less than 1,000 m. It is composed of the Alaska Range and the Coast Mountains, the Rocky Mountains, and the Mackenzie Mountains that comprise the system of Cordillera. The central region includes pediment that is bounded on the west by the coastline of the Hudson Bay and the Arctic Archipelago. The eastern region, where annual precipitation is 500–1,000 mm, adjoins the Atlantic Ocean and includes the Labrador Peninsula, the Newfoundland Island, and the St. Lawrence Basin.

Long-term changes of characteristics of snow cover are demonstrated by the example of February that is the snowiest month. Generally, greater snow depths characterize the northern part of North America. Mean average snow depth in February (1966–1996) is greater in Canada (62 cm) than in the Northern Eurasia (39 cm) by a factor of 1.5. Snow water equivalent differs in the same manner; it is 162 mm in the north of the North America and 103 mm in the Northern Eurasia.

Such differences were also found in some regions of North America. Maximum values of snow water equivalent (282 mm) are typical for the western region because of the Aleutian low pressure system affecting this area associated with the mountainous relief there. The Icelandic low pressure system is responsible for high values of snow water equivalents (152 mm) in the eastern region. The central, most continental part is characterized by less snow water equivalent (83 mm).

Maximal snow water equivalent in Northern Eurasia is observed in Western Siberia and East European Plain (133 and 110 mm accordingly). This situation is caused by the cyclones coming in via the East European Plain. Minimal snow water equivalent is registered on the Turan Plain and in Kazakhstan (68 mm), where the amount of precipitation is rather small. In the Eastern Siberia and Far East regions, the snow water equivalent is ca. 90 mm.

The long-term trends of snow water equivalents are radically different for the considered continents. Trends of snow water equivalents for the north of North

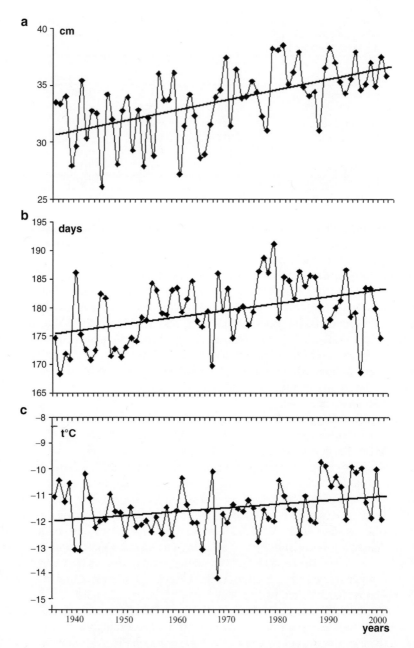

Fig. 5.14 Long-term variability in Northern Eurasia of (**a**) snow depth, (**b**) number of days with snow cover >50%, and (**c**) winter temperature (November–May). *Solid lines* indicate linear trends

America are negative and make up −1.11 mm/year (the western region) and −1.29 mm/year (the eastern region). The central region is characterized by an insignificant negative trend that makes up to −0.12 mm/year. A long-term change of snow water equivalent at the western coast of North America reflects negative long-term changes of pressure of the Aleutian low system, an impact which abates in the central part of the continent. The long-term changes of snow water equivalent at the eastern coast of the North America reflect changes in the local atmospheric circulation.

General increase of snow water equivalent is typical for Northern Eurasia. Greatest rate of changes (+0.58 mm/year) characterizes the Far East. The Aleutian low pressure system has the opposite effect on formation of snow cover in comparison with the Pacific coast of Canada. Rather high positive trends of snow water equivalents of the East European Plain, Ural, and Siberia (+0.35, +0.41, and +0.26 mm/year) may be caused by intensification of cyclonic activity in these regions.

5.2.3.3   Conclusions

On the basis of observational data, the spatial variability and long-term trends of snow depth, snow water equivalent, and number of days with snow coverage >50% for Northern Eurasia and for North America are estimated. As a whole, for Northern Eurasia there has been an increase of snow depth and duration of snow cover during 1936–2000. Also the winter temperature has increased in this period. This situation is consistent with the tendencies of the present warming, where the increase of air temperature causes an increase of annual precipitation and winter precipitation in particular, and consequently also an increase in the snow water equivalent (Førland and Hanssen-Bauer 2000; Kitaev 2002).

Although there is an universal long-term increase of winter air temperature in the region, the duration of snow cover increases in the north of the studied region and decreases in the south. This may be connected to zonal increase of air temperature from the north to the south at weak change of precipitation. The conclusion about connection of increase snow cover duration with negative trends of autumn air temperature is revealed also.

The exception is the Scandinavian Peninsula, where there is a long-term tendency of decrease in number of days with stable snow cover and a change of precipitation structure (increase of amount of mixed precipitation and decrease of amount of solid).

5.2.3.4   Problems

Observations over snow cover and snow depth in different countries are performed by different techniques, complicating common calculations. Metadata are incomplete or absent in general. The spatial distribution of weather stations is not uniform; the network is sparse in Arctic regions and denser in middle latitudes.

5.2.4   Atmospheric Moisture Flux, P–E Estimates

5.2.4.1   The Significance of the Atmospheric Moisture Flux

The atmospheric moisture flux arises from the three-dimensional distribution of moisture and wind speed. These may be observed with remote sensing techniques, such as horizontal wind speed estimates from cloud tracking or microwave-based measurement of moisture. Yet, the most direct method of measuring upper air humidity and (horizontal) wind speed – and the one with the longest history – are the radiosonde observations.

Atmospheric moisture content and storage are directly assimilated into atmospheric (re-)analyses, which provide a physically consistent image of the atmosphere. This direct link constitutes a significant difference to other modeled components of the atmospheric moisture budget such as precipitation P (which is forecasted using parameterizations of cloud microphysics) or evapotranspiration E (which is diagnosed using parameterizations of near-surface turbulence). Thus, the atmospheric moisture flux and its convergence are the part of the reanalyzed hydrological cycle which is most closely related to observations.

The budget equation for the atmospheric moisture reads:

$$\frac{\partial}{\partial t}Q + \int_V \vec{\nabla}(\rho q \vec{v})dV \approx \overline{E} - \overline{P} \qquad (5.4)$$

Here, $Q = \dfrac{\partial}{\partial t}\displaystyle\int_V \rho q dV$ is the integrated water vapor, ρ is air density, q is specific humidity, $\vec{v}$ is the horizontal wind speed, and the overbars denote that precipitation P and evapotranspiration E are integrated over the surface under the volume V. The derivation of the equation includes several approximations (such as storage and horizontal advection of cloud water being neglected); thus the equality is only approximate. For common use with spatial scales of hundreds of kilometers, it may readily be considered as exact.

Often, the components of Eq. 5.4 are only considered as temporal averages over one or more years. In this case the rate of change of stored water vapor $\partial Q/\partial t$ is negligible and the budget equation reduces to:

$$\int_V \vec{\nabla}(\rho q \vec{v})dV \approx \overline{E} - \overline{P} \qquad (5.5)$$

In other words, the net water flux into the surface $(\overline{E} - \overline{P})$ is – in the long-term average – equal to the integrated convergence of moisture flux, i.e., the water vapor flux into the volume above the surface.

5.2.4.2   Observational and Computational Problems

The analysis of atmospheric moisture flux convergence, outlined here, concentrates on the evaluation of radiosonde data. This brings up the question of how reliably this instrument type may measure moisture and horizontal wind speed. Unfortunately, this problem may not be tackled here satisfactorily; for further discussion, the reader is referred to, e.g., Wang et al. (2002). In the present context, moisture and wind speed measurements of radiosondes are considered as unbiased, at least for the relatively warm lower atmosphere where the main part of water vapor is present.

Computation of the moisture flux convergence from radiosonde data brings along two severe problems:

- The irregular distribution of the original data makes it necessary to derive an appropriate method for computing the derivatives. The seemingly obvious (and easiest) way is to first interpolate the data to some regular grid – a method which produces serious errors in the analyzed fields due to the sparsity of observations (cf. Doswell and Caracena 1988). The advisable alternative is to do the calculations directly on the observational grid.
- The continuity equation for dry air enters the derivation of Eq. 5.4. It is well known (cf., e.g., Ehrendorfer et al. 1994) that the use of a wind field which does not fulfill the continuity equation (that is not "mass consistent") produces serious errors which cannot be expected to vanish in spatial or temporal means. The way out of this problem is a variational approach to modify the wind field minimally making it mass consistent.

Keeping this in mind, the recommended method to incorporate radiosonde data in the computation of moisture flux convergence is a combination of a variational approach with the so-called finite-element method described in Göber et al. (2003) and Hagenbrock (2003).

5.2.4.3 Data and Calculations

The Historical Arctic Rawinsonde Archive (HARA) (Serreze and Shiotani 1997) is, to the knowledge of the author, the most comprehensive collection of radiosonde data for the Arctic region. It includes data for the period 1973 to mid-1996 north of 50°N with an average of about 180 stations reporting per day.

The most common number for comparison is the moisture flux convergence, integrated over the Arctic atmosphere north of 70°N (given as precipitation equivalents in mm/day). Via Gauss' law this is closely related to the integrated moisture flux over the 70th parallel, sometimes referred to as "vapor flux" in the literature. Still, the data enables to derive more regional statements such as for river catchment areas or lat/lon grids.

5.2.4.4 Comparison of Different Results

The first estimate of the net water vapor flux across the 70th parallel from radiosonde data was calculated by Peixoto and Oort (1983) for the years 1963–73. Their estimate of approx. 0.33 mm/day is rather small compared to budgets by Overland and Turet (1994) with 0.59 mm/day and by Serreze et al. (1995) with 0.45 mm/day

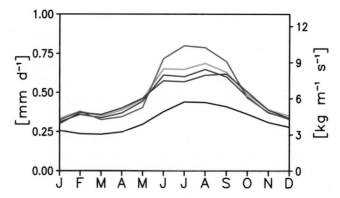

Fig. 5.15 Mean (1979–93, except for the Peixoto/Oort estimate) annual cycle of the moisture input into the polar cap north of 70°N, based on radiosonde data (*blue*: Göber et al. (2003), *purple*: Serreze et al. (1995), *black*: Peixoto and Oort (1983)) and on the ERA15 reanalysis with full resolution (*red*: Cullather et al. (2000)) and computed on the radiosonde grid (*green*: Göber et al. (2003))

(but for different time periods). A large uncertainty with respect to this key number based on observations is obvious. It may be hypothesized that this is due to different time intervals investigated and that it stems from inadequate calculation techniques in view of the problems mentioned above.

The uncertainty gets even more complex when results based on reanalyses are also taken into account. Here, a severe discrepancy becomes visible in the depiction of the form of the annual cycle: Reanalysis estimates show a much more pronounced summer maximum. It has been supposed that this difference is caused by the very different horizontal resolutions of the data entering the investigations. To further investigate this, an additional calculation has been performed with reanalysis data thinned out to the radiosonde grid, therefore avoiding the influence of different sampling errors.

A synopsis of different estimates of the net water vapor input into the Arctic is shown in Fig. 5.15. The estimate by Göber et al. (2003) is similar to the result of Serreze et al. (1995), which, basically, is based on the same data. The difference between reanalysis and radiosonde-based estimates is strongly reduced by performing the calculations on the same grid.

In Fig. 5.16 different depictions of the interannual variability are shown. Clearly, the form and amplitude of interannual variations become much more similar between reanalysis and radiosonde estimates when performing the calculations on the same grid. The difference becomes even smaller in the second half of the period, the reason for this being unclear. The estimate based on the calculations by Serreze et al. (1995) shows, for many years, a somewhat different behavior. This calculation seems to be incapable of describing interannual variability, probably due to the lack of mass consistency.

Finally, Fig. 5.17 shows two estimates of the horizontal structure of the net water gain of the Arctic atmosphere: one based on ERA15 reanalyses data (calculated by

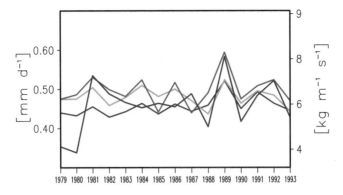

Fig. 5.16 Same as Fig. 5.15, but synopsis of annual means

the Climate Analysis Section of the NCAR 2002) and one based on the calculations by Göber et al. (2003), smoothed to a T42 resolution to enhance comparability (cf. Hagenbrock 2003). The latter calculation is only based on radiosonde data and thus devoid of uncertainties induced by moisture parameterizations and other inaccuracies of an underlying weather forecast model. This is – apart from the analysis by Peixoto and Oort (1992), which is focused on other parts of the globe – the only database to validate this quantity. Certain characteristics are shown in both figures, such as the maximum water input over southeast Greenland and Norway or the minimum in the Barents Sea, yet also large differences remain.

5.2.4.5 Summary and Conclusion

During the ACSYS period, the analysis of the atmospheric moisture and its convergence in the Arctic region has made a major step forward. A new analysis method has been developed which overcomes two obstacles in the evaluation of radiosonde data, namely, (1) the appropriate calculation of derivatives based on irregularly distributed data and (2) the necessity of the wind field to comply with the continuity equation. This gives an estimate of the average net water gain of the Arctic atmosphere of 0.46 mm/day, consistent with a previous estimate by Serreze et al. (1995). In contrast to this, the interannual variations and the horizontal structure may also be evaluated. The gap between radiosonde- and reanalysis-based estimates can be attributed largely to different sampling errors.

Calculations based on radiosonde data are still assumed a low estimate of the "true" value as (with respect to the radiosonde network) "sub-grid" fluxes are expected to be directed northward on average. The representation of orography, especially over Greenland, remains a problem.

All results obtained from the HARA data set $\left(\bar{\nabla}(\rho q \bar{v}), \quad Q, \quad \partial Q / \partial t\right)$ are available as monthly means for the years 1973 to mid-1996 on a regular $0.5° \times 0.5°$ grid north of 65°N and may be obtained from the author.

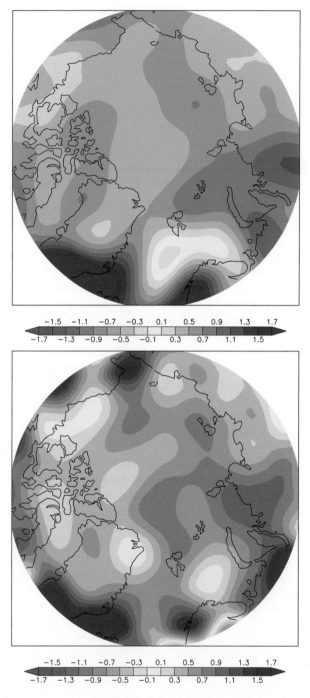

Fig. 5.17 Long-term average (1979–93) of the vertically integrated moisture flux convergence based on full ERA15 reanalysis data (*upper panel*) and on radiosonde data (*lower panel*). Units are mm/day

5.3   Evaluation of Reanalysis Results and Observations

5.3.1   Description of the Reanalyses

Global numerical weather predictions require global initial fields that represent the present state of the atmosphere. These fields are provided by operational analyses and comprise a data assimilation suite combining observations, previous forecasts, and model assumptions about the evolution of different meteorological variables. Since operational analyses are an estimate of the actual weather situation, long time series of these analyses should give an adequate description of the evolution of weather patterns, and their average would describe the climate.

However, the individual analyses are influenced by changes in the model, analysis technique, data assimilation, and the use of observations, which are an essential product of research and development at a numerical weather forecast center. Thus apparent changes of atmospheric conditions may occur in long time series of analysis fields which are caused only by changes in the corresponding analysis system. This led to the implementation of the reanalysis projects, in which a fixed analysis/forecast system is used to assimilate past observations over a long period of time. (Certain inconsistencies are still present, however, since the amount of available observations varies for different time periods.) For more detailed information on these topics, see, for example, Uppala (1997) and Kållberg (1997).

In this study about the Arctic Ocean catchment, we consider the two reanalyses of the European Centre for Medium-Range Weather Forecasts (ECMWF), the recent 40 years reanalysis, ERA40 (Simmons and Gibson 2000), and the previous 15 years reanalysis, ERA15 (Gibson et al. 1997), as well as the reanalysis of the National Centers for Environmental Prediction (NCEP) NRA (Kalnay et al. 1996). Here we will mainly focus on ERA40 as the hydrological cycle at the land surface of ERA15, and NRA has already been discussed before, e.g., by Hagemann and Dümenil Gates (2001) and Dai and Trenberth (2002).

5.3.1.1   ERA15

ERA15 was conducted for the years 1979–1993 (Gibson et al. 1997) at a spectral T106 resolution that corresponds to about 1.1°. In this study, the daily averages of two 24-h forecasts (ERA15–24; at 0000 and 1,200 UT) were used. In Sect. 5.3.2.1 (Table 5.10), also annual averages of the 6-h forecasts (ERA15–6) are considered. Stendel and Arpe (1997) have found that in most circumstances, ERA15–24 has a better representation of precipitation than ERA15–6, which is related to spin-up problems in the 6-h forecasts. Therefore, only ERA15–24 is considered elsewhere.

Table 5.10 Catchment water balances (P precipitation, E evaporation, R runoff) for the years 1979–1993 in km³/a. The observed discharge is a climatological value

Catchment	Six largest Arctic rivers				Full Arctic catchment			
	P	E	P–E	R	P	E	P–E	R
ERA40	4,609	3,299	1,310	1,713	7,002	4,580	2,422	3,027
ERA40-SL		2,514	2,095	2,291		3,205	3,797	3,709
ERA15-6	3,634	3,193	441	520	5,341	4,375	966	1,117
ERA15-24	4,146	3,103	1,043	520	6,071	4,230	1,841	1,117
NRA	6,261	5,332	929	5,093	8,629	7,000	1,629	8,884
CMAP	3,921	–	–	–	5,698	–	–	–
CRU	3,988	–	–	–	5,862	–	–	–
GPCP	4,808	–	–	–	7,341	–	–	–
Observed discharge	–	2,423[a]	1,975	1,975	–	–	–	–

[a]Estimated from the mean precipitation of CRU and GPCP minus the observed discharge

5.3.1.2 ERA40

ERA40 was conducted for the years 1958–2001 (Simmons and Gibson 2000) at a spectral T159 resolution whereas all grid point space variables (such as all surface variables) are available at the same T106 grid as in ERA15. The whole ERA40 time period can be divided into three parts: the satellite period 1989–2001 when a lot of satellite data were assimilated into the ERA40 system, the pre-satellite period 1958–1972 when no satellite data were available, and the transition period 1973–1988 when the amount of satellite data that were assimilated increases with time. These three periods correspond also to the three streams which were produced separately during the ERA40 production timeframe. For ERA40, only the 6-h forecasts are considered as monthly means; the 24-h forecasts were not available at the time of the study.

5.3.1.3 NCEP Reanalysis

The NCEP reanalysis (NRA) was conducted with the numerical weather forecast model of the NCEP (Kalnay et al. 1996). A spectral horizontal resolution of T62 was used, which corresponds to a resolution of about 1.9°. The reanalysis extends from the year 1948 to the present. Due to the known deficiencies in the NRA hydrological cycle (Hagemann and Dümenil Gates 2001), we consider NRA values only for comparisons with regard to the ERA15 period.

5.3.1.4 Additional Tools

As in Hagemann and Dümenil Gates (2001), the combination of the simplified land surface (*SL*) scheme (Hagemann and Dümenil Gates 2003) and the hydrological discharge (*HD*) model (Hagemann and Dümenil 1998) was used to simulate river

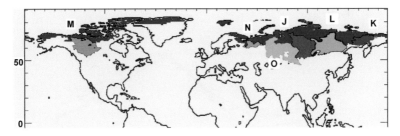

Fig. 5.18 The Arctic Ocean catchment is represented by the six largest rivers draining into the Arctic Ocean (*J* Yenisei, *K* Kolyma, *L* Lena, *M* Mackenzie, *N* Northern Dvina, *O* Ob)

discharges from daily reanalysis data of precipitation and 2-m temperature. The application of the SL scheme was necessary since neither of the reanalyses is suitable for a direct application of a global-scale discharge simulation that requires surface runoff and drainage as input fields (cf. Hagemann and Dümenil Gates 2001). For instance, the ERA15 drainage from the soil (at 3-m depth) is larger than its surface runoff by 2 orders of magnitude at the majority of the land points. This is unrealistic for most parts of the world. Even though this ratio has decreased in ERA40, here also almost the whole total runoff consists of drainage.

5.3.2 The Hydrological Cycle over the Arctic Ocean Catchment

The Arctic Ocean catchment is represented by the catchment area (about 9.7 million km^2) of its six largest rivers (Yenisei, Kolyma, Lena, Mackenzie, Northern Dvina, Ob) which cover an area of about 65% of the whole Arctic Ocean catchment (Fig. 5.18). More detailed information about catchments and observed discharges can be obtained from Dümenil Gates et al. (2000).

5.3.2.1 ERA15 Period 1979–1993

Table 5.10 shows the annual mean components of the hydrological cycle over the catchment of the six largest Arctic rivers and the full Arctic Ocean catchment for the ERA15 period 1979–1993. Observations comprise CMAP (Xie and Arkin 1997), CRU (New et al. 2000), and GPCP (Huffman et al. 1997) precipitation data. Both CMAP and CRU precipitation data are not corrected for the systematic undercatch of precipitation gauges which is especially significant for snowfall. For GPCP data, a correction has been applied which is known to be too large by a factor of about 2 (Rudolf, 2001, personal communication) so that the actual precipitation amounts are expected to be in between GPCP and CMAP/CRU.

Due to soil moisture nudging conducted in all three reanalyses, the land surface water balance is not closed, and consequently the total runoff does not correspond

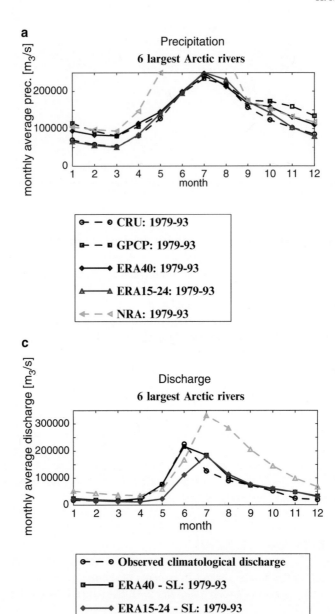

Fig. 5.19 Annual cycles of (**a**) precipitation, (**b**) difference of 2-m temperature to CRU data, (**c**) discharge over the six largest Arctic rivers

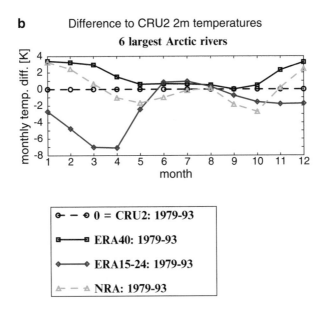

Fig. 5.19 (continued)

to precipitation minus evapotranspiration (P–E) in the long-term mean. In ERA40, the effect of soil moisture nudging is generally less severe than in ERA15 or NRA, where the water balance at the land surface is not closed for many large catchments (Hagemann and Dümenil Gates 2001). A detailed description of the different nudging methods used in ERA15 and NRA is given, for example, by Roads and Betts (2000). Thus, only P–E was considered to represent ERA40 runoff in Sect. 5.3.2.2.

NRA largely overestimates the precipitation over the Arctic Ocean catchment. This overestimation occurs mainly in the summer where the NRA precipitation is about twice as large as observed (Fig. 5.19). This applies to most land parts of the Northern Hemisphere (Hagemann and Dümenil Gates 2001). ERA15 underestimates the precipitation, especially in the 6-h forecasts, which is mainly to a winter dry bias (Fig. 5.19a). This dry bias has been eliminated in ERA40 where the precipitation agrees well with the observations.

The discharge (Fig. 5.19c) simulated with the SL scheme and the HD model is consistent with the precipitation comparison. The late discharge peak for ERA15–24-SL is connected to a severe ERA15 cold bias in high latitudes during the winter and spring (Fig. 5.19b) which causes a late snowmelt in the SL scheme. This cold bias has been corrected (Viterbo et al. 1999; Viterbo and Betts 1999) in ERA40 where now a smaller warm bias has evolved. The latter may be related to a too low snow albedo in forested regions (Betts et al. 2003).

Both ERA15 and NRA have several deficiencies in the treatment of snow and snowmelt over land (see Hagemann and Dümenil Gates 2001) so that they are not considered in this study. In ERA40 the snow analysis has changed completely compared to ERA15. In data-sparse regions there will be a relaxation to the Foster and

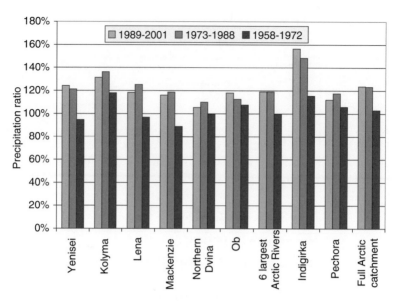

Fig. 5.20 Precipitation ratio: ERA40/observed precipitation

Davy (1988) climatology, and snowfall gauge observations, known to be biased low by as much as 30%, are not used. Consequently, the annual cycle of the ERA40 snowpack over the Arctic catchment agrees well with the Foster and Davy (1988) climatology in the ERA15 period (not shown).

5.3.2.2 Hydrological Cycle for Several Arctic Rivers in the Three ERA40 Subperiods

Figure 5.20 shows the precipitation ratio of ERA40 data against observations for all three ERA40 periods. Here, GPCC data (Rudolf et al. 1996) are used as observations for the latest period, and CRU data (New et al. 2000) are used for the two other periods. This precipitation ratio gives an idea of the precipitation bias which varies within the three periods. For almost all Arctic rivers, the ERA40 precipitation is close to the observations only in the earliest period. In the two later periods, it is larger than the observations by about 20% of the average. The fact that the precipitation bias is not the same in all three periods leads to the conclusion that the different observing systems available in the three periods are influencing the quality of the ERA40 precipitation over land. Thus, no conclusions on trends in precipitation over land can be drawn from the ERA40 data.

As GPCC and CRU data are uncorrected for precipitation undercatch (cf. Sect. 5.3.2.1), the ERA40 precipitation seems to be adequate in the two later periods while it is underestimated in the earliest period where the underestimation occurs mainly in the summer (see Fig. 5.21a). This is supported by the validation of the

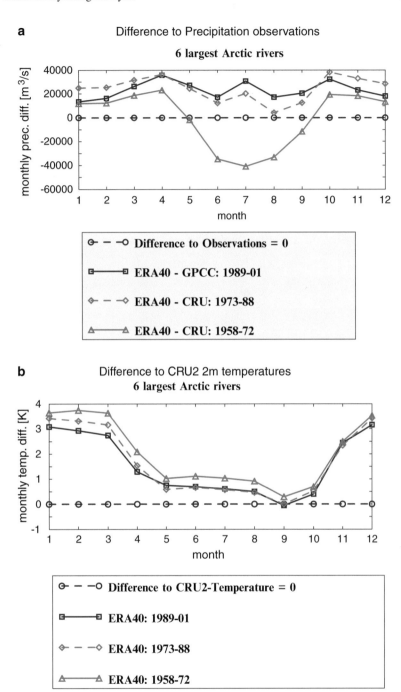

Fig. 5.21 Annual cycles in the 3 ERA40 subperiods for the six largest Arctic rivers: (**a**) difference of precipitation to observations, (**b**) difference of 2-m temperature to CRU2 data, (**c**) discharge simulated with the HD model, (**d**) accumulated snowpack

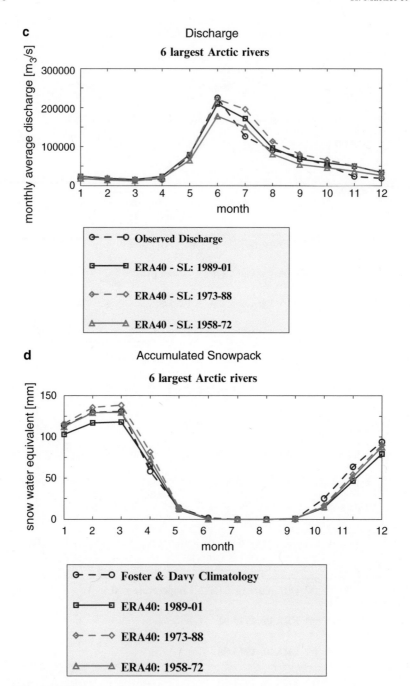

Fig. 5.21 (continued)

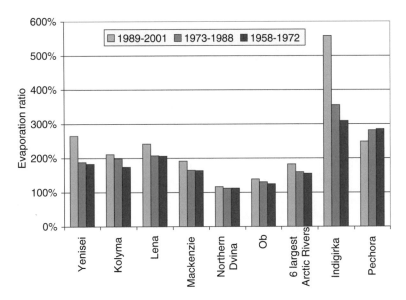

Fig. 5.22 Evapotranspiration ratio ERA40/estimated observations

simulated discharge. Figure 5.21c shows that the seasonal cycle of discharge is simulated reasonably well in the two later periods, but in the earliest period, the discharge is slightly underestimated. As the SL scheme evaporation (ERA40-SL) is much closer to the observational estimates (cf. Table 5.10) than the ERA40 evaporation, we gain confidence that the simulated discharge is consistent with the ERA40 precipitation, i.e., deficiencies in the ERA40 precipitation become visible in the simulated discharge.

Figure 5.22 shows an estimate of the ERA40 evaporation ratio. Here, the observed evaporation was estimated by subtracting the climatological observed discharge (Dümenil Gates et al. 2000) from the observed precipitation. For most of the catchments, evaporation is largely overestimated in ERA40. As for precipitation, the bias varies between the 3 periods for some of the catchments. Here, the evaporation ratio in the latest period tends to be larger than in the two earlier periods. Consequently, the runoff represented by P–E is underestimated for most of the Arctic rivers as shown in Fig. 5.23. Here, the runoff ratio is calculated as the ERA40 P–E divided by the climatological discharge. For the Mackenzie River, the runoff ratio is even negative in the earliest period which is directly related to the underestimation of precipitation. Here, the soil moisture nudging (cf. Sect. 5.3.2.1) acts as a source of water.

A comparison of the ERA40 2-m temperature with CRU2 surface air temperatures (New et al. 2002) shows that the ERA40 warm bias in the winter tends to be larger in the earliest period for most of the Arctic rivers (Fig. 5.21b). As for the ERA15 period (cf. Sect. 5.3.2.1), there is generally a very good agreement between ERA40

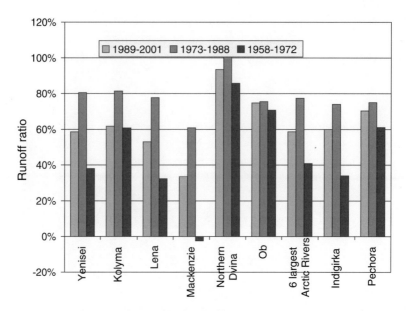

Fig. 5.23 Runoff (P–E) ratio ERA40/observed runoff

snowpack and the snow data climatology of Foster and Davy (1988). Only for the latest period, the snowpack is fairly underestimated in ERA40 for most of the Arctic rivers (Fig. 5.21d).

5.4 Synopsis of the Observed Arctic Hydrological Cycle: Where Are We Now?

During the ACSYS decade (1994–2003), important progress has been made in the Arctic research resulting in an extension of the observational data basis and reexamination of the main components of the Arctic hydrological cycle.

At GPCC the Arctic Precipitation Data Archive (APDA) was established, which has significantly extended and updated monthly precipitation records. The number of available precipitation stations in the Arctic was increased nearly by a factor of 10 during the ACSYS decade. Based on the quality-controlled, prolonged, and homogenized monthly time series of total precipitation, a gridded precipitation climatology (in a 100-×100-km EASE grid) for the Arctic drainage system spanning the period 1950–2000 was compiled. This precipitation climatology constitutes a new improved observational data set for studies of energy and water cycle as well as for validation of satellite and model data in the Arctic region. It is published on the homepages of the Global Precipitation Climatology Centre (gpcc.dwd.de). In addition, GPCC starts also collection of observed snow data.

GRDC (in Koblenz, Germany) established the Arctic Runoff Data Base (ARDB). This discharge data collection originates from the networks of the National Hydrological Services of the countries within the Arctic drainage basin. It contains 2,404 stations with monthly runoff records of a mean length of about 30 years. Daily records are available for 1,023 stations. Based on these data the fresh water flux into the Arctic Ocean as well as its changes were estimated (cf. Sect. 5.2.2).

During the ACSYS decade the snow cover and snow depth data sets from the Former Soviet Union were updated for the period 1991–2000. The snow depth and snow cover climatology consists of observations from more than 1,400 stations from CIS countries and 100 Fennoscandian stations; the winter air temperature is derived from 223 stations (cf. Sect. 5.2.3).

Based on radiosonde data the atmospheric moisture flux and transport (P–E) are reexamined by a new analysis method and compared with estimates of other authors and with ERA15 reanalysis (cf. Sect. 5.2.4). The new analysis gives an estimate of the mean net water input to the Arctic atmosphere of 0.46 mm/day, which is consistent with an earlier assessment by Serreze et al. (1995).

The extended and updated database enables now an assessment of long-term changes in precipitation, runoff, snow cover, and snow depth for large parts of the Arctic drainage basin. The comparison of the yearly precipitation and runoff variations in the three Eurasian catchment areas of North Dvina, Lena, and Yenisei indicates that no significant changes in both components have been occurred in the period 1950–2000 (cf. Sect. 5.2.1).

The analysis of the annual discharge of the seven largest rivers draining to the Arctic Ocean in the period 1979–1993 suggests that no significant trends can be identified. Slightly negative and partly significant trends, however, are observed in runoff maximum. This indicates that the runoff maximum occurs few days earlier (earlier snowmelt and earlier runoff), which is consistent with observed warming in high latitudes (Sect. 5.2.2).

In Sect. 5.2.3 the spatial and temporal variability of snow cover and snow depth in Eurasia and Fennoscandia are discussed in connection with the dynamic and variability of the North Atlantic Oscillation. While in Fennoscandia the snow cover duration and snow depth decrease, they increase in Eurasia accompanied by an increasing winter time temperature in the period 1936–2000.

The atmospheric moisture flow did not show any tendency in the examined period 1979–1993 (Sect. 5.2.4).

More comprehensive information about the global hydrological cycle can be obtained from the 4-dim. global reanalysis (ERA15, ERA40, and NCEP), which contain not or rarely observable parameters such as water equivalent of snow, snow temperature, soil water, or latent and sensible heat flux.

However, the reliability of the estimated hydrological cycle components is partly questionable, especially in high latitudes. In Sect. 5.3 the evaluation of the reanalysis and observations are discussed in more detail. Table 5.11 summarizes the results and compares the averaged hydrological cycle components throughout

Table 5.11 Catchment water balances (P precipitation, E evaporation, R runoff) for the years 1979–1993 in km³/a (for details see Table 5.10)

Catchment	Six largest Arctic rivers				Full Arctic catchment			
	P	E	P–E	R	P	E	P–E	R
Reanalysis	4,663	3,488	1,164	2,027	6,761	4,678	2,131	3,571
Observations	4,239	2,376	1,863	1,863	6,300	3,992	2,609	2,609
Mean (line (1+2)/2)	4,451	2,932	1,514	1,945	6,568	4,222	2,370	3,090

the three reanalyses with observations for the 6 largest Arctic rivers and full Arctic catchment area. As can be seen, the reanalyses overestimate the observed values of precipitation by 7%, the evaporation by 17%, and runoff by about 37%. In contrast, the reanalyses underestimate the observed P–E by 22%. Thus the water balance in the Arctic catchment area is not closed (cf. Sect. 5.3). Because of inhomogeneities in the reanalysis data, resulting from changing number and type of the assimilated observations, although the Numerical Weather Prediction Model (NWP) was not modified, these data sets are not applicable for an assessment of long-term trends in hydrology.

The progress in the ACSYS program was only possible with the scientific progress in several others areas. Some of these should be mentioned here, which had essentially contributed to an improved assessment of the different components of the global and high-latitude hydrological cycle. In this context, special attention is received to such activities which result in a compilation of new/updated observational data sets for the high latitudes in the Northern Hemisphere.

Using of satellite-derived precipitation for the oceans and data-sparse land areas is one of the milestones in the collection of an extended database.

- In 1995 the version 1 of the GPCP merged satellite-gauge monthly data set (without high latitudes) for the period 1988–1994 was published (Huffman et al. 1997).
- Monthly precipitation estimates from TOVS for high latitudes were released in 1999. This was the version 2 of the merged satellite-gauge data set which covers the period 1979–1999 (Adler et al. 2003).
- In the following year, GPCP provided a satellite-based, daily gridded (1° resolution) precipitation data set for the period 1997–1999. For high latitudes it contains precipitation estimated from TOVS (Huffman et al. 2001).
- The second important precipitation data set CMAP (CPC Merged Analysis of Precipitation), compiled by Xie and Arkin (1997), merges gauge measurements, satellite observations, and NCEP reanalysis.
- Snow cover data, derived from visible satellite imagery at NOAA/NESDIS, are operationally available since November 1966 in a 190-km and 7-day resolution on an 89×89 polar stereographic grid for the Northern Hemisphere.

- Since February 1997 a daily snow/ice analysis exists in an improved resolution (~25 km). They use new as well as existing satellite and surface imagery products. In early 2004, the spatial resolution was again improved to 4 km.

The WMO has published in 1998 the results of the Solid Precipitation Measurement Intercomparison (Goodison et al. 1998). Based on these findings, the appropriate correction factors for individual precipitation events can be evaluated. This type of correction is more reliable than those derived from climatological conditions. The transfer of this "on-event" correction to application on global and monthly data sets is still in development.

The US Weather Service and the Meteorological Service of Canada made great effort to digitize their own historic daily climatic data and made them available for the public on CD-ROM. The data set Nordklim (Tuomenvirta et al. 2001) includes monthly observations of 12 different climate elements from more than 100 stations in the Nordic region for more than 100 years.

An overview of the Arctic climate is given in the Arctic Meteorology and Climate Atlas, which is a cooperative work of the US and Russian scientists (Arctic Climatology Project, Fetterer and Radionov 2000). This atlas contains monthly mean gridded fields of the two-meter air temperature, sea level pressure, precipitation, cloud cover, snow depth, and radiation from 65 Russian and 24 western coastal, island, and drifting stations for the period 1950–1990 and is available from NSIDC on CD-ROM.

In addition, noteworthy is the compilation of reanalysis data products, at several data centers (e.g., ECMWF, NCEP/NCAR) which enable a 3-dim. view on global weather and climate. Their features, especially in regard to the hydrological cycle components at high latitudes, are discussed in Sect. 5.3.1.

5.4.1   Updated Precipitation Time Series

The Global Precipitation Climatology Centre (GPCC) computes regularly global monthly precipitation maps based on land surface stations. This data can be analyzed for the Arctic. Figure 5.24 shows seasonal precipitation time series for the period 1951–2009 north of 55°N. In all four seasons the precipitation in the first decade of the twenty-first century has increased. This is even more clearly displayed by the annually accumulated anomalies shown in Fig. 5.25. The anomalies are calculated from the long-term monthly sums of the period 1951–2000. The accumulated values start each hydrological year in November (first bar), the second bar represents the sum of the anomalies for November and December, and the twelfth bar is the annual anomaly from November to October. The twenty-first century shows most of the time positive precipitation anomalies.

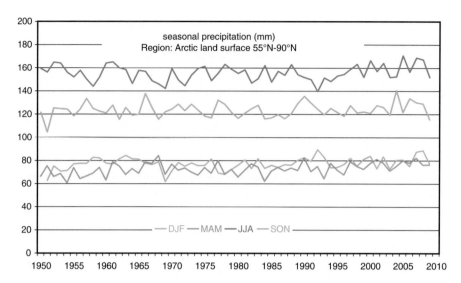

Fig. 5.24 Land-based precipitation of seasonal totals for the period 1951–2009 north of 55°N (DJF for December, January, and February; MAM for March, April, May …)

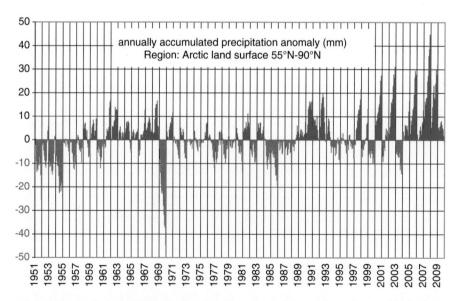

Fig. 5.25 Annually accumulated precipitation anomalies (1951–2009) north of 55°N, reference period 1951–2000 (for details see text)

5.5 Open Questions, Future Tasks

The Arctic hydrological cycle is in many aspects a challenge to the science. The transition between the liquid and solid phase affects all processes in the Arctic. Despite the enormous progress, which is made in the collection of the hydrological cycle–related data and their analysis, many problems are still unresolved. The highest priority should be given to the improvement of the availability and quality/reliability of the observational data basis. This can be achieved, for example, by means of:

- Acceleration of the data flow from the national services to the global data centers in the era of internet and automation of observations.
- Enhanced accuracy of the observed data.
- Collection of metadata. For the adjustments of the catch deficiencies (cf. Sect. 5.2.1.1), the global data centers need readily retrievable metadata information for all of the stations, including, in the case of climate and precipitation stations, location and instrumentation characteristics and their changes. This can be done by updating the instrument parameter catalog compiled by Sevruk and Klemm (1989).
- Enhanced data basis through digitizing of historical observations in their best time resolution as was already began by many National Weather Services (see previous section), and to make them available for scientists as required by the WMO Data Rescue Program, DARE.
- Additional incorporation of radar observation to the existing gridded data sets.
- Extension of the measurement programs. In the future, there will be a need of additional observed parameters (like significant weather, frozen ground, soil moisture, etc.) to make the historic observations comparable to the modern measurements (e.g., precipitation from radar or remote sensing snow depth).
- Application of different correction techniques to increase the data reliability. An interpolation scheme for the monthly bias corrections from synoptic stations to precipitation-only stations as a routine has to be implemented. The assessment of the accuracy of the new correction method by error analysis based on variation of the input parameters has to be realized. To increase the accuracy of the satellite-based precipitation estimates, a temporal interpolation between satellite overflights has to be incorporated.
- Development of homogenization procedures for daily data and homogenization that should be done by the national services.
- Cross-validation of the observed data, e.g., snow depth data derived from remote sensing need to be validated by surface measurements.
- Improvements of measurements for winter discharge of frozen rivers.

Because the surface observational networks are under severe cost pressures and automation leads to loss and degradation of key data, it is to hope that the conventional measurements could be replaced by new remote sensing monitoring programs such as Global Precipitation Mission (GPM) based on TRMM experience.

References

Adler RF, Huffman GJ, Chang A, Ferraro R, Xie P-P, Janowiak J, Rudolf B, Schneider U, Curtis S, Bolvin D, Gruber A, Susskind J, Arkin P, Nelkin E (2003) The version-2 global precipitation climatology project (GPCP) monthly precipitation analysis (1979-present). J Hydrometeorol 4:1147–1167

Baumgartner A, Reichel E (1975) The world water balance – mean annual, global, continental and maritime precipitation, evaporation and runoff. Elsevier, Amsterdam

Betts AK, Ball JH, Viterbo P (2003) Evaluation of the ERA-40 surface water budget and surface temperature for the Mackenzie River basin. J Hydrometeorol 4:1194–1211

Chen M, Xie P, Janowiak JE, Arkin PA (2002) Global land precipitation: a 50-yr monthly analysis based on gauge observations. J Hydrometeorol 3:249–266

Climate Analysis Section (2002) Atmospheric mass, moisture and energy budget products from ECMWF 15-Year reanalysis at tapered T42. http://www.cgd.ucar.edu/cas/catalog/outline.html. Accessed 26 Mar 2010

Cullather RI, Bromwich DH, Serreze MC (2000) The atmospheric hydrologic cycle over the Arctic Basin from reanalyses. Part I: comparison with observations and previous studies. J Clim 13:923–937

Dai A, Trenberth KE (2002) Estimates of freshwater discharge from continents: latitudinal and seasonal variations. J Hydrometeorol 3:660–687

de Couet T, Maurer T (2009) Surface freshwater fluxes into the world oceans. Global Runoff Data Centre, Koblenz, Federal Institute of Hydrology (BfG), http://www.bafg.de/cln_016/nn_294838/GRDC/EN/02__Services/02__DataProducts/FreshwaterFluxes/freshflux__node.html. Accessed 26 Mar 2010

Doswell CA III, Caracena F (1988) Derivative estimation from marginally sampled vector point functions. J Atmos Sci 45:242–253

Dümenil Gates L, Hagemann S, Golz C (2000) Observed historical discharge data from major rivers for climate model validation, report 307, Max-Planck-Inst für Meteorol, Hamburg

Ehrendorfer M, Hantel M, Wang Y (1994) A variational modification algorithm for the three-dimensional mass flux non-divergence. Q J R Meteorol Soc 120:655–698

Fetterer F, Radionov V (2000) Arctic climatology project: environmental working group Arctic meteorology and climate atlas. National Snow and Ice Data Center, Boulder, CD-ROM

Førland EJ, Hanssen-Bauer I (2000) Increased precipitation in the Norwegian Arctic: true or false? Clim Change 46:485–509

Førland EJ, Allerup P, Dahlström B, Elomaa E, Jónsson T, Madsen H, Perälä J, Rissanen P, Vedin H, Vejen F (1996) Manual for operational correction of Nordic precipitation data. Norwegian Meteorological Institute, Oslo

Foster DJ, Davy RD (1988) Global snow depth climatology, USAFETAC/TN-88/006. Scott Air Force Base, Scott AFB

Gibson JK, Kållberg P, Uppala S, Hernandez A, Nomura A, Serrano E (1997) Era description, ECMWF Reanal. Project Report Series 1. European Centre for Medium-Range Weather Forecasting, Geneva

Gleason BE (2002) National Climatic Data Center. Data documentation for data set 9101, Global Daily Climatology Network, V1.0. NCDC. http://lwf.ncdc.noaa.gov/oa/climate/research/gdcn/GDCN_V1_0.pdf. Accessed 26 Mar 2010

Göber M, Hagenbrock R, Ament F, Hense A (2003) Comparing mass consistent atmospheric budgets on an irregular grid: an Arctic example. Q J R Meteorol Soc 129:2383–2400

Goodison BE, Louie PYT, Yang D (1998) WMO solid precipitation measurement intercomparison. Final report. WMO/TD-No. 872. Instruments and observing methods report no. 67, Geneva

Grabs WE, Portmann F, de Couet T (2000) Discharge observation networks in Arctic regions: computation of the river runoff into the Arctic Ocean, its seasonality and variability. In: Lewis et al (eds) The freshwater budget of the Arctic ocean. Proceedings of the NATO advanced research workshop, Tallinn, 27 Apr–1 May 1998, Kluwer Academic, Dordrecht

Hagemann S, Dümenil L (1998) A parameterization of the lateral waterflow for the global scale. Clim Dyn 14:17–31

Hagemann S, Dümenil Gates L (2001) Validation of the hydrological cycle of ECMWF and NCEP reanalyses using the MPI hydrological discharge model. J Geophys Res 106:1503–1510

Hagemann S, Dümenil Gates L (2003) Improving a subgrid runoff parameterization scheme for climate models by the use of high resolution data derived from satellite observations. Clim Dyn 21:349–359

Hagenbrock R (2003) Der Feuchtehaushalt der arktischen Troposphäre aus Radiosonden-Messungen (The moisture budget of the Arctic atmosphere based on radiosonde measurements). PhD thesis, University of Bonn, Bonn (in German)

Huffman GJ, Adler RF, Arkin A, Chang A, Ferraro R, Gruber A, Janowiak J, Joyce RJ, McNab A, Rudolf B, Schneider U, Xie P (1997) The global precipitation climatology project (GPCP) combined precipitation data set. Bull Am Meteorol Soc 78:5–20

Huffman GF, Adler RF, Morrissey MM, Bolvin DT, Curtis S, Joyce R, McGavock B, Susskind J (2001) Global precipitation at one-degree daily resolution from multi-satellite observations. J Hydometeorol 2:36–50

Kållberg P (1997) Aspects of the re-analysed climate. ECMWF reanalysed Project Report Series 2, European Centre for Medium-Range Weather Forecasting, Geneva

Kalnay E et al (1996) The NCEP/NCAR 40-year re-analysis project. Bull Am Meteorol Soc 77:437–471

Kitaev LM (2002) Spatial-temporal variability of snow cover depth in the northern hemisphere. Russ Meteorol Hydrol 5:28–34

Kitaev L, Kislov A, Krenke A, Razuvaev V, Martuganov R, Konstantinov I (2002) The snow cover characteristics of northern Eurasia and their relationship to climatic parameters. Boreal Environ Res 7:437–446

Kopanev ID (1971) Methods for snow cover studying. Gidrometeoizdat, Leningrad (In Russian)

Korzoun VI, Sokolov AA, Voskresensky KP et al (1977a) World water balance and water resources of the earth. UNESCO Press, Paris

Korzoun VI, Sokolov AA, Voskresensky KP, Kalinin GP, Konoplyantsev AA, Korotkevich ES, Lvovich MI (1977b) Atlas of world water balance. UNESCO Press, Paris

Legates DR (1987) A climatology of global precipitation, Publ Climatol, vol XL, no 1, Newark, Delaware

MacKay DK, Løken OH (1974) Arctic hydrology. In: Ives JD, Barry RG (eds) Arctic and alpine environments. William Clowes and Sons, London

Manual for weather stations and posts (1985) vol. 3, part I and II, Gidrometeoizdat, Leningrad, Russian Federation, 164 p. (In Russian)

Masuda K (1990) Atmospheric heat and water budgets of polar regions: analysis of FGGE data. Proc NIPR Symp Polar Meteorol Glaciol 3:79–88

McKay GA, Gray DM (1981) The distribution of snow cover. In: Gray DM, Male DH (eds) Handbook of snow: principles, processes, management and use. Pergamon Press, Toronto

New M, Hulme M, Jones P (1999) Representing twentieth-century space-time climate variability. Part I: development of a 1961–90 mean monthly terrestrial climatology. J Clim 12:829–856

New M, Hulme M, Jones P (2000) Representing twentieth-century space-time climate variability. Part II: development of 1901–96 monthly grids of terrestrial surface climate. J Clim 13:2217–2238

New M, Lister D, Hulme M, Makin I (2002) A high-resolution data set of surface climate over global land areas. Clim Res 21:1–25

Overland JE, Turet P (1994) Variability of the atmospheric energy flux across 70°N computed from the GFDL data set. The polar oceans and their role in shaping the global environment, The Nansen centennial volume, Geophysical Monograph no 85. American Geophysical Union, Washington, DC, pp 313–325

Peixoto JP, Oort AH (1983) The atmospheric branch of the hydrological cycle and climate. In: Street-Perrott A, Beran M, Ratcliffe R (eds) Variations in the global water budget. D. Reidel, Dordrecht

Peixoto JP, Oort AH (1992) Physics of climate. Am Institute of Physics/Springer, New York

Peterson BJ, Holmes RM, McClelland JW et al (2002) Increasing river discharge to the Arctic ocean. Science 298:2171–2173

Roads J, Betts A (2000) NCEP-NCAR and ECMWF reanalysis surface water and energy budgets for the Mississippi River basin. J Hydrometeorol 1:88–94

Rubel F, Hantel M (1999) Correction of daily rain gauge measurements in the Baltic Sea drainage basin. Nordic Hydrol 30:191–208

Rudolf B, Hauschild H, Rüth W, Schneider U (1996) Comparison of rain gauge analyses, satellite-based precipitation estimates and forecast model results. Adv Space Res 7:53–62

Schmidlin TW (1995) Automated quality control procedure for the "water equivalent of snow on the ground" measurement. J Appl Meteorol 34:143–151

Sellers WD (1965) Physical climatology. The University of Chicago Press, Chicago/London

Serreze MC, Shiotani S (1997) Historical Arctic rawinsonde archive. Technical report, National Snow and Ice Data Center, Boulder, CD-ROM

Serreze MC, Barry RG, Walsh JE (1995) Atmospheric water vapor characteristics at 70°N. J Clim 8:719–731

Sevruk B (ed) (1986) Correction of precipitation measurements. ETH/IAHS/WMO workshop of precipitation measurements, Zurich, 1–3 Apr 1985. ETH Geographisches Institut, Zurich

Sevruk B (ed) (1990) Precipitation measurement. WMO/IAHS/ETH workshop of precipitation measurements, St. Moritz, 3–7 Dec 1989. ETH Geographisches Institut, Zurich

Sevruk B, Klemm S (1989) Catalogue of national standard precipitation gauges. World meteorological organization, instruments and observing methods Report, WMO/TD-no. 313

Sherwood K, Idso C (2003) Will global warming shut down the thermohaline circulation of the world's oceans? CO2 Sci Mag 6(8), editorial commentary http://www.co2science.org/edit/v6_edit/v6n8edit.htm. Accessed 25 Mar 2010

Shiklomanov IA, Shiklomanov AI, Lammers RB, Peterson BJ, Vorosmarty CJ (2000) The dynamics of river water inflow to the Arctic Ocean. In: Lewis et al (eds) The Freshwater budget of the Arctic Ocean, Proceedings of the NATO advanced research workshop, Tallinn, 27 Apr–1 May 1998, Kluwer Academic, Dordrecht

Simmons AJ, Gibson JK (2000) The ERA-40 project plan, ERA-40 Project Report Series 1, European Centre for Medium-Range Weather Forecasting, Reading

Stendel M, Arpe K (1997) Evaluation of the hydrological cycle in re-analyses and observations. Report 228, Max-Planck-Institut for Meteorology, Hamburg

Tuomenvirta H, Drebs A, Førland E, Ole Tveito E, Alexandersson H, Vaarby Laursen E, Jónsson T (2001) Nordklim data set 1.0 – description and illustrations. DNMI report no. 08/01, Norwegian Meteorological Institute, Oslo, Norway

Uppala S (1997) Observing system performance in ERA, ECMWF Reanalysed Project Report Series 3, European Centre for Medium-Range Weather Forecasting, Reading

Viterbo P, Betts AK (1999) Impact on ECMWF forecasts of changes to the albedo of the boreal forests in the presence of snow. J Geophys Res 104:27803–27810

Viterbo P, Beljaars ACM, Mahfouf J-F, Teixeira J (1999) The representation of soil moisture freezing and its impact on the stable boundary layer. Q J R Meteorol Soc 125:2401–2426

Wang J, Cole HL, Carlson DJ, Miller ER, Beierle K (2002) Corrections of humidity measurement errors from the Vaisala RS80 radiosonde – Application to TOGA COARE data. J Atmos Ocean Technol 19:981–1002

Xie P, Arkin P (1997) Global precipitation: a 17-year monthly analysis based on gauge observations, satellite estimates and numerical model outputs. Bull Am Meteorol Soc 78:2539–2558

Chapter 6
Interaction with the Global Climate System

T.A. McClimans, G.V. Alekseev, O.M. Johannessen, and M.W. Miles

Abstract The Arctic is part of the global climate system. To address the issue of climate, the fluxes of heat, salt, and fresh water must be considered. One of the most speculated reasons for rapid climate change in the subarctic North Atlantic, and the global conveyor belt, is a breakdown of the thermohaline circulation (THC) due to an increased fresh water supply. Whitehead's (Estuaries 21:281–293, 1998) one-box dynamic model is used to show how multiple states and catastrophe can occur in the Arctic Mediterranean with variable freshening and cooling. The broader question is how this interacts with the global climate. In this chapter, we focus on the oceanic aspects of the arctic climate system, discuss processes, review the data, and speculate on the role this part of the globe has in the greater context of global climate. The interaction with the global system comprises the outflow of freshwater and ice, and deeper, freshened, and cooled seawater into the subarctic North Atlantic, via the Labrador Sea. An example of significant climate variability in the twentieth century is presented.

T.A. McClimans (✉)
SINTEF Fisheries and Aquaculture, 7465 Trondheim, Norway
e-mail: thomas.mcclimans@sintef.no

G.V. Alekseev
AARI, St. Petersburg, Russia
e-mail: alexgv@aari.nw.ru

O.M. Johannessen
NERSC, 5059 Bergen, Norway
e-mail: ola.johannessen@nersc.no

M.W. Miles
Bjerknes Centre for Climate Research, 5007 Bergen, Norway and Environmental Systems Analysis Research Center, Boulder, USA

ESARC, Boulder, CO, USA
e-mail: martin.miles@geo.uib.no

P. Lemke and H.-W. Jacobi (eds.), *Arctic Climate Change: The ACSYS Decade and Beyond,* Atmospheric and Oceanographic Sciences Library 43, DOI 10.1007/978-94-007-2027-5_6, © Springer Science+Business Media B.V. 2012

List of Symbols and Abbreviations

Symbols

A	Albedo
G_i	Mode of convection
M	Advection term
M_a	Atmospheric advection
M_o	Ocean advection
q_T	Cooling
Q	Ventilation of saline water in AM
Q_f	Fresh water supply
S	Short wave radiation; Salinity
S_o	Solar constant
t	Time
T	Temperature
T_0	Surface air temperature
U	Outgoing long wave radiation
z	Vertical coordinate
α	Variation of density with temperature (pressure dependent)
β	Variation of density with salinity
θ	Potential (adiabatic) density
ρ	Density
ρ_o	Reference density
σ	Density anomaly ($\rho - 1000$ kg/m^3)
σ_M	RMS fluctuation in heat advection
σ_t	Density anomaly ignoring pressure effects
σ_T	RMS fluctuation in air temperature
ψ	Local, temporal solar elevation

Abbreviations

AM	Arctic Mediterranean
AO	Arctic Oscillation Index (SLP at North Pole)
AW	Atlantic Water
BODC	British Oceanographic Data Centre
BSBW	Barents Sea Branch Water
ECHAM	European Centre Hamburg Model (global climate model)
EWG	Environmental Working Group
FSBW	Fram Strait Branch Water
GSR	Greenland–Scotland Ridge
MAIA	Monitoring Atlantic Inflow toward the Arctic

MOC   Meridional Overturning Circulation
NAO   North Atlantic Oscillation Index (SLP at Lisbon – SLP at
                Reykjavik)
NPR   North Polar Region
psu   Practical salinity units (= ppt)
PW   Polar Water
R/V   Research vessel
SAT   Surface air temperature
SLA   Sea level anomaly
SLP   Sea level pressure
Sv   Sverdrups (= 10^6 m^3/s)
THC   Thermohaline circulation
TRACTOR   **TRA**cer and **C**irculation in **T**he **NOR**dic Seas Region
WSC   West Spitsbergen Current

6.1   The Arctic in the Global Climatic System

The Arctic is the northern polar part of the global climate system, but its boundary does not coincide everywhere with the Polar Circle. The most typical climatic features of the Arctic are the polar night (day) in the winter (summer) and a constant presence of the cryosphere in the form of ice sheets, glaciers, multiyear sea ice, and solid permafrost. The southern limit of permafrost on land, multi-year sea ice in the ocean, and the northern polar circle, in the regions where typical arctic climate indicators are absent to the north of it, are taken as the arctic climate zone boundary.

The arctic climate results from a much smaller heat flux from the sun compared to the non-polar regions. This deficit is partly replenished by the internal thermodynamic processes occurring both in the climate system in general and in its polar portion (Fig. 6.1). The greatest contribution to the arctic heat transport, compared to fluxes that would have been observed with an immobile atmosphere and ocean, is made by the poleward heat advection resulting from the atmosphere and ocean circulations. Due to these processes, the arctic climate becomes warmer by several tens of degrees compared to a climate in the absence of advection, whereas the much publicized greenhouse effect increases the surface temperature by a much smaller amount (e.g., Bobylev et al. 2003).

Advection (M) comprises the overwhelming portion of the energy balance for the earth – atmosphere climate system, $U=S+M$, at high latitudes of the Northern Hemisphere (Table 6.1). In the table, U is the outgoing long-wave radiation flux at the upper boundary of the atmosphere, including a significant heat release during the freezing of seawater, $S=S_o(1-A)\Psi/4$ is the intensity of the absorbed short-wave radiation flux from the sun to the upper boundary of the atmosphere, where S_o is the solar constant, A is the system albedo, and Ψ is the local, temporal solar elevation,

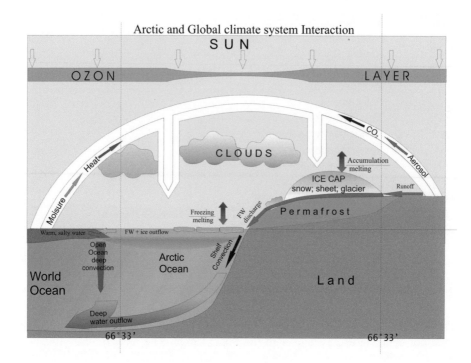

Fig. 6.1 Fluxes in the arctic climate system (Alekseev 2003)

Table 6.1 Mean latitudinal, annual average energy flux components of the earth-atmosphere system, W/m² (Khrol 1992)

Northern latitude	A	S	U	M
Pole to 85°	0.62	68	184	117
85–80°	0.60	72	187	115
80–75°	0.57	80	188	109
75–70°	0.52	94	192	98
70–65°	0.49	106	195	89
90–65°	0.53	93	193	100

and $M = M_a + M_o$ is the advective energy flux, where M_a is the advection in the atmo-sphere and M_o is the advection in the ocean. All fluxes are given as normalized unit fluxes for the zones in W/m².

It follows from the table that more than half of the annual average energy flux to the North Polar Region (NPR) is a result of advection from lower latitudes. During the polar night, the contribution of advection increases to 100% (Table 6.2).

The negative values of M in the summer are due to heat absorption by ice and snow melting in the polar ocean, and to heating of the upper freshened layer. In the winter, this heat will comprise one of the main components of the influx from the

Table 6.2 Estimates of the NPR energy flux components from the results of measurements of the radiation balance components from satellites, W/m² (Marchuk et al. 1988)

Northern latitude	year			January			July		
	A	U	M	A	U	M	A	U	M
90–80°	0.67	177	118	–	165	165	0.69	207	61
80–70°	0.57	179	97	–	157	157	0.50	212	−18
70–60°	0.46	191	72	0.78	165	158	0.39	224	−52

ocean. Though the oceanic transport of heat to the Arctic Ocean at 80°N is comparable to the atmospheric transport at that latitude (Sect. 4.3.3.1) the average annual advection M_o to the NPR is much less than M_a.

During the polar night, when the heat flux from the sun is absent, the surface air temperature fluctuations are determined by fluctuations in M. Using a simplified energy budget, we can obtain the standard deviation of air temperature fluctuations

$$\sigma_T \cong \frac{T_0}{4} \frac{\sigma_M}{M} \tag{6.1}$$

where T_0 is the absolute surface temperature (~248°K) and σ_M is the standard deviation of fluctuations in M: i.e., the temperature variability at night depends primarily on the variability of the heat advection. The root-mean-square air temperature deviation for the Northern Polar Area in January is approximately 2°, giving σ_M/M ~ 0.025. It is suggested that this estimate applies to *this interdependence* also in other seasons when (6.1) does not apply due to insolation. Many studies (e.g., Thompson and Wallace 1998; Walsh and Portis 1999; Dickson et al. 2000) indicate that changes in intensity of heat and moisture transfer to the Arctic are proxied by atmospheric circulation indices (NAO and AO). The heat and moisture transfer to the Arctic intensifies and air temperature increases during high indices, and vice versa.

The influence of oceanic circulation is clearly seen in the oscillations of the warm Atlantic water inflow to the Nordic Seas and then to the Arctic Ocean (*see also* Chap. 4). Observations from icebreakers in the Arctic Basin in the 1990s revealed unusually high temperatures in the intermediate layer of Atlantic origin (Carmack et al. 1995; Aagaard et al. 1996). Their comparison with earlier observations conducted by the Soviet "Sever" expeditions (Alekseev et al. 1997, 1998) and with observations of F. Nansen in 1894 (Alekseev et al. 1999) show that, since 1989, the temperature in this water layer increased in several regions of the Arctic Basin. Figure 6.1 depicts the Arctic as a sensitive climate system with direct forcings, interactions between the arctic system components and advective exchanges with its subarctic neighbors and associated changes in the global climate system.

While the *global impact* on the arctic climate system is primarily made through the atmosphere, an inverse influence of the Arctic on global climate is realized through the ocean, in particular through its most active components – the upper freshened layer and sea ice. In this part of the climate system, freshwater accumulates

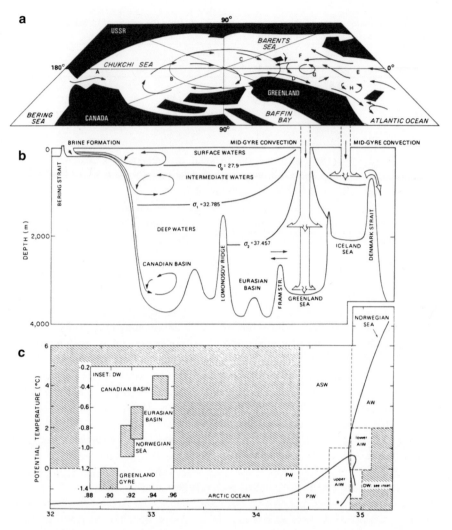

Fig. 6.2 Sketches of the Arctic Mediterranean circulation and water properties (Aagaard et al. 1985)

from river runoff and precipitation, the brackish surface inflow from the Pacific, through the Bering Strait, and separation from seawater by the freezing and melting processes at the surface, predominantly in the arctic shelf seas. These processes comprise a peculiar internal freshwater cycle, which redistributes freshwater in the upper layer between the solid and liquid phases. Interannual oscillations of the conditions of summer melting and winter freezing can be one of the causes of significant variations in freshwater content and correspondingly, in salinity of its upper layer (Alekseev et al. 2000). The sea-ice mass growth in the winter is greater than the summer melting, and with this, the main component in the winter heat sink balance is evident. Excessive ice is exported through Fram Strait with freshened water in the upper layer, with the largest amount in the summer. This freshwater and sea-ice outflow from the Arctic Basin is subjected to significant interannual oscillations that influence the winter convection processes and sea-ice spreading in the Greenland Sea and over the North Atlantic (Dickson et al. 1988; Häkkinen 1993; Mysak et al. 1990; Vinje 2001; Smedsrud et al. 2008).

Another impact of the Arctic on global climate is the generation of denser deep and near-bottom water masses comprising a return branch of the global vertical ocean circulation. The conditions of the development of these processes are controlled by competing flows of saline water of Atlantic origin, freshened arctic water, and winter heat fluxes and freshwater flows under the action of atmospheric circulation. Studies initiated by Nansen (1909) and continued in the 1950s–1990s (Aagaard 1968; Carmack and Aagaard 1973; Worcester et al. 1993; Johannessen et al. 1996; Meincke et al. 1997) have revealed that in the central Greenland Sea, deep and near-bottom waters are formed by deep winter convection. The latter develops due to the inflow and subsequent cooling of saline Atlantic water (AW) in the convective gyre and is hindered by the inflow of freshened arctic water and ice. The icon of the circulation and water mass transformations within the basins is shown in Fig. 6.2 (Aagaard et al. 1985). Figure 6.2c shows the water properties as a function of the potential temperature (θ) and salinity (S). The values of σ in Fig. 6.2b are in situ density anomalies (density minus 1,000 kg/m^3) including the effect of pressure. To understand these values for overflow considerations, the authors show that the equivalent σ_t values (ignoring the effects of pressure) are, nominally, $\sigma_{t0} = 27.9$, $\sigma_{t1} = 28.05$ and $\sigma_{t2} = 28.08$ kg/m^3. This shows how weak the gravitational stability of the water below sill depth is and how little energy is needed to homogenize these masses. However, during the last decades of the twentieth century, cardinal changes in the structure of water masses have occurred here, testifying to an almost complete termination of deep convective mixing (Bonish et al. 1997; Budeus et al. 1998; Alekseev et al. 2001; Drange et al. 2005).

6.2 Thermohaline Convection and the "Great Conveyor Belt"

Although M_o accounts for only a small part of the annual average heat advection to the Arctic, it plays an important role for the THC in the North Atlantic and the resulting Meridional Overturning Circulation (MOC). The AW inflow into the Arctic is the final northern stretch of the global oceanic heat transport before subduction (Fig. 6.3). Upper layer water from the North Atlantic breaches the Greenland–Scotland Ridge (GSR), spreads over the Nordic Seas and flows into the Arctic Basin where it intrudes at depths of 50–800 m and joins the circulation system of water and ice (Timofeyev 1960; Gorshkov 1980; EWG 1997; Bobylev et al. 2003). See Chap. 4 for details. AW is the important heat supply and practically the only source of salt for the Arctic Ocean, which experiences a constant cooling and freshening. Cooling increases its density ρ while freshening reduces ρ of the AW.

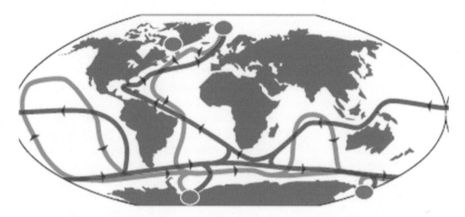

Fig. 6.3 The oceanic conveyor belt carries heat northward into the North Atlantic. The *purple spots* represent the four major downwelling zones: the Greenland, Labrador, Weddell, and Ross seas (Hansen et al. 2004). *Red lines* are surface flows and *blue paths* are deep return flows

Its presence leads to mixing, including deep convection, formation of new water masses and the increase of bio productivity of the upper mixed layer in the Nordic Seas. Weak, but constant heat flux from the Atlantic layer to the upper layer of the Arctic Basin limits the winter growth of ice. In this connection, the AW inflow is an important climate process in the Arctic climate system, and its changes influence not only the Arctic climate but the "Great conveyor belt" as well.

Atlantic water penetrates into the Nordic Seas through the Faeroe–Shetland Strait, over the Faeroe–Iceland Ridge and to a less degree through the Denmark Strait around Iceland (Fig. 6.4). The water is cooled significantly on its way to the Arctic Basin and the Barents Sea (Mauritzen 1996a). The changes of AW characteristics in the Norwegian and Greenland Seas can be seen from the mean near-surface temperatures that vary from 9°C in the Faeroe–Shetland Strait to 1–3°C in the Fram Strait due to heat loss to the atmosphere and mixing with the surrounding water. The average salinity decreases 1.2 psu and dissolved oxygen content increases approximately by 1.2 ml/l. Its maximum salinity varies from 35.3 psu near the Faeroe–Shetland Strait to 35.0 psu in the Fram Strait at depths from 50 to 200–300 m. The core of the Norwegian Atlantic Current with maximum salinity values is seen at depths of 100–200 m. In the Greenland Sea, the depth of salinity maximum exceeds 1,000 m (*see* Chap. 4).

Temperatures, typical for the AW in the North Atlantic disappear in the Norwegian Sea due to exchange with the atmosphere and mixing with arctic water. The transformation process is mostly seen in the zone of influence of the East-Icelandic Current. To the north of 66°, the temperature changes are less, but they continue as the AW spreads further north. Salinity from the North Atlantic changes little from the Faeroe–Shetland Strait to the Fram Strait, showing its important role connecting the subpolar circulation system of the North Atlantic to the Norwegian and Greenland Seas.

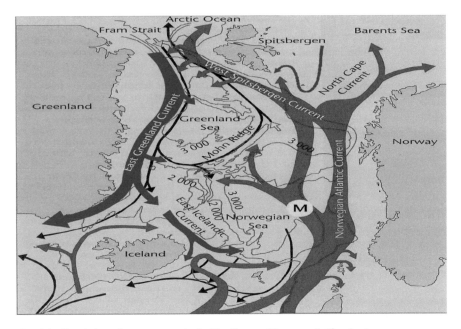

Fig. 6.4 Circulation of water masses in the Nordic seas (Courtesy, S. Østerhus)

The Greenland Sea is one of the most climatically active regions in the Nordic Seas. Here, the warm and salt AW meets the cold and fresh arctic water and sea ice, as far north as the Fram Strait. The proximity to the Greenland ice sheet promotes the cooling of the sea surface followed by either the formation of deep convection mixing of the underlying water, or intense ice formation on the surface, depending on the dominant waters of the upper layer in the central part of the sea.

AW contributes sufficiently to the formation of the upper layer in the central part of the Greenland Sea. According to Carmack and Aagaard (1973), 90% of the water mass volume in this layer comprises AW while polar water (PW) comprises only 1–2% of the upper layer volume. Both water masses form the whole spectrum of secondary water masses in the upper layer in this part of the Greenland Sea, varying from each other by the relative maintenance of the original water masses, persistence and efficiency of the atmospheric impact, formation time, and position in the water mass.

AW inflow into the central part of the Greenland Sea first leads to the formation of its unique vertical structure, favorable for the development of deep winter convection. A water mass with an AW dominance would have a salinity of 34.80 psu or more while water of less salinity could be considered as the product of a dominant impact of PW and the local freshening. A salinity of the surface water in the Greenland Gyre exceeding 34.80‰ was measured on numerous expeditions, conducted first by F. Nansen. The presence of more salty water at the surface of the Greenland Gyre, compared to PW, occurs in two ways: (1) AW in the West

Spitsbergen Current (WSC) intruding from the north and northeast, and (2), upwelling of saltier and warmer intermediate water created by winter convection and horizontal spreading. The latter circulates under the polar waters, flowing with the Jan Mayen Current, and intrudes under the *locally* freshened surface water masses.

Oceanographic observations in the central part of the Greenland Sea, conducted in the winter often revealed more saline and warmer water at the surface. The latter is more typical for the northern part of the region, located closer to the AW supply. For example, surface observations in February 1994 from R/V "Hakon Mosby" (Johannessen et al. 1996), near 77°30′N, 2°W showed temperatures from 0.6°C to 1.2°C and salinities of 34.90–34.96 psu, and the observations in February 1995 showed the presence of water with $S > 34.88$ psu and temperatures close to 0°C, to the west of 75°N, 2°W. Thus, the main source of salinity increase of the upper layer in the gyre is the horizontal advection of AW toward the center of the gyre, mainly from the north, northeast or northwest, i.e., from regions where AW spreads from the WSC. Moreover, the viewed examples of AW intrusions into the center of the Greenland Sea suggest that the AW inflow into this region occurs from the surface to the depth of their extension and only afterward are these waters covered by lighter PW, becoming thus the intermediate layer (Alekseev et al. 2001).

While approaching the center of the circulation, the AW gets freshened due both to mixing with the retreating PW and precipitation, which is significant here, and a water mass is formed with a rather wide range of salinities and temperatures. The formation of ice during the winter may contribute additionally to the salinity increase of the upper layer as new ice has been observed on the gyre surface in the absence of the pack and young drifting ice (Nagurniy et al. 1985). The absence of drifting ice, noticed in all known cases of winter convection, is connected to the fact that ice and PW are pushed away by northern winds, and they are replaced by saltier water of Atlantic origin. Cooling of such high salinity water close to the freezing point will produce water that is denser than the underlying water. Calculations of the possible depth of convection based on observed salinity and constant cooling at the surface showed that the minimum salinity at which the convection could reach the bottom is 34.82 psu (Alekseev et al. 1995). This is due in part to thermobaric convection, typical for the central part of the Greenland Sea (Aagaard and Carmack 1989) (*see* Sect. 6.4). Exceeding the limiting salinity does not mean a guaranteed convection to the bottom, unless followed by the intense cooling at the surface. Favorable conditions for the formation of deep convection in the Greenland Sea were registered in the 1960s and in separate episodes until the beginning of the 1980s. In the 1990s the convection depth did not reach over 2,000 m (Alekseev et al. 2001). As a consequence, the structure of the water mass in the central part of the sea during these periods differed greatly (Fig. 6.5).

Freshening of the upper layer of the Norwegian and Greenland seas and the decrease of the area, occupied by the AW in the 1970s, led to a weakening of the vertical exchange in the Norwegian Sea and in the area of the deep water formation in the Greenland Sea. The reduced production of Greenland Sea deep water led to the increase of the temperature of the intermediate layer, the deep water, and the outflow to the North Atlantic, as well as to the decrease of the density gradients

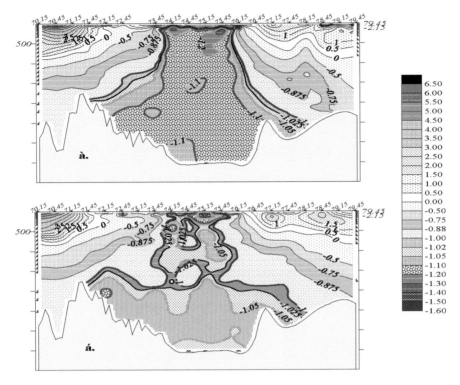

Fig. 6.5 Temperature T [°C] on the section along $2°15'$ W in the Greenland Sea averaged for periods (*top*) with high (1965–1980) and (*bottom*) low (1981–1998) deep water production rates (Alekseev et al. 2003)

responsible for the Atlantic conveyor efficiency. Intensification of the closed cyclonic circulation in the Norwegian and Greenland seas occurred as well as in the East Islandic Current that promotes the maintenance of the low temperatures and water salinity in the western and central parts of the basin. Nevertheless, since the late 1980s, the AW temperatures in the Norwegian and Greenland seas increased together with the AW inflow into the Arctic Basin.

In the northeastern part of the Norwegian Sea, the AW inflow separates into two branches. The western branch of relatively warm and saline water of the WSC penetrates the Arctic Basin through the Fram Strait, moving along the continental slope eastward and developing the cyclonic gyres above the deep basins of the Arctic Ocean (Fig. 4.6, Chap. 4). It is this current that transports the heat into the subsurface layer of the Arctic Basin due to its isolation from the atmosphere and weak mixing with the surrounding waters.

The eastern branch is formed by AW being cooled and freshened as it flows through the Barents and Kara Seas to reach the Arctic Basin *via* the St. Anna Trough (Schauer et al. 2002). This branch, called Barents Sea Branch Water (BSBW) transports about 2 Sv (Blindheim 1989; Loeng et al. 1993) and is less salty and colder than the western branch of AW penetrating the Eurasian part of the Arctic Basin through the Fram Strait, called Fram Strait Branch Water (FSBW). Mauritzen (1996b) computed a volume flux of 3.4 Sv for the FSBW. The first significant trans- formation (cooling and freshening) of the flow of AW through the Fram Strait occurs after passing the St. Anna Trough, where it meets the cooler, fresher BSBW. Here, on the continental slope of the Nansen Basin BSBW is mixed with FSBW and the current forms the intermediate layer that deepens gradually eastward and is consid- ered to be the main part of the conveyor belt of the Arctic Basin circulation (Rudels et al. 1994, 2005). Sometimes, it is an important source of the deep water ventilation.

The analysis of the oceanographic data, obtained from the Arctic Ocean up to the 1980s, did not cast doubts on the robust climatic regime of the AW, but in the begin- ning of the 1990s, there were observed unusually high temperatures (compared to the climatic values) in the AW layer north of the Kara Sea (Quadfasel 1991). In summer 1993, the observations from R/V "Polarstern" confirmed the increase of temperature of the AW in the northern part of the Laptev Sea (Schauer et al. 1995), and from the Canadian R/V "Henry Larsen," north of Wrangel Island (76° 57′ N, 174° 08′ E). Here the temperature was 1.4°C (Carmack et al. 1995) with a "climatic" value of 0.6°C (Gorshkov 1980). In July–August 1994 the expeditions aboard R/V "Polar Sea" and RV "Louis S. St-Lourent," conducting the section from the Bering Strait to the Fram Strait, discovered a warming of 0.5–1°C in the AW layer north of the Chuckchi Sea and above the Lomonosov Ridge (Aagaard et al. 1996). Around Alfa Rise (85°N, 155°W) during the summer of 1998, 0.85°C was measured. At 88°N, 102°W it remained close to normal (0.44°C). Observations from R/V "Polarstern" in the summers of 1995 (Rudels et al. 1997), 1996 and 1999 (Schauer 2000), and 1998, showed that the heat anomaly continued in the 1990s. Observations from R/V "Larsen" in 1993, R/V "Louis S. St-Laurent" in 1994 and R/V "Akademik Fedorov" in 2000 revealed also its spread into new regions. At 180° longitude a 2000 temperature of 1.4°C (0.8°C above the normal) in the AW layer was registered at 77°N. In 1994 a temperature of 1.1°C (0.5°C above normal) was registered at 80°N, and in 2000, 1.3° (0.7° above normal) at 82°N.

Comparisons with the previous observational data showed that the most signifi- cant changes of temperature in the Atlantic layer occurred in the sector 90–180°E (Alekseev et al. 1997). Changes were less noticeable to the north of Svalbard and St. Josef Land. Probably such spatial dissimilarity in temperature changes is connected not only with the inflow of warmer-than-usual AW, but with the increase of their volume and deeper penetration into the Arctic Ocean. This is seen in the thickening of the AW layer by 50–100 m at different locations. The compilation of the time series by the normalized anomalies method shows that the 1990s are characterized by large positive anomalies.

The previous period of positive anomalies for the AW temperatures occurred from the mid 1920s until the end of the 1940s (*see* Sect. 6.6). Time series, covering both periods in the Fram Strait and northward from Arkticheskiy Cape, show greater positive anomalies compared to the 1990s. Large negative anomalies were regis- tered around the turn of the *last* century. The second period of relatively low

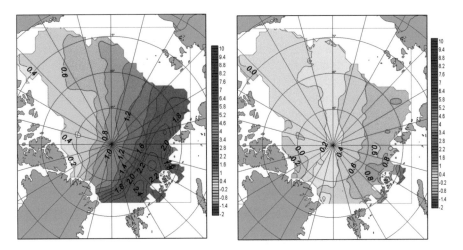

Fig. 6.6 Maximum AW temperature in the 1990s (*left*) and anomalies relative to the 1970s (*right*) (Alekseev et al. 2009)

Table 6.3 MOC overturning transports in the North Atlantic (Sv) (Lumpkin and Speer 2003)

Section	Flow (Sv)
GSR	4.8 ± 0.8
Cape Farwell-Ireland	12.0 ± 1.6
Great Banks-France	14.1 ± 1.2
23°N	15.5 ± 1.2

temperatures in the AW in the Arctic Basin is the period 1950–1980s. A comparison of the mean temperatures for the 1970s and 1990s (Fig. 6.6) witnesses the overall temperature increase of the AW in the Arctic Basin. It is worth noticing that observations in 2004, at the end of the ACSYS Decade, show the inflow of the warmer AW conveyor into the Nordic Seas, Fram Strait (Schauer et al. 2004) and further east to the Laptev Sea (Polyakov et al. 2002).

The "separator" role of the freezing–melting process causes an export of cold water, both at the surface and at the bottom overflows. The fresher surface outflow, together with a more than 1 Sv contribution from the Canadian Archipelago, preconditions deep water convection in the Labrador Sea. The cold overflows at sill depths of the Denmark Strait and the Faroe Bank Channel have enough potential energy to entrain large amounts of AW as they plunge to the abyss. The entrainment in the plunging flows more than doubles the volume flux to the deeper, equatorward return flow. Using inverse modeling, Lumpkin and Speer (2003) estimated MOC transports in the North Atlantic section of the conveyor belt. Their results for different latitudes are shown in Table 6.3. Clearly, the most significant increase occurs where the Arctic overflow entrains upper AW to large depth to the south of the GSR. Fischer et al. (2003) measured a value of 14 ± 3.3 Sv. The Labrador Gyre accounts for an increase of less than 2 Sv. The GSR overflows and the Labrador Gyre are two of the four major source regions that pump the conveyor belt (Fig. 6.3). The other two are the Weddell and Ross seas in the Southern Ocean.

The overall model results above are in reasonable agreement with the estimates of Hansen and Østerhus (2000), who give 9 Sv inflow to the Arctic, including the inflow through the Bering Strait, and an equal outflow to the Atlantic, of which 4 Sv, including the flow through the Canadian Archipelago is surface outflow. These numbers are typical for the ACSYS Decade and appear to be quite stable. The overflow to the Atlantic is surprisingly stable in spite of the variability, suggesting a buffering due to the large reservoir of deeper water and the topographic hydraulic control at the sill. However, changes in the height of the deeper water in the Nordic Seas are associated with changes in the hydrography and there is a significant recirculation of cooled and freshened AW (Mauritzen 1996a,b) in the Nordic Seas. The water mass characteristics of the overflowing water change more than the volume flux (Hansen and Østerhus 2000, p. 198).

Mauritzen (1996b) and Lumpkin and Speer (2003) used diagnostic box models to compute the fluxes. This technique is good for assessing the present state, but a prognostic model with known, or assumed, forcing is needed to predict future states. The focus should be on processes. There are many climate models that attempt to predict the future and these will be exploited in other chapters of this volume. We will now look at the physical ocean processes that must be accounted for in the models.

6.3   The Arctic Mediterranean System and Bimodal Catastrophe

The Arctic Ocean and the Nordic Seas are often referred to as the Arctic Mediterranean (AM). This is a semi-enclosed estuary with a throttle at the GSR. The simplest analytic *dynamical* model of this basin is the one-box model by Whitehead (1998). This assumes a balanced salt budget and lumps the atmospheric and ocean fluxes of freshwater and heat to independent forcing conditions.

Wind and cooling over the AM cause mixing of the surface brackish water with the more saline water that flows in from the North Atlantic. The present situation at the GSR is an upper layer inflow of saline Atlantic water from the North Atlantic and a fresher, colder, deeper outflow over the sills to the global ocean. It is the interaction between these domains that is of interest for ACSYS.

There have been several 2-box models and more complicated models of climate dynamics following Stommel (1961) who showed multiple equilibrium solutions. Whitehead (1998) realized that only one box was necessary when he designed a hydraulic laboratory test. The independent cooling (q_T) and freshwater (Q_f) forcings are the key to obtaining different states. With further constraints on the outflow, like sills, width, and rotation, there are parameter ranges of multiple equilibria, obtained in the following way.

Consider the AM to be open to a saline reservoir through a constriction at the GSR. The torque of the buoyancy of the water (negative or positive) will drive a circulation through this opening. The buoyancy torque depends on the relative strengths of (positive) freshwater flux and (negative) cooling that changes the density ρ. The amount of warm, saline water (Q) that is admitted to the region to satisfy the

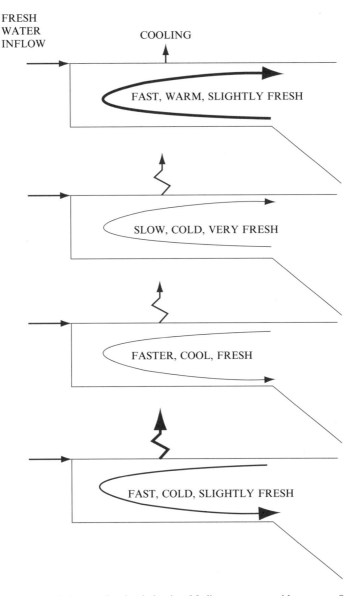

FRESH
WATER
INFLOW

COOLING

FAST, WARM, SLIGHTLY FRESH

SLOW, COLD, VERY FRESH

FASTER, COOL, FRESH

FAST, COLD, SLIGHTLY FRESH

Fig. 6.7 Schematic solutions to the circulation in a Mediterranean sea with constant Q_f and variable cooling q_T (Whitehead 1998)

heat loss is restricted through a so-called overmixing condition (Stommel and Farmer 1953). It is a double hydraulic control condition, restricting the amount of inflow and the key to obtaining a solution to the problem using a one-box model. The different flow regimes with increased cooling, but the same Q_f, are sketched in Fig. 6.7. The relative contributions of cooling and freshwater flux to the buoyancy,

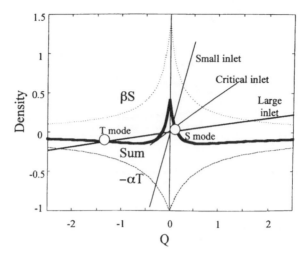

Fig. 6.8 Solutions to Whitehead's (1998) coupled equations

and their sum, can be viewed as a function of the flow of seawater admitted through the control, the so-called ventilation Q. For given values of Q_f and q_r, there may be one or two stable solutions that match the topographic constraint as shown in Fig. 6.8. Here, β is the density change with salinity S and α is the density change with temperature T. The GSR is a large inlet in this connection. The theory of Whitehead shows that the convection-driven temperature mode, which is today's situation, has much more ventilation than the estuarine mode, or salt mode (driven by the freshwater buoyancy).

This single box model illustrates the essence of the general climate behavior that more complex models have shown. As the freshwater forcing is changed (upper curve in Fig. 6.8), the solution curves will cross the topography (control) condition at different values of Q. Above a critical value of Q_f there will be only one solution (*see* Fig. 6.9), a salt mode, where the flow will reverse and the surface layer will exhibit an outflow of cold, brackish water with entrained seawater (probably covered with ice during a greater part of the year). This represents the speculated shutdown of the THC. In the context of this simple model, this transition will be abrupt and is referred to as a catastrophe.

There are no feedback loops in this model. These will cause oscillations that depend on the time scales of the system. Even the "catastrophe" will take decades to occur as the system adjusts to the new equilibrium state, since the model requires complete mixing of the water masses. The constant-salinity external ocean does not take into account the dynamics of evaporation-mixing-driven circulation in this outer domain. More elaborate numerical models considering the climate system with feedbacks are presented in the following chapters of this book.

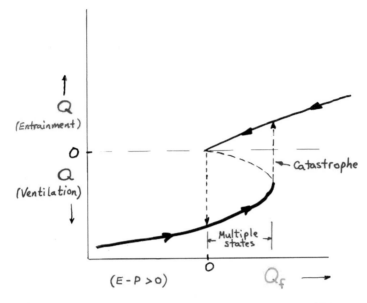

Fig. 6.9 Schematic of solutions with variable Q_f (Adapted from Whitehead 1995)

6.4   Physical Processes Within the AM

The single box model assumes that the entire water mass is fully mixed at all times. The AM is quite different. In Fig. 6.2, large convection regions were depicted in the Iceland and Greenland seas. Although the Greenland Sea also had a mode of deep convection (well below sill depth of the AM), this is not so important for the climate discussion above. However, the brine rejection during freezing on the shelf seas is of equal importance for the production of intermediate water at overflow depths (600–800 m). Thus, there are three, nearly equal and independent regions of ventilation within the AM. Mauritzen (1996a) emphasized the significant buoyancy loss in the Norwegian Sea due to *cooling of the inflowing Atlantic water*, contributing to the total picture.

The double control that regulates the exchange with the global ocean in the south is affected by the rotation of the earth. The major inflows of AW follow the bottom slopes to the right, forming so-called "fast track" slope jets that transport heat quite rapidly to higher latitudes. Likewise, the deeper overflows to the Atlantic are affected by the rotation of the earth. Østerhus et al. (2001) showed a section of the Faroe Bank Channel overflow which seems to be quite stable (Fig. 6.10) with a seasonal fluctuation. The velocity distribution seems to obey the conservation of potential vorticity, as expected. The overflow in the Denmark Strait appears to be quite variable, but lacks a significant seasonal signal.

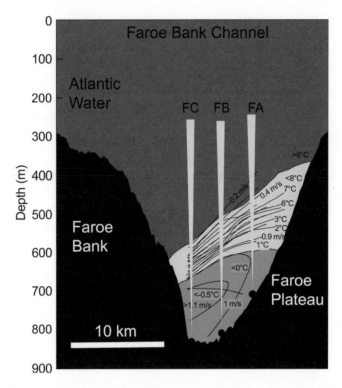

Fig. 6.10 Velocity and temperature distributions in the outflow through the Faroe Bank Channel (Østerhus et al. 2000)

The distribution of inflows between the three major branches (Shetland slope Current, Faroe Current, and North Icelandic Irminger Current to the west of Iceland, in Fig. 6.4) is likely related more to the wind-stress curl than the NAO, as many investigators have suggested. Orvik and Skagseth (2003) showed a good correlation between this and direct measurements at the Svinøy section off the coast of Norway, with a 15 month delay. It appears that the wind-stress curl imparts vorticity to the water masses during their passage in the North Atlantic Drift Current and that vorticity determines the relative strengths of the inflows to the east. A positive vorticity enhances the fast-track slope current that is being monitored.

There are two major routes for the inflowing AW in the Nordic Seas. The slope current is predominantly barotropic, while the off-slope frontal jet is predominantly baroclinic (McClimans et al. 1999; Orvik et al. 2001). However, the notion of stream tube flows must be abandoned because there is an exchange between these branches both in time (season) and in latitude. The cooling in the Nordic Seas renders the inflow more barotropic toward the Arctic Ocean (Walin et al. 2004).

To the extent the flow is barotropic and geostrophic, it can be monitored by bottom pressure gradients. This was explored in MAIA (BODC 2003), and appears to hold for the slope currents along the Norwegian continental slope. However, the attempt to use *coastal water level* measurements to monitor the flow has to take into account that the barotropic transport is proportional to the sea level rise across the flow while the baroclinic transport is proportional to the square of the rise. Thus, a combined baroclinic + barotropic transport needs more information. An attempt to use satellite altimetry (SLA) showed that the seasonal changes on the shelves and in the deep basins are out of phase, and that the gradients do indeed agree with the seasonal behavior of the slope currents (Jakobsen et al. 2003). The interannual variations in the SLA do not have a clear relation to the NAO. This is what makes Orvik and Skagseth's correlation interesting.

The slope current is the so-called "fast track" for the heat transport to the Arctic. This becomes quite clear from the ice fronts in the Barents Sea and the Yermak Plateau. Häkkinen and Cavalieri (1989) showed that the most intensive "hot spots" in the ocean during the winter are the slope regions where the inflowing Atlantic water meets the polar front (Murman Current, Warm Core Current toward Hopen Deep and WSC). It is the locations of these hot spots that determine the latitudes of the maximum heat transfer to the atmosphere and the transformations of AW on its way to the Arctic halocline (Rudels et al. 1994). These transports are by and large topographically steered. A hydrographic cross section of the Nansen Basin (Anderson et al. 1989) shows the narrow, high-salinity core flowing eastward at about 400–600 m depth past the northern slope of Svalbard (*see* Sect. 6.2).

The paradigm of deep water formation from brine rejection during freezing on the marginal seas has been invoked by many investigators since Aagaard et al. (1985). Direct measurements of dense plumes on the slope are few, and it appears that, due to entrainment, the depth of penetration may be somewhat less than 2,000 m. However, this is much deeper than sill depth at the GSR, and suffices for "deep convection." In spite of the salinity excess from the brine rejection, mixing along the entire path, through intermediate layers, causes the dense water plumes to be less saline than the inflowing, AW in the halocline. Details of this process were shown by Quadfasel et al. (1988) off Storfjord on Svalbard.

6.4.1 Convection

There have been many studies on convection at geophysical scales where the rotation of the earth may be important. Maxworthy and Narimousa (1994) showed that brine rejection on *shelf seas* may cause mesoscale eddies to form at the bottom, containing large volumes of denser fluid in local pockets. These move around the basin and cause highly time-dependent, intense outflows in troughs. This is a challenge for field studies, but such events have been observed. For the deeper basins, there is a more diffuse intrusion of the denser water masses, and the variability may be less. Indeed, when several sources throughout the basin are pooled together, the variability of the integral is less event-full. There is less variability of the overflow in the Faroe Bank Channel, indicating that this location is far from the convection events. On the other hand, there is a significant variability in the overflow through the Denmark Strait, implying nearby convective sources and/or more energetic atmospheric forcing. The normal cyclonic ocean circulation around basins suggests that this outflow is "upstream" from the Faroe overflow, although the actual flow routes are not well mapped. The Denmark Strait is also the outer, more variable branch of the Atlantic inflow to the Nordic Seas. Results from the EU project TRACTOR have revealed some of these details (Eldevik et al. 2005).

There are several types of convection. Brine rejection has been mentioned. Although this is thermally generated through freezing of seawater, surface density increase due to cooling is more important in the total picture (Mauritzen 1996a). Chu (1991) expressed a simple criterion for convection invoking 5 main forms of forcing, in which there are some subdivisions:

$$\frac{\partial}{\partial t} \frac{1}{\rho_o} \frac{\partial \rho}{\partial z} = \sum_{i=1}^{5} G_i < 0, \tag{6.2}$$

where
G_1 = buoyancy flux
G_2 = potential instability
G_3 = double diffusion
G_4 = thermobaric
G_5 = cabbeling
Here, ρ_o is a reference density and z is the vertical coordinate.

Both brine rejection and thermal convection belong to G_1. G_2 accounts for changes in stability as currents transport water masses to regions of variable static stability. G_3 accounts for the difference in molecular diffusion of heat and salt. Above the Atlantic water intrusion to the halocline, we expect layers of double-diffusive convection, whereas below this salinity maximum, we expect salt fingers. Thermobaric convection arises due to the pressure-dependent thermal compressibility of water. Aagaard and Carmack (1989) showed how stable water masses with the same density profiles, but different salinities, can become dense enough to reach to much greater depths when subjected to cooling. Cabbeling is due to the non-linear equation of state $\rho(T, S)$. When equally dense water masses of different $\theta - S$ (or $T - S$) properties mix, the product is a water mass that is denser.

To this list, we can add dynamic processes that produce vertical motions and mixing: baroclinic instability, Kelvin–Helmholtz instability, and (wave-induced) Langmuir rolls. The baroclinic instability is a particularly interesting dynamical process. This was invoked by Killworth (1979) to account for the *preconditioning* of what he called "chimneys"; i.e., small mesoscale, intensive cyclones that convect large amounts of surface water to large depth. His analysis is based primarily on G_1. He concluded that brine rejection is far more efficient in producing deep convection than thermal convection in the deeper waters near Antarctica (Weddell Sea Gyre). For the Arctic, Johannessen et al. (1983) suggested that ice-edge upwelling during the winter causes deeper, denser water to reach the surface exposing it for intense cooling and subsequent deep convection.

6.5   Problems of Monitoring Climate Change in the AM

These small cyclonic features including "plumes" with 100–200 m horizontal scales in the centers, and their erratic nature make them hard to find, but chimneys have been found and studied by, among others, Gascard (1991) and Johannessen et al. (1991, 2005). The vertical velocities indicate strong vertical mass fluxes in local events. For climate studies, this becomes a game of statistics and we return to the overall picture of how much of the total flux in the MOC is due to convection in the three regions of the AM: the marginal shelf seas, the Greenland Sea, and the Iceland Sea, not to forget the preconditioning cooling in the Norwegian Sea. Aagaard and Carmack (1994) presented scenarios for increased and decreased freshwater supply, but the same thermal forcing (as in the example of Fig. 6.9). Although their scenarios showed deep convection in the Greenland Sea and the Arctic Ocean, it is only necessary to have convection to below sill depth (800 m) to run the thermal mode.

The single box model assumes complete mixing, or integrated properties throughout the basin. Average properties are impossible to *measure* in nature. Due to the small scale of the important transport routes and their temporal variability, they are difficult to monitor, but the focus should be here. The episodic nature of both slope jets and overflows, especially at the Denmark Strait, is a challenge. If you are not there at the right time, you may miss the action. Knowledge of the processes is essential for interpreting field observations, constructing useful benchmarks for model validations and for planning useful field studies.

6.6   Early Twentieth Century Warming: An Example of Global Interaction?

The warming in the early twentieth century was a long-lasting event that commenced about 1920 and reached its maximum some 20 years later. The warming, though evident in hemispheric and even global averages of surface air temperature (SAT), was actually confined to the high northern latitudes. The largest warming occurred in the Arctic (60–90°N) and the average for the 1940s was some 1.7°C (2.2°C for the winter half-year) above the long-term SAT (Johannessen et al. 2004; Bengtsson et al. 2004). The early twentieth century warming trend in the Arctic was nearly as large as the warming trend for the last part of the century, such that they could be considered to be part and parcel of the same natural low-frequency oscillation (e.g., Polyakov et al. 2002).

However, recent investigations (Johannessen et al. 2004; Bengtsson et al. 2004) have been able to distinguish and better understand the early and late twentieth century warming events. Johannessen et al. (2004) found that a spatial comparison of SAT trends in these periods (and the intervening colder period) reveals key differences in their patterns (Fig. 6.11). The 20-year SAT trends for the 1920s–1930s warming period (Fig. 6.11a, b) and the subsequent cooling period (Fig. 6.11c, d)

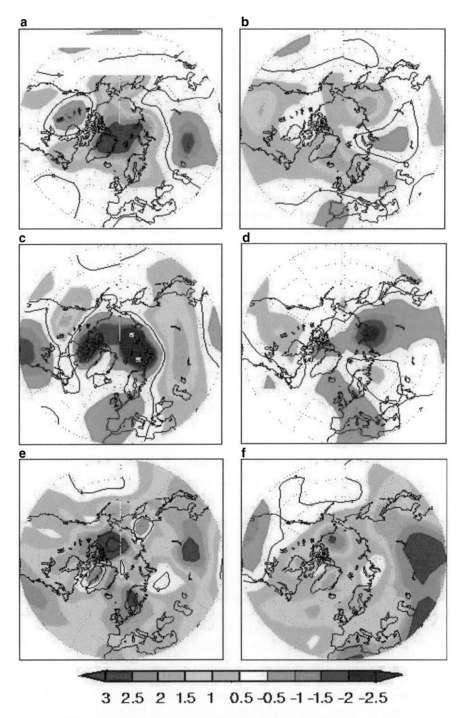

3 2.5 2 1.5 1   0.5 -0.5 -1 -1.5 -2 -2.5

Fig. 6.11 Observed surface air temperature (*SAT*) trends north of 30°N in the winter (November–April) and summer (May–October) half-years for 20-year periods representing warming, cooling and warming in the twentieth century: (**a–b**) 1920–1939, winter and summer, respectively, (**c–d**) 1945–1964, winter and summer, respectively, and (**e–f**) 1980–1999, winter and summer, respectively (Johannessen et al. 2004)

have similar patterns, thus suggesting similar underlying processes. In the winter half-year, the high-latitude warming (Fig. 6.11a) and cooling (Fig. 6.11c) patterns are organized symmetrically around the pole, while in the summer half-year, the warming (Fig. 6.11b) and cooling (Fig. 6.11d) appear to reflect the positions of the latitudinal quasi-stationary wave structure, predominantly wavenumber three and four. However, the warming trend for the last 20 years is more widespread and has a different pattern from the earlier periods in both winter (Fig. 6.11e) and summer (Fig. 6.11f). Both the early and late twentieth century warming events were most pronounced in the Arctic and during the winter. In addition, in the latter period there is a pronounced warming in the Eurasian mid-latitudes, especially in the summer.

The underlying *cause* of the early twentieth century warming is a decades-old mystery. Four possible mechanisms, individually or in combination, may be responsible for the warming: (1) increased greenhouse gases, (2) increased solar irradiation, (3) reduced volcanic activity, and (4) internal variability within the climate system. This last-mentioned aspect is now considered in more detail, as recent modeling and observational research suggests its importance.

Delworth and Knutson's (2000) modeling study suggested that the 1920s–1930s warming anomaly was due to natural processes, rather than external forcing. Johannessen et al. (2004) found a similar high-latitude anomaly, albeit less extreme and of a somewhat shorter duration, in a 300-year control run (without anthropogenic forcing) using the ECHAM4 model implemented at the Max-Planck Institute for Meteorology, Hamburg. The anomaly occurred after 150 years of integration and lasted for some 15 years. That this simulation is able to produce – without anthropogenic forcing – an anomaly similar to the observed high-latitude warming in the 1920s–1930s strongly supports Delworth and Knutson's (2000) contention that the warming event represents primarily natural variability within the climate system. The most plausible explanation in this case is the multi-decadal oscillation related to North Atlantic Ocean circulation (e.g., Delworth and Mann 2000).

Based on a combination of observational and modeling evidence, Bengtsson et al. (2004) theorize that a particular sequence of events and interactions in the coupled atmosphere– ocean system may explain the early twentieth century warming. They propose a mechanism involving a steadily increasing transport of warm water into the Barents Sea, driven by increasing SW–W winds between Spitsbergen and the northern Norwegian coast, i.e., the Barents Sea opening. Between 1920 and 1940, the SLP gradient across the Barents Sea opening increased by 8 mb, corresponding to a geostrophic wind anomaly of 6 m/s. (It is notable that the increased winds were not due to the NAO – which actually weakened during the 1920s and remained so during the warm anomaly – in contrast to the recent influence of the NAO on the AM (e.g., Dickson et al. 2000)). This would lead to a massive transport of warm water into the Arctic basin, with a major reduction of ice in the Barents Sea, which is the region where the largest SAT anomalies were observed.

Bengtsson et al. (2004) argue further, based on observations and atmospheric modeling experiments, that the reduced sea-ice coverage was the main process for the amplified SAT increase in the region. The ECHAM4-simulated SAT increases

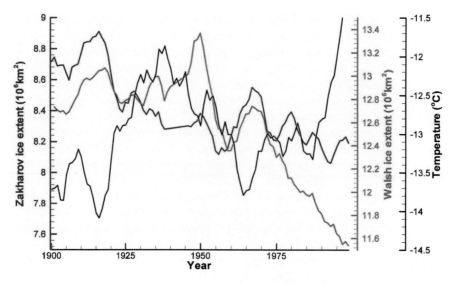

Fig. 6.12 Annual sea-ice extent derived from the "Zakharov" sea-ice dataset (*red*), Northern Hemisphere sea-ice extent from the widely used "Walsh" dataset (*green*) and zonal (70–90°N) mean annual SAT (*black*) since 1900. The time series shown are 5-year running means (Johannessen et al. 2004)

in the area of reduced sea-ice extent is related to stronger oceanic heat flux to the atmosphere, exceeding 150 W m^{-2} in the areas of largest decreases in sea ice.

In order to assess the importance of ocean–ice–atmosphere feedbacks, coupled model control experiments were performed using ECHAM4/OPYC3. It was found that the sensitivity of arctic (60–90°N) SAT to sea-ice changes is about −1.4°C per 10^6 km^2. A close link between arctic sea ice and SAT variability has also been shown (Fig. 6.12) using century-scale sea-ice observations that include the early twentieth century warming period (Johannessen et al. 2004) where the Zakharov ice extent data set covering 75% of the Arctic Ocean showed a significant decrease in response to the warming (not shown in the Walsh ice data set). Regional sea-ice variability is determined largely by atmospheric conditions, through driving ocean currents and wind-driven sea-ice advection, each of which are particularly important in the Barents Sea region. The modeled correlation between volume inflow through the Barents Sea opening and Barents Sea ice extent is −0.77. This is linked to enhanced SAT increases in the region, both as observed during the early twentieth century warming (Bengtsson et al. 2004; Johannessen et al. 2004) and in the model experiments, which further suggest a positive feedback involving increases in latent and sensible heat fluxes, decreases in SLP, and increases in vorticity in the region. Thus the AM and in particular the Barents Sea appears to be another key area for interactions both *within* the region and with the global climate system.

6.7   Summary Statement

The present chapter focused on the ocean link to the Arctic climate system, emphasizing the various parts that must be considered in constructing benchmarks to develop reliable coupled models. Sea ice was the focus of Chap. 3 and a comprehensive account of the oceanography was presented in Chap. 4. At the end of the ACSYS decade there were several parts of the system that were still poorly known. However, a recent paper by Eldevik et al. (2009) analyzing the hydrography of the Nordic Seas from 1950 to 2005, which includes the ACSYS decade, supports much of the revised circulation scheme suggested by Mauritzen (1996a, b) and shows a robust system; in particular, the recirculation of AW within the basin and more complete routes of the freshened, cooled seawater in the GSR overflows to the MOC in the North Atlantic. The ocean segment of the interaction between the Arctic and the global climate system is important and there are feedback mechanisms that must be considered, as in Sect. 6.6. These mechanisms are not accounted for in simple box experiments. The following chapters in this volume will address various forms of more complete systems using other advanced models.

References

Aagaard K (1968) Temperature variations in the Greenland Sea deep-water. Deep-Sea Res 15:281–296

Aagaard K, Carmack EC (1989) The role of sea ice and other fresh water in the arctic circulation. J Geophys Res 94:14485–14498

Aagaard K, Carmack EC (1994) The Arctic Ocean and climate: a perspective. In: Johannessen OM, Muench RD, Overland JE (eds) The polar oceans and their role in shaping the global environment, AGU geophysical monograph 85. American Geophysical Union, Washington, DC, pp 5–20

Aagaard K, Swift JH, Carmack EC (1985) Thermohaline circulation in the Arctic Mediterranean seas. J Geophys Res 90:4833–4846

Aagaard K, Barrie LA, Carmack EC, Jones EP, Lubin D, Macdonald RW, Swift JH, Tucker WB, Wheeler PA, Whrither USRH (1996) Canadian researchers explore Arctic Ocean. EOS 77(22):209–213

Alekseev G (2003) Research of the Arctic climate change in 20th century. Trudy AARI 446:7–17, In Russian

Alekseev G, Ivanov V, Korablev A (1995) Interannual variation of deep convection in the Greenland Sea. Oceanology 35:45–52

Alekseev GV, Bulatov LV, Zakharov VF, Ivanov VV (1997) Inflow of unusually warm Atlantic waters to the Arctic Basin. Rep Acad Sci 356(3):401–403

Alekseev GV, Bulatov LV, Zakharov VF, Ivanov VV (1998) Heat expansion of Atlantic waters in the Arctic Basin. Meteorol Gidrol 7:69–78

Alekseev GV, Bulatov LV, Zakharov VF, Ivanov VV (1999) To the change in the thermal state of Atlantic water in the Arctic Basin for the last 100 years. Problemy Arktiki i Antarktiki 71:179–183

Alekseev GV, Bulatov LV, Zakharov VF (2000) Freshwater melting/freezing cycle in the Arctic Ocean. In: Lewis EW et al (eds) The freshwater budget of the Arctic Ocean, NATO science series. Kluwer Academic, Boston, pp 589–608

Alekseev GV, Johannessen OM, Korablev AA, Ivanov VV, Kovalevsky DV (2001) Interannual variability in water masses in the Greenland Sea and adjacent areas. Polar Res 20(2):201–208

Alekseev GV, Johannessen OM, Korablev AA, Proshuntinsky A (2003) Arctic Ocean and sea ice. In: Bobylev LP, Kondratyev KYa, Johannessen OM (eds) Arctic environment variability in the context of global change. Springer-Praxis, Berlin

Alekseev GV, Pnyushkov AV, Ivanov NE, Ashik IM, Sokolov VT (2009) Assessment of climate change in the marine Arctic with IPY 2007/08 data. Problemy Arktiki i Antarctiki 1(81):7–14

Anderson LG, Jones EP, Koltermann KP, Schlosser P, Swift JH, Wallace DWR (1989) The first oceanographic section across the Nansen Basin in the Arctic Ocean. Deep-Sea Res 36:475–482

Bengtsson L, Semenov VA, Johannessen OM (2004) The early century warming in the Arctic – a possible mechanism. J Clim 17:4045–4057

Blindheim J (1989) Cascading of Barents Sea bottom water into the Norwegian Sea. Rapports et Procès-verbaux Reunion Conseil Internationale Exploration de la Mer 188:49–58

Bobylev L, Kondratyev KYa, Johannessen OM (2003) Arctic environment variability in the context of global change. Springer-Praxis, Berlin

BODC: Monitoring Atlantic inflow toward the Arctic. MAIA CD-ROM, © NERC (2003)

Bonish B, Blindheim J, Bullister JL, Schlosser P, Wallace DWR (1997) Longterm trends of temperature, salinity, density and transient tracers in the central Greenland Sea. J Geophys Res 102:18553–18571

Budeus G, Schneider W, Krause G (1998) Winter convective events and bottom water warming in the Greenland Sea. J Geophys Res 103:18513–18527

Carmack E, Aagaard K (1973) On the deep water of the Greenland Sea. Deep-Sea Res 20:687–715

Carmack EC, Macdonald RW, Perkin RG, MacLaughlin FA, Pearson RJ (1995) Evidence for warming of Atlantic water in the southern Canadian Basin of the Arctic Ocean: results from the Larsen – 93 expedition. Geophys Res Lett 22:1061–1064

Chu PC (1991) Geophysics of deep convection and deep water formation in oceans. In: Chu PC, Gascard JC (eds) Deep convection and deep water formation in the oceans, Proceedings of the international Monterey colloquium on deep convection and deep water formation in the oceans, Elsevier oceanography series 57. Elsevier, Amsterdam/New York, pp 3–16

Delworth TL, Knutson TR (2000) Simulation of early 20th century global warming. Science 287:2246–2250

Delworth TL, Mann ME (2000) Observed and simulated multidecadal variability in the Northern Hemisphere. Clim Dyn 16:661–676

Dickson RR, Meincke J, Malmberg S-A, Lee AJ (1988) The "great salinity anomaly" in the Northern North Atlantic 1968–1982. Prog Oceanogr 20:103–151

Dickson RR, Osborn TJ, Hurrell JW, Meincke J, Blindheim J, Adlandsvik B, Vinje T, Alekseev G, Maslowski W (2000) The Arctic Ocean response to the North Atlantic oscillation. J Clim 13:2671–2696

Drange H, Dokken T, Furevik T, Gerdes R, Berger W (eds) (2005) The Nordic seas: an integrated perspective, AGU monograph 158. American Geophysical Union, Washington, DC

Eldevik T, Straneo F, Sandø AB, Furevik T (2005) Pathways and export of Greenland Sea water. In: Drange H, Dokken T, Furevik T, Gerdes R, Berger W (eds) The Nordic seas: an integrated perspective, AGU monograph 158. American Geophysical Union, Washington, DC, pp 89–103

Eldevik T, Nilsen JEØ, Iovino D, Olsson KA, Sandø AB, Drange H (2009) Observed sources and variability of Nordic seas overflow. Nat Geosci 2:406–410

Environmental Working Group (EWG) (1997) Joint U.S. Russian atlas of the Arctic Ocean: oceanography atlas for the winter period. National Snow and Ice Data Center, Boulder

Fischer J, Schott FA, Dengler M (2003) Boundary circulation at the exit of the Labrador Sea. J Phys Oceanogr 34:1548–1570

Gascard JC (1991) Open ocean convection and deep water formation revisited in the Mediterranean, Labrador, Greenland and Weddell seas. In: Chu PC, Gascard JC (eds) Deep convection and deep water formation in the oceans, Proceedings of the international Monterey colloquium on deep convection and deep water formation in the oceans, Elsevier oceanography series 57. Elsevier, Amsterdam/New York, pp 157–181

Gorshkov SG (ed) (1980) Atlas of the oceans. Arctic Ocean. USSR Ministry of Defense, VMF, GUNIO. (GUNIO = Glavnoe Upravlenie navigazii i okeanografii). In English: Main Administration on Navigation and Oceanography, 199 pp

Häkkinen S (1993) An Arctic source for the great salinity anomaly: a simulation of the Arctic Ice-Ocean system for 1955–1975. J Geophys Res 98:16397–16410

Häkkinen S, Cavalieri DJ (1989) A study of oceanic surface heat fluxes in the Greenland. Norwegian and Barents seas. J Geophys Res 94:6145–6157

Hansen B, Østerhus S (2000) North Atlantic-Nordic seas exchanges. Prog Oceanogr 45:109–208

Hansen B, Østerhus S, Quadfasel D, Turrell W (2004) Already the day after tomorrow? Science 305:953–954

Jakobsen PK, Ribergaard MH, Quadfasel D, Schmith T, Hughes CW (2003) Near-surface circulation in the northern North Atlantic as inferred from Lagrangian drifters: variability from the mesoscale to interannual. J Geophys Res. doi:10.1029/2002JC001554

Johannessen OM, Johannessen JA, Morison J, Farrelly BA, Svendsen EA (1983) Oceanographic conditions in the marginal Ice Zone North of Svalbard in early fall 1979 with emphasis on Mesoscale Process. J Geophys Res 88:2755–2769

Johannessen OM, Sandven S, Johannessen JA (1991) Eddy-related winter convection in the Boreas Basin. In: Gascard JC, Chu PC (eds) Deep convection and deep water formation in the oceans, Elsevier oceanographic series. Elsevier, New York, pp 87–105

Johannessen OM, Muench R, Overland JE (eds) (1994) The polar oceans and their role in shaping the global environment, The Nansen centennial volume, Geophysical monograph 85. American Geophysical Union, Washington, DC. ISBN 0–87590–042–9

Johannessen OM, Lygre K, Samuel AJ, Samuel P (1996) Observations of convective chimneys in the Greenland Sea in late winter 1994 and 1995. NERSC technical report, 12, June 30, 34 pp

Johannessen OM, Bengtsson L, Miles MW, Kuzmina SI, Semenov VA, Alekseev G, Zakharov VF, Nagurnyi AP, Bobylev LP, Pettersson LH, Hasselmann K, Cattle H (2004) Arctic climate change – observed and modeled temperature and sea ice variability. Tellus 56A:328–341

Johannessen OM, Lygre K, Eldevik T (2005) Convective chimneys and plumes in the Northern Greenland Sea. In: Drange H, Dokken T, Furevik T, Gerdes R, Berger W (eds) The Nordic seas: an integrated perspective. Geophysical Monograph Series No 158, American Geophysical Union, Washington, DC, pp 251–272

Khrol VP (ed) (1992) Atlas of the energy budget of the northern polar region. Gidrometeoizdat, St.-Petersburg, 72 pp

Killworth P (1979) On "chimney" formations in the ocean. J Phys Oceanogr 9:531–554

Loeng H, Sagen H, Ådlandsvik B, Ozhigin V (1993) Current measurements between Novaya Zemlya and Frans Josef Land September 1991–September 1992 data report. Institute of Marine research report no 2 – 1993, 23 pp

Lumpkin R, Speer K (2003) Large-scale vertical and horizontal circulation in the North Atlantic Ocean. J Phys Oceanogr 33:1902–1920

Marchuk GI, Kondratyev KYa, Kosoderov VV (1988) Earth radiation budget: key aspects. Nauka, Moscow, 216 pp

Mauritzen C (1996a) Production of dense overflow waters feeding the North Atlantic across the Greenland–Scotland Ridge. Part 1: evidence for a revised circulation scheme. Deep-Sea Res 1 Oceanogr Res Pap 43:769–806

Mauritzen C (1996b) Production of dense overflow waters feeding the North Atlantic across the Greenland–Scotland Ridge. Part 2: an inverse model. Deep-Sea Res 1 Oceanogr Res Pap 43:807–835

Maxworthy T, Narimousa S (1994) Unsteady, turbulent convection into a homogeneous, rotating fluid, with oceanographic applications. J Phys Oceanogr 24:865–887

McClimans TA, Johannessen BO, Jenserud T (1999) Monitoring a shelf edge current using bottom pressures or coastal sea level data. Cont Shelf Res 19:1265–1283

Meincke J, Rudels B, Friedrich HJ (1997) The Arctic Ocean – Nordic seas thermohaline system. ICES J Mar Sci 54:283–299

Mysak LA, Manak DK, Marsden RF (1990) Sea ice anomalies observed in the Greenland and Labrador seas during 1901–1984 and their relation to an interceded Arctic climate cycle. Clim Dyn 5:111–132

Nagurniy A, Bogorodsky P, Popov A, Svyaschennikov P (1985) Forming of cold deep water on the surface of the Greenland Sea. Doklady AN USSR 284(2):478–480

Nansen F (1909) The oceanography of North Polar Basin. The Norwegian North polar expedition 1893–1896. Sci Res 3(9):390 pp

Orvik KA, Skagseth Ø (2003) The impact of the wind stress curl in the North Atlantic on the Atlantic inflow to the Norwegian Sea toward the Arctic. Geophys Res Lett 30(17):1884. doi:10.1029/2003GL017932

Orvik KA, Skagseth Ø, Mork M (2001) Atlantic inflow to the Nordic seas: current structure and volume fluxes from moored current meters, VM ADCP and SeaSoar CTD observations, 1995–1999. Deep-Sea Res 48:937–957

Østerhus S, Turrell WR, Hansen B, Lundberg P, Buch E (2001) Observed transport estimates between the North Atlantic and the Arctic Mediterranean in the Iceland–Scotland region. Polar Res 20:169–175

Polyakov IV, Johnson MA, Colony RL, Bhatt U, Alexseev GV (2002) Observationally based assessment of polar amplification of global warming. Geophys Res Lett 29:1878–1881

Quadfasel D (1991) Warming in the Arctic. Nature 350:385

Quadfasel D, Rudels B, Kurz K (1988) Outflow of dense water from a Svalbard fjord into the Fram Strait. Deep-Sea Res 35:1143–1150

Rudels B, Jones EP, Anderson LG, Kattner G (1994) On the intermediate depth waters in the Arctic Ocean. In: Johannessen OM, Muench RD, Overland JE (eds) The polar oceans and their role in shaping the global environment, Geophysical monograph 85. American Geophysical Union, Washington, DC, pp 33–46

Rudels B, Darnell C, Gunn J, Zakharchuck E (1997) CTD observations. In: Scientific Gruise report of the Arctic expedition ARK – XI/1 of RV "Polarstern" in 1995. Berichte zur Polarforschung 226:22–25

Rudels B, Björk G, Nilsson J, Winsor P, Lake I, Nohr C (2005) The interaction between waters from the Arctic Ocean and the Nordic seas north of Fram Strait and along the East Greenland Current: results from the Arctic Ocean-02 Oden expedition. J Mar Syst 55:1–30

Schauer U (2000) The expedition ARKTIS XV/3 of the research vessel "Polarstern" in 1999. Rep Polar Res 350:63 pp, Alfred-Wegener-Institute, Bremerhaven

Schauer U, Rudels B, Muench RD, Timokhov L (1995) Circulation and water mass modification along the Nansen Basin slope. Berichtezur Polarforschung 176:94–98

Schauer U, Loeng H, Rudels B, Ozhigin VK, Dieck W (2002) Atlantic water flow through the Barents and Kara Seas. Deep-Sea Res I 49:2281–2298

Schauer U, Fahrbach E, Osterhus S, Rohardt G (2004) Arctic warming through the Fram Strait: oceanic heat transport from 3 years of measurements. J Geophys Res 109:C06026. doi:10.1029/2003JC001823 (2004)

Smedsrud LH, Sorteberg A, Kloster K (2008) Recent and future changes of the Arctic sea – ice cover. Geophys Res Lett. doi:10.1029/2008GL034813

Stommel H (1961) Thermohaline convection with two stable regimes of flow. Tellus 13:224–230

Stommel H, Farmer HG (1953) Control of salinity in an estuary by a transition. J Mar Res 12:13–20

Thompson DWJ, Wallace JM (1998) The Arctic oscillations signature in the wintertime geopotential height and temperature fields. Geophys Res Lett 25:1297–1300

Timofeyev VT (1960) Water masses of the Arctic Basin. Gidrometeoizdat, Leningrad

Vinje T (2001) Anomalies and trends of sea ice extent and atmospheric circulation in the Nordic seas during the period 1864–1998. J Clim 14:255–267

Walin G, Broström G, Nilsson J, Dahl O (2004) Baroclinic boundary currents with downstream decreasing buoyancy: a study of an idealized Nordic seas system. J Mar Res 62:517–543

Walsh JE, Portis DH (1999) Relationship between the atmospheric circulation of the Central Arctic and the North Atlantic. In: 5th AMS Conference on polar meteorology and oceanography, Dallas, 10–15 Jan 1999

Whitehead JA (1995) Thermohaline ocean processes and models. Ann Rev Fluid Mech 27:89–113

Whitehead JA (1998) Multiple T-S states for estuaries, shelves and marginal seas. Estuaries 21:281–293

Worcester PF, Lynch JF, Morawitz MWL, Pawlowicz R, Sutton PJ, Cornuelle BD, Johannessen OM, Munk WH, Owens WB, Schuchman R, Spindel RC (1993) Evolution of the large-scale temperature field in the Greenland Sea during 1988–89 from tomographic measurements. Geophys Res Lett 20:2211–2214

Part II
Modelling

Chapter 7
Mesoscale Modelling of the Arctic Atmospheric Boundary Layer and Its Interaction with Sea Ice

Christof Lüpkes, Timo Vihma, Gerit Birnbaum, Silke Dierer,
Thomas Garbrecht, Vladimir M. Gryanik, Micha Gryschka, Jörg Hartmann,
Günther Heinemann, Lars Kaleschke, Siegfried Raasch, Hannu Savijärvi,
K. Heinke Schlünzen, and Ulrike Wacker

Abstract This chapter summarises mesoscale modelling studies, which were carried out during the ACSYS decade until 2005. They were aiming at the parameterisation and improved understanding of processes in the Arctic boundary layer over the open ocean and marginal sea ice zones and over the Greenland ice sheet. It is shown that progress has been achieved with the parameterization of fluxes in strong convective situations such as cold-air outbreaks and convection over leads. A first step was made towards the parameterization of the lead-induced turbulence for high-resolution, but non-eddy resolving models. Progress has also been made

C. Lüpkes (✉) • G. Birnbaum • V.M. Gryanik • J. Hartmann • U. Wacker
Alfred Wegener Institute for Polar and Marine Research, Postfach 120161, D-27515
Bremerhaven, Germany
e-mail: christof.luepkes@awi.de; gerit.birnbaum@awi.de; vladimir.gryanik@awi.de;
jorg.hartmann@awi.de; ulrike.wacker@awi.de

T. Vihma
Finnish Meteorological Institute, POB 503, FIN-00101 Helsinki, Finland
e-mail: timo.vihma@fmi.fi

S. Dierer • K.H. Schlünzen
Meteorological Institute, Centre for Marine and Climate Research, University of Hamburg,
Bundesstr. 55, D-20146 Hamburg, Germany
e-mail: sdierer@web.de; heinke.schluenzen@zmaw.de

T. Garbrecht
OPTIMARE Sensorsysteme GmbH & Co. KG, Am Luneort 15a, D-27572
Bremerhaven, Germany
e-mail: thomas.garbrecht@optimare.de

M. Gryschka • S. Raasch
Institute of Meteorology and Climatology, Leibniz University of Hannover,
D-30419 Hannover, Germany
e-mail: gryschka@muk.uni-hannover.de; raasch@muk.uni-hannover.de

P. Lemke and H.-W. Jacobi (eds.), *Arctic Climate Change: The ACSYS Decade and Beyond,* Atmospheric and Oceanographic Sciences Library 43,
DOI 10.1007/978-94-007-2027-5_7, © Springer Science+Business Media B.V. 2012

with the parameterization of the near-surface atmospheric fluxes of energy and momentum modified by sea ice pressure ridges and by ice floe edges. Other studies brought new insight into the complex processes influencing sea ice transport and atmospheric stability over sea ice. Improved understanding was obtained on the cloud effects on the snow/ice surface temperature and further on the near-surface turbulent fluxes. Finally, open questions are addressed, which remained after the ACSYS decade for future programmes having been started in the years after 2005.

Keywords Polar atmospheric boundary layer • Atmosphere-ice-ocean interaction • Sea ice • Leads • Cold-air outbreaks • Surface roughness

7.1   Introduction

The Arctic atmospheric boundary layer (ABL) exhibits several characteristics, which have a large influence on the transport processes of mass, energy and momentum and their exchange between the atmospheric, oceanic and cryospheric parts of the Arctic climate system. The spatial scale of these processes is mostly smaller than that resolved by large-scale climate models. So, the application of mesoscale models can help to obtain a better understanding and quantification of the processes and to develop and test their parameterization for models using different grid sizes. Examples of characteristic features of the Arctic ABL, which need further consideration with respect to (climate-) modelling and parameterization and which have been addressed during ACSYS, are the following:

The first one concerns the ABL thickness. Over the huge pack-ice region, it is mostly below 400 m. But during off-ice flow regimes with cold-air outbreaks, the height increases over the open ocean with increasing distance from the ice edge to values of sometimes more than 2,000 m in a distance of several 10 km off the ice edge.

The second peculiarity of the Arctic ABL concerns its stratification. According to radiosonde sounding data from Russian ice stations, surface-based inversions prevail during winter. They extend to a height of approximately 1,200 m, with a

G. Heinemann
Department of Environmental Meteorology, University Trier, D-54286 Trier, Germany
e-mail: heinemann@uni-trier.de

L. Kaleschke
Institute of Oceanography, University of Hamburg, Bundesstr. 55,
D-20146, Hamburg, Germany
e-mail: lars.kaleschke@zmaw.de

H. Savijärvi
Department of Physics, University of Helsinki, P.O. Box 64, 00014, Helsinki, Finland
e-mail: hannu.sarvijarvi@helsinki.fi

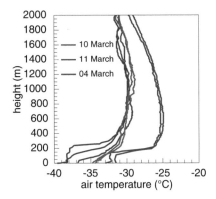

Fig. 7.1 Temperature profiles north of the Fram Strait marginal sea ice zone during cold conditions in March 1993, measured during the campaign REFLEX (Kottmeier et al. 1994) with drop sondes released from the aircraft Polar 4 of the Alfred Wegener Institute

typical temperature increase of 10–12K (Serreze et al. 1992b). Ship and aircraft data from the Atlantic sector of the Arctic Ocean show that in winter and early spring, especially during low temperatures around −30°C, strong surface-based inversions can exist also in this region. But at higher temperatures, the inversions are mostly elevated, and below the inversion, a slightly stable ABL prevails of typically 100–250 m thickness (Fig. 7.1). Modelling studies presented in this section are mostly based on the latter conditions. In summer, a slightly stable or near-neutral stratification prevails in the ABL over sea ice with a capping inversion of variable thickness and temperature increase.

The stable stratification over sea ice is caused by various reasons. These are the generally negative radiation balance over snow and ice and, during on-ice flow regimes, also the advection of warm air from the open ocean over the cold ice surface (e.g. Overland and Guest 1991). The performance of models for these situations was not well known before the ACSYS decade.

A third peculiarity of the Arctic ABL is caused by the inhomogeneities in the ice distribution ranging from large transitional zones between the open ocean and pack ice (marginal ice zones) to regions in the pack ice with open water patches and/or thin-ice areas (leads and polynyas). Leads exist during the whole year even in regions with thick pack ice. Due to divergent sea ice drift, leads are formed with lengths of up to hundreds of kilometres extending over the Arctic. Over leads, strong convective plumes can develop even when they are covered with thin ice. These plumes penetrate into the stably stratified ABL while generating gravity waves (Mauritsen et al. 2005). Due to the small width of leads, ABL processes in the vicinity of leads cannot be resolved by climate models, but have to be parameterized.

Both drift and deformation of sea ice depend on the vertical transport of momentum within the ABL. This transport is strongly influenced by the heterogeneous roughness of the sea ice surface due to inhomogeneous distribution of sea ice pressure ridges and ice floes. Mesoscale models can be used to study the effect of a heterogeneous sea ice surface on both the ABL and in turn on the drift and deformation of sea ice. During ACSYS, the influence of ridges and floe edges on the momentum transport has been investigated in detail.

In the ACSYS decade, mesoscale models were also used to study ABL processes over the open ocean, especially during conditions of cold-air outbreaks. Due to their high frequency of occurrence and their significance on the energy budget of ocean and atmosphere, it is important that cold-air outbreaks are investigated and that the parameterization of turbulent fluxes within the ABL is examined for this special type of flow regime. Furthermore, cold-air outbreaks influence huge regions over the Nordic Seas and are sometimes triggering the development of polar lows. Sometimes, they extend from the marginal sea ice zone near Svalbard far to the mid-latitudes while interacting with low-pressure systems over the Nordic Seas. Therefore, a correct treatment of cold-air outbreaks in models is of high relevance for air and ship traffic.

During ACSYS, also katabatic wind systems over the sloped ice sheet of Greenland have been modelled, which have a strong influence on the transport of momentum and energy and hence also on the mass balance of the ice sheet.

In this chapter, we give a review on mesoscale modelling studies of Arctic ABL processes during the ACSYS decade. Moreover, some case studies carried out during the ACSYS period are summarised, which are related to measuring campaigns (REFLEX, ARTIST, KABEG, WARPS).

Most of the modelling studies presented in this chapter have been carried out with the 2D hydrostatic mesoscale model UH of the University of Helsinki (Alestalo and Savijärvi 1985; Savijärvi and Matthews 2004; Vihma et al. 2005) and with the 3D non-hydrostatic mesoscale transport and fluid model METRAS developed at the University of Hamburg. METRAS is described by Schlünzen (1990) and Schlünzen et al. (1996) and a 2D version by Lüpkes and Schlünzen (1996).

7.2   Cold-Air Outbreaks

One of the most striking atmospheric phenomena in the Arctic is the formation of cold-air outbreaks with strong convection developing under off-ice wind conditions downstream of the pack-ice region. Satellite images document that between late autumn and early spring, this type of convection, visible by the typical cloud patterns, covers a large part of the Arctic ice-free ocean. Sometimes, cold-air outbreaks originating from the northern Fram Strait reach even to mid-latitudes, and often, cold-air outbreaks interact with low-pressure systems over the Nordic Seas.

Cold-air outbreaks have been studied in much detail during the ACSYS decade by observations and mesoscale modelling. Several aircraft-based campaigns have been carried out over the northern Fram Strait (Brümmer et al. 1992; Brümmer 1997, 1999; Hartmann et al. 1992, 1999; Kottmeier et al. 1994) to investigate the boundary layer structure in a region of about 300 km extent around the marginal sea ice zone. From such observations, we know that over the Fram Strait pack-ice region, the atmospheric flow during cold-air outbreaks is characterised by a shallow slightly stable boundary layer capped by a strong inversion with a base at about 100–300 m height (Brümmer 1996). During cold-air outbreaks, the air temperatures north of the sea ice edge range from −20°C, as observed by Hartmann et al. (1992), down to −35°C, as observed by Kottmeier et al. (1994). Downstream of the ice edge, the thickness of the convective layer increases gradually up to values between 900 and 2,200 m, and air temperatures are between −15°C and −5°C after 300 km fetch over the open water (Brümmer 1996). In most cases of cold-air outbreaks, the convection is organised into rolls accompanied with cloud streets and, at a distance of several hundreds of kilometres downstream of the ice edge, into hexagonal cells (Etling and Brown 1993; Atkinson and Zhang 1996). In the last decade, numerous studies of roll convection have been carried out with LES models (e.g. Chlond 1992; Raasch 1990; Hartmann et al. 1997; Müller et al. 1999; Gryschka and Raasch 2005; and many others). Liu et al (2004) presented the modeling of a cold air outbreak using a nonhydrostatic model with 500 m horizontal grid size and resolving the larger sized convection rolls. For shortness, we concentrate here, however, only on mesoscale modelling studies.

Due to the large differences between the temperatures of the open water surface and the near-surface air, extremely large heat fluxes with several 100 Wm⁻² occur over a very large region. Brümmer (1996) found from an analysis of ten cold-air outbreaks that the sensible and latent heat fluxes were the dominant terms in the surface energy budget over the open water. Values ranged between 200 and 700 Wm⁻². In contrast, the net longwave radiation was the dominating term over the pack ice, where the turbulent heat fluxes amounted only to about 10 Wm⁻². Owing to the considerable amount of heat and moisture transported from the open ocean to the atmosphere and the related potential for the development of polar lows, it is important for Arctic weather and climate models to accurately simulate the development of the convective ABL.

An accurate modelling of energy and momentum transport is especially difficult in the marginal sea ice zone. During cold-air outbreaks, the wind forcing on sea ice drift tends to generate a diffuse ice edge with the ice concentration decreasing gradually over a long distance. Hence, during cold-air outbreaks, there is usually a significant heat flux from the open leads upstream of the ultimate ice edge. Parameterization of this subgrid-scale heat flux is one of the key aspects for a successful modelling of the ABL heat budget during a cold-air outbreak (Vihma and Brümmer 2002). Moreover, the high number of ice floe edges cause a large form drag, which has been found critical for an accurate parameterization of the surface momentum fluxes (Birnbaum and Lüpkes 2002; Lüpkes and Birnbaum 2005). This topic is addressed in Sect. 7.5.2.

In a time scale of hours, well-developed cold-air outbreaks are often quasi-stationary in the sense that the mixed-layer height and the mean fields of temperature and humidity downstream of the ice edge change only slightly in time, since cold-air advection from north and the strong heating from the surface

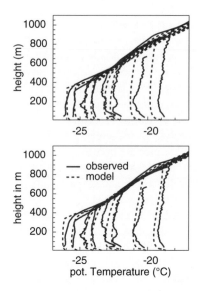

Fig. 7.2 Modelled and observed potential temperature during a cold-air outbreak, modelled without clouds (*top*) and with clouds (*bottom*). The first three profiles on the left hand side are above the MIZ; the profile on the right end of the figure is 114 km downstream of the ice edge (From Augstein et al. 2000; a similar case is discussed in Kaleschke et al. 2001)

balance each other. This stationarity is the reason why cold-air outbreaks were considered during the ACSYS decade and already before for the testing and derivation of turbulence closures in 2D models. Lüpkes and Schlünzen (1996) tested several first-order closures. They used the non-hydrostatic model METRAS, which was initialised with observed profiles over pack ice and a geostrophic wind according to the observations. For a case of extreme convection observed during a cold-air outbreak by Kottmeier et al. (1994), it was shown that only schemes allowing very efficient turbulent mixing were able to reproduce the typical temperature structure of the well-mixed convective ABL. According to their study and a similar one by Chrobok et al. (1992), a strong mixing can be realised by closures allowing large eddy diffusivities with maxima in the order of several $100 \ m^2 s^{-1}$. Furthermore, in a simple nonlocal closure as that of Troen and Mahrt (1986) or of Lüpkes and Schlünzen (1996), countergradient transport of heat and humidity contributes to the mixing.

The study of Lüpkes and Schlünzen (1996) was carried out with 4 km horizontal grid size. Nevertheless, their nonlocal closure can be used also with much larger grid sizes being typical for regional high-resolution climate or forecast models. As an example, results obtained from METRAS with 8 km horizontal grid size and with the nonlocal closure of Lüpkes and Schlünzen (1996) are shown in Fig. 7.2. It contains modelled and observed profiles of potential temperature for a cold-air outbreak on 5 April 1998 in the Fram Strait, which was studied during the campaign

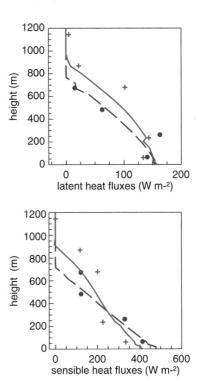

Fig. 7.3 Modelled and observed heat fluxes referring to a cold-air outbreak, observed during the ARTIST campaign in the Fram Strait (From Augstein et al. 2000)

ARTIST (Hartmann et al. 1999). As described in detail by Kaleschke et al. (2001), the use of satellite-based sea ice concentration data in the marginal sea ice zone was important to obtain a good agreement between model results and observation. The differences between modelled and observed profiles are smaller, when clouds are accounted for (treated in this model run via the Kessler (1969) parameterization). At a first glance, the agreement between modelled and observed fluxes of sensible and latent heat (Fig. 7.3) seems to be only fair; however, the error of the measurement is about 25% of the observed values, so there should not be too much emphasis on the differences.

Bearing in mind that the used closure of Lüpkes and Schlünzen (1996) does not consider explicitly any impact of roll vortices on the vertical transport of energy and momentum, the good quality of agreement between modelled and observed profiles as in Fig. 7.2 is not self-evident. Note that in previous studies by others (see Hartmann et al. 1997 and the review article by Etling and Brown 1993), it was found that roll vortices can contribute 25% to the transport of energy. However, Fig. 7.2 proves that there exist at least cases where this additional effect of transport does not seem to play a large role. More case studies should be carried out in future to confirm this finding and to investigate whether any additional effect of roll vortices has to be taken into account in turbulence closures used, e.g. in climate models. Besides that, Schlünzen and Katzfey (2003) found that the above nonlocal closure is also applicable to cases with convection over land surfaces, when no roll vortices exist.

Also, 3D studies have been performed with mesoscale models. Pagowski and Moore (2001) used the Polar MM5 to model an extreme cold-air outbreak over the Labrador Sea. They found that an accurate representation of the inhomogeneities in sea ice cover within the marginal sea ice zone is necessary to reproduce the observed boundary layer evolution over the open sea. Furthermore, they point to the strong impact of the initial and boundary conditions in regional models used for the simulation of processes. Due to inaccuracies and to the limited resolution of the used analysed fields supplying the initial and boundary conditions, they were not able to correctly simulate the interaction of a cold-air outbreak with the formation of a polar low.

A similar conclusion on the importance of a high-quality analysis was drawn by Wacker et al. (2005), who presented 3D simulations of the same cold-air outbreak episode over Fram Strait as shown in Fig. 7.2 as a result from the 2D version of the model METRAS. They used the operational mesoscale weather prediction model Lokal-Modell (LM) (Doms and Schättler 2002) of the German Weather Service (DWD). Differences between model results and observations could be traced back mainly to differences between the observed ice conditions and those resulting from the analysis data. A sensitivity study showed that physical parameterizations as the treatment of the surface fluxes above water had a significant effect on the results. Another result was that the model tends to underestimate the convective boundary layer thickness, especially in large distance (~200 km) from the ice edge. Since a similar problem occurred also in a simulation by Lüpkes and Schlünzen (1996), Wacker et al. (2005) speculated that in mesoscale models of this type, some physical processes, e.g. the penetration of plumes into the inversion layer, are not represented well enough. It can be expected that a similar problem would occur also in climate models unless mixing processes (entrainment) near the inversion are not explicitly parameterized.

The development of clouds during cold-air outbreaks was also addressed by Wacker et al. (2005). In contrast to the METRAS simulation discussed above, the cloud parameterization scheme includes also the ice phase. Wacker et al. (2005) could not find a significant influence of the ice phase on the evolution of the cold-air outbreak. Model runs with and without a consideration of ice particles in clouds resulted in only small differences, e.g. in the fields of temperature and cloud water content. Figure 7.4 contains a segment of the vertical north-south cross section along 5°E through the LM model domain, considered by Wacker et al. (2005). It covered an area of about $1,000 \times 1,500$ km^2 over the North Atlantic. According to the figure (not shown in Wacker et al. 2005), most of the condensate is in liquid form, although the temperatures are about 15–20K below the freezing point. The maximum ice content is found close to the ice edge in cold air. Light precipitation falls out of the clouds. The amount of condensate, the dominance of liquid water, and presence of precipitation are in agreement with observations.

Also, Vihma and Brümmer (2002) considered the role of clouds during cold-air outbreaks. They found that it was essential for an accurate modelling of the specific humidity field to account for deposition growth of cloud ice. Olsson and Harrington (2000) conclude from their sensitivity study with a high-resolution 2D model that the amount of cloud condensate strongly influences the ABL structure due to radiation effects.

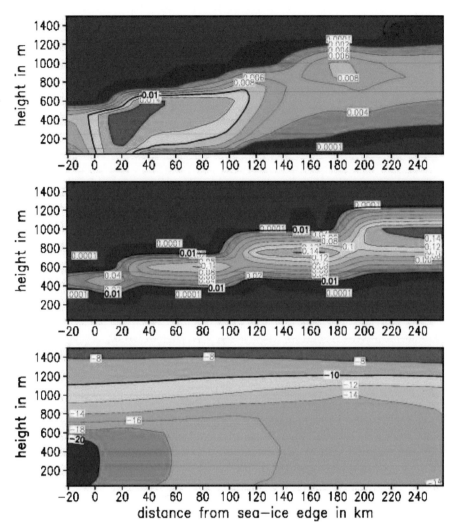

Fig. 7.4 Modelled potential temperature (*bottom*), liquid water content (*middle*) and ice content (*top*) in g/kg during a cold-air outbreak observed during the ARTIST campaign in the Fram Strait. Bottom Figure modified from Wacker U, Potty KVJ, Lüpkes C, Hartmann J, Raschendorfer M (2005) A case study on a polar cold air outbreak over Fram Strait using a mesoscale weather prediction model. Boundary-Layer Meteorol 117(2):301–336, Figure 6c, with kind permission by Springer Science+Business Media B.V.

Summary and Conclusions

In the ACSYS decade, several studies with mesoscale models have been carried out successfully to study cold-air outbreaks. Such model runs and the comparison of their results with observations improved our understanding of the effect of different parameterizations on the development of the convective ABL, and they helped to quantify the deficiencies of the model results. Remaining uncertainties, which should be subject to future research, concern the role of entrainment and cloud physical processes on the growth of the convective ABL.

7.3   On-Ice Airflow and Influence of Ridges

Especially in the marginal sea ice zone, the ABL is often influenced by warm-air advection from the open ocean. Such cases of on-ice flow are usually related to cyclones, and the effects of heat advection can sometimes be felt several 100 km downstream of the ice edge. During the springtime campaigns, ARTIST (Hartmann et al. 1999) and WARPS (Schauer and Kattner 2004) cold-air outbreaks over the Fram Strait were often accompanied by strong warm-air advection with on-ice flow over Storfjorden and the Barents Sea. This was due to low-pressure systems moving from southwest to northeast along the eastern side of Svalbard.

During on-ice flow, even in winter over the open ocean, the temperature of the advected air masses is close to the freezing point. While crossing the ice edge, the flow has to adjust to changed surface conditions: its properties are modified, and a shallow internal boundary layer develops, which is characterised by weak turbulent heat fluxes directed downwards. Due to the low heat conductivity of the snow, the snow surface temperature is rapidly affected by the overflowing warm air, and a large temperature gradient develops in the uppermost snow and ice layers (Cheng and Vihma 2002). Hence, on-ice flow is of importance not only for the state of the Arctic ABL but also for the sea ice thermodynamics.

Before the ACSYS decade, first aircraft observations of on-ice flows over the Fram Strait were made during summer by Fairall and Markson (1987). They found that the near-surface stress was reduced by a factor of two due to stability effects during on-ice flow. In at least one of their observed cases, the near-surface temperature decreased by 2.5 K over a distance of 70 km. A review of Guest et al. (1995) addressed the wind and momentum flux fields during various flow conditions over the marginal sea ice zone. Springtime observations (May at 80°N) of an on-ice airflow are presented by Brümmer et al. (1994). In their observations and in an aircraft-based study over the Baltic sea by Vihma and Brümmer (2002), the temperature differences between the air, ice and open sea were small, and so, the air-mass modification also remained small (a few tenths of a degree over horizontal distances of 100–300 km). The next aircraft-based study was carried out by Brümmer and Thiemann (2002). They document a case with strong warm-air advection along a 300 km flight section from the Fram Strait MIZ north-west of Svalbard towards north-westerly direction across the pack ice. Over this distance, the internal ABL grew to a height of 145 m, and the near-surface air temperature decreased from about −2°C over the open ocean to about −10°C at 100 km distance from the ice edge. The increase of ABL depth was not monotonic, but influenced by various factors such as cloud formation and inhomogeneous surface temperature. A strong temperature gradient of 12 K per 10 km, related to a synoptic-scale front, was observed 270 km off the ice edge.

The first modelling study of on-ice flow was carried out before ACSYS by Kantha and Mellor (1989). They presented 2D simulations with different geostrophic wind. In case of on-ice flow, they described the development of a shallow boundary layer combined with a low-level jet. An analogous modelling study by Glendening (1994) concentrated mainly on ice-parallel flow regimes. Neither study included a comparison with data since sparse data existing at that time could not be used for a validation of models.

The first comparison of mesoscale model results with aircraft data on the spatial ABL variability was made by Vihma and Brümmer (2002), but in their case of on-ice flow over the Baltic Sea, the change in ABL structure was dominated by the changes in surface roughness and temperature at the coastline. A comparison between model results and data of on-ice flow was carried out by Vihma et al. (2003). This study is explained in the remaining part of this section in more detail.

As a part of the campaign ARTIST (Hartmann et al. 1999), aircraft observations were carried out in spring 1998 over the Storfjord (Svalbard) during conditions of on-ice airflow. Low-level flights were performed at about 30 m height, and standard meteorological parameters, turbulent fluxes and sea ice topography were measured, the latter with a laser altimeter along a distance of about 150 km along the mean wind. Due to the on-ice flow regime, the sea ice concentration was close to 100% over the entire flight distance.

Two 2D models were used: the hydrostatic UH model and the non-hydrostatic model METRAS. Both were initialised at the inflow boundary over water with the observed profiles of temperature and humidity. Based on observations, the same geostrophic wind was used for the model runs with a horizontal grid size of 2 km. Figure 7.5 shows observed and modelled profiles of wind and temperature along the mean wind in the ABL. South of the pack-ice edge, there was an area of about 50 km width with thin pancake ice, and on its southern end, the first profile was flown. There, a surface-based inversion was observed of about 300 m thickness. The boundary layer depth increased to 150 m at 162 km distance from the pack-ice edge. Obviously, both models were able to reproduce the boundary layer growth with a slight underestimation of thickness and cooling at 162 km. However, the latter could be traced back to the non-stationarity of the observed flow during the observation period. Neither model was able to produce the weak low-level jet observed at 300 m height at 51 km. The very good agreement between the solutions of the two models with different turbulence closures (METRAS: TKE-based closure, UH: first-order mixing length scheme) is once more shown in the contour plots of potential temperature and wind in Fig. 7.6. Also, the effect of the hydrostatic approximation in the UH model does not generate problems in this type of flow.

Vihma et al. (2003) included a sensitivity study, in which they investigated the role of different assumptions mainly concerning the turbulence closures. They obtained the largest effect on the boundary layer by changing the parameterization of surface roughness. In their reference runs, they used a parameterization of surface roughness by Garbrecht et al. (2002), which includes the effect of sea ice pressure ridges by splitting the entire drag into contributions of skin drag of the plane sea ice and of form drag caused by ridges. This parameterization could be used since ridge heights and their distances could be derived from the measured surface

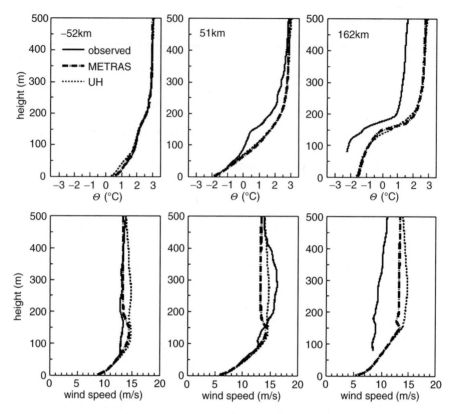

Fig. 7.5 Modelled and observed profiles of potential temperature and horizontal wind during on-ice flow over the Storfjord at different distances from the edge of the thick ice pack. The position −52 km is over thin (pancake) ice, whose thickness increased towards the ice edge. Reproduced from Vihma T, Hartmann J, Lüpkes C (2003) A case study of an on-ice air flow over the Arctic marginal sea-ice zone, Boundary Layer Meteorol. 107:189–217, Figure 3, with kind permission by Springer Science+Business Media B.V.

topography. With this parameterization, a very good agreement was obtained between model results and observed near-surface fluxes (Fig. 7.7) in the region up to 100 km distance from the ice edge. At larger distances, the agreement was worse, which is due to the inability of the models to reproduce an observed increase of wind speed, which was probably not related to the ice surface structure.

Figure 7.7 shows that both observed and modelled momentum fluxes have a maximum at $x=0$, where x is the distance from the pack-ice edge. At $x<0$, there was a region of pancake ice, and then, ice thickness and roughness increased towards $x=0$, which resulted in an increase of the drag coefficients and thus in increasing momentum fluxes. For $x>0$, two effects led to a decrease of fluxes. The first was the occurrence of smooth ice with lower observed drag coefficients, and the second one was stabilisation of the near-surface layers due to a lower surface temperature of thicker ice. As a result, in the considered case, the fluxes show an almost symmetric behaviour around $x=0$, but probably, this is not always the case over the ice edge zone.

Fig. 7.6 Potential temperature (°C) obtain with two different mesoscale models (UH and METRAS) for on-ice flow (wind from *left* to *right*). From Vihma T, Hartmann J, Lüpkes C (2003) A case study of an on-ice air flow over the Arctic marginal sea-ice zone, Boundary Layer Meteorol. 107:189-217, Figure 4, with kind permission by Springer Science+Business Media B.V.

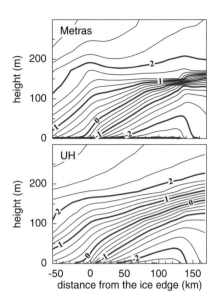

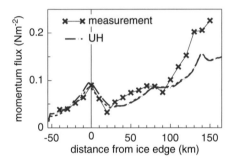

Fig. 7.7 Modelled and observed momentum flux at 30 m height for on-ice flow. Reproduced from Vihma T, Hartmann J, Lüpkes C (2003) A case study of an on-ice air flow over the Arctic marginal sea-ice zone, Boundary Layer Meteorol. 107:189-217, Figure 6a, with kind permission by Springer Science+Business Media B.V

The effect of ridges is illustrated in Fig. 7.8 showing the difference between the METRAS model run with the form drag parameterization and with a parameterization that uses a constant roughness length for momentum ($z_0 = 10^{-3}$ m). Differences occur in the entire ABL since the modified roughness affects the ABL depth. Especially, close to its top at about 150 m height, the effect of a modified parameterization is large since the ABL height is influenced.

Tests with different horizontal grid sizes were also made by Vihma et al. (2003); a vertical resolution of more than seven layers below 500 m was required for a reasonable modelling of the strong inversion, and a horizontal resolution higher than 20 km was necessary. Also, experiments applying six regional-scale climate models, validated against the SHEBA data, demonstrated that the model with highest vertical resolution yielded best results for the near-surface variables (Tjernström et al. 2005).

Fig. 7.8 Differences between two METRAS runs obtained with different surface roughness (see text). All other parameters remained unchanged. Modified from Vihma T, Hartmann J, Lüpkes C (2003) A case study of an on-ice air flow over the Arctic marginal sea-ice zone, Boundary Layer Meteorol. 107:189-217, Figure 8, with kind permission by Springer Science+Business Media B.V.

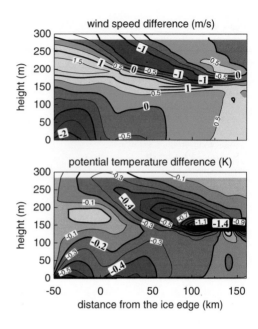

Conclusions

During ACSYS, some case studies of on-ice flow with warm-air advection have been performed with mesoscale models; the evolution of the stably stratified boundary layer over the ice has been well reproduced. Sensitivity studies under slightly stable stratification have shown that the different turbulence closures used above the surface layer have a smaller effect on the modelled boundary layer structure than the used parameterizations of surface roughness. The effect of sea ice ridges on the surface fluxes can influence wind and temperature in the entire ABL, which is due to their impact on the ABL depth. Due to the shallow boundary layer and small horizontal scales of changes in roughness and boundary layer depth, the application of a higher horizontal and vertical resolution would most probably improve the performance of climate models in the Arctic even far from topography.

7.4 Convective Processes Over Ice-Covered Oceans

Convection in the ABL over ice-covered oceans can be generated by several factors including (1) heat fluxes from leads, polynyas and areas with thin ice; (2) strong downward longwave radiation from clouds; (3) cloud-top radiative cooling; (4) cold-air advection and (5) summertime solar radiation on an ice surface with a reduced albedo. We shall discuss the first four factors.

7.4.1 Convection Over Leads

The importance of leads and polynyas in generating localised convection was well known already before the ACSYS decade. Among the first observations of the lead effect on the polar boundary layer were those carried out during AIDJEX (Paulson and Smith 1974) with measurements both in the upwind and downwind sides of leads (Andreas et al. 1979). These data clearly showed the influence of leads on the downstream profiles of wind speed and temperature. Parameterizations of the sensible heat flux were proposed as a function of bulk quantities. Further studies of leads were made by Schnell et al. (1989), who found by Lidar detection that plumes originating from leads may penetrate the Arctic inversion while transporting heat and moisture into the troposphere. According to Serreze et al. (1992a), however, such penetration is a rare event, as it requires an atypical combination of conditions (leads or polynyas at least 10 km wide, weak surface winds, low surface temperature and a weak temperature inversion).

During ACSYS, additional studies were carried out, resulting in more quantitative information on the fluxes over leads (Ruffieux et al. 1995; Alam and Curry 1997; Persson et al. 1997; Pinto et al. 2003). Andreas and Cash (1999) analysed three data sets: the AIDJEX data, data from a Russian drifting station over a refrozen polynya (Makshtas 1991), and data from a polynya in the Canadian archipelago (Smith et al. 1983). They found that smaller leads and polynyas transport sensible heat more efficiently than large ones, which might be explained by the combined effect of forced and free convection. An algorithm was proposed to calculate the turbulent heat fluxes over leads in fetch-limited convective conditions.

Attempts to model the flow over leads have been undertaken during the ACSYS decade applying LES models (Glendening and Burk 1992; Alam and Curry 1997; Weinbrecht and Raasch 2001; Zulauf and Krueger 2003). With such models, the number of cases, which can be considered, was restricted due to the models' high resolution and computational costs. Simpler models can, however, be used to simulate the characteristic features of the ABL above leads. Dare and Atkinson (1999, 2000) used a high-resolution mesoscale model to simulate the flow over polynyas 2–50 km wide. Mauritsen et al. (2005) used the non-hydrostatic atmospheric component of the Coupled Ocean-Atmosphere Prediction System (COAMPS) to simulate a flow over leads in idealised summer and winter conditions. Their primary focus was on the internal gravity waves generated by leads; in summer conditions, the gravity waves were related to the development of the internal boundary layer. In winter conditions, secondary circulations and intermittent wave breaking occurred.

A limitation in many studies has been that the in situ measurements were made over sea ice only. This was circumvented by Lüpkes et al. (2004) during the campaign WARPS (Winter Arctic Polynya Study, Schauer and Kattner 2004), where turbulence measurements were obtained over leads from a mast mounted at the bow of RV Polarstern. Instruments were run, while the ship was drifting slowly across the lead with several stops, similarly as described earlier by Garbrecht et al. (1999) for the drift towards an ice ridge. The helicopter-borne turbulence measuring system

Helipod (Bange et al. 2002) was used for additional measurements. Over a lead, covered by nilas of 8 cm thickness, the sensible heat flux amounted to about 70 W m^{-2} and increased to 100 $W m^{-2}$ when the nilas was destroyed by Polarstern. This demonstrates that a very accurate knowledge of the sea ice thickness is necessary for the calculation of the energy transport over a lead.

Two mesoscale modelling studies related to WARPS are discussed in the following lines: First, an application of the UH model with high resolution is explained; the model was used to simulate the convective boundary layer over leads observed during WARPS.

A 10 km-wide region was considered with two leads occurring over the northern Fram Strait on 3 April 2003. Both leads were oriented perpendicular to the wind direction. Their widths were 1 and 2 km with a 4 km-wide pack-ice patch in between. Flights with Helipod were conducted along the mean wind on eight levels below 300 m with the lowermost level at about 30 m. Several vertical soundings were flown at both ends of the horizontal flight sections. A well-mixed layer over the lead was found reaching a height of 150 m. The maximum sensible heat flux observed over the leads at the lowermost flight level was about 400 $W m^{-2}$. In the UH model, a horizontal grid spacing of 250 m was used with 50 layers in the lowermost 3 km. The initial $\theta(z)$ and $q(z)$ for the model were set according to the upwind Helipod profile up to 400 m and above according to Polarstern rawinsonde sounding data. The geostrophic wind was set according to observed winds above the ABL to 10 $m s^{-1}$. At the lateral boundaries, zero-gradient boundary conditions for $\theta(z)$, $q(z)$ and $V(z)$ were used. The surface temperature was prescribed on the basis of data obtained from a radiation thermometer at the helicopter. The model was run for 2 h.

The observed and modelled sensible and latent heat fluxes as well as potential temperatures are shown in Figs. 7.9 and 7.10. The vertical distribution of the sensible heat flux is well reproduced, but the modelled temperature increase at 30 m height caused by the leads is somewhat larger than observed. Further, there is a horizontal shift of ~1 km between the observed and modelled peak values of the turbulent fluxes at the height of 30 m.

This comparison of the hydrostatic model with the observations indicates that a non-eddy resolving model reproduces to some degree the structure of observed plumes, although the simple local turbulence closure applied was originally developed for horizontally homogeneous flow. This conclusion, however, has to be treated with some caution, since the spatial resolution of measurements was restricted, so they can give only an approximate picture of the real topology of temperature fields and fluxes over the leads.

In another modelling study (Lüpkes and Gryanik 2005; Lüpkes et al. 2008) (abbreviated by L0508) on the convection over leads, results obtained with the LES model PALM (Raasch and Schröter 2001) were compared with those of the non-hydrostatic model METRAS (Schlünzen 1990). The LES model used grid sizes of 10 m, which was small enough to resolve lead-induced convection. METRAS, however, was run with a non-eddy resolving horizontal grid length of 200 m. Two different turbulence closures were used in METRAS: a local mixing length closure, similar

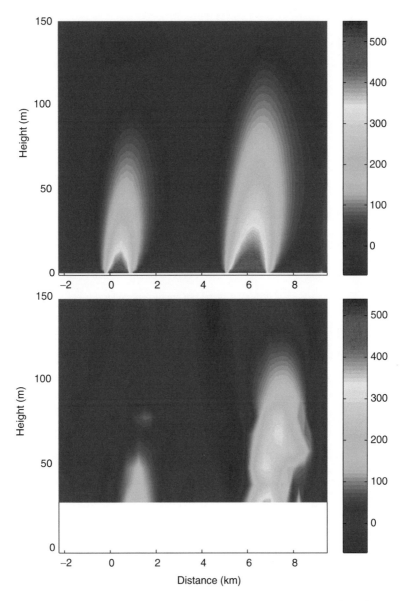

Fig. 7.9 Observed (*bottom*) and modelled (*top*) sensible heat flux over two leads (at 0–1 km and 5–7 km) in the Fram Strait pack-ice region (units: W m^{-2}). Note that there were no observations available from the layer below 30 m height

to the one used for the result shown in Fig. 7.9, and a new nonlocal turbulence closure, which accounts for the inhomogeneous nature of the flow over leads and allows gradient and countergradient fluxes of heat. Dominating external parameters for the closure are the buoyancy flux over the lead surface, the wind speed at the

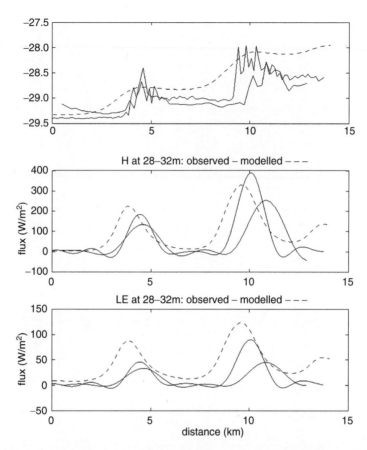

Fig. 7.10 Observed (*solid lines*) and modelled (*dashed line*) air potential temperature (*top*), sensible heat flux and latent heat flux at the height of 30 m over a 12 km-wide sea ice region with two leads (centred at 3.5 and 9.5 km). Two low-level flights were made, and data are shown from both

upstream lead edge and the ABL height on the upstream side of a lead. The closure includes a parameterization equation for the internal boundary layer developing on the downstream side of the leads during lead-orthogonal flow.

L0508 considered the flow over two leads of 1 km width separated by 10 km sea ice from each other. A neutral boundary layer was prescribed initially at the inflow boundary of the first lead with a strong capping inversion at 300 m height. In Fig. 7.11, results of both models are shown. Only by application of the nonlocal closure to METRAS, it became possible to reproduce the distribution of potential temperature as obtained from the LES model. Obviously, in both model results, the potential temperature decreases with height only in a small region over the lead. However, in a much larger region downstream of the lead, it increases slightly with height. Also, the simulated plume inclinations agree well in both model results.

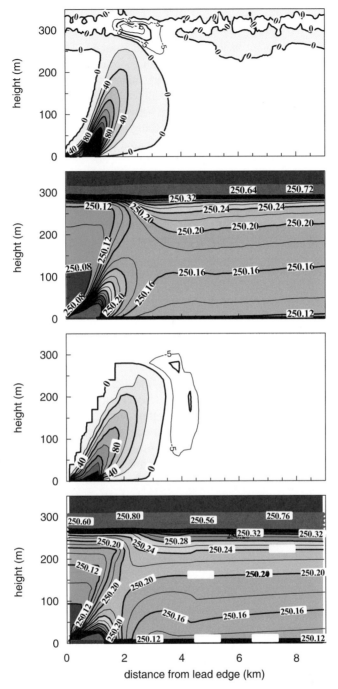

Fig. 7.11 Potential temperature (K) and sensible heat flux (W m^{-2}) obtained from the LES model (two *top* figures) and from METRAS (two *bottom* figures). The lead position is between distance = 0 and distance = 1 km; surface wind of about 5 m s^{-1} is directed in this case from left to right. Similar model results are described in From Lüpkes C, Gryanik VM, Witha B, Gryschka M, Raasch S, Gollnik T (2008) Modelling convection over leads with LES and a non-eddy-resolving microscale model. J Geophys Res C09028, Figures 3 and 6, Copyright American Geophysical Union (2008), modified by permission of American Geophysical Union

Below 80 m, turbulent heat fluxes obtained from METRAS are smaller than those of the LES. However, this does not hint to a failure of the closure. The reason is a too weak resolution of the LES at the upstream lead edge, which produces an overshooting of the plume further downstream over the lead (for details, see Lüpkes et al. 2008). A similar quality of agreement between both models was obtained for nine other cases differing by the external forcing parameters.

With the local closure, an increase of potential temperature with height on the downstream side of the lead could not be obtained. Thus, as for homogeneous convection, also for the lead situation, the physical nature of the turbulent heat fluxes could not be reproduced since a local closure accounts only for gradient fluxes. Nevertheless, the temperatures, averaged in vertical direction over the boundary layer depth, differed only slightly when the different closures were used. A difference in the order of 10 W m^{-2} existed between the heat fluxes obtained with the local and the nonlocal closure. In the upper part of the ABL, fluxes were underestimated with the local closure.

Summary and Conclusions

The main results of the lead modelling studies can be summarised as follows: Localised shallow convection in a slightly stable background stratification, which is a typical situation over Arctic winter leads, can be qualitatively simulated with a high-resolution hydrostatic model using a local turbulence closure. The more detailed structure of the heat plumes originating from leads can, however, only be resolved either with LES models or with non-eddy-resolving models using a nonlocal closure. The hydrostatic assumption seems to work well enough for conditions as presented here. Hence, also, results from hydrostatic models with a high horizontal resolution can form a certain basis for the derivation of flux parameterizations over large leads and polynyas for climate models. In the case of narrow leads, additional detailed observations and large eddy simulations are, however, needed to clarify the flow regimes. More studies are needed on the effects of leads in generating gravity waves since the latter are very important for the physics of the stable Arctic ABL.

7.4.2   Convection Over Sea Ice Caused by a Combination of Factors

During the ACSYS decade, various studies, e.g. Persson et al. (2002), have documented that the ABL stratification over Arctic sea ice can be slightly unstable even in winter, and the effects of various individual factors in generating the unstable stratification have been addressed. The heat budget of the ABL over sea ice is affected by turbulent, radiative and advective processes. During warm-air advection from the open ocean to sea ice, the stratification is typically stable (Sect. 7.3), but

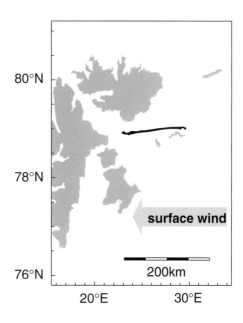

Fig. 7.12 Horizontal flight legs (*solid lines*) of the AWI Polar 2 aircraft over sea ice east of Spitsbergen on 27 March 1998. Islands are marked in grey and the sea ice in white. The flight track was completed by profile flights at the upstream and downstream end of the flight legs and by another one at about 24.5°E. From Vihma T, Lüpkes C, Hartmann J, Savijärvi H (2005) Observations and modelling of cold-air advection over Arctic sea ice. Boundary Layer Meteorol 117(2): 275-300, Figure 1, modified with kind permission by Springer Science+Business Media B.V.

during cold-air advection, slightly unstable stratification can be generated also over sea ice.

In winter, unstable stratification over a thick, compact (nearly 100% ice concentration) Arctic sea ice cover is less common, but may occur in overcast conditions with enhanced downward longwave radiation from clouds (Intrieri et al. 2002). This was one of the important findings during the ACSYS decade. Further, Pinto (1998) and Wang et al. (2001) found out that cloud-top radiative cooling may cause top-down mixing and generate a convective ABL even without any major heating from the surface. In the study of Vihma and Brümmer (2002), a slightly convective ABL over sea ice upwind of the ice edge was related to cold-air advection.

In summer, with a larger incoming solar radiation and a lower surface albedo, the ABL over sea ice can be slightly unstable even without the presence of the above-mentioned factors (although a cloud cover is usually present). The SHEBA results indicated that a well-mixed boundary layer is common in summer, although the upward sensible heat flux was never large (Persson et al. 2002).

In the following lines, a modelling study of Vihma et al. (2005) is described in detail. Its objective was to identify and quantify various processes contributing simultaneously to the unstable stratification observed over sea ice in early spring. The model simulations were based on aircraft observations in the ABL over Arctic sea ice east of Spitsbergen (Fig. 7.12). Non-stationary conditions of cold-air advection prevailed with a cloud-covered area, whose edge was retreating through the study region. Four horizontal legs of 100 km length were flown back and forth in upwind and downwind directions at an altitude of ~40 m. In addition, four pairs of vertical profiles were measured. In the following figures (Figs. 7.13 and 7.14), the

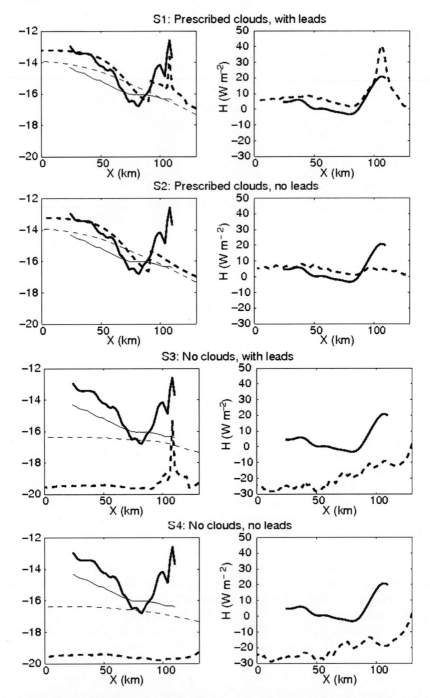

Fig. 7.13 Observed (*continuous lines*) and modelled (*dashed lines*) horizontal distributions from east (*left*) to west (*right*) of the potential temperature in °C (*left column*) of the surface (*thick lines*) and air (*thin lines*) and the sensible heat flux averaged over the four low-level flight legs for the sensitivity tests S1–S4. From Vihma T, Lüpkes C, Hartmann J, Savijärvi H (2005) Observations and modelling of cold-air advection over Arctic sea ice. Boundary Layer Meteorol 117(2):275-300, Figure 10, with kind permission by Springer Science+Business Media B.V.

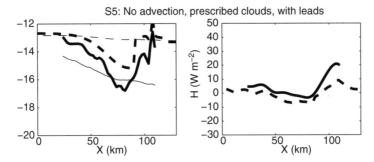

Fig. 7.14 As Fig. 7.13, but for sensitivity test, S5 without the effect of cold-air advection. From Vihma T, Lüpkes C, Hartmann J, Savijärvi H (2005) Observations and modelling of cold-air advection over Arctic sea ice. Boundary Layer Meteorol 117(2):275-300, Figure 11, with kind permission by Springer Science+Business Media B.V.

x-coordinate is defined, increasing eastwards with $x=0$ at 23°E, where the westernmost vertical profile was flown (Fig. 7.12). The turbulent fluxes of momentum, sensible heat and latent heat were calculated from the instantaneous observations. The sea ice concentration, roughness and ABL stratification varied along the flight track. In the west, the ice field was compact, but in the east, there was a region almost 30 km wide with small leads covering 5% of the surface area.

The observed case was modelled, applying the UH 2D hydrostatic boundary layer model equipped with a good-quality (modified ECMWF) radiation scheme (see also Sects. 7.3 and 7.4.1). To better understand the reasons for the spatial variations and for the existence of a well-mixed boundary layer with unstable stratification above the surface, different model runs were carried out. Five of these runs are described as follows:

Run S1: The surface temperature was modelled with an eight-layer scheme for the heat conduction in the snow and ice. The surface of each grid interval was divided into subsections of snow-covered ice and leads on the basis of the observations. The grid-averaged surface fluxes of heat and moisture were then calculated as area averages of the local fluxes, applying the mosaic method (Vihma 1995). Time-dependent vertical profiles of potential temperature and specific humidity at the inflow boundary, the cloud cover and the surface roughness were prescribed according to the observations. The flow was forced by the geostrophic wind speed based on the observations.

Run S2: As S1, but no leads were included.

Run S3: As S1, but no clouds were included.

Run S4: As S1, but no clouds and no leads were included.

Run S5: As S1, but instead of the prescribed $\partial\theta(z)$, t at the inflow boundary, zero-gradient boundary conditions were applied. Accordingly, in S5, there was no large-scale cold-air advection into the model domain.

In S1, the results (Fig. 7.13) are close to the observations, although the sensible heat flux is overestimated in the lead region, as is also the air temperature in the middle of the study region. The success of S1 was naturally mostly based on the prescribed clouds. However, the same cloud condensate contents and cloud fractions based on observations were used in all other model runs with clouds included (S2 and S5). This approach is supposed to yield reliable results on the relative importance of clouds, leads and advection.

In S2, with prescribed clouds, but without leads, the surface temperatures are too cold in the lead region at the western part of the domain, and the sensible heat flux does not show any peak there. In S3, with leads but without clouds, θ_s is far too low, resulting in a downward sensible heat flux. In S4, with no clouds and no leads, the situation is almost the same as in S3, except for the surface temperature peak.

On the basis of the results of S1–S4, it appeared that, under cold-air advection conditions, the thick clouds were responsible for the upward sensible heat flux in the downwind region, while leads together with thin clouds caused the upward flux in the upwind region. The effects of leads and clouds interacted non-linearly with the ABL turbulence and advection: in the presence of clouds, the lead effect on the ABL appeared stronger than in the absence of clouds (Fig. 7.13).

To understand the importance of the imposed cold-air advection, we look at the model results without it (S5, Fig. 7.14). $\theta_s(x)$ is warmer than in S1; $\theta_{40m}(x)$ is much warmer, and it is constant downstream. The sensible heat flux displays only a weak positive peak in the upwind region. The results accordingly suggest that the observed upward sensible heat flux was a combined effect of clouds, leads and cold-air advection.

Vihma et al. (2005) also made 5-day simulations to study the quasi-steady-state structure of the cloudy ABL over the Arctic sea ice. The simulations demonstrated that the evolution towards a deep, well-mixed ABL may take place via a formation of two mixed layers: one related to mostly shear-driven surface mixing and the other to buoyancy-driven top-down mixing due to the cloud-top radiative cooling. These mixed layers may gradually merge. The results were comparable to those of Wang et al. (2001). The combined effects of shear-driven turbulence and cloud-top radiative cooling over Arctic sea ice were addressed also in the LES modelling study by Inoue et al. (2005).

Summary and Conclusions

The effects of several factors acting simultaneously on the ABL development have received little attention in earlier studies. Studies during the ACSYS decade demonstrated that the important factors influencing the surface layer stratification are downward longwave radiation from clouds, upward heat fluxes from leads and cold-air advection. Qualitatively similar results have been derived from analyses of the Russian ice station data (Vihma and Pirazzini 2005). From the point of view of the ABL temperatures and stratification, clouds are usually more important than leads since they can strongly shape the temperature profile via both cloud-top radiative cooling (generating top-down mixing) and radiative heating of the snow surface (generating turbulent surface flux and bottom-up mixing).

7.5   Mesoscale Studies with Coupled Atmosphere-Sea Ice Models

A detailed understanding of the atmosphere-sea ice interaction requires the application of models on different scales including the mesoscale. Before the ACSYS decade, studies with coupled models were mainly restricted to large-scale models or to 1D models, and only very few investigations were made with multidimensional mesoscale models. One of the first studies was carried out by Wefelmeier (1992) (see also Wefelmeier and Etling 1991), who considered ice drift in the Fram Strait and its interaction with the ABL. The ice model did not include thermodynamic processes. During ACSYS, more complex coupled mesoscale atmosphere-sea ice models have been developed. A two-class dynamic-thermodynamic coupled atmosphere-sea ice model has been applied for studies of coastal polynyas (see Sect. 7.6). The results of the present section are based on the METRAS/MESIM model, which has been developed at the University of Hamburg in cooperation with the Alfred Wegener Institute, Bremerhaven. It consists of the atmospheric mesoscale model METRAS (see the previous subsections of Chap. 7) and the mesoscale sea ice model MESIM (Birnbaum 1998; Dierer et al. 2005). MESIM simulates the drift as well as freezing and melting of sea ice for several ice categories.

During the ACSYS decade, the model has been applied by Birnbaum (1998), by Dierer et al. (2005) and by Dierer and Schlünzen (2005). In the latter two articles, the impact of a cyclone passage on sea ice is studied. A similar case is considered in the following Sect. 7.5.1, and another application of the model system, based on Birnbaum (1998), is described in Sect. 7.5.2. In these studies, the thermodynamic effects are neglected in respect of the short time scales considered. The ice model is directly coupled to METRAS by using a flux aggregation scheme according to von Salzen et al. (1996). It includes the form drag effect of floe edges as in Birnbaum and Lüpkes (2002) (see also Sect. 7.5.2). A partial sea ice cover is allowed in each model grid cell. In each time step, the atmospheric forcing is used to calculate the ice drift, and in return, the new ice distribution influences the atmosphere by changing, for example, roughness length and heat flux. The influence of the ocean current is considered for the ice drift as well, but the changes in ocean currents by atmospheric forcing are not considered.

7.5.1   Influence of Polar Mesoscale Cyclones on Fram Strait Sea Ice Export – Case Studies with the Mesoscale Model System METRAS/MESIM

Every 5–6 days, a low-pressure system is passing the Fram Strait region (Affeld 2003). Based on ECMWF reanalysis data, Affeld (2003) additionally shows that 13% of the 1,374 lows that occurred between 1979 and 2000 in the Fram Strait were moving in northerly direction and thus passing the marginal sea ice zone, where

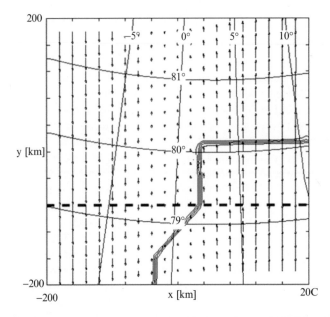

Fig. 7.15 Ocean current (*vectors*) and ice edge (*grey line*) in the Fram Strait. The region south and east from the ice edge is ice free in cases A and B. The dashed line denotes the budgeting line for calculating ice export and average wind

they strongly influence the energy exchange between ocean and atmosphere as well as the drift of sea ice. A study with METRAS/MESIM has been carried out to gain a better understanding of the atmosphere-sea ice interaction in such conditions and to quantify the modification of the ice export induced by a mesoscale low.

METRAS/MESIM was set up for the Fram Strait with the model domain completely over water and ice without any land surfaces included (Fig. 7.15). The passage of a mesocyclone was simulated for springtime meteorological conditions with ABL temperatures between −10°C and −30°C depending on the underlying surface type. The prescribed large-scale pressure gradient forced a cyclone track from south to north. The centre of the cyclone passed the model domain across its centre. This means that the atmospheric flow was from south to north in the eastern part of the domain and in the western part vice versa, when the cyclone centre was above the domain centre. Different scenarios were considered. The characteristics of the basic case A can be summarised as follows: ocean flow and ice coverage were used as shown in Fig. 7.15. A southward ocean current was prescribed in the western and a northward current in the eastern part of the model domain. This type of ocean current is typically found in the Fram Strait. The ice coverage amounted to 97% in the ice-covered sector of the domain and to 75.7% on average in the entire domain. Case A corresponds to a situation that was observed during the campaign ARKTIS'93 (Brümmer and Höber 1999). The sea ice distribution is characteristic for the northern Fram Strait region and is found in the climatological average.

Table 7.1 Parameters for run A and parameters used in the sensitivity studies B and C

Case	Changed parameter
A	Ice coverage in the ice-covered region 97%, total ice coverage 75.7%
B	No ocean flow, ice field as in A
C	Homogenous ice coverage of 75% in the entire domain

Fig. 7.16 West-east averaged modelled wind speed *ff* and direction *dd* at the budgeting line for cases A–E

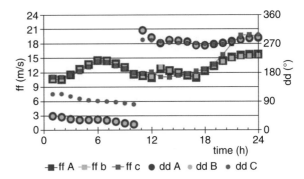

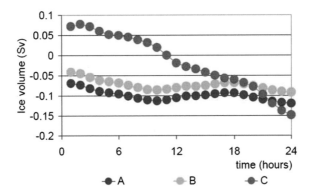

Fig. 7.17 West-east averaged ice volume transport at the budgeting line. Positive values denote ice transport to the north (import to the Arctic); negative values indicate ice transport to the south (export from the Arctic)

Two further model runs were carried out. In case B, the ocean flow was assumed as zero; the sea ice remained unchanged. In case C, a homogenous sea ice cover of 75% was prescribed in the entire domain (Table 7.1); the ocean flow was as in case A.

The modelled wind speed as well as the wind direction at the budgeting line (Fig. 7.15) varies considerably during the simulation (Fig. 7.16). The wind speed maximum amounts to 16 m/s at the end of the runs, and the minimum is about 10 m/s at the beginning. The centre of the cyclone passes the budgeting line at $t = 10$ h. At that time, the average wind at the budgeting line turns by about 40° to the east, and the wind speed drops from a previous maximum value of 15 m/s to about 12 m/s. The net ice export across the budgeting line is shown in Fig. 7.17.

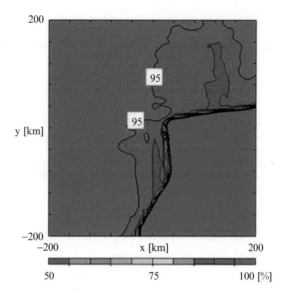

Fig. 7.18 Ice coverage at $t = 24$ h for case A

The largest difference exists between scenario C and the others. For scenario C, there is a net northward transport of ice before the passage of the cyclone's centre and a net southward transport after it. This is in contrast to the other cases showing always a transport towards the south. The initial northward transport in case C is a result of the initial ocean and atmospheric flow. Both, ocean current and atmospheric flow, are northward in the eastern part of the domain, where less ice exists in the cases A and B than in case C. A comparison of wind speed and ice export for cases A and B shows the dependence of ice export on wind. The ice drift export has a saddle point between $t = 11$ and $t = 18$ h. This is approximately the same time, when the wind speed is at a more or less constant low value. When it increases again, the ice export increases too. Obviously, the sea ice drift follows quite rapidly changes in atmospheric forcing.

The wind direction is the other important impact factor: south-easterly winds before the cyclone passage advect ice towards the ice-covered west. This ice is loosened after the cyclone passage by the north-westerly winds, which advect ice to the open water in the south-east. The ice distribution after the cyclone passage (Fig. 7.18) reflects the large impact of the cyclone. Close to the ice edge, the ice coverage is reduced by up to 10% (case A) and by about the same amount in the centre of the formerly homogeneous ice cover in case C. During the cyclone passage, the modelled ratio ice drift/wind speed increases, which is confirmed by observations of Brümmer and Hoeber (1999). This is most probably caused by the loosening of the ice. Hence, the increase in ice export in all cases is a combined effect of the increase of the ice drift/wind speed ratio, the changed wind direction and the larger wind speeds after the cyclone passage.

The total effect of the different scenarios for the initial ice distribution and ocean flow can be derived from the ice export across the budgeting line averaged over 24 h

Fig. 7.19 Mean ice export
(Sv) at the budgeting line
(79.5°N) for scenarios *A*, *B*
and *C*

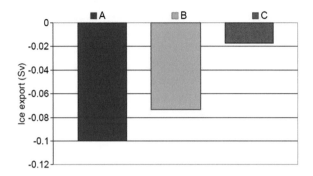

(Fig. 7.19). The neglect of the ocean current (case B) reduces the ice export by about 27% compared to case A. The smallest ice export was found for case C. As can be seen from Fig. 7.17, in case C, the ice import before the cyclone passage and the export thereafter more or less compensate each other, resulting in a small export of only 0.017 Sv for the 24 h integral. Obviously, the small-scale ice distribution is crucial to properly simulate ice export.

Summary and Conclusions

The influence of atmospheric forcing and sea ice distribution on sea ice export through Fram Strait has been investigated during the ACSYS period with the coupled atmosphere-sea ice model METRAS/MESIM. The sea ice was found to react within hours on the atmospheric forcing. The model results have shown that mesoscale heterogeneities in atmospheric forcing and sea ice cover can considerably change the sea ice export. To simulate it properly, a high resolution (here with 7 km horizontal grid length) needs to be employed. It is the distribution of ice, not its total amount in the model domain (here 400×400 km^2), which mainly determines differences in ice export. In the considered case, the export is considerably lower for a homogeneous ice distribution within the model domain without any clear ice edge. This is due to compensating effects of ice import and export in the homogeneous case. The latter result is especially important for climate modelling since it shows that models with large grid sizes in the order of 100 km can produce serious errors in modelling the ice export.

7.5.2   Modelling of Sea Ice Drift in the Marginal Sea Ice Zone

The interaction of atmosphere, ocean and sea ice is particularly strong in areas with heavily ridged ice or with a broken ice cover. Such conditions are typical for the marginal sea ice zone (MIZ), which is the transition zone between the pack ice and the open ocean region. In this zone, floes exist with typical diameters between a few and several hundreds of metres. Its horizontal extension mostly depends on the prevailing wind direction. In case of on-ice flows, there is sometimes a sudden change from the open ocean to the pack ice with very large floes. Sometimes, relatively small floes are collected in front of the pack ice. During off-ice flows, the MIZ spreads out, and the floes are surrounded by open water. Hence, their edges are exposed to the wind and ocean currents. Thus, off-ice flows are ideal to investigate the possible effect of floes on the momentum exchange between atmosphere, ocean and sea ice (Anderson 1987; Andreas et al. 1984; Claussen 1991; Guest and Davidson 1987; Fairall and Markson 1987; Overland 1985). Birnbaum (1998) studied the interaction between sea ice and atmosphere in the MIZ with the coupled model system METRAS/MESIM (see Sect. 7.5.1). The goal was to investigate the role of the different stress terms in the momentum balance equation for sea ice on the sea ice drift. In the following lines, this study is summarised including some additional figures not shown in the original work:

The atmospheric drag in the MIZ is calculated in METRAS as the sum of skin drag and form drag caused by floe edges and ridges at the edges of floes. This parameterization was developed during the ACSYS decade by Birnbaum and Lüpkes (2002) as well as Lüpkes and Birnbaum (2005). It is based on aircraft observations of turbulent fluxes above the MIZ during the campaign REFLEX (Kottmeier et al. 1994; Mai et al. 1996). The total drag on sea ice is parameterized in the model as the sum of the atmospheric skin drag and form drag, the oceanic skin drag and form drag, and the wave radiation stress. A similar concept as for the parameterization of atmospheric drag is used in MESIM for the calculation of oceanic stress. It is based on Steele et al. (1989), Perrie and Hu (1997) and Hehl (1997). The oceanic skin drag at the bottom of the ice is calculated similarly to McPhee (1979), whose approach is widely used in ice modelling (Lemke et al. 1990; Kreyscher et al. 2000). In the studies presented, the geostrophic ocean current is set to zero. As argued by Gallee (1997) and used in other coupled models (Lynch et al. 1995), this assumption is acceptable for short-time simulations.

Following Steele et al. (1989), the wave radiation stress is parameterized as a function of the surface wind stress over open water. It is assumed to point into the same direction as the near-surface wind. Internal ice forces are calculated with the most widely used viscous-plastic rheology (Hibler 1979; Lemke et al. 1990).

An off-ice flow across the Fram Strait MIZ close to Svalbard is investigated, which was also considered by Birnbaum and Lüpkes (2002) using METRAS without the sea ice model. The meteorological conditions are similar to cases typically observed during the campaign REFLEX.

The 2D model domain of 300 km extension is oriented in north-south direction from the pack-ice region across a large MIZ of 76 km length to the ice-free ocean. In the MIZ, initially, the ice concentration decreases linearly towards the open ocean (Figure 7.20a). Ice thickness and floe length are prescribed initially as a function of the sea ice concentration as proposed by Mai (1995) (see also Lüpkes and Birnbaum 2005).

In the model simulations, the geostrophic wind is set to $10 \, \mathrm{m \, s^{-1}}$ from north. This direction is perpendicular to the model's east-west oriented ice edge. Sea water is

Fig. 7.20 (a) Ice concentration A in the initial state and after 1, 2 and 3 days (b) Absolute value of ice drift velocity v_i in the initial state and after 3 days as a function of distance from the northern (*inflow*) boundary. The results refer to case 1

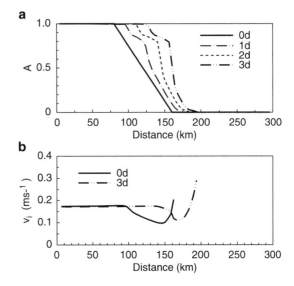

assumed to be at its freezing point of −1.8°C. The ice surface temperature is prescribed constant to −34°C.

Before the coupled simulation starts, the atmospheric model is run into a quasi-stationary state using the initial ice distribution. The duration of the coupled simulation is 3 days, which is a typical period for cold-air outbreaks.

In the following, results of five cases are discussed, which differ by the drag parameterization.

Case 1: First, a model run is analysed, where all five drag terms mentioned above are taken into account in the sea ice momentum balance equation. Since the ocean current is neglected, the wind causes the ice to drift with the main component to the south and a small component to the west. During the 3 days of simulation, the ice distribution within the MIZ changes considerably (Fig. 7.20a). In the central part of the MIZ, a very sharp gradient of ice concentration evolves.

The new distribution of ice results from a spatially inhomogeneous ice drift velocity field, which is shown in Fig. 7.20b. In the initial MIZ, the drift velocity decreases with decreasing ice concentrations for 20%<A<80%. Hence, in this zone, the ice drift is convergent, and the north-south gradient of ice concentration in this central part of the MIZ increases. In the region with A<20%, the drift velocity increases strongly with decreasing A. This pattern of sea ice drift can be explained as a result of the various stress terms. The atmospheric drags and the wave radiation stress accelerate the ice drift, whereas the oceanic drags have a decelerating effect, since in this study, the ocean current is assumed to be zero. The oceanic form drag increases considerably with decreasing A, except for very small concentrations. The reason is the dependence of form drag on the floe size. It is smaller for an ensemble of larger floes close to the pack ice than for the smaller floes further downstream, an

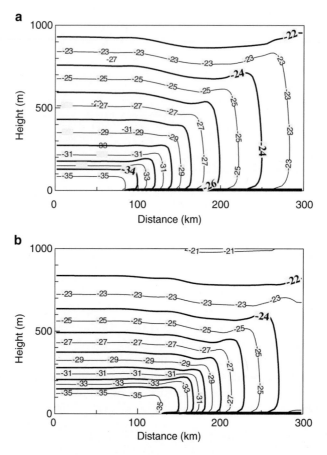

Fig. 7.21 Potential temperature (°C) **a** in the initial state and **b** after 3 days of the coupled simulation as a function of height and distance from the northern (*inflow*) boundary. The results refer to case 1

effect already described by Steele et al. (1989). Therefore, the larger floes drift with a higher velocity than the smaller floes, and the former collect the smaller floes.

At the southern edge of the MIZ, the strong wave radiation stress and wind stress cause the highest ice drift velocities found in the entire MIZ (Fig. 7.20b). Hence, after 3 days, a zone of more than 10 km width with very small ice concentrations (between 1% and 5%) forms the southernmost part of the MIZ.

As evident from Fig. 7.21, the atmospheric flow rapidly adjusts to the changes in the underlying ice cover. Since after 3 days a larger part of the model domain is covered with sea ice than in the beginning, the heat input from the ocean into the atmosphere is reduced. This leads to a considerable decrease of the ABL height over the MIZ, which is after 3 days 200 m lower at the outflow boundary than in the initial state.

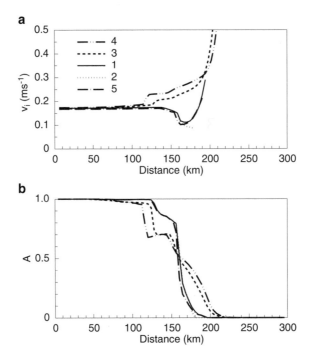

Fig. 7.22 (**a**) Absolute value of ice drift velocity v_i and (**b**) ice concentration A after 3 days as a function of distance from the northern (*inflow*) boundary for case 1 (all stress terms included), case 2 (no radiation stress), case 3 (no form drag), case 4 (no oceanic form drag) and case 5 (no atm. form drag)

Case 2: In this run, the wave radiation stress is neglected. It turns out that in general, wave radiation stress has a negligible influence on the ice drift velocity. This was also found by Steele et al. (1989), who showed with an ice-ocean model that wave radiation stress has only a minor influence on the sea ice drift. However, for very low ice concentrations as at the southern edge of the MIZ, its influence is large. There, the drift velocity is only one third of that in case 1 (see Fig. 7.22).

Case 3: Only wave radiation stress and the frictional forces at the top and the bottom of the ice (skin drag) are taken into account. Hence, the atmospheric and oceanic form drag is neglected. This leads to a considerably different ice drift velocity pattern in the MIZ (Fig. 7.22). The drift velocity increases in the entire region downstream of the pack ice. Furthermore, the ice concentration gradient in the centre of the MIZ is not as pronounced as in case 1.

In case 4, the oceanic form drag is neglected; in case 5, the atmospheric form drag is neglected.

Comparing all cases, it can be emphasised that besides skin drag, the most important stress is the oceanic form drag, whereas the atmospheric form drag has only a small influence. Nevertheless, Birnbaum and Lüpkes (2002) show that its influence on the atmospheric fluxes is of large importance. The highest drift velocities occur in case 4 because there only a decelerating stress term, the oceanic form drag, is switched off. The second largest velocities result if, additionally, the accelerative atmospheric form drag is neglected (case 3). Conversely, the drift velocities are generally much lower if the oceanic form drag is taken into account. If only the two skin drags are accounted for, the ice drift velocity is significantly overestimated in most parts of the MIZ. As shown in Fig. 7.22b, after 3 days, the position of the leading edge of the MIZ differs by about 30 km between case 4 and case 5.

Summary and Conclusions

During ACSYS, the ice–atmosphere interaction was studied in the marginal sea ice zone with the coupled ice–atmosphere model METRAS/MESIM. The results demonstrated that the ice drift velocity is very sensitive to the various stress terms. A neglect of oceanic form drag has a strong effect on both the drift velocity and the sea ice distribution in the MIZ.

In this investigation, the ocean current was neglected. Results could be different, especially in case of an accelerating oceanic flow. Furthermore, this analysis was restricted to the relative influence of the different stress terms only for conditions as in the MIZ. Future studies with mesoscale models should also concentrate on the inner Arctic.

7.6   Stable Boundary Layer Over the Greenland Ice Sheet

7.6.1   Simulations of Katabatic Wind Dynamics

The stable boundary layer (SBL) represents a quasi-permanent phenomenon for most parts of the year over the ice sheets of Greenland and Antarctica. As a result of the slopes of the ice sheets over large areas, katabatic winds develop, and the Coriolis force is important because of the relatively large horizontal scale of the wind system. The katabatic wind system over the sloped ice sheet of Greenland (Fig. 7.23) is very important for the near-surface climate and the air/snow energy and momentum exchange, which in turn plays an important role in questions of the mass balance of the ice sheet. The flow over the interior regions is relatively homogeneous, and hydrostatic mesoscale models with resolutions of 25–40 km have proven to reproduce the structure of the SBL in good agreement with observations, provided that the vertical resolution of the models is high in the ABL and clouds are forecasted correctly (Klein et al. 2001a; Bromwich et al. 2001). The structure of the katabatic wind system becomes more complex in the coastal areas, where the terrain

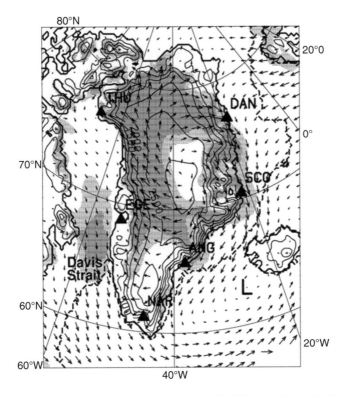

Fig. 7.23 Monthly mean wind field from a simulation with 25 km resolution for January 1990 (only every fourth wind *vector plotted*, orography isolines every 500 m). Directional constancy is represented by *shading*: *light shading* 0.8–0.9, *dark shading* larger than 0.9.The centre of the mean low-pressure system southwest of Iceland is marked by '*L*'. Greenland radiosonde stations are indicated by *triangles* (From Heinemann and Klein 2002)

is steepest. As a consequence, katabatic winds and associated air/ice interactions are also strongest near the ice margin, and high-resolution, non-hydrostatic models are needed (Klein et al. 2001b). Especially in the transition region between tundra and inland ice along the western coast of Greenland, the surface structures are quite complicated, consisting of steep glaciers, small lakes and fjords with a length scale of a few km.

During the ACSYS decade, our knowledge about the SBL structure over polar ice sheets has increased significantly in both the observational and the modelling point of view. The KABEG experiment (Heinemann 1999) has provided an unprecedented high-quality data set for the 3D structure of mean and turbulent quantities in the katabatic SBL (Heinemann 2002).

The KABEG data set has also been used for a couple of comparisons with mesoscale model simulation studies. As an example, the simulated and observed vertical structure of the SBL is displayed in Fig. 7.24 for a case study on 22 April

KA3 (22 April 1997):

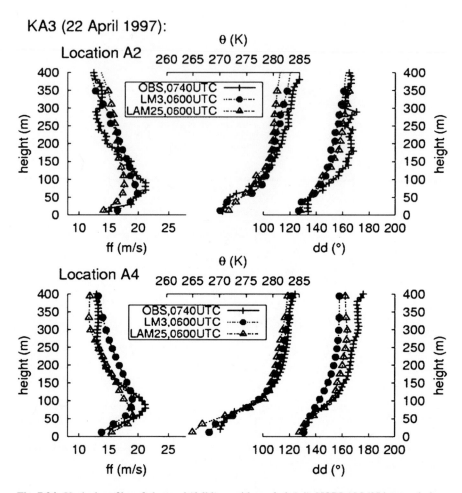

Fig. 7.24 Vertical profiles of observed (*full lines* with *symbols* '+'), NORLAM (25 km resolution, 18 h forecast, *full lines* with *open triangles*) and LM (2.8 km resolution, 6 h forecast, *full lines* with *filled circles*) simulated wind speed, potential temperature and wind direction at the locations A2 (*upper panel*) and A4 (*lower panel*) of the KABEG experiment (From Klein et al. 2001b, www.borntraeger-cramer.de)

1997. Vertical profiles of wind speed, potential temperature and wind direction are shown for two locations over the ice sheet of West Greenland, which coincide with the positions of two surface stations northeast of Kangerlussuaq during KABEG (Klein et al. 2001b). Location A2 has a height of about 800 m and lies in an area of increased slope near the edge of the ice sheet. Location A4 has a height of about 1,600 m and is situated at a distance of about 80 km from the ice edge. The three profiles shown in Fig. 7.24 are the hydrostatic NORLAM simulation with 25 km resolution, the non-hydrostatic LM simulation with 2.8 km resolution (both valid at 0600 UTC, 22 April 1997) and measured profiles obtained by KABEG aircraft measurements (at about 0740 UTC). In general, the modelled vertical profiles of both NORLAM and LM agree well with the measured boundary layer profiles, which have been averaged horizontally over 25 km. At site A4, the aircraft data show a well-developed katabatic wind system with a low-level jet (LLJ) of more than 22 m/s at a height of about 100 m. The wind speed at heights above 200 m is relatively high (13 m/s), which reflects the strong synoptic forcing for this case. The profile of the potential temperature shows a strong inversion of more than 10K per 100 m. The aircraft profiles show a veering of the wind direction with height from 130° near the surface (40° right to the local fall line) to southerly directions above the stable boundary layer. The boundary layer structure is captured well by both models, and NORLAM is slightly superior with respect to the wind profile. In the homogeneous area near A4, the higher resolution of the LM has obviously no advantage, and also, the different turbulence schemes of the models (first-order closure in NORLAM, TKE closure in LM) seem to have no major impact on the quality of the simulation of mean quantities. However, at site A2, which is also over the ice but close to the ice margin, the LM results are superior to the NORLAM results since the topography gradient is underestimated in the coarse resolution of NORLAM, and a much weaker LLJ is simulated.

7.6.1.1 Interaction Between the Katabatic Wind and the Formation of Coastal Polynyas

On several occasions, extreme katabatic storms occur when the katabatic wind system is enforced by synoptic processes. A well-known (and feared) phenomenon is the so-called piteraq, which is a strong synoptically enforced katabatic wind at the south-east Greenlandic coast (Fig. 7.23). Near the margins of the Greenland and the Antarctic ice sheet, scale interactions between katabatic storms and other phenomena occur, including the formation of mesocyclones and coastal polynyas. The effect of katabatic storms on the formation of coastal polynyas and associated feedback processes has been studied by coupling a two-class dynamic-thermodynamic sea ice model with a mesoscale non-hydrostatic atmospheric model (Heinemann 2003).

As an example of studies with the coupled mesoscale atmospheric/sea ice model with 12 km resolution, results of idealised studies for the Angmagssalik region of Greenland (near radiosonde station ANG in Fig. 7.23) are given as follows: In the Angmagssalik/Tasiilaq area, several pronounced valleys are present, and the channelling of the katabatic wind by these valleys together with the synoptic forcing by a cyclone near Iceland leads to the occurrence of extreme piteraq events in that area (Klein and Heinemann 2002; Rasmussen 1989). It is also one of the few regions of Greenland where the katabatic wind can extend to the coast. Heinemann (2003) shows that synoptically forced katabatic winds can result in a fast formation of a coastal polynya within 24 h in that area (Figs. 7.25 and 7.26). For a synoptic forcing of 8 m s^{-1} and north-westerly flow for the

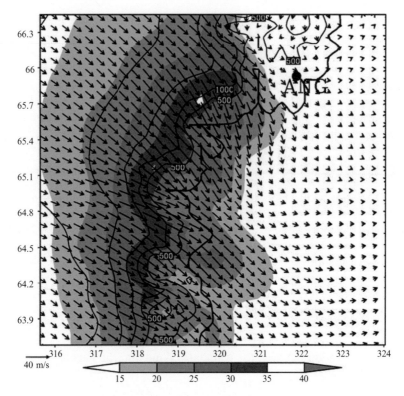

Fig. 7.25 Wind vectors (scale in the *lower left*) and wind speed (*shaded*) at 15 m after 14 h simulation for a subsection of the 12-km domain of the coupled FOOT3DK/sea ice model (no polynya and 100% sea ice coverage at the initial stage). The orography is shown as solid isolines every 500 m. The insert in the *upper left* corner indicates the synoptic forcing (From Heinemann 2003, Taylor & Francis Ltd., http://www.tandfonline.com)

Angmagssalik area, which corresponds to the synoptic situation of a stationary synoptic cyclone over Iceland, the near-surface wind field of the 12 km model after 14 h simulation (Fig. 7.25) shows the development of a strong synoptically forced katabatic wind in the Angmagssalik area. Maximum near-surface wind speeds are about 30–35 m s^{-1} over the steepest slope areas, and the katabatic wind is channelled by the large valleys. The signal of an orographically generated lee trough can be seen in the sea ice-covered eastern part of the model domain, which supports the extent of the katabatically generated airflow from the ice sheet over the sea ice for relatively larger distances. This large-scale lee trough moves southeastwards during the further development.

The development of a coastal polynya in the Angmagssalik area is displayed by the simulation of the frazil ice coverage in Fig. 7.26. After 6 h simulation time of the coupled model (not shown), some few isolated areas of frazil ice with 20–40% concentration (equivalent to a reduction of the consolidated ice

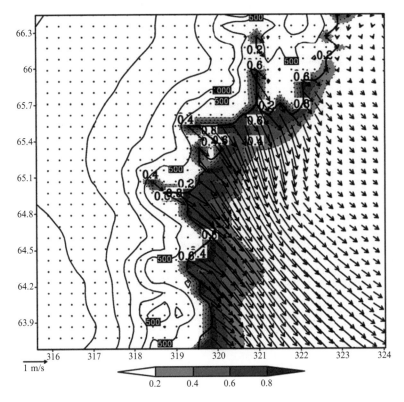

Fig. 7.26 As Fig. 7.25, but for the frazil ice coverage (*shaded* and *dotted* isolines every 0.2) and ice velocity of the consolidated ice (vector scale in the *lower left*), after 22 h simulation time

of the same amount) have formed at the exit regions of the large valleys. The formation of the polynya proceeds, and after 22 h simulation time, extended areas with only 20% consolidated ice are present near the coast. The thinning of the consolidated ice under the influence of intense katabatic winds (see Fig. 7.25) results in relatively large ice velocities. Sea ice advection is the main process during the first 12 h, but the production of frazil ice and the subsequent conversion to consolidated ice become important at later stages. Simulations for a scenario of an eastward-moving cyclone being typical for piteraq events in the Angmagssalik area reproduce the main features known from observational and realistic modelling studies, such as the development of the lee trough, its interaction with the katabatic wind and a development of a low-level mesoscale cyclone in the bay-like area southwest of Angmagssalik. The simulations of the coupled model were performed for the Greenland area, but because of their idealised setup, the results are also applicable for the conditions of the Antarctic.

7.7 Summary

During the ACSYS decade, a lot of studies have been carried out applying mesoscale models aiming at an improved understanding of the most relevant processes in the Arctic boundary layer and their parameterization. The studies addressed many different topics, such as the modelling of cold-air outbreaks, stable on-ice airflows and convective processes caused by leads, clouds and cold-air advection. Furthermore, the impact of ridges and floe edges on surface roughness was investigated. Finally, the interaction of the boundary layer with sea ice was studied in detail during different weather situations such as the passing of mesoscale cyclones. Most studies addressed the boundary layer processes over sea ice. It was shown in the last subsection, however, that also progress has been achieved in the mesoscale modelling of the stable boundary layer over the Greenland ice sheet.

In most of the model runs presented, the initial and boundary conditions were based on observational data obtained from aircraft. Comparison of model results with data mainly focussed on the representation of turbulent fluxes and on the boundary layer structure (thickness and stratification). It was shown that progress has obviously been achieved with the parameterization of fluxes in strong convective situations such as cold-air outbreaks and convection over leads. Remaining uncertainties in the modelling of cold-air outbreaks, which should be subject to future research, concern the role of entrainment and of cloud physical processes for the growth of the convective ABL. A first step was made towards the parameterization of the lead-induced turbulence for high-resolution but non-eddy resolving models. Future work is, however, necessary to derive such parameterizations for climate models.

Based on the modelling of on-ice airflow, it has been concluded that the parameterization of the impact of ridges on surface roughness is of large importance. It was shown that ridges influence the wind and temperature distribution not only close to the surface but also in the entire ABL. Progress has been made with the parameterization of the near-surface fluxes modified by ridges and by ice floe edges. In case of floes of up to 1 km width, the drag coefficients can be parameterized according to Lüpkes and Birnbaum (2005) as a function of sea ice concentration. However, in the inner Arctic regions with complete sea ice cover, the drag parameterizations require the knowledge of the surface topography. This forms a large challenge for remote sensing since the present resolution of satellite sensors is not high enough for the detection of small ridges.

One of the studies considered the effects of several factors acting simultaneously on the flow over a region covered almost completely with sea ice. It was shown that the surface layer stratification was influenced mainly by downward longwave radiation from clouds, upward heat fluxes from leads and cold-air advection. The latter produced upward fluxes over the sea ice, which was not often described in the literature for wintertime conditions.

The influence of atmospheric forcing and sea ice distribution on sea ice export through Fram Strait has been investigated during the ACSYS period with the coupled atmosphere-sea ice model METRAS/MESIM. The model results have shown that mesoscale heterogeneities in atmospheric forcing and sea ice cover can considerably change the sea ice export.

Another study with the same model system demonstrated that in the MIZ, the ice drift velocity is very sensitive to the various stress terms. A neglection of oceanic form drag has at least in the MIZ a considerable effect on both the drift velocity and the sea ice distribution. Future studies should investigate the role of the different stress terms also in the more inner Arctic regions with larger floes and a larger impact of ridges than floe edges.

Fortunately, it was shown that the hydrostatic assumption works reasonably well in mesoscale models not only for the stably stratified boundary layer over sea ice but also for cases with shallow convection over leads. Parameterizations of the energy and momentum fluxes in the ABL or the correct treatment of clouds seem to have a larger impact on the results compared with the effect of the hydrostatic assumption. Application of a higher horizontal and vertical resolution would most probably improve the performance of climate models in the Arctic, which are in most cases hydrostatic ones.

Although many different topics were addressed by the mesoscale modelling and observational studies during ACSYS, there are still many open questions to be addressed in future programmes. It is beyond the scope of the present chapter to develop a complete future strategy of mesoscale research. But a few points can be addressed here.

Most of the studies carried out during ACSYS were focussing on the ice edge zones and considered processes during winter and spring. As already mentioned, future studies should include the inner pack-ice regions and also summer conditions in both the MIZ and the central Arctic. This seems to be important also in the light of the dramatic changes, which occurred in the years after ACSYS such as the strong sea ice melting in summer 2007. Promising activities are the use of data from drifting stations and campaigns focussing on the melting processes of sea ice. Recent publications have demonstrated that a better understanding of the Arctic climate processes also requires a consideration of the processes above the ABL and their interaction with ABL processes. Furthermore, observations used for modelling studies during ACSYS were mostly carried out under fair weather conditions. So, finally, more attempts should be made in the future to include observations during severe weather conditions, when the turbulent fluxes of energy and momentum are largest and when we can expect a strong mixing between the ABL and the upper layers.

Acknowledgements Part of the presented research was conducted in the frame of the German ACSYS joint programme funded by the Federal Ministry of Education and Research (FKZ 03PL034A and FKZ 03PL034C). Other parts were funded by the Deutsche Forschungsgemeinschaft under grants He 2740/1 and WA 1334/2–1 and within SFB512. The project ARTIST was funded by the EU (grant no. ENV4-CT97–0487). Finalisation of this work was supported by the EC 6th Framework Programme DAMOCLES. The model simulations were performed on the DKRZ computers within the University of Hamburg contingent. KABEG was an experiment of the Meteorologisches Institut der Universität Bonn (MIUB) in cooperation with the Alfred-Wegener-Institut (AWI). The ECMWF provided the analyses taken as initial and boundary conditions for simulations presented in Sects. 7.3 and 7.6. SSM/I data used for the derivation of the sea ice coverage for KABEG and ARTIST were provided by the Global Hydrology Resource Center (GHRC) at the Global Hydrology and Climate Center (Huntsville, Alabama, USA).

The numerical models were provided by the Institut für Geophysik und Meteorologie, Universität zu Köln, the Norwegian Meteorological Institute (DNMI, Oslo), the German Meteorological Service (DWD, Offenbach) and the University of Hamburg.

References

Affeld B (2003) Zyklonen in der Arktis und ihre Bedeutung für den Eistransport durch die Framstraße. Dissertation, FB Geowissenschaften, Universität Hamburg

Alam A, Curry JA (1997) Determination of surface turbulent fluxes over leads in Arctic sea ice. J Geophys Res 102:3331–3344

Alestalo M, Savijärvi H (1985) Mesoscale circulations in a hydrostatic model: coastal convergence and orographic lifting. Tellus 37A:156–162

Anderson RJ (1987) Wind stress measurements over rough sea ice during the 1984 marginal ice zone experiment. J Geophys Res 92(C7):6933–6941

Andreas EL, Cash BA (1999) Convective heat transfer over wintertime leads and polynyas. J Geophys Res 104(C11):25721–25734

Andreas EL, Paulson CA, Williams RM, Lindsay RW, Businger JA (1979) The turbulent heat flux from Arctic leads. Boundary Layer Meteorol 17:57–91

Andreas EL, Tucker WB, Ackley SF (1984) Atmospheric boundary-layer modification, drag coefficient, and surface heat flux in the Antarctic marginal ice zone. J Geophys Res 89(NC1): 649–661

Atkinson BW, Zhang JW (1996) Mesoscale shallow convection in the atmosphere. Rev Geophys 34:403–432

Augstein E, Lüpkes C, Hartmann J (2000) Atmospheric boundary layer investigations over the Arctic Ocean, in: Final Report on The Arctic Radiation and Turbulence Interaction Study (ARTIST), Contract Nr. ENV4-CT97-0497-0487 (DG12-ESCY), Bremerhaven, pp 5–21

Bange J, Buschmann M, Spiess T, Zittel, Vorsmann P (2002) Umrüstung der Hubschrauber-schleppsonde Helipod, Vorstellung eines einzigartigen meteorologischen Forschungssystems. Deutscher Luft- und Raumfahrtkongress, Stuttgart, 23–26 Sept 2002

Birnbaum G (1998) Numerical modelling of the interaction between atmosphere and sea ice in the Arctic marginal sea ice zone. Reports on polar research, vol 268. Alfred Wegener Institute, Bremerhaven

Birnbaum G, Lüpkes C (2002) A new parameterization of the surface drag in the marginal sea ice zone. Tellus 54A:107–123

Bromwich DH, Cassano JJ, Klein T, Heinemann G, Hines KM, Steffen K, Box JE (2001) Mesoscale modeling of katabatic winds over Greenland with polar MM5. Mon Weather Rev 129:2290–2309

Brümmer B (1996) Boundary-layer modification in wintertime cold-air outbreaks from the Arctic sea ice. Boundary Layer Meteorol 80:109–125

Brümmer B (1997) Boundary-layer mass, water, and heat budgets in wintertime cold-air outbreaks from the Arctic sea ice. Mon Weather Rev 125:1824–1837

Brümmer B (1999) Roll and cell convection in wintertime Arctic cold-air outbreaks. J Atmos Sci 56:2613–2636

Brümmer B, Höber H (1999) A mesoscale cyclone over the Fram Strait and its effect on sea ice. J Geophys Res 104(D16):19085–19098

Brümmer B, Thiemann S (2002) The atmospheric boundary layer in an Arctic wintertime on-ice air flow. Boundary Layer Meteorol 104:53–72

Brümmer B, Rump B, Kruspe G (1992) A cold air outbreak near Spitsbergen in springtime - boundary-layer modification and cloud development. Boundary Layer Meteorol 61:13–46

Brümmer B, Busack B, Hoeber H (1994) Boundary-layer observations over water and Arctic sea ice during on-ice air flow. Boundary Layer Meteorol 68:75–108

Cheng B, Vihma T (2002) Modelling of sea ice thermodynamics during warm-air advection. J Glaciol 48:425–438

Chlond A (1992) Three-dimensional simulation of cloud street development during a cold air outbreak. Boundary Layer Meteorol 58:161–200

Chrobok G, Raasch S, Etling D (1992) A comparison of local and nonlocal turbulence closure methods for the case of a cold air outbreak. Boundary Layer Meteorol 58:69–90

Claussen M (1991) Local advection processes in the surface layer of the marginal ice zone. Boundary Layer Meteorol 54:1–27

Dare RA, Atkinson BW (1999) Numerical modelling of atmospheric response to polynyas in the Southern Ocean sea ice zone. J Geophys Res 104(D14):16691–16708

Dare RA, Atkinson BW (2000) Atmospheric response to spatial variations in concentration and size of polynyas in the Southern Ocean sea-ice zone. Boundary Layer Meteorol 94:65–88

Dierer S, Schlünzen KH (2005) Influence parameters for a polar mesocyclone development. Meteorol Z 14:781–792

Dierer S, Schlünzen KH, Birnbaum G, Brümmer B, Müller G (2005) Atmosphere-sea ice interactions during a cyclone passage investigated by using model simulations and measurements. Mon Weather Rev 133(12):3678–3692

Doms G, Schättler U (2002) A description of the nonhydrostatic regional model LM. Part I: dynamics and numerics. http://www.cosmo-model.org/content/model/documentation/core/. Accessed 10 Mar 2010

Etling D, Brown RA (1993) Roll vortices in the planetary boundary layer: a review. Boundary Layer Meteorol 65:215–248

Fairall CW, Markson R (1987) Mesoscale variations in surface stress, heat fluxes, and drag coefficient in the marginal ice zone during the 1983 marginal ice zone experiment. J Geophys Res 92:6921–6932

Gallee H (1997) Air-sea interactions over Terra Nova Bay during winter: simulation with a coupled atmosphere-polynya model. J Geophys Res 102:13835–13849

Garbrecht T, Lüpkes C, Wamser C, Augstein E (1999) Influence of a sea ice ridge on low-level air flow. J Geophys Res 104(D20):24499–24507

Garbrecht T, Lüpkes C, Hartmann J, Wolff M (2002) Atmospheric drag coefficients over sea ice – validation of a parameterization concept. Tellus 54A:205–219

Glendening JW (1994) Dependence of boundary layer structure near an ice-edge coastal front upon geostrophic wind direction. J Geophys Res 99:5569–5581

Glendening JW, Burk SD (1992) Turbulent transport from an Arctic lead: a large eddy simulation. Boundary Layer Meteorol 59:315–339

Gryschka M, Raasch S (2005) Roll convection during a cold air outbreak: a large eddy simulation with stationary model domain. Geophys Res Lett 32:L14805. doi:10.1029/2005GL022872

Guest PS, Davidson KL (1987) The effect of observed ice conditions on the drag coefficient in the summer East Greenland marginal sea ice zone. J Geophys Res 92(C7):6943–6954

Guest PS, Glendening JW, Davidson KL (1995) An observational and numerical study of wind stress variations within marginal ice zones. J Geophys Res 100:10887–10904

Hartmann J, Kottmeier C, Wamser C (1992) Radiation and Eddy Flux Experiment (REFLEX I). Reports on Polar Research, vol 105. Alfred Wegener Institute, Bremerhaven

Hartmann J, Kottmeier C, Raasch S (1997) Roll vortices and boundary layer development during a cold air outbreak. Boundary Layer Meteorol 84:45–46

Hartmann J, Albers F, Argentini S et al (1999) Arctic Radiation and Turbulence Interaction Study (ARTIST). Reports on Polar Research, vol 305. Alfred Wegener Institute, Bremerhaven

Hehl O (1997) Die Bestimmung der hydrodynamischen Rauhigkeit der Unterseite polaren Meereises - Numerische Simulationen. Berichte des Institutes für Meteorologie und Klimatologie der Universität Hannover, vol 52. University of Hannover, Hannover

Heinemann G (1999) The KABEG'97 field experiment: an aircraft-based study of the katabatic wind dynamics over the Greenlandic ice sheet. Boundary Layer Meteorol 93:75–116

Heinemann G (2002) Aircraft-based measurements of turbulence structures in the katabatic flow over Greenland. Boundary Layer Meteorol 103:49–81

Heinemann G (2003) Forcing and feedback mechanisms between the katabatic wind and sea ice in the coastal areas of polar ice sheets. Glob Atmosphere-Ocean Sys 9:169–201

Heinemann G, Klein T (2002) Modelling and observations of the katabatic flow dynamics over Greenland. Tellus 54A:542–554

Hibler WD III (1979) A dynamic thermodynamic sea ice model. J Phys Oceanogr 9(4):815–846

Inoue J, Kosovic B, Curry JA (2005) Evolution of a storm-driven cloudy boundary-layer in the Arctic. Boundary Layer Meteorol 117:213–230

Intrieri JM, Fairall CW, Shupe MD, Persson POG, Andreas EL, Guest PS, Moritz RE (2002) An annual cycle of Arctic surface cloud forcing at SHEBA. J Geophys Res. doi:10.1029/2000JC000439

Kaleschke L, Lüpkes C, Vihma T, Haarpaintner J, Bochert A, Hartmann J, Heygster G (2001) SSM/I sea ice remote sensing for mesoscale ocean-atmosphere interaction analysis. Can J Remote Sens 27(5):526–537

Kantha LH, Mellor GL (1989) A numerical model of the atmospheric boundary layer over a marginal ice zone. J Geophys Res 94:4959–4970

Kessler E (1969) On the distribution and continuity of water substance in atmospheric circulations. Metrol Monogr 32:1–84

Klein T, Heinemann G (2002) Interaction of katabatic winds and mesocyclones at the eastern coast of Greenland. Meteorol Appl 9:407–422

Klein T, Heinemann G, Bromwich DH, Cassano JJ, Hines KM (2001a) Mesoscale modeling of katabatic winds over Greenland and comparisons with AWS and aircraft data. Meteorol Atmos Phys 78:115–132

Klein T, Heinemann G, Gross P (2001b) Simulation of the katabatic flow near the Greenland ice margin using a high-resolution nonhydrostatic model. Meteorol Z (NF) 10:331–339

Kottmeier C, Hartmann J, Wamser C, Bochert A, Lüpkes C, Freese D, Cohrs W (1994) Radiation and Eddy Flux Experiment 1993 (REFLEX II), Reports on polar research, vol 133. Alfred Wegener Institute for Polar and Marine Research, Bremerhaven

Kreyscher M, Harder M, Lemke P, Flato GM (2000) Results of the sea ice model intercomparison project: evaluation of sea-ice rheology schemes for use in climate simulations. J Geophys Res 105:11299–11320

Lemke P, Owens WB, Hibler WD III (1990) A coupled sea ice-mixed layer-pycnocline model for the Weddell Sea. J Geophys Res 95:9513–9525

Liu AQ, Moore GWK, Tsuboki K, Renfrew IA (2004) A high-resolution simulation of convective roll clouds during a cold-air outbreak. Geophys Res Lett 31:L03101. doi:10.1029/2003GL018530

Lüpkes C, Birnbaum G (2005) Surface drag in the arctic marginal sea ice zone – a comparison of different parameterisation concepts. Boundary Layer Meteorol 117:179–211

Lüpkes C, Gryanik VM (2005) The effect of sea ice on regional atmospheric processes in the Arctic, in Final report ACSYS, coordinated by A. Hense, Grant 03 PL034, Universität Bonn, Bonn, pp 11–21

Lüpkes C, Schlünzen KH (1996) Modelling the arctic convective boundary-layer with different turbulence parameterisation. Boundary Layer Meteorol 79:107–130

Lüpkes C, Hartmann J, Birnbaum G et al. (2004) Convection over Arctic leads (COAL). In: Schauer U, Kattner G. (ed) The expedition ARKTIS XIX a,b and XIX/2 of the Research Vessel 'Polarstern' in 2003. Reports on Polar Research, vol 481. Alfred Wegener Institute for Polar and Marine Research, Bremerhaven, pp 47–62

Lüpkes C, Gryanik VM, Witha B, Gryschka M, Raasch S, Gollnik T (2008) Modelling convection over leads with LES and a non-eddy-resolving microscale model. J Geophys Res C09028. doi:10.1029/2007JC004099

Lynch AH, Chapman WL, Walsh JE, Weller G (1995) Development of a regional climate model of the western Arctic. J Clim 8:1555–1570

Mai S (1995) Beziehungen zwischen geometrischer und aerodynamischer Oberflächenrauhigkeit arktischer Meereisflächen. M.S. thesis, University of Bremen

Mai S, Wamser C, Kottmeier C (1996) Geometric and aerodynamic roughness of sea ice. Boundary Layer Meteorol 77:233–248

Makshtas AP (1991) The heat budget of Arctic ice in the winter. International Glaciological Society, Cambridge

Mauritsen T, Svensson G, Grisogono B (2005) Wave flow simulations over arctic leads. Boundary Layer Meteorol 117:259–273

McPhee MG (1979) The effect of the oceanic boundary layer on the mean drift of pack ice: application of a simple model. J Phys Oceanogr 9:388–400

Müller G, Brümmer B, Alpers W (1999) Roll convection within an Arctic cold-air outbreak: interpretation of in situ aircraft measurements and spaceborne SAR imagery by a three-dimensional atmospheric model. Mon Weather Rev 127(3):363–380

Olsson PQ, Harrington JY (2000) Dynamics and energetics of the cloudy boundary layer in simulations of off-ice flow in the marginal ice zone. J Geophys Res 105(D9):11889–11899

Overland JE (1985) Atmospheric boundary layer structure and drag coefficients over sea ice. J Geophys Res 90:9029–9049

Overland JE, Guest PS (1991) The Arctic snow and air temperature budget over sea ice during winter. J Geophys Res 96:4651–4662

Pagowski M, Moore GWK (2001) A numerical study of an extreme cold-air outbreak over the Labrador sea: sea ice, air-sea interaction, and development of polar lows. Mon Weather Rev 129(2):47–72

Paulson CA, Smith JD (1974) The AIDJEX lead experiment. AIDJEX Bull 23:1–8

Perrie W, Hu Y (1997) Air-ice-ocean momentum exchange. Part II: ice drift. J Phys Oceanogr 27:1976–1996

Persson POG, Ruffieux D, Fairall CW (1997) Recalculations of pack ice and lead surface energy budgets during the Arctic leads experiment (LEADEX) 1992. J Geophys Res 102:25085–25089

Persson POG, Fairall C, Andreas EL, Guest PS, Perovich DK (2002) Measurements near the atmospheric surface flux group tower at SHEBA: near-surface conditions and surface energy budget. J Geophys Res 107:8045. doi:10.1029/2000JC000705

Pinto JO (1998) Autumnal mixed-phase cloudy boundary layers in the Arctic. J Atmos Sci 55:2016–2038

Pinto JO, Alam A, Maslanik JA, Curry JA, Stone RS (2003) Surface characteristics and atmospheric footprint of springtime Arctic leads at SHEBA. J Geophys Res 108:8051. doi:10.1029/2000JC000473

Raasch S (1990) Numerical simulation of the development of the convective boundary layer during a cold air outbreak, Boundary Layer Meteorol 52:349–375

Raasch S, Schröter M (2001) PALM – a large eddy simulation model performing on massively parallel computers. Z Meteorol 10:363–372

Rasmussen L (1989) Den dag, Angmagssalik naesten blaeste i havet. Vejret, 2, Danish Meteorological Society, pp 3–14

Ruffieux D, Persson POG, Fairall CW, Wolfe DE (1995) Ice pack and lead surface energy budgets during LEADEX 1992. J Geophys Res 100:4593–4612

Savijärvi H, Matthews S (2004) Flow over small heat islands: a numerical sensitivity study. J Atmos Sci 61:859–868

Schauer U, Kattner G (2004) The expedition ARKTIS XIX a,b and XIX/2 of the Research Vessel 'Polarstern' in 2003. Reports on Polar Research, vol 481. Alfred Wegener Institute for Polar and Marine Research, Bremerhaven

Schlünzen KH (1990) Numerical studies on the in-land penetration of sea breeze fronts at a coastline with tidally flooded mudflats. Beitr Phys Atmos 63:243–256

Schlünzen KH, Katzfey JJ (2003) Relevance of sub-grid-scale land-use effects for mesoscale models. Tellus 55A:232–246

Schlünzen KH, Bigalke, K, Lüpkes, C, Niemeier U, v. Salzen K (1996) Concept and realization of the mesoscale transport and fluid model ‚METRAS'. Technical Report 5, METRAS, Meteorologisches Institut, Universität Hamburg

Schnell RC, Barry RG, Miles MW, Andreas EL, Radke LF, Brock CA, McCormick MP, Moore JL (1989) Lidar detection of leads in Arctic sea ice. Nature 339:530–532

Serreze MC, Maslanik JA, Rehder MC, Schnell RC, Kahl JD, Andreas EL (1992a) Theoretical heights of buoyant convection above open leads in the winter Arctic pack ice cover. J Geophys Res 97:9411–9422

Serreze MC, Kahl JD, Schnell RC (1992b) Low-level temperature inversions of the Eurasian Arctic and comparisons with Soviet drifting station data. J Clim 5:615–629

Smith SD, Anderson RJ, den Hartog G, Topham DR, Perkin RG (1983) An investigation of a polynya in the Canadian Archipelago 2, structure of turbulence and sensible heat flux. J Geophys Res 88:2900–2910

Steele M, Morison JH, Untersteiner N (1989) The partition of air-ice-ocean momentum exchange as a function of ice concentration, floe size, and draft. J Geophys Res 94:12739–12750

Tjernström M, Zagar M, Svensson G, Cassano JJ, Pfeifer S, Rinke A, Wyser K, Dethloff K, Jones C, Semmler T, Shaw M (2005) Modelling of the Arctic boundary layer: an evaluation of six ARCMIP regional-scale models using data from the SHEBA project. Boundary Layer Meteorol 117:337–381

Troen IB, Mahrt L (1986) A simple model of the atmospheric boundary-layer: sensitivity to surface evaporation. Boundary Layer Meteorol 37:129–148

Vihma T (1995) Subgrid parameterization of surface heat and momentum fluxes over polar oceans. J Geophys Res 100:22625–22646

Vihma T, Brümmer B (2002) Observations and modelling of on-ice and off-ice air flows over the northern Baltic Sea. Boundary Layer Meteorol 103:1–27

Vihma T, Pirazzini R (2005) On the factors controlling the snow surface and 2-m air temperatures over the Arctic sea ice in winter. Boundary Layer Meteorol 117:73–90

Vihma T, Hartmann J, Lüpkes C (2003) A case study of an on-ice air flow over the Arctic marginal sea ice zone. Boundary Layer Meteorol 107:189–217

Vihma T, Lüpkes C, Hartmann J, Savijärvi H (2005) Observations and modelling of cold-air advection over Arctic sea ice. Boundary Layer Meteorol 117(2):275–300

von Salzen K, Claussen M, Schlünzen KH (1996) Application of the concept of blending height to the calculation of surface fluxes in a mesoscale model. Meteorol Z 5(2):60–66

Wacker U, Potty KVJ, Lüpkes C, Hartmann J, Raschendorfer M (2005) A case study on a polar cold air outbreak over Fram Strait using a mesoscale weather prediction model. Boundary Layer Meteorol 117(2):301–336

Wang S, Wang Q, Jordan RE, Persson POG (2001) Interactions among longwave radiation of clouds, turbulence, and snow surface temperature in the Arctic: a model sensitivity study. J Geophys Res 106:15323–15333

Wefelmeier C (1992) Numerische Simulation mesoskaliger, dynamischer Wechselwirkungen zwischen Atmosphäre, Eis und Ozean, vol 42. Berichte des Instituts für Meteorologie und Klimatologie der Universität Hannover

Wefelmeier C, Etling D (1991) The influence of sea ice distribution on the atmospheric boundary layer. Z Meteorol 41(5):333–342

Weinbrecht S, Raasch S (2001) High-resolution simulations of the turbulent flow in the vicinity of an Arctic lead. J Geophys Res 106(C11):27035–27046

Zulauf MA, Krueger SK (2003) Two-dimensional cloud-resolving modelling of the atmospheric effects of Arctic leads based upon midwinter conditions at the surface heat budget of the Arctic ocean ice camp. J Geophys Res 108(D10):4312. doi:10.1029/2002JD002643

Chapter 8
Arctic Regional Climate Models

K. Dethloff, A. Rinke, A. Lynch, W. Dorn, S. Saha, and D. Handorf

Abstract In this chapter, we provide an overview of current applications of regional climate models (RCMs) to the Arctic. There are increased applications of RCMs to present-day climate simulations and process parameterisations. Any advances in regional climate modelling must be based on analysis of physical processes in comparison with observations. In data-poor regions like the Arctic, this approach may be completed by a collaborative analysis of several research groups. Within the ARCMIP (Arctic Regional Climate Model Intercomparison Project), simulations for the SHEBA year 1997–1998 have been performed by several Arctic RCMs. The use of high resolution RCMs can contribute to a better description of important regional physical processes in the ocean, cryosphere, atmosphere, land and biosphere including their interactions in coupled regional model systems. This is based on identifying and modelling of the key processes and on an assessment of the improved understanding in the light of analysis of instrumental as well as paleoclimatic and paleoenvironmental records. The main goal is to address the deficiencies in understanding the Arctic by developing improved physical descriptions of Arctic climate feedbacks in atmospheric and coupled regional climate models and to

K. Dethloff (✉) • A. Rinke • W. Dorn • D. Handorf
Alfred Wegener Institute for Polar and Marine Research, Research Unit Potsdam,
Telegrafenberg A43, D-14473 Potsdam, Germany
e-mail: Klaus.Dethloff@awi.de; Annette.Rinke@awi.de; Wolfgang.Dorn@awi.de;
Doerthe.Handorf@awi.de

A. Lynch
School of Geography and Environmental Sciences, Monash University,
Melbourne, VIC 3800, Australia
e-mail: Amanda.Lynch@arts.monash.edu.au

S. Saha
C & H Division, Indian Institute of Tropical Meteorology, Pune 411008, India
e-mail: Subodh@tropmet.res.in

P. Lemke and H.-W. Jacobi (eds.), *Arctic Climate Change: The ACSYS Decade and Beyond,* Atmospheric and Oceanographic Sciences Library 43, DOI 10.1007/978-94-007-2027-5_8, © Springer Science+Business Media B.V. 2012

implement the improved parameterisations into global climate system models to determine their global influences and consequences for decadal-scale climate variations. A further aim is to model the main feedbacks correctly to arrive at a more reliable estimate of future changes due to the coupling between natural and anthropogenic effects.

8.1   Introduction

In 1994, the World Climate Research Programme (WCRP) published the Arctic Climate System Study (ACSYS) Implementation Plan. The main scientific goal of ACSYS was to ascertain the role of the Arctic in the global climate. Concerning the Arctic atmosphere and modelling research programme, the main idea was described as following: 'The Arctic atmosphere provides the dynamic and thermodynamic forcing of the Arctic Ocean circulation and sea ice. Through the cloud and radiation fields, it also provides an essential link for the climate system, so maintaining the large-scale global temperature gradient and providing a source of locked polar air masses, important for the atmospheric general circulation. Correct representation in coupled models of the interactions between the troposphere, the atmospheric boundary layer and the sea ice and upper ocean and especially of the role of radiation and clouds is essential for proper simulation of climate and climate change'. These ideas formulated in 1994 were very visionary and belong still today to the most important research topics.

The ACSYS modelling programme focused on improving the representation of polar processes in coupled general circulation models (GCMs). One main motivating factor was the enhanced polar warming in the GCM projections for future climate change. The modelling strategy based on quantitative evaluation of individual model components and process parameterisations, assessment of model sensitivities and investigation of coupling strategies and representation of coupled processes. In order to improve the parameterisations of polar processes in atmospheric general circulation models, the representation of high-latitude processes and their potential impact of climate simulations, the ACSYS modelling programme suggested the use of regional climate models (RCM). Observational field and process studies which should lead to improved parameterisations of Arctic-specific processes are carried out on a much finer scale than current GCM can resolve. The adaptation of this mesoscale information to a global-scale parameterisation is a complex and difficult topic. To address these problems, a limited-area climate system model is a more efficient tool for testing and verifying physical parameterisations. The first development of a nested regional climate simulation has been undertaken by Dickinson et al. (1989) for the western part of the American continent.

Walsh et al. (1993) and Lynch et al. (1995) developed a limited-area climate system model (Arctic Regional Climate System Model or ARCSyM) and applied it to the western Arctic. The primary objective of the ARCSyM effort was to use a high-resolution regional model to test and improve those aspects of Arctic process formulation that have limited the validity of global model simulations of the Arctic. The ARCSyM was based on the NCAR regional climate model RegCM2 developed by Giorgi (1991), Giorgi et al. (1993) and Giorgi and Shields (1999) with substantial modifications to the cloud physics. The primary conclusion of this effort was that the complex interactions and feedbacks taking place include many processes that are poorly understood, sparsely measured and rarely modelled and that the physically based RCM approach has the potential for promising results in a diverse set of applications.

Dethloff et al. (1996) applied the RCM HIRHAM at 50km resolution to a pan-Arctic integration domain and were successful in simulating the climate of the Arctic region. They found that the model reproduces the ECMWF – analysed monthly mean circulation rather closely, but the model underestimates the vertical transports of heat and moisture in the planetary boundary layer in winter time. Jürrens (1999) applied the RCM REMO to a European–Arctic integration domain. Laprise et al. (1998) and Caya and Laprise (1999) described results of the Canadian RCM application for the Arctic. The dynamical core of this model, based on the fully elastic non-hydrostatic equations and was solved by semi-implicit and semi-Lagrangian schemes. The physical parameterisation package of the CRCM was taken from the second-generation Canadian GCM (McFarlane et al. 1992).

When regional climate research began in the late 1980s, GCMs at roughly 300–500-km resolution were considered inadequate for producing climate information needed for assessing regional impacts of climate change and variability. RCMs now fill the gap between global climate modelling and regional application. Whereas in middle latitudes, the main research focus was on dynamical downscaling of climate scenarios with GCMs, the main research topic in the Arctic was mainly on improvements in the uncertain physical parameterisations.

In recent years, the nested regional climate modelling technique, which consists of using initial conditions, time-dependent lateral meteorological conditions and surface boundary conditions from a GCM to drive high-resolution RCMs, has been well established in the Arctic climate research community. It has been shown by Rinke et al. (1997, 1999a, b), Box and Rinke (2003) and Dethloff et al. (2001, 2002) that e.g. the HIRHAM model is able to simulate spatial patterns of important climate variables in reasonable agreement with observations in the vertical structure, the annual cycle and interannual variability for the pan-Arctic integration domain. Antic et al. (2004) evaluated the downscaling ability of a one-way nested regional climate model over a region subjected to strong surface forcing: the west of North America. The sensitivity of the results to the horizontal resolution jump and updating frequency of the lateral boundary conditions are also evaluated. In order to accomplish this, a perfect-model approach has been applied. This so called Big-Brother Experiment over the west coast of North America showed that complex topography and coastline have a strong impact on the downscaling ability of the one-way nesting technique. The best downscaling ability is obtained when the ratio of spatial resolution between the nested model and the nesting fields is less than 12 and when the update frequency is more than twice a day. Lorenz and Jacob (2005) investigated the influence of regional-scale information on the global circulation through the development and application of the so-called two-way nesting technique. The two-way nesting climate model system consists of a spectral atmospheric general circulation model and a grid point regional atmospheric model.

Within the last decade, the applications of Arctic RCMs increased strongly. In this chapter, we try to provide an overview of current applications of RCMs to the Arctic. There are increased applications of RCMs to present-day climate simulations and process parameterisations. Any advances in regional climate modelling must be based on analysis of physical processes in comparison with observations. In data-poor regions like the Arctic, this approach may be completed by a collaborative analysis of several research groups. Within the ARCMIP (Arctic Regional Climate Model Intercomparison Project), simulations for the SHEBA year 1997–1998 have been performed by several Arctic RCMs as described by Rinke et al. (2006), Tjernström et al. (2005) and Wyser et al. (2007).

Another important application of Arctic RCMs deals with time-slice projections of GCM scenarios as in Kiilsholm et al. (2003). They conducted high-resolution climate change simulations for an area covering Greenland with the regional climate model HIRHAM. Three 30-year time-slice experiments were conducted for periods representing present day (1961–1990) and the future (2071–2100) in two climate scenarios. Due to a much better representation of the surface topography in the RCM, the simulated geographical distribution of present-day accumulation represents a substantial improvement compared to the driving GCM. Estimates of the regional net balance are also better represented by the RCM. In the future climate, the net balance for the Greenland ice sheet is reduced in all the simulations. In both scenarios, the estimated melt rates are larger in HIRHAM than in the driving model. Dorn et al. (2003) investigated regional magnitudes and patterns of Arctic winter climate changes in consequence of regime changes of the North Atlantic Oscillation (NAO) using the regional atmospheric climate model HIRHAM. The regional model has been driven with data of positive and negative NAO phases from a control simulation as well as from a time-dependent greenhouse gas and aerosol scenario simulation. Both global model simulations include a quite realistic interannual variability of the NAO with pronounced decadal regime changes and no or rather weak long-term NAO trends. The results indicate that the effects of NAO regime changes on Arctic winter temperatures and precipitation are regionally significant over most of northwestern Eurasia and parts of Greenland.

A similar important aspect is the application of RCMs to paleoclimatic time slices as in Rivers and Lynch (2004). Sensitivity experiments with respect to different parameterisations of the Arctic planetary boundary layer schemes have been described by Dethloff et al. (2001). The influence of changed topography of Greenland has been investigated by Box and Rinke (2003) and Dethloff et al. (2004). Currently, the main focus is on the development of coupled Arctic RCMs. They comprise atmospheric RCMs coupled to other subsystem models of the climate system such as sea-ice and ocean models, aerosol models and land and soil models, as in Maslanik et al. (2000), Rinke et al. (2003, 2004a, b), and Saha et al. (2006). Beside the application issue of RCMs, there are unsolved problems, which still need further investigations. Running pairs of RCM simulations driven by large-scale analyses and GCM outputs allow the evaluation of error sources arising from model internal components versus lateral boundary conditions. The sensitivity to the model domain size and location depends on whether large-scale forcings originate from inside or outside the regional integration domain. An outstanding technical issue with RCMs is related to the treatment of lateral boundary conditions, the one-way or two-way interaction and the influence of horizontal and vertical resolution on climate variability. The higher spatial resolution may lead to improvements in some aspects of the climate simulations and degradation in others.

In the EU-project GLIMPSE (Global Implications of Arctic Climate Processes and Feedbacks) several RCM and GCM groups from Europe, US and Canada used a hierarchy of atmospheric and coupled RCMs to improve Arctic climate process parameterisations in close interaction with observational studies carried out during the SHEBA year 1997–1998 and to implement the improved parameterisations into coupled AOGCMs, as described by Dethloff et al. (2006). By means of simulations with a global coupled AOGCM, it was shown that the improved parameterization of Arctic sea-ice and snow albedo can trigger changes in the Arctic and North Atlantic Oscillation patterns with strong implications for the European climate.

8.2 Regional Climate Models of the Arctic Atmosphere

The development of Arctic regional atmospheric climate models (RCMs) begun with the work of Walsh et al. (1993), Lynch et al. (1995) and Dethloff et al. (1996). Within the frame of the Arctic Regional Climate Model Intercomparison Project – ARCMIP, several other groups have started to apply their regional models on Arctic domains as described in Rinke et al. (2006). The models use the primitive form of the dynamic and thermodynamic equations describing the atmosphere. The basic equations of the models are the general conservation equations for momentum, mass, thermodynamic energy, water vapour and cloud water and/or cloud ice. For making the equation system complete, one needs additionally the state equation of ideal gases. Additionally source terms describe friction and diffusion, diabatic processes, e.g. radiation heating and cooling, latent heat of condensation and evaporation and microphysical sources and sinks due to condensation and evaporation which are determined by the model parameterisations. Most of the models use the hydrostatic approximation. For the handling of the horizontal diffusion (which is a horizontal scale-selective damping formulation to prevent energy accumulation in the smallest resolved scales), different schemes are used, e.g. non-linear second order scheme or linear fourth order scheme. To solve the basic equations numerically, most of the models apply the method of finite differences. A mesh grid is used which is rectangular (or for some, polar stereographic) on the transformed coordinates. The majority of the grid-point models use horizontally staggered grids. It is convenient for the vertical discretisation to use either pressure, sigma or hybrid pressure–sigma coordinates.

8.2.1 Spatial Resolution

The models applied to Arctic or Nordic sub-areas have a horizontal grid distance between 12 and 100 km and differ regarding the vertical resolution between 19 and 30 levels. The Arctic atmosphere is characterised by some specific properties. One is the extreme static stability of lower troposphere connected with long-lasting inversions caused by the wintertime absence of solar radiation, strong surface long-wave cooling and semi-permanent snow and sea ice coverage. Seventy-five percent of the time, the Arctic planetary boundary layer (PBL) is stratified stably (Persson et al. 2002), and the PBL height is often low. Thus, it is crucial to have a high enough vertical resolution in the PBL as well as a low enough placed first model level in Arctic RCMs. Very high resolution experiments are still very rare; practically, only one single study with a 12-km resolution has been published over a Russian tundra domain (Christensen and Kuhry 2000). In addition, a 7-km resolution study was conducted by Lynch et al. (1997). While the time period simulated was limited to a single polynya (ice-opening) event, it did extend beyond typical numerical weather prediction timescales. The RCM solution depends on the driving boundary fields. The mismatch between the scales in the driving coarse-resolution model and the high-resolution RCM does not cause fundamental problems if proper boundary condition procedures are applied, as demonstrated by Denis et al. (2002). The maximum acceptable spatial resolution jump between the driving and the nested models is six to twelve fold, i.e. T60 to T30 resolution of the coarse driving model for a 45km resolution RCM (Denis et al. 2003; Antic et al. 2004). These results have been found for a North American 45km RCM and need still to be investigated in Arctic climatic region.

8.2.2 Domain Choice and Lateral Boundary Forcing

The proper choice of the appropriate model domain is not trivial. The RCM simulations are sensitive to both the size and the position of the domain chosen for calculations. One fundamental question to be considered is the strength of the internal solution dynamics affected by the forcing from the lateral boundary conditions (LBC). The LBC can have a significant impact on the evolution of the predicted atmospheric fields through the propagation of boundary errors into the interior of the domain. LBC can produce transient gravity–inertia modes in the domain. Additionally, an ill-posed mathematical formulation of the LBC can cause spurious interactions between the model solution and the lateral boundaries. Marbaix et al. (2003) present an evaluation of the boundary relaxation methods used in state-of-the-art RCMs. All domains in Arctic RCMs are small enough to keep the large-scale circulation features similar to those of the driving model. In the large pan-Arctic domain, the lateral boundary control is weaker causing the geopotential root-mean-square difference between daily simulated and observed 500 hPa geopotential heights to be larger than for smaller Arctic sub-domains (Rinke and Dethloff 2000). However, more freedom benefits the testing of the sensitivity to internal model processes because the model solution is more free to respond to variations in internal parameters (Seth and Giorgi 1998). The determination of the magnitude of the internal variability of the Arctic RCMs is a prerequisite to assess the significance of climate sensitivity signals. The internal variability is caused by the non-linearity in the model physics and dynamics and can modulate, or even mask, physically forced signals in the model. A few studies have specifically addressed the uncertainty and/ or internal variability in Arctic RCMs (Rinke and Dethloff 2000; Lynch and Cullather 2000; Wu et al. 2004; Caya and Biner 2004; Rinke et al. 2004a). Pan-Arctic RCMs tend to have a large internal variability, triggered by uncertainties in LBC and/or initial conditions (IC). Both small changes in LBC and IC can result in a model divergence from the driving large-scale fields. The extent to which this occurs depends on the control of the model by the LBC forcing and is largely independent on the magnitude as well as on the kind or origin of the perturbation. However, some geographical key regions (Barents and Kara Seas, Fram Strait, Canadian archipelago) have been identified as main contributors to the internal variability. The locations may be connected with orographic features and/or sea-ice distribution with their ability to trigger the spatial response pattern of the model, but this hypothesis needs further investigations. Two regimes in the internal variability are found, a regime of large variability in autumn/winter and a regime of smaller variability in summer. The atmospheric circulation and planetary wave activity are weaker during summer than during winter. Therefore, the response of the model to LBC and/or IC perturbations is developing (amplifying) at a slower rate in summer due to the reduced model-generated wave activity and instability. The temporal classification of the two regimes of the variability in a pan-Arctic RCM is completely converse to those found by Caya and Biner (2004) in a RCM covering a Canadian domain. They detected the largest internal variability in their RCM during summer due to a weaker zonal circulation and the stronger importance of local processes (e.g. convection). The comparison of ensemble simulations of two pan-Arctic RCMs (HIRHAM and ARCSyM) showed that each individual RCM has its own internal variability patterns (Rinke et al. 2000). But, as the internal variability is large, ensemble runs are required for each individual model in order to be able to assess significant simulation responses e.g. to changes in physical parameterization changes. The lateral boundary data are generally provided either by global models or observational data analyses. Both have resolutions coarser than those used in current Arctic RCMs and do not allow for specific Arctic process parameterizations. The accuracy of the commonly used data analyses (ECMWF or NCEP) is uncertain over the data-sparse Arctic ocean and land areas and has been evaluated for a variety of Arctic parameters (e.g. Bromwich et al. 2000; Cullather et al. 2000; Serreze and Hurst 2000; Francis 2002).

8.2.3 Lower Boundary Forcing

At the lower boundary of atmospheric RCMs, the sea surface temperatures (SSTs) and sea-ice fraction have to be provided. In the several Arctic RCM studies, both are taken either from observational data analyses (e.g. from ECMWF) or from satellite products (e.g. sea ice concentrations from Nimbus-7 SSMR, DMSP SSM/I and AVHRR surface temperatures). There are different ways to handle sea ice in RCMs. One way is to diagnose it from the prescribed SST (i.e. if the SST in a grid cell is below the freezing point of sea water, the whole grid cell is covered by ice, and a constant sea ice thickness is prescribed). A more advanced method takes partial sea ice cover as well as variable sea ice thickness into account. Several Arctic RCM experiments have been performed using different lower boundary conditions. The results underline very clearly the effect of sea ice on regional atmospheric processes in the Arctic. The influence of the different description of sea ice on the Arctic atmospheric RCM simulations has been investigated by Rinke and Dethloff (2000), Rinke et al. (2006) and Semmler et al. (2004). It was shown that there is a clearly better agreement (biases reduced by ~50%) between measured and modelled surface and boundary layer temperatures and humidities when a more advanced sea ice scheme is used. Only with the accurate satellite-derived sea ice data, the models are able to reproduce anomalous atmospheric pressure, patterns, as in summer 1990. It was also shown that the sea-ice thickness plays an important role in the simulation of ice temperatures and mean sea-level pressure and that it impacts the whole tropospheric air column. As the Greenland ice sheet is the largest orographic feature of the Arctic, and mountains of such a scale together with surface heating or cooling anomalies exert a strong influence on the atmospheric circulation, it is necessary to prescribe the orography in Arctic RCMs realistically. Most of the models use the GTOPO30 elevation data produced by the United States Geological Survey's EROS Data Center (Gesch 1994; Gesch and Larson 1996). However, Box and Rinke (2003) identified certain systematic model biases, caused in particular by inaccurate GTOPO30 elevation data over Greenland (180 m lower on average, with errors of up to -840 m over 50 km grid cells as compared to a new elevation model for Greenland by Ekholm (1996) and Bamber et al. (2001)). The resulting warm biases due to the wrong elevation data enhance a negative albedo bias which, in turn, leads to positive net shortwave radiation biases.

8.2.4   Physical Parameterisations

Due to the physical parameterisations, the source/sink terms of the model equations are determined. The physical processes which contribute to these terms are basically sub-grid-scale processes which have to be parameterized in terms of model variables on the resolvable scales. The following processes are generally taken into account: vertical diffusion in the planetary boundary layer (PBL), radiation, stratiform and convective condensation, soil and land surface processes and gravity wave drag. Concerning the turbulent flux parameterizations, most of the models use the Monin-Obuchov-similarity for the surface layer scheme and a prognostic turbulent kinetic energy scheme for the PBL scheme. The complexity of the radiation and cloud schemes varies among the different Arctic RCMs. For the land surface and soil processes, the RCMs use multi-layer (3–6 levels) soil models with surface energy and water budget calculations or simplified bucket model for hydrology. According to Curry et al. (1995), the PBL, the aerosol–cloud interactions and the surface albedo can be considered as the most Arctic specific and complex physical processes which need Arctic specific parameterisations. Additionally, the land surface and soil processes seem to be necessarily handled more carefully as the Arctic is characterised by permafrost thawing and freezing processes and different soil types. Numerous Arctic RCM experiments have been conducted to study the influence of different description of these processes: e.g. representation of momentum, heat and moisture exchange in the PBL (Dethloff et al. 2001; Mirocha et al. 2005; Tjernström et al. 2005), surface albedo (Køltzow et al. 2003), cloud–radiation interaction (Pinto et al. 1999; Cassano et al. 2001; Bromwich et al. 2001; Girard and Blanchet 2001a; Morrison et al. 2003; Jones et al. 2004) and aerosol (Rinke et al. 2004b; Girard and Blanchet 2001b).

8.2.5 *Applications*

The application of Arctic atmospheric RCMs started with the work of Lynch et al. (1995) and Dethloff et al. (1996). Lynch et al. (1995) applied the ARCSyM model over a Western Arctic domain, while simulations for pan-Arctic model domains were conducted by Maslanik et al. (2000) with ARCSyM, Dethloff et al. (1996) with HIRHAM and, recently, Semmler et al. (2004) with REMO. Many other groups have applied RCMs with a more southerly region of interest, but included substantial parts of the North Atlantic, Scandinavia and Greenland (Christensen et al. 1997; Rummukainen et al. 2001), and also, the Canadian RCM includes parts of the Arctic (Denis et al. 2002). For the first ARCMIP experiment, eight RCMs have been applied on a small Beaufort Sea domain. The Arctic RCMs have been used for a lot of different kinds of climate applications which can be grouped by the scientific aim of the application as follows:

1. Studies to optimise the model configuration (parameterisation schemes, influence of boundary forcing, nesting strategy). The downscaling ability and the sensitivity to the spatial resolution and temporal updating frequency of the lateral boundary conditions have been investigated by Denis et al. (2002, 2003), Antic et al. (2004) and Dimitrijevic and Laprise (2005). Rinke and Dethloff (2000) and Rinke et al. (2004a) investigate the influence of the domain, lower and lateral boundaries. The internal variability has been investigated by Christensen et al. (2001), Caya and Biner (2004) and Rinke et al. (2004a).
2. Analysis of field experiments with the aim of process studies and to validate and develop parameterisations. Greenland climate is one of the areas a lot of work has been focussed on, see Bromwich et al. (2001), Cassano et al. (2001), Box and Rinke (2003), Dethloff et al. (2002, 2004) and Kiilsholm et al. (2003).
3. Regional climate studies (interannual variability, interaction atmosphere–land, hydrological cycle, climate change). The impact of land surface schemes and

properties has been extensively studied within the ARCSyM model (Lynch et al. 1998, 1999; Wu and Lynch 2000). A limited number of Arctic RCM experiments of multi-year type have been conducted (Rinke et al. 1999a, b, 2004a, b; Dorn et al. 2000, 2003; Dethloff et al. 2002; Box and Rinke 2003; Kiilsholm et al. 2003). Only few climate change projection simulations have been conducted in the Arctic (Kiilsholm et al. 2003; Dorn et al. 2003; Saha et al. 2006).

4. Regional integrated assessment and socio-economic applications. The Polar MM5 is relatively new to regional climate applications but is participating in the ARCMIP project. This model is also being used as the basis for an integrated assessment of the impacts of climate change, and more specifically, extreme cyclones, flooding and erosion, on the community of Barrow, Alaska (Lynch et al. 2004).

8.2.6   Arctic Regional Climate Model Intercomparison Project (ARCMIP)

ARCMIP (Curry and Lynch 2002), http://curry.eas.gatech.edu/ARCMIP, has been organised under the auspices of the WCRP GEWEX Cloud System Studies Working Group on Polar Clouds and the ACSYS Numerical Experimentation Group. Funding for coordination of ARCMIP activities has been provided by the International Arctic Research Consortium and the Global Implications of Arctic Climate Processes and Feedbacks (GLIMPSE), http://www.awi-potsdam.de/www-pot/atmo/glimpse/index.html project (funded by the European Union). The primary ARCMIP activities have focused on coordinated simulations by different Arctic RCMs and their evaluation using observations from satellites and field measurements. The combination of model intercomparison and evaluation using observations allows strengths and weaknesses of model structures, numerics and parameterisations to be assessed. The simulation experiments are carefully designed so that each of the models is operating under the same external constraints (e.g. domain, boundary conditions). The ARCMIP experiment has been conducted for the 1997–1998 period of the Surface Heat Budget of the Arctic Ocean project (SHEBA; Uttal et al. 2002), which included extensive field observations and accompanying satellite analyses. Simulations of eight different Arctic RCMs (ARCSyM, COAMPS, CRCM, HIRHAM, RegCM, PolarMM5, RCA and REMO) have been performed for the SHEBA period. Each of the models employed the same domain covering the Beaufort Sea (~70×55 grid points), the same horizontal resolution of 50 km and the same atmospheric lateral boundary (ECMWF analyses) and the same ocean/sea-ice lower boundary forcings (AVHRR, SSM/I). The surface temperatures over land and glaciers are not prescribed but calculated individually by each model using their own energy balance calculations. The models differ in the vertical resolution as well as in the treatments of dynamics and physical parameterisations. Tjernström et al. (2005), Rinke et al. (2006) and Wyser et al. (2007) discussed the results of this first intercomparison experiment: First, compared with the SHEBA observations, the

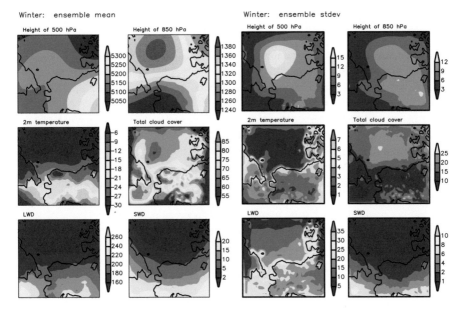

Fig. 8.1 Winter model ensemble mean (*left*) and across-model scatter (*right*) of the 850 and 500 hPa geopotential heights (m), 2 m temperature (°C), total cloud cover (%), longwave and shortwave downward radiation (W m^{-2}). Calculations are based on the eight ARCMIP models from Rinke et al. (2006)

modelled near surface variables (e.g. surface pressure, temperature, wind, radiation, etc.) agree well with the observations. Two examples are given: The model ensemble mean bias in net radiation is −10 W/m^2 and less than 1 m/s in wind. Although the mean turbulent heat flux bias is also small, the models differ strongly from each other and reveal some common bias features. They share a common large-scale flow bias in all seasons (an underestimation of the geopotential height by the models) and a common seasonal bias in temperature and humidity profile (models are colder in lower levels in the transition periods and warmer elsewhere, relatively dry in the near surface layers and wet in the free troposphere) compared to ECMWF analyses. Even using a very constrained experimental design (small integration domain, specified lower boundary condition for ocean and sea ice) and specified 'perfect' horizontal boundary conditions from data analyses, there is considerable scatter among the different RCMs. The largest across-model scatter is found in the 2-m air temperature over land (up to 5 K), in the surface radiation fluxes (up to 55 W/m^2) and in the cloud cover (5–30%) and presented in Fig. 8.1. This is not surprising given the very complex and individually different land surface and radiation–cloud schemes within the models. The quantified scatter between the individual models highlights the magnitude and seasonal dependency of the disagreement and unreliability for current regional climate simulations.

8.3   Coupled Regional Models of the Arctic Climate System

Understanding the seasonal and long-term variability of the Arctic climate requires the decoding of the interactions between the climate subsystems atmosphere, hydrosphere, cryosphere, biosphere and pedosphere. In particular, sea ice plays an important role within the Arctic climate system, forced by and modulating, in turn, atmospheric and oceanic processes and circulations due to its impact on atmosphere–ocean exchanges of heat, moisture and momentum. It is also generally accepted that the sea-ice-albedo feedback effect is an important factor in the amplification of climate change in the Arctic (Ingram et al. 1989; Curry et al. 1995). During the ACSYS decade, several efforts have been done to couple regional high-resolution atmosphere models with regional high-resolution ocean and sea-ice models for Arctic domains. A short overview about existing model systems is given in Table 8.1.

In Table 8.1, CF ice dynamics refers to dynamics with a cavitating fluid rheology according to Flato and Hibler (1992), and EVP ice dynamics refers to dynamics with an elastic–viscous–plastic rheology according to Hunke and Dukowicz (1997). The land–soil component is an integral part of the atmospheric component, and only the number of layers currently used for calculating soil temperature (t), soil water (w) and snow (s) is specified in brackets.

8.3.1   Model Dynamics and Physics

8.3.1.1   Atmosphere

The model dynamics of the atmosphere components is consistently based on the hydrostatic, primitive equations on a rotating sphere. Prognostic equations exist for horizontal wind components, temperature, specific humidity, cloud water, surface pressure and associated diagnostic equations for vertical wind and geopotential. The horizontal resolutions used for the atmosphere components vary between 50 and 100 km, while the vertical resolution of the atmosphere consists of a different number of unequally spaced levels (from 19 to 24 levels) between the surface and about 10 hPa in terrain-following (sigma) or hybrid sigma–pressure coordinates. The physical parameterisations for sub-grid-scale processes are generally adapted from global atmosphere models. For instance, HIRHAM and REMO use the physical parameterisation package of ECHAM4 (Roeckner et al. 1996) that includes parameterisations for radiation, vertical diffusion, land surface processes (see Sect. 8.3.1.4), cumulus convection, stratiform clouds and gravity wave drag. Since most of these parameterisations are adapted for global and mid- latitude climate simulations, they are not always sufficient for the specific Arctic climate conditions, for example, for the vertical diffusion in a shallow stable boundary layer. Several efforts have been done to develop improved process descriptions for Arctic climate simulations. New or modified parameterisations for radiation, clouds, sea-ice and snow albedo and the

Table 8.1 Coupled regional atmosphere–ocean–ice and coupled atmosphere–land–soil models applied to the entire Arctic Basin

Model	Atmosphere component	Ocean–Ice component	Ice dynamics	Land–Soil component	Reference
ARCSyM	Extensively modified from RegCM2	Mixed-layer ocean	CF (or EVP)	NCAR LSM ($t6+w6+s1$)	Lynch et al. (1995, 1998, 2001a)
RCAO	RCA	RCO	EVP	($t5+w2+s1$)	Döscher et al. (2002)
HIRHAM/NAOSIM	HIRHAM	NAOSIM	EVP	ECHAM4 ($t5+w1+s1$) NCAR LSM ($t6+w6+s1$)	Rinke et al. (2003), Dorn et al. (2007), Saha et al. (2006)
HIRHAM/MICOM/ MI-IM	HIRHAM	MICOM/MI-IM	EVP	ECHAM4 ($t5+w1+s1$)	Debernard et al. (2003)
REMO/MPI-OM	REMO	MPI-OM (global)	Viscous–plastic	ECHAM4 ($t5+w5+s2$)	Mikolajewicz et al. (2004)

planetary boundary layer have been tested in the stand-alone atmosphere models. An implementation into the coupled model systems has already partly taken place or is under way.

8.3.1.2 Ocean

Except ARCSyM, which comprises the one-dimensional mixed-layer ocean model of Kantha and Clayson (1994), the ocean components of the coupled models are also based on the hydrostatic, primitive equations on a rotating sphere. In this case, there are prognostic equations for horizontal velocity components, (potential) temperature and salinity (and in addition, sea surface elevation, in case of a free surface like in RCO and MPI-OM) and diagnostic equations for vertical velocity, pressure and density (or depth, in case of an isopycnal vertical coordinate model like MICOM). The prognostic equation for sea surface elevation in ocean models is the equivalent to the prognostic equation for surface pressure in atmosphere models. The difference between a traditional z-coordinate ocean model and an isopycnic model is, simply stated, that the former predicts water density changes at fixed depth levels, whereas the latter predicts the depths at which certain density values are encountered. Thus, the traditional roles of water density and depth as dependent and independent variables are reversed. Although all models are applied to the entire Arctic Basin, they differ partly from each other in domain choice, horizontal (20–100 km) and vertical resolution (30–59 levels), bottom topography and open boundary treatment. As example, we focus here on NAOSIM, a MOM2-based regional model for the northern North Atlantic, the Nordic Seas and the Arctic Ocean. This ocean model uses a rotated spherical grid where the model grid poles lie on the geographical equator and the geographical longitude of 30°W lies on the equator of the model grid. In the coupled model set-up, a horizontal resolution of 0.25° (~25 km) in longitude and latitude is used. In the vertical, the model has 30 unequally spaced levels. Bottom topography is based on the Etopo5 data set of the National Geophysical Data Center. The southern model boundary of NAOSIM approximately runs along 50°N. Here, an open boundary condition has been implemented following Stevens (1991). All other boundaries are treated as closed walls. Especially, there is no inflow through Bering Strait in the current model set-up. Freshwater influx from rivers is not explicitly included. To account for river run-off and diffuse run-off from the land, as well as to include the effect of flow into the Arctic through Bering Strait on the salinity, a restoring flux (with an adjustment timescale of 180 days) is added to the surface freshwater flux. In contrast to all other ocean models, MPI-OM is a global ocean model with a spatially variable resolution, which is strongly increased in the coupling region (approximately 20 km), and with 40 levels in the vertical. Accordingly, there are no open boundaries, only a gradual transition from the fine-resolution into the coarse-resolution model domain, and the Bering Strait is open for in- and outflow of water and ice. The latter might be the major advantage of using a global ocean model since the inflow of warmer ocean water through the Bering Strait is likely to account for an amplified melting of sea ice in the East Siberian, Chukchi and Beaufort Seas during summer. The physical process descriptions depend strongly on the ocean model used and include parameterisations for horizontal and vertical viscosity, vertical and isopycnal diffusivity, eddy-induced mixing, convection, bottom boundary layer slope transport and salt restoring flux.

8.3.1.3 Sea Ice

Although sea ice is usually incorporated in the ocean models (except for MICOM which is separately coupled to the sea-ice flux model MI-IM), it represents an important interface between atmosphere and ocean and requires the inclusion of additional equations. For realistic simulations of the Arctic climate, both thermodynamic and dynamic processes connected with sea ice have to be taken into account by sophisticated schemes. The coupled regional models for the Arctic include thermodynamic sea-ice schemes in which thermodynamic growth and melting of sea ice is derived from an energy balance equation following Semtner (1976) or Parkinson and Washington (1979). Prognostic equations for ice concentration and ice thickness are usually based on the widely used approach by Hibler (1979) with two ice thickness categories (thick ice and thin ice/open water). More complex sea-ice schemes use multiple sub-grid-scale ice thickness categories instead of the uniform distribution between zero and twice the mean thickness of the thick ice, and in particular, a snow layer which is described by a prognostic equation for the snow thickness, similar to that for the ice thickness. Sea-ice dynamics is considered by a two-dimensional momentum balance equation. Kreyscher et al. (2000) have shown that the atmospheric wind stress, the ocean current stress and the strength of the ice are the crucial forces for ice motion. The latter is represented by the internal stress tensor which is determined by the given rheology. In most coupled regional models, an elastic–viscous–plastic (EVP) sea-ice rheology according to Hunke and Dukowicz (1997) is used (in the meantime also in ARCSyM). This EVP ice dynamic model reduces to the viscous–plastic standard model by Hibler (1979) at long timescales but uses elastic waves to improve the response on short timescales and increase computational efficiency.

8.3.1.4 Land Surface and Soil

A typical feature of Arctic land surfaces is the existence of large permafrost areas that are temporary covered with snow or permanently covered with ice (glaciers). The exchange of heat and moisture between soil, vegetation and atmosphere is radically affected by both the land surface type and the seasonal cycle of vegetation and snow cover. In general, the models distinguish between multiple predetermined land cover types and diagnostically between snow and no snow. The land surface characteristics of these types, like surface albedo, vegetation type, leaf area index or water holding capacity of the soil, are either given as fixed constants based on a seasonal climatology or use more advanced schemes. This means that, so far, there is no interaction with the biosphere. The parameterisation of the land surface processes comprises prognostic equations for the evolution of the soil temperatures, the soil hydrology and the snow pack over land. The coupled model systems use here different soil schemes with different numbers of soil layers for calculating the vertical profiles. Rather similar is the calculation of the temperature profile in the soil (or in the glacier if present) that is derived from a numerical solution of the heat conduction equation in a multiple layer model where the thickness of the individual layers increases with depth. At the lower boundary, a zero heat flux condition is imposed. In the presence of a snow pack over land, the heat conduction equation is solved for one or two extra layers with snow characteristics. Soil water is considered in different ways, either by a simple soil moisture bucket model or by solving a conservation equation for the soil water content in multiple soil layers as for the soil temperatures. The latter and more advanced scheme allows for calculating phase transformations in the soil that are of prime importance for the surface energy balance in permafrost areas. Therefore, the more advanced NCAR land surface model (LSM; Bonan 1996), which includes multiple soil layers and some other improvements, has been successfully incorporated into ARCSyM and recently also into a coupled HIRHAM/LSM version. A transfer into both coupled model systems with HIRHAM components is planned in the near future.

8.3.2   Coupling Strategy

Since all components of the coupled models have their origin usually in stand-alone models, a definite strategy for the coupling is required. There are different methods to couple two or more models: (a) integration of all components in one individual model system, (b) two-way exchange of all required variables needed to drive the stand-alone versions, (c) using an external interface that communicates with the individual model components (the so-called coupler). All these methods are used to couple the regional atmosphere–ocean–ice models. The only model that uses method (a) is ARCSyM, but also the coupled atmosphere–land surface model HIRHAM/LSM uses this method. Here, all model components have been incorporated in one individual modular model system. Method (b) is applied by HIRHAM/NAOSIM and HIRHAM/MICOM/MI-IM, while method (c) is used by RCAO and REMO/MPI-OM. In the current configuration of HIRHAM/NAOSIM, the component models for atmosphere and ocean/ice use different domains, different grids and different resolutions because there is long experience with the particular model setup from the stand-alone versions. The coupling procedure of this model therefore uses a common grid on which the exchange variables have to be extra- and interpolated from the respective model component grids. The interpolation is realised by the use of three times three sub-grid points which are each assigned to a grid cell of the common grid. Subsequently, the values are averaged on this grid cells having regard to the number and origin of sub-grid points. Because there is no separate coupler between both component models, the required surface fluxes have to be calculated in one of the models, in this case in the atmosphere model, while the grid-to-grid interpolation is done in the ocean–ice model. The procedure for exchange between the components of HIRHAM/MICOM/MI-IM is an integral part of the ice model MI-IM. The atmosphere and ocean/ice models are coupled on a grid with a common orientation, but with different horizontal resolutions. In RCAO, the ocean model shares the grid of the atmosphere model. The OASIS coupler is used and has been configured to transfer energy and state variables between ocean and atmosphere. REMO/MPI-OM also uses the OASIS coupler. Here, the limited-area atmosphere and the global ocean/ice model consequently use quite different domains with different grid orientations and resolutions. The strategy of coupling between the atmosphere and the soil has been done by updating and exchanging the key surface variables in every time step. Soil moisture, skin moisture, surface temperature and snow water equivalent are transferred from the land–soil model to the atmosphere model HIRHAM, and the simulated atmospheric driving fields are transferred back to the LSM.

8.3.3   Validation and Applications

A general problem in the evaluation of Arctic climate simulations is the scarcity of area-wide observations in the Arctic. The SHEBA project provides a good dataset, but for a single point only. During the ACSYS decade, new satellite data became available, but to this day, there are no reliable area-wide datasets for sea-ice thickness and surface fluxes. Multi-decadal runs have been performed only with the coupled REMO/MPI-OM driven with NCEP re-analyses for the period 1958–2001 as discussed by Mikolajewicz et al. (2005). All other coupled models have been focused on decadal runs for the period 1990–1999 or case studies for single years. A collective validation of RCAO, HIRHAM/NAOSIM and HIRHAM/MICOM/ MI-IM has recently begun using the extensive dataset of the SHEBA period 1997– 1998. The ARCSyM model has been validated in several different contexts over both a western and a whole Arctic domain. Case studies with respect to the anomalous Arctic sea-ice extent during summer 1990 have been conducted by Maslanik et al. (2000) and Lynch et al. (2001a) as well as by Rinke et al. (2003) with the coupled HIRHAM/NAOSIM. These case studies suggest a positive feedback between ice dynamics and ice melt that contributed to the strong ice anomaly in this particular year. Dorn et al. (2007) conducted sensitivity experiments using a coupled regional atmosphere–ocean–ice model of the Arctic in order to identify the requirements needed to reproduce observed sea-ice conditions. The ability of the coupled model to reproduce the observed summer ice retreat depends largely on a quasi-realistic ice volume at the beginning of the melting period, determined by the relationship between winter growth and summer decay of ice. Summer ice decay is strongly affected by the parameterisation of the sea-ice albedo. Winter ice growth depends significantly on the parameterisation of lateral freezing. A large uncertainty

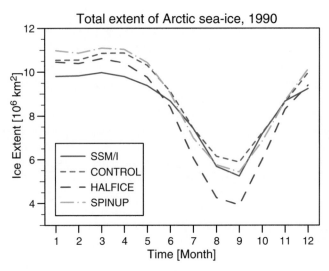

Fig. 8.2 Total sea-ice extent from HIRHAM-NAOSIM simulation for the year 1990, described by Dorn et al. (2007)

in the model relates to the simulation of long-wave radiation most likely as a result of overestimated cloud cover. The results suggest that uncertainties in the descriptions for Arctic clouds, snow and sea-ice albedo and lateral freezing and melting of sea ice, including the treatment of snow, are responsible for large deviations in the simulation of Arctic sea ice in coupled models. Model biases in polar regions must be further reduced in order to enhance the credibility of future climate change projections. The ice-albedo feedback is important in allowing the continued formation of the ice anomaly throughout the summer. The sea-ice conditions, in turn, affect the regional atmospheric circulation, and thus, the ice drift. Therefore, a realistic reproduction of the sea-ice extent in coupled models is linked to a realistic reproduction of the regional atmospheric circulation and vice versa. For this reason, coupled models have more difficulties to reproduce the observed climate than stand-alone models driven with observational data analyses. A comprehensive validation of all variables related to the key processes is essential and requires high-quality observational datasets for the entire Arctic. As an example, Fig. 8.2 shows the computed total sea-ice extent from the HIRHAM-NAOSIM simulation for the year 1990, carried out by Dorn et al. (2007). This run shows the seasonal freezing and thawing of the averaged Arctic sea-ice cover for the control run. The control run started on January 1 of 1990. Two sensitivity experiments have been carried out where the model started in May 1999 (SPINUP) and with a halved initial sea-ice thickness distribution (HALFICE) compared to the control run. These sensitivity runs are in qualitative agreement with the observations from NSIDC.

8.4 Past Climate Simulations of Arctic Climate

The energy exchange between the tropical source and polar sink regions as the main key regions of the earth system is not only essential for the atmospheric variability on decadal-scale climate variability but also for past climate changes. Current climate models are not able to reproduce the observed phases of low-frequency variability and the right speed for transitions and changes compared to real observations. The models show a tendency to respond more slowly than indicated by data. This indicates that important feedbacks and energy cascades in the models are not described in agreement with observations. It is also essential to identify the feedbacks basic to the model-to-model variation of climate sensitivity. Interaction of aerosols, clouds and water vapour gives rise to most relevant but also highly uncertain feedback mechanisms with high implication for the net response in future climate projections. Essential climate feedback mechanisms operate on different temporal and spatial scales. These interactions must be modelled in a realistic way to understand pathways of future change in the climate system. One approach to improve our understanding of the modern system is to use knowledge of past environments. In this way, a set of 'natural experiments', representing different past climatic scenarios, can be used to evaluate strengths and weaknesses of any computer model to be used in simulating possible twenty-first century climates. However, to make this evaluation using paleodata, it is important that the model spans similar spatial scales to the data. It is well known that global climate models exhibit most skill at larger spatial scales: the IPCC Third Assessment Report (IPCC 2001) suggested a minimum scale of 107 km^2 for reliable reproductions of climate features. This is a significant limitation to the evaluation of paleoclimatic experiments since most proxy data represent a significantly smaller footprint.

To address this problem, several research groups have used regional climate models to perform a dynamical downscaling of global paleoclimate simulations (Renssen et al. 2003). The efficacy of the approach was demonstrated in contemporary climate simulations in the 'big-brother' experiment of Denis et al. (2002). This technique has matured in the last decade and is now being applied to the paleoclimate problem. Probably, the earliest example of the use of regional climate models in paleoclimate applications is Hostetler et al. (1994) in which a regional climate model was used to study the interactions between the midlatitude atmosphere and Lakes Lahontan and Bonneville during the Pleistocene. It was found through sensitivity studies that local lake–atmosphere interaction may explain differences in the relative sizes of these lakes 18,000 years ago. However, at this time, only short ensembles of single months were possible. Since that time, there have been numerous studies of various eras using regional climate models, primarily in midlatitudes and the tropics (e.g. Diffenbaugh and Sloan 2004). An important, and potentially confounding, issue in paleoclimatic simulations is interannual variability. This issue is compounded in high-latitude regions. In particular, unless relatively long simulations (e.g. 25–50 years) or representative ensembles are conducted, the potential exists for interannual variations in the simulated climate to overwhelm the effects of the changed boundary conditions. In other words, the potential exists for: (a) particular configurations of present and paleoatmospheric circulation patterns to be indistinguishable from one another (e.g. Masson and Joussaume 1997), or (b) extreme phases of present circulation modes to differ as much as present typical circulation patterns differ from the paleoclimate modes (and vice versa) (e.g. Kageyama et al. 1999).

In conducting regional model simulations, there are several potential strategies for diminishing this issue: (1) run very long simulations of both the GCM and regional model; (2) select a specific interval from a long GCM simulation to use for the regional model boundary conditions (often, this is done in a subjective fashion, guided by analyses of the GCM simulations, e.g. Hostetler et al. 2000); (3) select a specific interval or construct a representative composite from a long GCM simulation in a more objective fashion or (4) run an ensemble of experiments with the regional model, and average them to produce the paleoclimatic simulation. The strategy chosen depends in large part on available resources, both labour and computational. As implied above, it is generally supposed that higher resolution models will perform better than GCMs in regional studies of paleoclimate. For example, Renssen et al. (2001) found that nesting a mesoscale model within a global model improved data–model agreement in Europe for the Younger Dryas anomalies. However, Alley (2003) comments that in this case, the sea-ice boundary conditions used were more extreme than likely, and yet simulated wintertime temperatures for western Europe still exceeded reconstructions by between 5 and 25 K. In another example, Felzer and Thompson (2001) performed an evaluation of a regional climate model for paleoclimate simulations of the North Atlantic and found that while their model was capable of capturing the correct temperature and precipitation patterns, grid-to-grid comparisons were not supported. This limited the applicability of the model to individual core evaluation. Finally, Haywood et al. (2004) found data/model inconsistencies in their simulated net moisture budget for the Early Cretaceous due to problems in the simulation of correct temperatures and hence evaporation in their limited area model. Hence, while the techniques have matured, the quality of the simulations remains an important question. Hence, a second major issue of concern for comparisons of model simulations and paleodata is the apparent circularity of using paleoecological data both to specify surface boundary conditions and to check the resulting simulations. The strongest instance of this issue is illustrated by the reconstruction of atmospheric patterns using local proxy data (e.g. Appenzeller et al. 1998). This method forms the basis of an approach to force models to incorporate specific proxy data-based information (von Storch et al. 2000). This practice, and even the straightforward use of proxy data for model evaluation, is mitigated by differences in the resolution, even at regional scales, of the paleoenvironmental data used in each step. For example, information used to describe this boundary condition mainly describes the coastline and topography, the structure of the vegetation (forest versus tundra) and extent of land and sea ice. In this way, the specified vegetation boundary conditions will be just one of a multiplicity of factors influencing the resulting climate simulation. In contrast, the paleoecological data that can be used to evaluate model output is at a much finer resolution of vegetation associations, plant functional types and species. At the temporal and spatial scales of most simulations, physical climate models are more sensitive to topography, coastline and soils than to vegetation distribution (e.g. Lynch et al. 2001a, b), making paleoecological data in particular ideal for model evaluation studies. An alternative to traditional model evaluation exercises is the use of limited area models for the explicit analysis of feedbacks in order to generate hypotheses consistent with the proxy record. Motivation for this approach is the result of studies such as Diffenbaugh and Sloan (2004), who found that paleosimulations driven only by orbital forcing of mean annual (precipitation minus evaporation) of the NW United States were in disagreement with proxy reconstructions from the Pacific coast suggesting that consideration of regional-scale climate system feedbacks is crucial for the application of limited area models to paleoclimate problems. An example of explicit analysis of feedbacks is shown in Rivers and Lynch (2004), who present a factorial experimental strategy for assessing the effects of the changing land surface on early Holocene climate in Beringia, a vast area between the Kolyma River in the Russian Far East to the Mackenzie River in the Northwest Territories of Canada. In that application, it was found that large-scale interannual variability has a greater impact on the local climate than changes in the land surface specification for temperature-based responses, but not for moisture availability. Vegetation transitions from tundra to boreal forest led to increased precipitation, winter snow depth and low cloud, leading to a delayed snowmelt in forested regions. This is in contrast to expectations based on albedo arguments alone. In western Beringia, a positive feedback is suggested in warming due to the presence of boreal forest, consistent with some observations that Siberian deciduous forest persisted longer than Alaskan deciduous forest in the context of large-scale cooling of the climate. The reverse was found to be true in eastern Beringia, where deciduous forests exhibited a cooling effect due to increases in albedo. The influence of Greenland's deglaciation on the atmospheric winter and summer circulation of the Arctic has been quantified with the high-resolution regional atmospheric model HIRHAM by Dethloff et al. (2004). Greenland's deglaciation exerts a pronounced influence on the atmospheric winter circulation of the Arctic. The land areas over Siberia and the Canadian archipelago are warmed up to 5 K, and parts of the Atlantic and Pacific Oceans are cooled up to 3 K. A north-eastward shift of the storm tracks occurs over the North Atlantic as well as an increase of synoptic activity over Alaska. The pronounced changes in the 'precipitation minus evaporation' differences during winter could have important consequences for the atmospheric freshwater input into the Arctic Ocean and the Nordic Seas with the potential to cause fundamental changes in the ocean circulation. The significant differences between simulations with and without Greenland's orography are linked to the blocking of cold air masses west of Greenland that result in a decrease of the geopotential height on the upstream side of the mountain and a dominant barotropic response of the Arctic atmosphere. These changes correspond to an enhanced winter polar vortex and stratospheric conditions more favourable for Arctic ozone losses as described by Rex et al. (2004).

8.5 Scenario Simulations of the Future Arctic Climate

A further application of regional climate models is the dynamical downscaling of global climate model simulations using different emission scenarios of greenhouse gases (GHG) and aerosols. As noted before, all climate simulations contain biases, and the impact of GHG and aerosols on future climate trends and tendencies contains uncertainties due to changes in the global circulation regimes under the influence of external forcings. Rummukainen et al. (2001) carried out dynamical downscaling experiments of two GCM climate scenario runs for the European Arctic. Kiilsholm et al. (2003) conducted transient climate change simulations for the Arctic from ECHAM4/OPYC coupled GCM runs and showed 30-year time slices for two scenarios for the years 2071–2100. Dorn et al. (2003) investigated the influence of changes in the NAO patterns under increased atmospheric carbon dioxide and aerosol content on the Arctic winter climate using a climate scenario run with the AOGCM ECHAM4/OPYC, described by Roeckner et al. (1999). The regional model has been driven with data of positive and negative NAO phases from a control simulation as well as from a time-dependent greenhouse gas and aerosol scenario simulation. The results indicate that the effects of NAO regime changes on Arctic winter temperatures and precipitation are regionally significant over most of north-western Eurasia and parts of Greenland. In this regard, mean winter temperature variations of up to 6 K may occur over northern Europe. Precipitation and synoptic variability are also regionally modified by NAO regime changes with a stronger synoptic variability during positive NAO phases. The climate changes associated with the NAO are, in some regions, stronger than those attributed to enhanced greenhouse gases and aerosols. This result indicates that the projected global changes of the atmospheric composition and internal circulation changes are competing with each other in their importance for the Arctic climate evolution in the near future. The knowledge of the future NAO trend on decadal timescales is vitally important for a regional assessment of climate scenarios for the Arctic.

Natural climate variability along with the anthropogenic changes in the climate system may lead to a change in future states of the climate. Demographic, socio-economic and the technological developments contribute to the greenhouse gas (GHG) and aerosol emissions and are accompanied partly by drastic land use changes. Different emission scenarios, based on projected future GHG and aerosol emissions and land use with consistent assumptions of future demographic, socio-economic and technological developments, have been prepared by a group of economists and social scientists in SRES (IPCC Special Report on Emission Scenario). Here, we focus on the B2 scenario, a world with an emphasis on local solutions to economic and environmental sustainability. Global B2 scenario simulations show for the last three decades of the twenty-first century (2071–2100), a change of 2.2 K (with a range of 0.9–3.4 K between the nine models used by TAR IPCC) in globally averaged surface air temperature relative to the period 1961–1990. However, the models differ significantly in the simulated temperature response in the Arctic, not only in the magnitude but also in regional aspects of the projected temperature change. A model with high horizontal resolution will be very useful to find out the regional

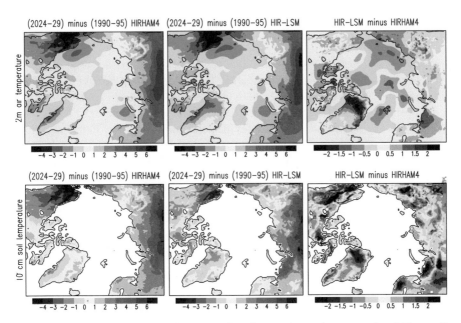

Fig. 8.3 Differences (°C) of mean winter (DJF) 2-m air temperatures (*upper panel*) and 10-cm soil temperatures (*lower panel*) between time slice 2024–2029 and time slice 1990–1995. *Left*: HIRHAM simulation; *middle*: HIRHAM-LSM simulation; *right*: difference between HIRHAM and HIRHAM-LSM simulations (Calculations are based on the simulations described by Saha et al. 2006)

aspects of Arctic climate changes in the context of global warming. A dynamical downscaling of a B2 scenario simulation of the coupled atmosphere–ocean model ECHO–G (ECHAM4/HOPE–G) was done with the regional atmospheric model HIRHAM over a pan-Arctic domain at a horizontal resolution of 50×50 km. Two 6-year-long time slices (1990–1995 and 2024–2029) were chosen for the dynamical downscaling of this scenario with the original HIRHAM as well as with the new HIRHAM-LSM (HIRHAM coupled with NCAR Land Surface Model) and are shown in Fig. 8.3.

The regions of warming and cooling during 2024–2029 winter (DJF) compared to 1990–1995 winter are similar for both models. With advanced vegetation and soil schemes, the coupled model shows a deviation from HIRHAM in 2-m air temperature by about ±1.5°C. In both scenario runs, there is an enhanced warming over the eastern hemisphere and parts of Northern America and a cooling over Alaska and Greenland. The difference plot shows that the impact of different soil schemes varies with a strong regional signature. The LSM reduces the anthropogenic warming over Siberia and enhances the warming over European Russia and Northern Canada. Both models show a similar warming and cooling trend in 2-m air and 10-cm soil temperatures at high latitudes, but the HIRHAM-LSM shows a stronger soil warming than HIRHAM. The anthropogenic impact is amplified by the use of a more advanced land–soil scheme in the Arctic. This would have strong implications for the additional release of methane from permafrost areas.

8.6   Global Implications of Arctic Climate Processes and Feedbacks

The Arctic region is one of the key areas for understanding how climate might change in the future because it is where the powerful ice-albedo feedback mechanism operates as discussed by Curry et al. (1995) and Holland and Bitz (2003). This feedback is the most important factor for the polar amplification of global warming. Possible future changes in Arctic sea-ice cover and thickness, and consequently changes in the ice-albedo feedback, represent one of the largest uncertainties in the prediction of future temperature change. Variability in the Arctic and polar amplification and feedbacks are recurrent themes in numerical climate modelling, but our knowledge of the mechanisms supposed to underpin this link is weak as shown by Moritz et al. (2002). Biases and across-model scatter are present in simulations of Arctic sea ice, either in the radiative forcing or in the parameterisations of surface melt and its influence on the absorption of shortwave radiation (Flato et al. 2004). Main causes of the interannual variability of the sea-ice cover in the Arctic are the year-to-year variations in the atmospheric fields of wind and temperature due to the high sensitivity of the Arctic sea-ice cover to atmospheric forcing as discussed by Arfeuille et al. (2000). Sea ice introduces additional feedbacks into the coupled climate system, making climate naturally more variable in polar regions. Surface albedo has long been recognised as one of the key surface parameters in climate models through its direct effect on the energy balance. By means of simulations with a global coupled AOGCM, Dethloff et al. (2006) investigated the influence of a more realistic sea-ice and snow albedo treatment on the energy balance and atmospheric circulation patterns. They showed that changes in the polar energy sink region can exert a strong influence on the mid- and high-latitude climate by modulating the strength of the midlatitude westerlies and storm tracks. It is found that a more realistic sea-ice and snow albedo treatment changes the ice-albedo feedback and the radiative exchange between the atmosphere and the ocean–sea-ice system. The planetary wave energy fluxes in the middle troposphere of midlatitudes between 30°N and 50°N are redistributed, which induces perturbations in the zonal and meridional planetary wave trains from the tropics over the midlatitudes into the Arctic. It is shown that the improved parameterisation of Arctic sea-ice and snow albedo can trigger changes in the Arctic and North Atlantic Oscillation patterns with strong implications for the European climate. The improved and validated Arctic sea-ice and snow albedo parameterisations have been introduced for present-day forcing conditions into the ECHO-G model, and unforced simulations over 500 years with fixed present-day greenhouse gases concentrations have been carried out. The averaged annual surface air temperature difference between the new sea-ice and snow albedo (C_NSS) run and the control (C) run for this time slice is characterised by a cooling of the Arctic regions as a result of the increased albedo and a warming in a zonal belt covering the midlatitudes. The annual cycle of Arctic sea-ice extent is found to be more realistic. The 500 hPa winter geopotential differences between the C_NSS-run and the C-run show pronouced changes over high and middle latitudes.

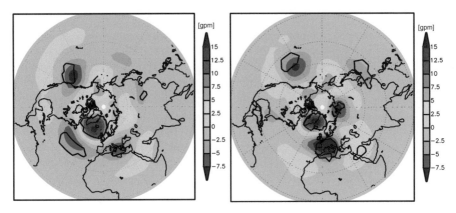

Fig. 8.4 Winter (DJF) geopotential difference (gpm) at 500 hPa 'New sea-ice and snow albedo run minus control run' (C_NSS-C), averaged over the first 250 years of the 500 year long run (*left*) and over the last 250 years of the 500 year long run (*right*). *Thick black contours* describe the 95% significance level (Dethloff et al., A dynamical link between the Arctic and the global climate system, Geophys. Res. Lett. 33:L03703, 2006. Copyright [2006] American Geophysical Union. Reproduced by permission of American Geophysical Union)

These results indicate that Arctic changes connected with sea-ice and snow cover changes induce significant anomalies in dynamical fields connecting the Arctic and midlatitudes through meridional and zonal planetary wave guides during winter. Arctic sea- ice and land surface snow perturbations can disturb the atmospheric wave propagation pathways between the Arctic and the tropics in the middle troposphere and in the latitudinal belt between 30°N and 50°N. These results provide a physical explanation on how regional sea-ice or snow anomalies in the Arctic can influence the stationary planetary waves and transient storm tracks on hemispheric scale and seasonal to interannual timescales. On centennial timescales, additional feedbacks between the atmospheric and oceanic circulation and the sea-ice cover develop. Therefore, the 500 hPa winter geopotential differences between the C_NSS simulations and the C-run have been examined over the whole 500-year-long simulations. Figure 8.4 shows the 500 hPa geopotential differences between the C_NSS and the C-run, averaged over the first 250 and the second 250 years of the 500-year-long run. Strongest differences occur over the whole northern hemisphere, which are statistically significant over the Pacific and the Atlantic Ocean, parts of the Arctic Ocean and the Mediterranean Sea. The difference pattern shows a strong similarity with the Arctic Oscillation (AO) pattern described by Thompson and Wallace (1998) and reveals a clear hint on triggering a North Atlantic Oscillation (NAO) like pattern.

The AO/NAO like patterns persist also over the second 250 years interval with shifts in the main centres of action over the Pacific Ocean and the Atlantic Ocean and Mediterranean Sea. These shifts are a result of the non-stationarity of the AO/NAO pattern and could be connected with changes in the oceanic circulation. The circulation differences which result from the C_NSS simulations project on the existing natural circulation modes of the climate system and trigger changes in the AO/NAO pattern with strong implications for the European climate. Rinke et al. (2003) showed in a coupled regional atmosphere–ocean–sea-ice model that the atmospheric forcing strongly affects the sea-ice retreat, but there is also a strong impact of the sea-ice conditions on the atmospheric circulation. Since the sea-ice distribution reacts very sensitively to the position and strength of baroclinic weather systems, the changes in the mean sea-level pressure pattern as a result of modified surface albedo parameterisations require a deeper understanding of the feedbacks acting in the coupled atmosphere–ocean–sea-ice system. The changed albedo feedback as well as the aerosol impact would lead to mean sea-level pressure changes which have the potential to influence large-scale teleconnection patterns and exert a global impact. This shows the way how current AOGCMs need to be improved in their performance over the Arctic.

8.7   Summary and Conclusions

There is a need to describe the most important physical and chemical mechanisms that control the atmospheric and oceanic circulation patterns in polar regions and on the global scale and their consequences for decadal-scale climate variability as a result of the non-linear dynamics of the coupled atmosphere–ocean–sea-ice-ecosystem under the influence of external forcing factors. The use of high-resolution RCMs can contribute to a better description of important regional physical processes in the ocean, cryosphere, atmosphere, land and biosphere including their interactions in coupled regional model systems. This is based on identifying and modelling of the key processes and through an assessment of the improved understanding in the light of analysis of instrumental as well as paleoclimatic and paleoenvironmental records.

Reliable projections of the climate state in the Arctic throughout the twenty-first century are crucially dependent on long-term and high-quality observations and improved regional and global model systems carefully evaluated against available instrumental observations, the synthesis between them and a better understanding of the basic operation of high northern latitude climate variability modes and the interaction between high and low latitudes. Therefore, the challenge is to identify and model the important components of the climate system and to explore this in order to assess and predict changes in the Arctic and Antarctic climate system caused by natural and anthropogenic effects.

Because of the complexity of the climate system, this research requires modelling and prediction studies that include interactions between the processes in the atmosphere, oceans, cryosphere, land surfaces and the biosphere, encompassing timescales from days to decades. This requires the improvement of the regional and global models for the Arctic and Antarctica, especially the physical parameterisations of radiation, planetary boundary layer, clouds, aerosols and stratospheric ozone. A main aim is to include the key dynamical–radiative–chemical interactions in the tropo- and stratosphere and to clarify the interactions between the Arctic with the global system via preferred atmospheric circulation modes. A key scientific focus is to clarify the robustness of Arctic climate change and the dynamic and thermal feedbacks processes responsible for the large climate variations. In this respect, it is necessary to better understand and quantify to what extent Arctic and Antarctic climate change is due to regional or global processes. The main goal is to address the deficiencies in understanding the Arctic by developing improved physical descriptions of Arctic climate feedbacks in atmospheric and coupled regional climate models and to implement the improved parameterisations into global climate system models to determine their global influences and consequences for decadal-scale climate variations. A further aim is to model the main feedbacks correctly to arrive at a more reliable estimate of future changes due to the coupling between natural and anthropogenic effects. With respect to the Arctic, we need investigations in the following directions:

1. Development of regional models of the coupled Arctic climate system, including the climate subsystems atmosphere, ocean, sea ice, land and permafrost, glaciers and interactions with terrestrial and marine ecosystems. Bridging hydrological models, regional climate models and permafrost dynamical models as in Stendel et al. (2007).
2. Understanding of controls over spatial and temporal variability of terrestrial processes in the Arctic that affect exchange of water and energy and greenhouse gases and feedbacks with vegetation changes as suggested by McGuire et al. (2003).
3. Improvement of the model performance in close collaboration of experimental and modelling groups during the IPY via atmospheric and oceanic measurements in different key regions of the Arctic.
4. Process and feedback understanding of Arctic climate variations and observed variability patterns for improved description of atmospheric processes on seasonal to decadal timescales in global climate models.
5. Determination of the impact of polar regions and regional polar feedbacks as key drivers for global climate changes and reduction of the uncertainties of future climate change scenarios and possible unexpected climate surprises.
6. In addition, the development of new dynamically adaptive Arctic climate models with dynamically adaptive cores and the ability of two-way feedbacks as in Läuter et al. (2004) is encouraged that can be easily implemented within Earth system models.

Ackowledgement We would like to thank Prof. Ernst Augstein and Prof. Dirk Olbers from AWI Bremerhaven for their support of our research since 1992.

References

Alley RB (2003) Palaeoclimatic insights into future climate challenges. Philos Transact R Soc Lond A Math Phys Eng Sci 361:1831–1848
Antic S, Laprise R, Denis B, de Elia R (2004) Testing the downscaling ability of a one-way nested regional climate model in regions of complex topography. Clim Dyn 23:473–493

Appenzeller C, Stocker TF, Anklin M (1998) North Atlantic oscillation dynamics recorded in Greenland ice cores. Science 282:446–449

Arfeuille G, Mysak LA, Tremblay L-B (2000) Simulation of the inter-annual variability of the wind-driven Arctic sea-ice cover during 1958–1998. Clim Dyn 16:107–121

Bamber J, Ekholm S, Krabill WB (2001) A new, high-resolution digital elevation model of Greenland fully validated with airborne laser altimeter data. J Geophys Res 106:33773–33780

Bonan GB (1996) A land surface model (LSM version 1.0) for ecological, hydrological, and atmospheric studies: technical description and user's guide, NCAR Tech. Note TN-417+STR. National Center for Atmospheric Research, Boulder, p 150

Box JE, Rinke A (2003) Representation of Greenland ice sheet surface climate in the HIRHAM regional climate model. J Clim 16:1302–1319

Bromwich DH, Cullather RI, Serreze MC (2000) Reanalyses depiction of the Arctic atmospheric moisture budget. In: Lewis EL (ed) The fresh water budget of the arctic ocean. Kluwer Academic, Dordrecht, pp 163–196

Bromwich DH, Cassano JJ, Klein T, Heinemann G, Hines KM, Steffen K, Box JE (2001) Mesoscale modeling of katabatic winds over Greenland with the Polar MM5. Mon Weather Rev 129:2290–2309

Cassano JJ, Box JE, Bromwich DH, Li L, Steffen K (2001) Evaluation of polar MM5 simulations of Greenland's atmospheric circulation. J Geophys Res 106:33867–33889

Caya D, Biner S (2004) Internal variability of RCM simulations over an annual cycle. Clim Dyn 22:33–46

Caya D, Laprise R (1999) A semi-lagrangian semi-implicit regional climate model: the Canadian RCM. Mon Weather Rev 127:341–362

Christensen JH, Kuhry P (2000) High-resolution regional climate modelvalidation and permafrost simulation for the East European Russian Arctic. J Geophys Res 105:29647–29658

Christensen JH et al (1997) Validation of present-day regional climate simulations over Europe: LAM simulations with observed boundary conditions. Clim Dyn 13:489–506

Christensen OB, Gaertner MA, Prego JA, Polcher J (2001) Internal variability of regional climate models. Clim Dyn 17:875–887

Cullather RI, Bromwich DH, Serreze M (2000) The atmospheric hydrological cycle over the Arctic Basin from reanalyses. Part I: comparison with observations and previous studies. J Clim 13:923–937

Curry JA, Lynch AH (2002) Comparing arctic regional climate models. EOS Trans Am Geophys Un 83:87

Curry JA, Schramm J, Ebert EE (1995) On the sea ice albedo climate feedback mechanism. J Clim 8:240–247

Debernard JM, Køltzow O, Haugen JE, Røed LP (2003) Improvements in the sea-ice module of the regional coupled atmosphere-ice-ocean model and the strategy for the coupling of the three spheres. In: Iversen T, Lystad M (eds) RegClim general technical report no. 7. Norwegian Meteorological Institute, Oslo, pp 59–69

Denis B, Laprise R, Caya D, Cote J (2002) Downscaling ability of one-way nested regional climate models: the big-brother experiment. Clim Dyn 18:627–646

Denis B, Laprise R, Caya D (2003) Sensitivity of a regional climate model to the resolution of the lateral boundary conditions. Clim Dyn 20:107–126

Dethloff K, Rinke A, Lehmann R, Christensen JH, Botzet M, Machenhauer B (1996) Regional climate model of the Arctic atmosphere. J Geophys Res 101:23401–23422

Dethloff K, Abegg C, Rinke A, Hebestad I, Romanov V (2001) Sensitivity of Arctic climate simulations to different boundary layer parameterizations in a regional climate model. Tellus 53A:1–26

Dethloff K, Schwager M, Christensen JH, Kiilsholm S, Rinke A, Dorn W, Jung-Rothenhäusler F, Fischer H, Kipfstuhl S, Miller H (2002) Recent Greenland accumulation estimated from regional climate model simulations and ice core analysis. J Clim 15:2821–2832

Dethloff K, Dorn W, Rinke A, Fraedrich K, Junge M, Roeckner E, Gayler V, Cubasch U, Christensen JH (2004) The impact of Greenland's deglaciation on the Arctic circulation. Geophys Res Lett 31:L19201. doi:10.1029/2004GL020714

Dethloff K, Rinke A, Benkel A, Koltzow M, Sokolova E, Saha SK, Handorf D, Dorn W, Rockel B, von Storch H, Haugen JE, Roed LP, Roeckner E, Christensen JH, Stendel M (2006) A dynamical link between the Arctic and the global climate system. Geophys Res Lett 33:L03703. doi:10.1029/2005GL025245

Dickinson RE, Errico RM, Giorni F, Bates GT (1989) A regional climate model for the western U.S. Clim Chang 15:383–422

Diffenbaugh NS, Sloan LC (2004) Mid-holocene orbital forcing of regional-scale climate: a case study of western North America using a high-resolution RCM. J Clim 17:2927–2937

Dimitrijevic M, Laprise R (2005) Validation of the nesting technique in a Regional Climate Model through sensitivity tests to spatial resolution and the time interval of lateral boundary conditions during summer. Clim Dyn. doi:10.1007/s00382-005-0023-6

Dorn W, Dethloff K, Rinke A, Botzet M (2000) Distinct circulation states of the Arctic atmosphere induced by natural climate variability. J Geophys Res 105:29659–29668

Dorn W, Dethloff K, Rinke A (2003) Competition of NAO regime changes and increasing greenhouse gases and aerosols with respect to Arctic climate estimate. Clim Dyn 21:447–458. doi:10.1007/s00382-003-0344-2

Dorn W, Dethloff K, Rinke A, Frickenhaus S, Gerdes R, Karcher M, Kauker F (2007) Sensitivities and uncertainties in a coupled regional atmosphere-ocean-ice model with respect to the simulation of Arctic sea ice. J Geophys Res 112:D10118. doi:10.1029/2006JD007814

Döscher R, Willen U, Jones C, Rutgersson A, Meier H, Hansson U (2002) The development of the coupled ocean-atmosphere model RCAO. Boreal Environ Res 7:183–192

Ekholm S (1996) A full coverage, high-resolution, topographic model of Greenland computed from a variety of digital elevation data. J Geophys Res 101:21961–21972

Felzer BA, Thompson SL (2001) Evaluation of a regional climate model for paleoclimate applications in the Arctic. J Geophys Res Atmos 106:27407–27424

Flato GM, Hibler WD (1992) Modeling pack ice as a caviating fluid. J Phys Oceanogr 22:626–651

Flato GM, Participating CMIP modelling groups (2004) Sea-ice and its response to CO_2 forcing as simulated by global climate models. Clim Dyn 23:229–241

Francis JA (2002) Validation of reanalysis upper-level winds in the Arctic with independent rawinsonde data. Geophys Res 29:1–4

Gesch DB (1994) Topographic data requirements for EOS global change research. U.S. Geological Survey Open-File Report 94–626, p 60

Gesch DB, Larson KS (1996) Techniques for development of global 1-kilometer digital elevation models. In: Pecora thirteen, human interactions with the environment – perspectives from space, Sioux Falls, http://edcdaac.usgs.gov/gtopo30/papers/geschd3.asp

Giorgi F (1991) Sensitivity of simulated summertime precipitation over western United States to different physics parameterization. Mon Weather Rev 119:2870–2888

Giorgi F, Shields C (1999) Tests of precipitation parameterizations available in latest version of NCAR regional climate model (RegCM) over continental United States. J Geophys Res 104:6353–6375

Giorgi F, Marinucci MR, Bates GT (1993) Development of a second generation Regional Climate Model (RegCM2). Part I: boundary layer and radiative transfer processes. Mon Weather Rev 121:2794–2813

Girard E, Blanchet J-P (2001a) Microphysical parameterizations of arctic diamond dust, ice fog and thin stratus for climate models. J Atmos Sci 58:1181–1198

Girard E, Blanchet J-P (2001b) Simulation of Arctic diamond dust and ice fog or thin stratus using an explicit aerosol-cloud-radiation model. J Atmos Sci 58:1199–1221

Haywood AM, Valdes PJ, Markwick PJ (2004) Cretaceous (Wealden) climates: a modelling perspective. Cretac Res 25:303–311

Hibler WD (1979) A dynamic thermodynamic sea ice model. J Phys Oceanogr 9:815–846

Holland MM, Bitz CM (2003) Polar amplification of climate change in the coupled model intercomparison project. Clim Dyn 21:221–232

Hostetler SW, Giorgi F, Bates GT, Bartlein PJ (1994) Lake-atmosphere feedbacks associated with paleolakes Bonneville and Lahontan. Science 263:665–668

Hostetler SW, Bartlein PJ, Clark PU, Small EE, Solomon AM (2000) Simulated influences of Lake Agassiz on the climate of central North America 11,000 years ago. Nature 405:334–337

Hunke EC, Dukowicz JK (1997) An elastic-viscous-plastic model for sea ice dynamics. J Phys Oceanogr 27:1849–1867

Ingram WJ, Wilson CA, Mitchell JFB (1989) Modeling climate change: an assessment of sea ice and surface albedo feedbacks. J Geophys Res 94:8609–8622

Jones CG, Wyser K, Ullerstig A, Willen U (2004) The Rossby Centre Regional Atmospheric Climate Model (RCA). Part II: application to the arctic climate, Ambio 33:211–220

Jürrens R (1999) Validation of surface fluxes in climate simulations of the Arctic with the regional model REMO. Tellus 51:698–709

Kageyama M, D'Andrea F, Ramstein G, Valdes PJ, Vautard R (1999) Weather regimes in past climate atmospheric general circulation model simulations. Clim Dyn 15:773–793

Kantha LH, Clayson CA (1994) An improved mixed layer model for geophysical applications. J Geophys Res 99:25235–25266

Kiilsholm S, Christensen JH, Dethloff K, Rinke A (2003) Net accumulation of the Greenland ice sheet: high resolution modeling of climate changes. Geophys Res Lett 30:1485. doi:10.1029/2002GL015742

Køltzow M, Eastwood S, Haugen JE (2003) Parameterization of snow and sea ice albedo in climate models. Research report no. 149, Norwegian Meteorological Institute, Oslo

Kreyscher M, Harder M, Lemke P, Flato GM (2000) Results of the sea ice model intercomparison project: evaluation of sea-ice rheology schemes for use in climate simulations. J Geophys Res 105:11299–11320

Laprise R, Caya D, Bergeron G, Giguère M, Bergeron G, Côté H, Blanchet J-P, Boer GJ, McFarlane NA (1998) Climate and climate change in western Canada as simulated by the Canadian Regional Climate Model. Atmos Ocean 36:119–167

Läuter M, Handorf D, Rakowsky N, Behrens J, Frickenhaus S, Best M, Dethloff K, Hiller W (2004) A parallel adaptive barotropic model of the atmosphere. J Comput Phys 223(2):609–628. doi:10.1016/j.jcp. 2006.09.029

Lorenz P, Jacob D (2005) Influence of regional scale information on the global circulation: a two-way nesting climate simulation. Geophys Res Lett 32:L18706. doi:10.1029/2005GL023351

Lynch AH, Cullather RI (2000) An investigation of boundary forcing sensitivities in a regional climate model. J Geophys Res 105:26603–26617

Lynch AH, Chapman W, Walsh JE, Weller G (1995) Development of a regional climate model of the western Arctic. J Clim 8:1555–1570

Lynch AH, Glueck MF, Chapman WL, Bailey DA, Walsh JE (1997) Remote Sensing and Climate Modelling of the St Lawrence Is. Polynya. Tellus 49A:277–297

Lynch AH, McGinnes DL, Baily DA (1998) Snow-albedo and the spring transition in a regional climate system model: influence of land surface model. J Geophys Res 103:29037–29049

Lynch AH, Bonan GB, Chapin FS III, Wu W (1999) The impact of tundra ecosystems on the surface energy budget and climate of Alaska. J Geophys Res 104:6647–6660

Lynch AH, Maslanik JA, Wu W (2001a) Mechanisms in the development of anomalous sea ice extent in the western Arctic: a case study. J Geophys Res 106:28097–28105

Lynch AH, Slater AG, Serreze MC (2001b) The Alaskan Arctic frontal zone: forcing by orography. Coastal contrast and the Boreal Forest. J Clim 14:4351–4362

Lynch AH, Curry JA, Brunner RD, Maslanik JA (2004) Towards an integrated assessment of the impacts of extreme wind events on Barrow, Alaska. Bull Am Meteorol Soc 85:209–221

Marbaix P, Gallee H, Brasseur O, van Ypersele JP (2003) Lateral boundary conditions in regional climate models: a detailed study of the relaxation procedure. Mon Weather Rev 131:461–479

Maslanik JA, Lynch AH, Serreze MC, Wu W (2000) A case study of regional climate anomalies in the Arctic: performance requirements for a coupled model. J Clim 13:383–401

Masson V, Joussaume S (1997) Energetics of the 6000 BP atmospheric circulation in boreal summer, from large scale to monsoon areas: a study with two versions of the LMD AGCM. J Clim 10:2888–2903

McFarlane NA, Boer GJ, Blanchet J-P, Lazare M (1992) The Canadian Climate Centre second generation general circulation model and its equilibrium climate. J Clim 5:1013–1044

McGuire AD, Sturm M, Chapin FS (2003) Arctic transitions in the land-atmosphere system (ATLAS): background, objectives, results and future directions. J Geophys Res 108:D2. doi:1029/2002JD002367

Mikolajewicz U, Sein DV, Jacob D, Kahl T, Podzun R, Semmler T (2005) Simulating Arctic sea ice variability with a coupled regional atmosphere–ocean–sea ice model. Meteorol Z 14:793–800

Mirocha JD, Kosovic B, Curry JA (2005) Vertical heat transfer in the lower atmosphere over the Arctic Ocean during clear-sky periods. B Lay Meteo 117:37–71. doi:10.1007/s10546-004-1130–3

Moritz RE, Bitz CM, Steig EJ (2002) Dynamics of recent climate change in the Arctic. Science 297:149–151

Morrison HC, Shupe M, Curry JA (2003) Evaluation of a bulk microphysical scheme using SHEBA data. J Geophys Res 108(D8):4255. doi:10.1029/2002JD002229

Parkinson CL, Washington WM (1979) A large-scale numerical model of sea ice. J Geophys Res 84:311–337

Persson P, Ola G, Fairall CW, Andreas EL, Guest PS, Perovich DK (2002) Measurements near the atmospheric surface flux group tower at SHEBA: near-surface conditions and surface energy budget. J Geophys Res 107(C10):8045. doi:10.1029/2000JC000705

Pinto JO, Curry JA, Lynch AH, Persson POG (1999) Modelling clouds and radiation for the November 1997 SHEBA using a column climate model. Special issue of J Geophys Res on New Developments and Applications with the NCAR Regional Climate Model 104:6661–6678

Renssen H, Isarin RFB, Jacob D, Podzun R, Vandenberghe J (2001) Simulation of the Younger Dryas climate in Europe using a regional climate model nested in an AGCM: preliminary results. Glob Planet Chang 30:41–57

Renssen H, Bracconot P, Tett SFB, von Storch H, Weber SL (2003) Recent developments in Holocene climate modelling. In: Battarbee RW, Gasse F, Stickley CE (eds) Past climate variability through Europe and Africa. Kluwer Academic, Dordrecht

Rex M, Salawitch RJ, von der Gathen P et al (2004) Arctic ozone loss and climate change. Geophys Res Lett 31:L04116. doi:10.1029/2003GL018844

Rinke A, Dethloff K (2000) On the sensitivity of a regional Arctic climate model to initial and boundary conditions. Clim Res 14:101–113

Rinke A, Dethloff K, Christensen JH, Botzet M, Machenhauer B (1997) Simulation and validation of Arctic radiation and clouds in a regional climate model. J Geophys Res 102:29833–29847

Rinke A, Dethloff K, Spekat A, Enke W, Christensen JH (1999a) High resolution climate simulations over the Arctic. Polar Res 18:143–150

Rinke A, Dethloff K, Christensen J-H (1999b) Arctic winter climate and its interannual variations simulated by a regional climate model. J Geophys Res 104:19027–19038

Rinke A, Lynch AH, Dethloff K (2000) Intercomparison of Arctic regional climate simulations: case studies of January and June 1990. J Geophys Res 105:29669–29683

Rinke A, Gerdes R, Dethloff K, Kandlbinder T, Karcher M, Kauker F, Frickenhaus S, Koeberle C, Hiller W (2003) A case study of the anomalous Arctic sea ice conditions during 1990: insights from coupled and uncoupled regional climate model simulations. J Geophys Res 108(D9):4275. doi:10.1029/2002JD003146

Rinke A, Marbaix P, Dethloff K (2004a) Internal variability in arctic regional climate simulations: case study for the SHEBA year. Clim Res 27:197–209

Rinke A, Dethloff K, Fortmann M (2004b) Regional climate effects of Arctic haze. Geophys Res Lett 31(16):L16202. doi:10.1029/2004GL020318

Rinke A, Dethloff K, Cassano J, Christensen JH, Curry JA, Haugen JE, Jacob D, Jones C, Koltzow M, Lynch AH, Pfeifer S, Serreze MC, Shaw MJ, Tjernstrom M, Wyser K, Zagar M (2006) Evaluation of an ensemble of Arctic regional climate models: Spatiotemporal fields during the SHEBA year. Clim Dyn 26:459–472. doi:10.1007/s00382-005-0095-3

Rivers AR, Lynch AH (2004) The influence of land surface transitions on the local climate of Beringia in the early Holocene. J Geophys Res 109:D21114. doi:10.1029/2003JD004213

Roeckner E et al. (1996) The atmospheric general circulation model ECHAM-4: model description and simulation of present-day climate. MPI Rep. 218, Max Planck Inst. for Meteorol., Hamburg, pp 90

Roeckner E, Bengtsson L, Feichter J, Levieved J, Rodhe H (1999) Transient climate change with a coupled atmosphere-ocean GCM including the troposperic sulfur cycle. J Clim 12:3004–3032

Rummukainen M, Räisänen J, Bringfelt B, Ullerstig A, Omstedt A, Willén U, Hansson U, Jones C (2001) A regional climate model for northern Europe: model description and results from the downscaling of two GCM control simulations. Clim Dyn 17:339–359

Saha SK, Rinke A, Dethloff K, Kuhry P (2006) The influence of a complex land surface scheme on Arctic climate simulations. J Geophys Res 111:D22104. doi:10.1029/2006JD007188

Semmler T, Jacob D, Schlünzen KH, Podzun R (2004) Influence of sea ice treatment in a regional climate model on boundary layer values in the Fram Strait region. Mon Weather Rev 132:985–999

Semtner B (1976) A model for the thermodynamic growth of sea ice in numerical investigations of climate. J Phys Oceanogr 6:379–389

Serreze MC, Hurst CM (2000) Representation of mean Arctic precipitation from NCEP- NCAR and ERA reanalyses. J Clim 13:182–201

Seth A, Giorgi F (1998) The effects of domain choice on summer precipitation simulation and sensitivity in a regional climate model. J Clim 11:2698–2712

Stendel M, Romanovsky VE, Christensen JH, Sazonova T (2007) Using dynamical downscaling to close the gap between global change scenarios and local permafrost dynamics. Glob Planet Chang 56:203–214

Stevens DP (1991) The open boundary condition in the United Kingdom Fine-Resolution Antarctic Model. J Phys Oceanogr 21:1494–1499

Thompson DWJ, Wallace JM (1998) The Arctic oscillation signature in the wintertime geopotential height and temperature fields. Geophys Res Lett 25:1297–1300

Tjernstroem M, Zagar M, Svensson G, Cassano J, Pfeifer S, Rinke A, Wyser K, Dethloff K, Jones C, Semmler T, Shaw M (2005) Modelling the Arctic Boundary layer: An evaluation of six ARCMIP regional-scale models using data from the SHEBA project. Bound-Layer Meteorol 117:337–381. doi:10.1007/s10546-004-7954-z

Uttal T et al (2002) Surface energy budget of the Arctic ocean. Bull Am Meteorol Soc 83:255–275

von Storch H, Cubasch U, González-Ruoco J, Jones JM, Widmann M, Zorita E (2000) Combining paleoclimatic evidence and GCMs by means of Data Assimilation Through Upscaling and Nudging (DATUN). In: 11th Symposium on Global Change Studies, American Meteorological Society, Long Beach, pp 28–31

Walsh JE, Lynch A, Chapman W, Musgrave D (1993) A regional model for studies of atmosphere-ice-ocean interaction in the western Arctic. Metrol Atmos Phys 51:179–194

Wu W, Lynch AH (2000) The response of the seasonal carbon cycle in high latitudes to climate anomalies. J Geophys Res 105:22897–22908

Wu W, Lynch AH, Rivers AR (2005) Estimating the uncertainty in a regional climate model related to initial and lateral boundary conditions. J Clim 18:917

Wyser K, Jones CG, Du P, Girard E, Willén U, Cassano J, Christensen JH, Curry JA, Dethloff K, Haugen J-E, Jacob D, Koltzow M, Laprise R, Lynch A, Pfeifer S, Rinke A, Serreze M, Shaw MJ, Tjernström M, Zagar M (2007) An evaluation of Arctic cloud and radiation processes during the SHEBA year: simulation results from eight Arctic regional climate models. Clim Dyn. doi:10.1007/S00382-007-0286-1

Chapter 9
Progress in Hydrological Modeling over High Latitudes: Under Arctic Climate System Study (ACSYS)

Dennis P. Lettenmaier and Fengge Su

Abstract We review achievements in hydrological modeling over high latitudes during ACSYS, including development and improvement of land surface schemes in representing cold processes, large-scale hydrological modeling over high-latitude river basins, and estimates of freshwater river inflow to the Arctic Ocean. ACSYS hydrological modeling efforts were closely linked to the GEWEX continental-scale experiments (CSEs) and to the Project for Intercomparison of Land-Surface Parameterization Schemes (PILPS). Results in this review are mainly from PILPS 2(e), MAGS, BALTEX, GAME-Siberia (the latter three of which are CSEs), and other studies related to ACSYS. Based on these achievements from the 10 years efforts, the ACSYS scientific strategy for hydrology, which included adaptation of macroscale hydrological modes developed in the framework of GEWEX to Arctic (high-latitude) climate conditions and development of physical (conceptual) or parametric mesoscale hydrologic models for selected river catchments within the Arctic region, was implemented more or less as envisaged in the ACSYS Implementation Plan. In spite of major advances in high-latitude hydrological modeling during the ACSYS era, there remain important problems in parameterization of snow, frost, and lake/wetlands cold processes within climate and hydrology models and in linkages between atmospheric and hydrological models.

D.P. Lettenmaier (✉)
Department of Civil and Environmental Engineering, University of Washington,
164 Wilcox Hall, Box 352700, Seattle, WA 98195-2700, USA
e-mail: dennisl@u.washington.edu

F. Su
Institute of Tibetan Plateau Research, Chinese Academy of Sciences,
No.18, Shuangqing Rd., Beijing 100085, China
e-mail: fgsu@itpcas.ac.cn

P. Lemke and H.-W. Jacobi (eds.), *Arctic Climate Change: The ACSYS Decade and Beyond,* Atmospheric and Oceanographic Sciences Library 43, DOI 10.1007/978-94-007-2027-5_9, © Springer Science+Business Media B.V. 2012

Abbreviations

ACSYS	Arctic Climate System Study
ALMA	Assistance for Land Modeling Activities
AORB	Arctic Ocean River Basin
ARW	A NATO Advanced Research Workshop
BALTEX	Baltic Sea Experiment
CGCM	Canadian General Circulation Model
CLASS	Canadian Land Surface Scheme
CliC	Climate and Cryosphere project
CRCM	Canadian Regional Climate Model
CSEs	Continental-Scale Experiments
ECMWF	European Centre for Medium-Range Weather Forecasts
GAME	GEWEX Asian Monsoon Experiment
GEWEX	Global Energy and Water Cycle Experiment
MAGS	Mackenzie GEWEX Study
MOSES	The Met Office Surface Exchange Scheme
NSE	Nash–Sutcliffe efficiency
PILPS	Project for Intercomparison of Land-Surface Parameterization Schemes
SAST	Snow Atmosphere Soil Transfer
SEWAB	Surface Energy and Water Balance land surface scheme
VIC	Variable Infiltration Capacity model
WCRP	World Climate Research Programme
WGNE	Working Group on Numerical Experimentation

9.1 Introduction

Significant changes have been observed over the pan-Arctic land domain (defined for the purposes of ACSYS as all of the land area draining to the Arctic Ocean) in recent decades. These include increases in winter and fall precipitation (Wang and Cho 1997), reduction in spring snow cover extent (Armstrong and Brodzik 2001; Brown 2000), upward trends in permafrost active-layer depth (Frauenfeld et al. 2004; Smith et al. 2005), and a later freezing and earlier breakup dates of ice on lakes and rivers (Magnuson et al. 2000). Strong trends have also been detected in seasonal and annual patterns of Arctic river discharge (e.g., Peterson et al. 2002; Yang et al. 2002; Ye et al. 2004). All of these changes are directly linked to the Arctic hydrologic cycle, which plays an important role in land, atmosphere, and oceans of the global climate system (Vorosmarty et al. 2001).

Runoff from Arctic drainage basins represents 50% of the net flux of freshwater to the Arctic Ocean (Barry and Serreze 2000) which makes the strong role of the land surface unique among the world's oceans. The total volume and temporal variability of freshwater discharge to the Arctic Ocean exerts a strong control on the salinity of the polar ocean and subsequently the thermohaline circulation of the World Ocean (Aagaard and Carmack 1989; Broecker 1997; Karcher et al. 2005); hence, changes in the amount and timing of runoff from the land surface are of concern climatically. Accurate estimation of freshwater inflow to the Arctic Ocean and the spatial and temporal variations of Arctic river runoff in both gauged and ungauged basins are therefore of considerable concern not only to the land surface system but also to the coupled land–ocean–sea ice–atmosphere system of the Arctic.

These ongoing changes, and the connectivity of the Arctic freshwater system to global climate, motivated the objectives of ACSYS hydrological programme (WCRP 1994), which were to:

- Develop mathematical models of the hydrological cycle under specific Arctic climate conditions suitable for inclusion in coupled climate models
- Determine the elements of the freshwater cycle in the Arctic region and their time and space variability
- Quantify the role of atmospheric, hydrological, and land surface processes in the exchanges between different elements of the hydrological cycle
- Provide an observational basin for the assessment of possible long-term trends of the components of the freshwater balance in the Arctic region under changing climate

To respond to these stated objectives, the ACSYS hydrological programme was structured to include two major components. The first was the development of regional databases for the main components of the freshwater balance of the Arctic region and the second was the development of hydrological models of selected Arctic river basins and their validation using appropriate observational data sets. This review focuses on the second component, hydrological modeling of the arctic region conducted under the auspices of ACSYS.

Early in the evolution of ACSYS, a need was recognized for coordination of ACSYS hydrology activities with the companion World Climate Research Programme (WCRP) Global Energy and Water Cycle Experiment (GEWEX) project. GEWEX hydrological activities were initially formulated around a set of continental-scale experiments (CSEs), among which three – MAGS (Mackenzie GEWEX Study), BALTEX (the Baltic Sea Experiment), and GAME (GEWEX Asian Monsoon Experiment)-Siberia – were all focused in part on the hydrology of portions of the pan-Arctic domain. Hence, rather than embarking on a potentially duplicative hydrological modeling effort, ACSYS hydrological modeling efforts were closely linked to the three above-mentioned GEWEX CSEs.

Concurrent with the evolution of ACSYS and GEWEX came increasing recognition of the sensitivity of high-latitude land areas to climate change and the need for better representation of cold region processes in the land surface schemes (LSSs) used in numerical weather prediction and climate models. The Project for the Intercomparison of Land-Surface Parameterization Schemes (PILPS), initially an activity of GEWEX and later of the WCRP Working Group on Numerical Experimentation (WGNE), was designed to provide common databases and protocols for testing of LSSs and, in so doing, to motivate improvements in the models.

A key aspect of the ACSYS hydrological programme was co-hosting, with GEWEX, of Phase 2(e) of PILPS (see Bowling et al. 2003 for details). We summarize the key contributions of PILPS 2(e) in Sect. 9.2.

This chapter reviews documented achievements in hydrological modeling over high latitudes under ACSYS, including development and improvement of LSSs, large-scale hydrological modeling, and estimates of freshwater river inflow to the Arctic Ocean. Results are mainly from PILPS 2(e), MAGS, BALTEX, GAME-Siberia, and other studies related to ACSYS. Summaries are then provided regarding what ACSYS did and what remains to be done in high-latitude hydrological modeling.

9.2   Evolution of Land Surface Schemes

It has been widely recognized that the continental land surface processes and their characterizations play an important role in the accuracy of global climate and numerical weather prediction models (Dickinson and Henderson-Sellers 1988; Beljaars et al. 1993; Xue and Shukla 1993; Gedney et al. 2000). LSSs describe the interaction of energy, momentum, and water flux between the surface and its overlying atmosphere. LSSs were originally developed to provide lower boundary conditions for general circulation models (GCMs) (Manabe 1969). In the past two decades, there has been a great deal of attention focused on development of LSSs and evolution of their complexity from the simple "bucket" model to the more complicated soil–vegetation–atmosphere transfer schemes. There has also been an expansion in their applications from their original role in coupled land–atmosphere models used for climate and numerical weather predication to off-line, stand-alone use for hydrology, agriculture, and ecosystem studies. Although different LSSs have evolved from different heritages, and details differ, their representation of soil hydrology and runoff generation are of great significance regardless of their origins as the scheme used to partition water and energy in hydrological, ecological, and coupled land–atmosphere models (Henderson-Sellers 1996). PILPS was designed to assess the performance of LSSs in representing the hydrology, energy, momentum, and carbon exchanges with the atmosphere and to achieve greater understanding of the capabilities and potential applications of LSSs in atmospheric models (Henderson-Sellers et al. 1993, 1995).

9.3   Arctic Hydrological Modeling During ACSYS

As noted above, early in the history of ACSYS, it was decided that hydrologic model development would be undertaken by the companion WCRP GEWEX programme and that ACSYS would primarily be a user of GEWEX-fostered hydrologic development. Hydrologic model development within the three GEWEX CSEs (MAGS, BALTEX, and GAME-Siberia) most relevant to ACSYS is discussed below.

9.3.1   MAGS

MAGS, the Canadian contribution to the GEWEX continental-scale experiments, was designed to understand and model the high-latitude water and energy cycles of the Mackenzie River basin (1.8×10^6 km^2) and to improve the ability to assess the climate changes to Canada's water resources (Stewart et al. 1998; Rouse 2000; Rouse et al. 2003). Development of hydrologic models, land surface schemes, and land–atmosphere coupled models, suitably adapted for northern conditions, was one of the initiatives of MAGS. Consequently, the outcomes of MAGS contribute directly to the objectives of the ACSYS hydrological programme.

A number of sites for intensive measurement were developed within MAGS to represent different biophysical facets of the Mackenzie River basin. Intensive hydrological process studies promoted the development of process models based on these field experimentations (Rouse 2000; Woo et al. 2000; Woo and Marsh 2005). Among the targets of these field activities were improvement of process-based models of snow accumulation, blowing snow, intercepted snow, and snowmelt infiltration on frozen soils (Pomeroy et al. 1997, 1998, 2002; Pomeroy and Li 2000; Hedstrom and Pomeroy 1998; Woo et al. 1998; Essery et al. 1999). Woo et al. (2000) and Woo and Marsh (2005) provided detailed reviews of Canadian research in snow, frozen soils, and permafrost hydrology (including results from a first phase of MAGS) during the years 1995–2002.

They concluded that until the date of their writing, few advances in comprehensive mathematical models that represent cold region hydrologic processes had been incorporated into Canadian hydrological and atmospheric models despite the development of new process-based algorithms.

The Canadian Land Surface Scheme (CLASS) (Verseghy 1991; Verseghy et al. 1993), which was developed as a LSS for the Canadian General Circulation Model (CGCM), has been a participant in most PILPS experiments (including PILPS 2(e)). Development of improved parameterizations for CLASS was the ultimate objective of much of the MAGS hydrological modeling activity. A major effort within MAGS was to improve the ability of CLASS to simulate land surface hydrologic variables (especially runoff). This was accomplished by merging the surface energy flux (and vegetation) algorithms within CLASS with the hydrologic routing algorithms from WATFLOOD, a flood forecasting model (Kouwen et al. 1993). A three-level framework for hydrological modeling in MAGS was built in stages by combining CLASS and WATFLOOD, as shown in Fig. 9.1 (Pietroniro and Soulis 2003; Soulis et al. 2005). The resulting model is termed WATCLASS (Soulis et al. 2000) and uses CLASS for vertical processes and the lateral algorithms from WATFLOOD. Offline simulations using WATFLOOD, CLASS, and WATCLASS, which are driven by measured or forecast fields of near-surface data, represent Level 0, Level 1, and Level 2 efforts within the MAGS modeling strategy. The full coupling between CGCM/CRCM (Canadian Regional Climate Model) and land surface hydrology model (CLASS/WATCLASS) represents Level 3. During the course of MAGS, Levels 0, 1, and 2 were achieved and the framework for the progressing with Level 3 was established.

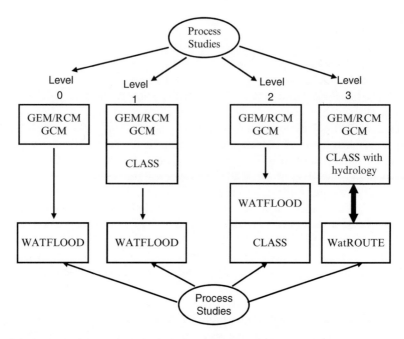

Fig. 9.1 Atmospheric-hydrological coupled modeling strategy for MAGS

Results from small research basins showed marked improvements in the simulation of streamflow when using WATCLASS with its enhanced hydrology in comparison to the original version of CLASS 2.6, with Nash–Sutcliffe efficiency (NSE) increased from <0 to 0.6 ~ 0.8 (Soulis and Seglenieks 2007). However, the preliminary application of WATFLOOD and WATCLASS in the Mackenzie basin indicated that the traditional hydrological model WATFLOOD was better able to simulate hydrographs than WATCLASS both in the timing and volume of the peak (Fig. 9.2) (Snelgrove et al. 2005). The worse performance of WATCLASS in Fig. 9.2b (especially for the Athabasca River station) suggests that the increase of model complexity degraded model capabilities with respect to the timing and magnitude of simulated streamflow. Snelgrove et al. (2005) suggested that future work was required to bridge gaps in the current theory in order to improve model simulations. These include implementation of theories for infiltration and liquid moisture flow through frozen ground and examination of snowmelt processes within WATCLASS.

Fassnacht and Soulis (2002) examined the effects of variations in snow representations in WATCLASS on water and heat fluxes and found that the inclusion of four enhancements to the CLASS snow process representations (occurrence of mixed precipitation, fresh snow density, maximum snowpack density, and canopy snowfall interception) strongly effected predicted heat fluxes, but had little impact on streamflow simulations. Davison et al. (2006) attempted to improve the simulation of the

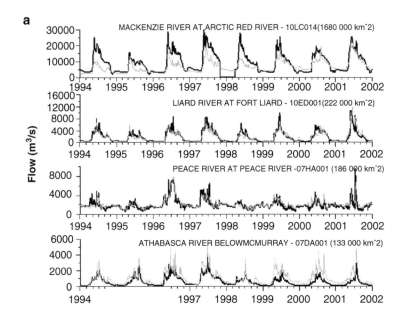

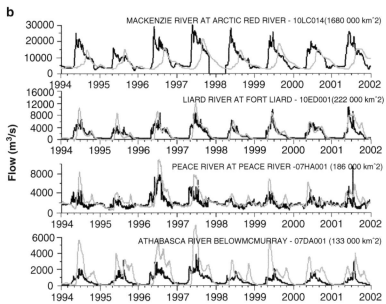

Fig. 9.2 Monthly streamflow simulations over different Mackenzie sub-basins with the WATFLOOD (**a**) and WATCLASS (**b**). *Black* is measured and *gray* is simulated (After Snelgrove et al. 2005)

spatial variability of snowmelt in WATCLASS by including wind-swept tundra and drift classes based on topography rather than the traditionally used vegetation land classes.

A second 5-year stage of MAGS, MAGS2 (which began in 2001), was aimed toward developing a fully coupled atmosphere/land surface/hydrologic modeling system (Level 3) based on the three primary models (CRCM, CLASS, and WATFLOOD). CLASS had previously been coupled with CRCM, which was the primary regional climate model used in MAGS. The coupled model (CRCM/CLASS) along with the high-resolution geophysical database was used to examine the mesoscale atmospheric circulations during the snowmelt period over the Mackenzie basin (MacKay et al. 2003a). Further evaluation of the coupled model was conducted by comparing the modeled surface water balance with observations during the water years 1998–1999 (MacKay et al. 2003b). The results demonstrated a plausible simulation of precipitation, temperature, and snow cover through the Mackenzie. However, streamflow was poorly simulated when the output was used to drive two off-line hydrologic models. MacKay et al. (2007) summarize the development and application of the version of CRCM used within MAGS (denoted CRCM-MAGS), which is essentially a developmental version of the CRCM. The emphasis of regional climate modeling in MAGS was largely on land surface processes and the interaction between the land surface and the atmosphere. The impact of lakes on regional climate is currently an active area of research within MAGS (Rouse et al. 2007). A one-dimensional thermal lake model (DYRESM) was being embedded within CLASS and was being tested over the Mackenzie River basin at the completion of MAGS2.

9.3.2   BALTEX

The Baltic Sea Experiment (BALTEX) is a European contribution to the investigation of the energy and water cycle over a large drainage system (Baltic Sea and related river basins) (Raschke et al. 2001). Developing coupled atmospheric, oceanographic, and hydrological models is a primary objective of BALTEX. The entire Baltic Sea drainage basin ranges from the subarctic climate in northern Finland (69°N) to the temperate and more continental climate in southern Poland (49°N). Although BALTEX, like MAGS, is not an ACSYS project, there are many interactions and common interests in modeling high-latitude hydrologic processes between these two projects (e.g., the PILPS 2(e) study area is part of BALTEX domain.).

The land surface scheme SEWAB (Surface Energy and Water Balance) (Mengelkamp et al. 1999), which was developed for use in BALTEX, is designed both for use in coupled land–atmospheric models and to be run off-line. It calculates the vertical energy and water fluxes between the land surface and the atmosphere and within the soil column. SEWAB's representation of runoff generation has been improved following its participation in PILPS Phases 2(a) and 2(c) (Chen et al. 1997; Wood 1998). A variable infiltration capacity approach was included for

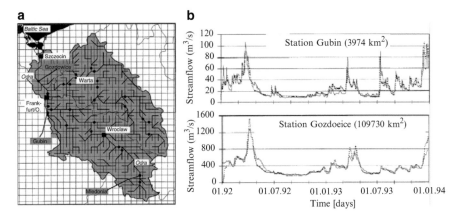

Fig. 9.3 (**a**) The Odra drainage basin overlaid by an equidistant grid of 18-km mesh size. The dots indicate locations of gauging stations; (**b**) daily observed (*solid*) and simulated (*dotted*) streamflow at gauging stations Gubin (river Nysa Lucycka) and Gozdowice (river Odra) for the time period 1992–1993 (From Mengelkamp et al. 2001)

surface runoff generation (Warrach et al. 1999), and a ponding at the surface was added to account for immediate streamflow response to precipitation events (Mengelkamp et al. 2001). The characteristic of seasonal snow cover and soil frost in the Baltic Sea drainage basin requires the inclusion of winter processes in the land surface model applied to this region. Soil freeze–thaw and improved snow accumulation and ablation representations were incorporated within SEWAB by Warrach et al. (2001). SEWAB overestimated the amount and duration of snow cover over the Torne–Kalix River basins in the PILPS Phase 2(e) experiments, although it produced good streamflow simulations (Nijssen et al. 2003). As a first attempt to include horizontal water processes in an atmospheric land surface scheme for studies of the water and energy cycle in the climate system, SEWAB was used to simulate runoff and streamflow at a regional scale in the Odra drainage basin with a drainage area of 1.19×10^5 km^2 (Fig. 9.3) (Mengelkamp et al. 2001) by linking a horizontal routing scheme which describes the transport of locally generated runoff into river systems (Lohmann et al. 1996).

The HBV model is a distributed conceptual hydrological model which was originally developed for flood forecasting (Bergstrom 1995). Snow accumulation and snowmelt in HBV are normally modeled by a degree-day approach based on air temperature observations. Within the framework of BALTEX, the large-scale hydrological model HBV-Baltic has been developed and used to simulate runoff of the entire Baltic basin (1.6×10^6 km^2) (Fig. 9.4) (Bergstrom and Graham 1998; Graham 1999), to evaluate the hydrological components of atmospheric models (Graham and Bergstrom 2001), and to assess climate change effects on river flow to the Baltic Sea (Graham 2004). These results suggest that continental-scale water balance modeling for the Baltic basin can be solved with the HBV-Baltic conceptual hydrological model. However, the lack of energy balance parameter-

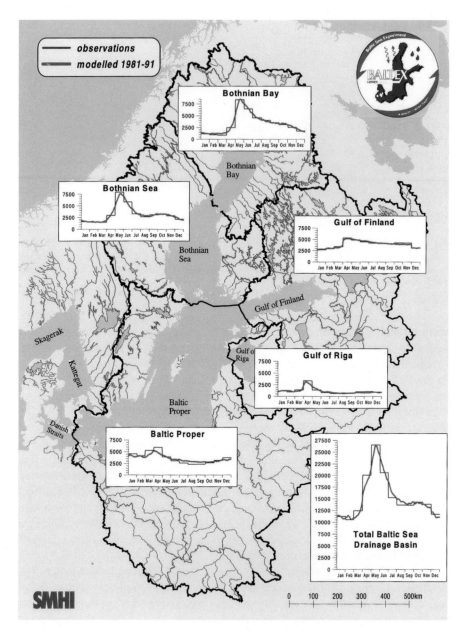

Fig. 9.4 Monthly averages (in $m^3 s^{-1}$) of freshwater flow into the major sub-basins of the Baltic Sea, calculated with the HBV model using meteorological input data. Note that major contributions are available during the melting season (Raschke et al. 2001)

izations within HBV is a major limitation, as the model is not appropriate for inclusion in coupled land–atmosphere models. In contrast, SEWAB is part of a coupled model system of land, atmosphere, and ocean developed in the context of BALTEX.

9.3.3   GAME-Siberia

The Lena River with a drainage area of about 2.43×10^6 km^2 is one of the largest rivers in the Arctic. Approximately 78–93% of the basin is underlain by permafrost (Zhang et al. 1999, 2000). The Lena River basin was chosen as a main field site within the GEWEX GAME-Siberia project. GAME-Siberia concentrates on observation and modeling of land surface processes, and regional analysis of energy and water cycle in permafrost region of eastern Siberia.

In GAME-Siberia, meteorological and hydrological observations were carried out at taiga and tundra areas within the Lena River basin during 1996–1998, followed by intensive observations during 2000 for various vegetation types (larch, pine, and grassland) (Ishii 2001; Ohta et al. 2001; Hamada et al. 2004). A one-dimensional land surface model was developed and improved to estimate water and energy fluxes in extremely cold regions based on the field data from GAME-Siberia (Yamazaki 2001; Yamazaki et al. 2004). Characteristics of snow cover and river runoff in a small watershed (5.5 km^2) in Arctic tundra near the mouth of the Lena River were studied by observation and a new land surface model simulation (Hirashima et al. 2004a, b). Ma et al. (1998) proposed a one-dimensional numerical model to estimate the heat transfer in permafrost regions by using the meteorological data obtained from observing sites in Lena River basin. These model implementations help to understand and explain observed land surface processes and seasonal flux variations; however, all these models were basically only conducted at point or small-basin scales.

A macroscale hydrological analysis of the Lena River basin was carried out to simulate snowmelt, evapotranspiration, runoff generation, and river flow by using a combined model which is composed of a soil–vegetation–atmosphere transfer model, runoff model, and river routing model (Ma et al. 2000). Two kinds of grid sizes were prepared for the combined model, in which a $1° \times 1°$ grid was used for the SVAT model and runoff model and a $0.1° \times 0.1°$ grid for the river routing model. Although this analysis was limited to only 1 year, it was one of the few macroscale modeling studies from GAME-Siberia.

9.4   NATO Arctic Freshwater Balance Workshop

A NATO Advanced Research Workshop (ARW) which focused specifically on the freshwater balance of the Arctic Ocean was held at Talinn, Estonia, in 1998. As a contribution to this workshop (results of which were published in a 2000 NATO Science Series volume, see Lewis (2000) for an overview), Bowling et al. (2000) made the first attempt to model in a comprehensive manner the land surface energy and water balance of the pan-Arctic drainage basin using the VIC (Variable Infiltration Capacity) LSS. VIC, as described by Liang et al. (1994, 1996), is a grid-based land surface scheme which parameterizes the dominant hydrometeorological processes taking place at the land surface–atmosphere interface. The model was designed both for inclusion in general circulation models (GCMs) as a land–atmosphere transfer scheme and for use as a stand-alone macroscale hydrologic model. The VIC model incorporates a two-layer energy balance snow model (Storck and Lettenmaier 1999; Cherkauer and Lettenmaier 1999) first used in the NATO study, which represents snow accumulation and ablation in both a ground pack and in an overlying forest canopy (if any).

In their contribution to the NATO Arctic Freshwater workshop, Bowling et al. (2000) report applications of the VIC to the Mackenzie and Ob River basins at 2° spatial resolution and daily temporal resolution to examine the space–time structure of runoff, evaporation, soil moisture, and snow water equivalent. The work suggested that the VIC macroscale hydrologic model was able to replicate the timing and variability of discharge to the Arctic Ocean from large northern rivers and also suggested the importance of the physical processes (e.g., sublimation from blowing snow, surface storage in lakes and wetlands, and infiltration limitation by frozen soils) which were not represented by the model generation at that time. These finding motivated many of the VIC model improvements that were tested in PILPS 2(e), as described in the next section.

9.5   PILPS 2(e)

A small working group was established by ACSYS in August 1998 for the purposes of planning an Arctic hydrology model intercomparison project (WCRP 1999). The resulting project, PILPS 2(e), was jointly sponsored by ACSYS and GEWEX. It was intended to evaluate the performance of land surface models in high latitudes (Bowling et al. 2003). The project tested 21 land surface models with respect to their ability to represent snow accumulation and ablation, soil freeze/thaw and permafrost, and runoff generation. PILPS 2(e) contributed to the goals of both GEWEX and ACSYS by providing a test bed for model modifications and improvements in representing high-latitude land processes and by providing information about the accuracy with which land schemes can be used to estimate runoff from ungauged areas draining to the Arctic Ocean.

The PILPS 2(e) experiment used BALTEX data from the 58,000-km^2 Torne–Kalix River system (65.5–69.5°N) in northern Scandinavia to evaluate the performance of LSSs in an off-line setting, meaning that prescribed atmospheric conditions were used to drive the LSSs and that there was no mechanism for representation of feedbacks from the land surface to the atmosphere. In the PILPS 2(e) experiment, two subcatchments of the Torne–Kalix system were selected for free calibration. Parameters were then transferred to two validation catchments and to the basin as a whole, using methods of the participants' choice. The purpose of the calibration experiment was to test the extent to which calibration could improve the performance of the LSSs and the extent to which parameter transfer from such calibrated catchments can improve the estimation of runoff from similar, ungauged basins. Results of the experiment indicated that those models that participated in a calibration experiment in which calibration results were transferred from small catchments to the basin at large had a smaller bias in their daily streamflow simulations than the group of models that did not (Bowling et al. 2003).

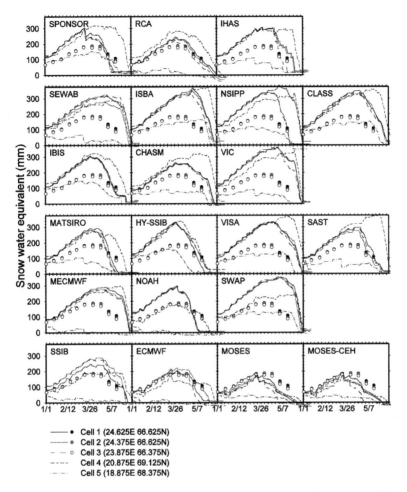

Fig. 9.5 Observed (*dots*) and simulated (*lines*) snow water equivalent for five locations during the first part of 1995 (no observations were available for cells 4 and 5) (From Nijssen et al. 2003)

Given the sparseness of observation sites in the Torne–Kalix River basin, the comparison of simulated results with observations largely focused on snow (extent, accumulation, and ablation) and streamflow (Nijssen et al. 2003). The results showed that in general, all 21 models captured the broad dynamics of snowmelt and runoff, but as shown in Figs. 9.5 and 9.6, there were large differences in snow accumulation, ablation, and streamflow. The greatest among-model differences in energy and moisture fluxes occurred during the spring snowmelt period, reflecting different model parameterizations of snow processes (e.g., fractional snow coverage, albedo, and land surface roughness). Nijssen et al. (2003) indicated that one important source of among-model differences in water and energy balances was the large differences in simulated snow sublimation, many of which resulted from differences in snow surface roughness parameterizations. The among-model differences in the phase and magnitude of the spring runoff were primarily attributed to differences in

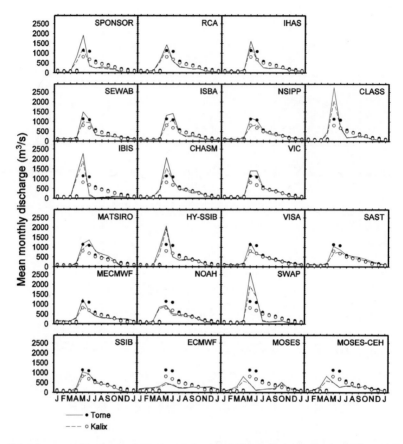

Fig. 9.6 Mean monthly observed (*dots*) and simulated (*lines*) discharge for the Kalix and Torne river basins (From Nijssen et al. 2003)

snow accumulation and melt as well as to differences in meltwater partitioning between runoff and infiltration. In a series of experiments in the Torne–Kalix River basin using the CHASM model, which can be operated with different complexities of the land surface energy balance, Pitman et al. (2003) demonstrated that the complexity of the representation of the land surface energy balance could not explain the difference between the PILPS 2(e) model results and instead attributed differences to variations in the hydrologic formulation incorporated in the participating models.

Results of the PILPS 2(e) experiment led to improvements in snow, frozen soil, and runoff process formulations in some of the participating models. These improvements were reported in part in a series of papers in a 2003 special issue of *Global and Planetary Change*. For instance, the snow–atmosphere–soil transfer (SAST) land surface scheme had difficulty in the PILPS 2(e) experiments in accurately simulating the pattern and amount of spring snowmelt runoff. As a result, Jin et al. (2003) updated the subsurface runoff parameterization of the SAST and obtained better hydrograph prediction. Habets et al. (2003) describe the two-layer frozen soil scheme and the three-layer explicit snow model used by ISBA in PILPS 2(e), as well as a new soil diffusion scheme of ISBA. Tests of ISBA using the PILPS 2(e) data showed that the new diffusion soil module performed well in terms of both hydrology and soil temperature, indicating a step forward in parameterizing frozen soils in ISBA.

Bowling et al. (2003) and Nijssen et al. (2003) concluded that the original ECMWF land scheme greatly overestimated sublimation, and underestimated spring snow accumulation, and hence snowmelt runoff. Van den Hurk and Viterbo (2003) reported improvements to the ECMWF land surface scheme in both timing and amount of runoff in the Torne–Kalix River basin that resulted from updates to the surface runoff scheme and the reduction of surface roughness. The Met Office Surface Exchange Scheme (MOSES) produced excessive sublimation and too early and too small peak runoff in the PILPS 2(e) simulations. Motivated by the PILPS 2(e) results, Essery and Clark (2003) improved the MOSES representations of snow processes in vegetation canopies and snow hydrology in the MOSES 2 land surface model. The modifications improved runoff simulations for two subcatchments used in the PILPS 2(e) experiment by reducing the amount of snow lost through sublimation and delaying the runoff of melt water.

PILPS 2(e) was used as a test site for evaluation of various changes in the VIC model. These include testing of a frozen soil/permafrost algorithm (Cherkauer and Lettenmaier 1999; Cherkauer et al. 2003) that represents the effects of frozen soils on the surface energy balance and runoff generation, a lake and wetland model (Bowling 2002) that represents the effects of lakes and wetlands on surface moisture and energy fluxes, and a parameterization of the effects of spatial variability in soil freeze–thaw state and snow distribution on moisture and energy fluxes (Cherkauer and Lettenmaier 2003). Although the flow attenuation by lakes and peat bogs in the Torne–Kalix basin appeared to be significant, predicted hydrographs using VIC with the lakes and wetlands algorithm did not differ much from those without lakes (Bowling et al. 2003). Bowling et al. (2003) also noted the apparent importance of sublimation during blowing-snow events in the Torne–Kalix basin, which was not represented by the version of VIC used in PILPS Phase 2(e). This finding motivated development of an algorithm that parameterizes the topographically induced sub-grid variability in wind speed, snow transport, and blowing-snow sublimation. The algorithm was designed to work within the structure of the existing VIC mass and energy balance snow model (Bowling et al. 2004). Subsequent testing on the Alaska North Slope demonstrated that the VIC macroscale algorithm was consistent with estimates from two different high-resolution blowing-snow algorithms (Liston and Sturm 1998; Essery et al. 1999) and with limited observations at Barrow, Alaska.

Most of those new cold updates to the LSSs will require further evaluation and validation with more observations at an appropriate spatial scale. The impact of frozen soil and blowing-snow schemes on large-scale simulation of water and energy terms (e.g., soil moisture, runoff, and evaporation) is still not clear. Models with frozen soil mostly assume that the presence of frozen water limits infiltration into the soil and changes the soil thermal fluxes through the dependence of soil thermal properties on soil water and ice content. Therefore, in Cherkauer and Lettenmaier (1999, 2003), for instance, the VIC model tended to produce higher spring peak flows and lower winter baseflow when the model was run with the frozen soil algorithm; however, the studies did not show apparent improvements (based on Nash efficiency) in streamflow simulations at basin scales. Results from PILPS 2(d), a previous cold regions experiment conducted at a grassland site in Valdai, Russia, indicated that models with an explicit frozen soil scheme produced better soil temperature simulations than those without a frozen soil scheme (Luo et al. 2003). However, the difference in soil moisture simulations from models with or without frozen soil physics was not clear in that experiment. An earlier study by Pitman et al. (1999) found that including a representation of soil ice in land surface models degraded runoff simulations in the Mackenzie River basin. Field studies in MAGS demonstrated that the infiltration is unlimited for organic materials even in permafrost areas (Woo and Marsh, 2005).

Overall, the PILPS 2(e) experiment offered valuable opportunities for the land surface modeling community to identify problems in representations of snow cover, frozen soil, surface runoff, and other physical processes in cold regions. It is worth noting that the 21 participating land surface schemes included LSSs from many well-known coupled models used for numerical weather and climate prediction. Furthermore, PILPS 2(e) was the first experiment to adhere to the ALMA (Assistance for Land Modeling Activities) data input and output protocols (based on NetCDF), one result of which is that the PILPS 2(e) data are available for future model testing. We are aware of one case in particular where the PILPS 2(e) data were used after the experiment to test three topography-based runoff schemes (Niu and Yang 2003), which did not exit at the time of the original experiment.

9.6   Freshwater Inflow to the Arctic Ocean

The goal of ACSYS hydrological programme was to determine the space–time variability of the Arctic hydrological cycle and the fluxes of freshwater to the Arctic Ocean. Numerous estimates have been made of the freshwater inflow to the Arctic Ocean based on available observed streamflow data (Prowse and Flegg 2000; Shiklomanov et al. 2000; Grabs et al. 2000; Lammers et al. 2001; Dai and Trenberth 2002). About 30% of the total drainage area to the Arctic is ungauged. Most of this area is along the Arctic coast downstream of the farthest gauging station in the major rivers and in the Canadian Archipelago. Most estimates of total discharge to the Arctic from ungauged areas are based on the assumption that runoff per unit area is equivalent for gauged and ungauged areas within each basin. Application of hydrologic models offers one option for providing consistent estimation of the discharge of both gauged and ungauged areas.

Although ACSYS ended in December 2003, the goal of its hydrological programme continues to motivate macroscale hydrologic model developments and applications over the Arctic regions and specifically to the problem of estimating total freshwater discharge. Recently, the VIC model with the cold land process updates described in Sect. 9.5 was applied to the entire pan-Arctic domain at a 100-km EASE-Grid system to evaluate the representation of Arctic hydrologic processes in the model and to provide a consistent baseline hydroclimatology for the region (Su et al. 2005). The model simulations of key hydrologic processes for the periods of 1979–1999 were evaluated using observed streamflow, snow cover extent, dates of lake freeze-up and break-up, and permafrost active-layer thickness. The pan-Arctic drainage basin was partitioned into 12 regions for model calibration and parameter transfer according to geographical definitions and hydroclimatology. Twenty-seven individual and sub-basins within different regions were chosen for model calibration and validation. Results indicated that the VIC model was able to reproduce the seasonal and interannual variations in streamflow quite well (for 19 basins out of 27 monthly Nash efficiency exceeded 0.75, and for 13 it exceeds 0.8). However, almost all the baseflow from January to April was underestimated, which was mostly due to the nature of the frozen soil algorithm in the VIC model. Although the primary purpose of the paper was to evaluate the ability of the model to reproduce hydrologic features of the major Arctic river basins, an evaluation was made of various estimates of freshwater discharge to the Arctic. In particular, the discharge simulated with the VIC model was used to estimate the total river inflow to the Arctic Ocean based on the farthest downstream outlets with the outflows to the Arctic Ocean in the 100-km river networks (Fig. 9.7). A 21-year average river inflow (1979–1999) to the Arctic Ocean from the AORB (Arctic Ocean River Basin) illustrated in Prowse and Flegg (2000) was estimated with the VIC model as 3,354 km^3/year, and 3,596 km^3/year with the inclusion of the Canadian Archipelago. The relationship between the inflow volume and contributing area resulting from various data sources and VIC simulations (Table 9.1) indicated that the VIC model was comparable to the previous estimates derived from the observed data (Fig. 9.8). More striking, however, was that a wide range of Arctic discharge estimates, when adjusted for differences in drainage areas, were shown to be closely equivalent – i.e., most of the differences in reported estimates of Arctic freshwater discharge can be attributed to differences in drainage areas used in the individual studies.

9.7 Conclusions: What Did the ACSYS Achieve?

During its 10-year history, ACSYS and related GEWEX projects motivated a number of advances in high-latitude hydrological modeling, particularly at large scales. The ACSYS scientific strategy for hydrology, which included adaptation of macroscale hydrological models developed in the framework of GEWEX to Arctic (high-latitude) climate conditions and development of physical (conceptual) or parametric mesoscale hydrologic models for selected river catchments within the Arctic region, was implemented more or less as envisaged in the ACSYS Implementation Plan (WCRP 1994). The following achievements can be attributed at least in part to ACSYS:

1. Improvement of land surface models in terms of their ability to represent high-latitude hydrologic processes, including snow accumulation and ablation, soil freeze/thaw and permafrost, and runoff generation (Specific examples include the ISBA,

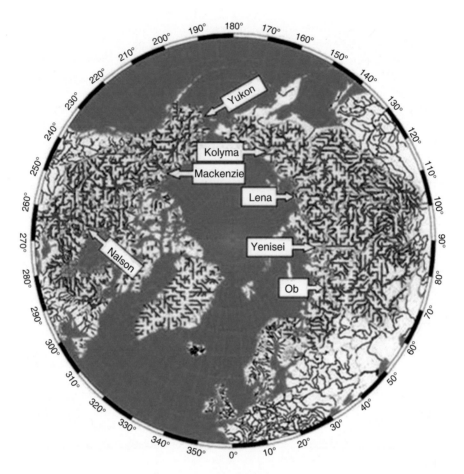

Fig. 9.7 Digital river networks for the pan-Arctic drainage basins at the 100-km resolution, showing the watershed boundaries of the Kolyma, Lena, Yukon, Yenisei, Ob, Mackenzie, and Nelson. Dots represent 200 the farthest downstream outlets with the outflows to the Arctic Ocean based on the river networks (From Su et al. 2005)

ECMWF, CLASS, and VIC LSMs, but there are almost certainly other model improvements that are less well documented.). These model improvements were motivated primarily by the PILPS 2(e) experiment in Torne–Kalix River basin.

2. Intensive field measurement under the GEWEX MAGS and GAME-Siberia projects promoted the development of improved process algorithms for snow accumulation, redistribution, and ablation, and water infiltration into frozen soil, and the development of one-dimensional land surface models for cold regions.

3. The VIC model and the macroscale hydrological models developed under the MAGS, BALTEX, and GAME-Siberia were used to simulate the surface water and energy balance of high-latitude river basins and (subsequent to ACSYS) to estimate the freshwater balance of the pan-Arctic land domain.

Table 9.1 Estimates of annual continental freshwater into the Arctic Ocean (After Su et al. 2005)

Basin definition	Contribution area (×1,000 km²)	Volume (km³/year)	Periods
"Arctic Ocean River Basin" in Prowse and Flegg (2000)[a]	11,045/15,504	2,338/3,299	1975–1984
"All Arctic Regions" in Shiklomanov et al. (2000)	18,875	4,300	1921–1996
"Arctic Ocean Basin" in Shiklomanov et al. (2000)	23,732	5,250	1921–1996
"Arctic Climate System" in Grabs et al. (2000)[a]	12,868/18,147	2,603/3,671	
AORB – Northern Greenland + Arctic Archipelago in Lammers et al. (2001)	16,192	3,302	1960–1989
The largest Arctic Rivers in Dai and Trenberth (2002)	16,850	3,658	
AORB – without Arctic Archipelago, VIC1	15,017	3,354	1979–1999
AORB – with Arctic Archipelago, VIC2	16,397	3,596	1979–1999

[a]The first area is the gauged area; the second area is the total contributing area in the definition; the second volume is the extrapolation over the total area

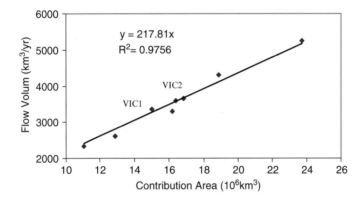

Fig. 9.8 Basin area annual flow volume relationship for different estimations in Table 9.1 (From Su et al. 2005)

4. Riverine freshwater fluxes to the Arctic Ocean have been better estimated (and the existing estimates shown to lie within reasonably tight error bars) through use of macroscale hydrological models.

What remains to be done in the post-ACSYS era? Despite major advances in high-latitude hydrological modeling during the ACSYS era, there remain important problems in parameterization of cold land hydrological processes within climate and hydrology models. Many of the key issues are identified in the Science Plan of the WCRP Climate and Cryosphere (CliC) project, which is the successor of ACSYS (Allison et al. 2001). These include:

1. The role of frozen soil moisture and blowing-snow parameterizations in the large-scale simulation of runoff, temperature, and evaporation are not completely clear.
2. Existing wetlands and lake models in land surface models need to be further improved and validated.
3. Many results from process investigations of snow and frost-related hydrological processes remain to be incorporated into large-scale hydrological models.
4. Continued development of hydrological models and linkages between atmospheric and hydrological models are needed in scientific studies of the interactions between climate, snow, and frost hydrology.

Acknowledgments This research was supported by NSF Grants 0230372 and 0629491 to the University of Washington.

References

Aagaard K, Carmack EC (1989) The role of sea ice and other fresh waters in the Arctic circulation. J Geophys Res 94(C10):14485–14498

Allison L, Barry RG, Goodison BE (2001) Climate and Cryosphere (CliC) project science and co-ordination plan, Version1. http://clic.npolar.no/introduction/science_plan.pdf. Accessed 24 Mar 2010

Armstrong RL, Brodzik MJ (2001) Recent northern Hemisphere snow extent: a comparison of data derived from visible and microwave sensors. Geophys Res Lett 28:3673–3676

Barry RG, Serreze MC (2000) Atmospheric components of the arctic ocean freshwater balance and their interannual variability. In: Levis EL et al (eds) The freshwater budget of the Arctic ocean. Springer, New York

Beljaars ACM, Viterbo P, Miller MJ, Betts AK, Ball JH (1993) A new surface boundary layer formulation at ECMWF and experimental continental precipitation forecasts. GEWEX News 3(3):15–18, 5–8

Bergstrom S (1995) The HBV model. In: Singh VP (ed) Computer models of watershed hydrology. Water Resources Publications, Highlands Ranch, Co

Bergstrom S, Graham LP (1998) On the scale problem in hydrological modeling. J Hydrol 211:253–265

Bowling LC (2002) Estimating the freshwater budget of high-latitude land areas. PhD dissertation. University of Washington, Seattle

Bowling LC, Lettenmaier DP, Matheussen BV (2000) Hydroclimatology of the Arctic drainage basin. In: Levis EL et al (eds) The freshwater budget of the Arctic ocean. Springer, New York

Bowling LC et al (2003) Simulation of high-latitude hydrological processes in the Torne-Kalix basin: PILPS phase 2(e): 1. Experiment description and summary intercomparisons. Glob Planet Change 38:1–30

Bowling LC, Pomeroy JW, Lettenmaier DP (2004) Parameterization of blowing snow sublimation in a macroscale hydrology model. J Hydrometeorol 5(5):745–762

Broecker WS (1997) Thermohaline circulation, the Achilles heel of our climate system: will man-made CO_2 upset the current balance? Science 28:1582–1588

Brown RD (2000) Northern Hemisphere snow cover variability and change, 1915–1997. J Clim 13:2339–2355

Chen T et al (1997) Cabauw experimental results from the Project for Intercomparison of Land-surface Parameterization Schemes (PILPS). J Clim 10:1194–1215

Cherkauer KA, Lettenmaier DP (1999) Hydrologic effects of frozen soils in the upper Mississippi River basin. J Geophys Res 104(D16):19599–19610

Cherkauer KA, Lettenmaier DP (2003) Simulation of spatial variability in snow and frozen soil field. J Geophys Res. doi:10.1029/2003JD003575

Cherkauer KA, Bowling LC, Lettenmaier DP (2003) Variable infiltration capacity cold land process model updates. Glob Planet Change 38:151–159

Dai A, Trenberth KE (2002) Estimates of freshwater discharge from continents: latitudinal and seasonal variations. J Hydrometeorol 3:660–3687

Davison B, Pohl S, Dornes P, Marsh P, Pietroniro A, MacKay M (2006) Characterizing snowmelt variability in land-surface-hydrologic model. Atmosphere-Ocean 44:271–287

Dickinson RE, Henderson-Sellers A (1988) Modelling tropical deforestation: a study of GCM land-surface parameterizations. Q J R Meteorol Soc 114(B):439–462

Essery R, Clark DB (2003) Developments in the MOSES 2 land-surface model for the PILPS 2e. Glob Planet Change 38:161–164

Essery R, Li L, Pomeroy J (1999) A distributed model of blowing snow over complex terrain. Hydrol Process 13:2423–2438

Fassnacht SR, Soulis ED (2002) Implications during transitional periods of improvements to the snow processes in the land surface scheme-hydrological model WATCLASS. Atmosphere-Ocean 40:389–403

Frauenfeld OW, Zhang T, Barry RG (2004) Interdecadal changes in seasonal freeze and thaw depths in Russia. J Geophys Res. doi:10.1029/2003JD004245

Gedney N, Cox PM, Douville H, Polcher J, Valdes PJ (2000) Characterizing GCM land surface schemes to understand their responses to climate change. J Clim 13:3066–3079

Grabs WE, Portman F, de Couet T (2000) Discharge observation networks in Arctic regions: computation of the river runoff into the Arctic ocean, its seasonality and variability. In: Levis EL et al (eds) The freshwater budget of the Arctic Ocean. Springer, New York

Graham LP (1999) Modeling runoff to the Baltic Sea. Ambio 28:328–334

Graham LP (2004) Climate change effects on river flow to the Baltic Sea. Ambio 33(4–5):235–241

Graham LP, Bergstrom S (2001) Water balance modelling in the Baltic Sea drainage basin-analysis of meteorological and hydrological approaches. Meteorol Atmos Phys 77:45–60

Habets F, Boone A, Noilhan J (2003) Simulation of a Scandinavian basin using the diffusion transfer version of ISBA. Glob Planet Change 38:137–149

Hamada S, Ohta T, Hiyama T, Kuwada T, Takahashi A, Maximov TC (2004) Hydrometeorological behavior of pine and larch forests in eastern Siberia. Hydrol Process 18:23–39

Hedstrom NR, Pomeroy JW (1998) Measurements and modelling of snow interception in the boreal forest. Hydrol Process 12:1611–1625

Henderson-Sellers A (1996) Soil moisture: a critical focus for global change studies. Glob Planet Change 13:3–9

Henderson-Sellers A, Yang Z-L, Dickinson RE (1993) The project for intercomparison of land surface parameterization schemes. Bull Am Meteorol Soc 74:1335–1350

Henderson-Sellers A, Pitman AJ, Love PK, Irannejad P, Chen TH (1995) The Project for Intercomparison of Land-surface Parameterization Schemes (PILPS): phase 2 and 3. Bull Amer Meteorol Soc 76:489–504

Hirashima H, Ohata T, Kodama Y Yabuki H, Sato N, Georgiadi A (2004a) Nonuniform distribution of tundra snow cover in Eastern Siberia. J Hydrometeorol 5(3):373–389

Hirashima H, Ohata T, Kodama Y, Yabuki H (2004b) Estimation of annual water balance in Siberian tundra region using a new land surface model. Northern Res Basins Water Balance 290:41–49

Ishii Y (2001) The outline of the field observation at the right bank of the Lena River. Activity report of GAME-Siberia 2000. GAME Publication 26:83–86

Jin J, Gao X, Sorooshian S (2003) Impacts of model calibration on high-latitude land-surface processes: PILPS 2(e) calibration/validation experiments. Glob Planet Change 38:93–99

Karcher M, Gerdes R, Kauker F, Koberle C, Yashayaev I (2005) Arctic ocean change heralds north Atlantic freshening. Geophys Res Lett. doi:10.1029/2005GL023861

Kouwen N, Soulis ED, Pietroniro A, Donald J, Harrington RA (1993) Grouping response units for distributed hydrologic modeling. J Water Res Manage Plan 119:289–305

Lammers RB, Shiklomanov AI, Vorosmarty CJ, Fekete BM, Peterson BJ (2001) Assessment of contemporary Arctic river runoff based on observational discharge records. J Geophys Res 106(D4):3321–3334

Lewis EL (2000) The freshwater budget of the Arctic Ocean – introduction. In: Levis EL et al (eds) The freshwater budget of the Arctic Ocean. Springer, New York

Liang X, Lettenmaier DP, Wood EF, Burges SJ (1994) A simple hydrologically based model of land surface water and energy fluxes for general circulation models. J Geophys Res 105(D17):14415–14428

Liang X, Wood EF, Lettenmaier DP (1996) Surface soil moisture parameterization of the VIC-2L model: evaluation and modifications. Glob Planet Change 13:195–206

Liston G, Sturm M (1998) A snow-transport model for complex terrain. J Glaciol 44:498–516

Lohmann D, Nolte-Holube R, Raschke E (1996) A large scale horizontal routing model to be coupled to land surface parameterization schemes. Tellus 48A:708–721

Luo L et al (2003) Effects of frozen soil on soil temperature, spring infiltration, and runoff: results from the POLPS 2 (d) experiment at Valdai, Russia. J Hydrometeorol 4:334–351

Ma X, Hiyama T, FukushimaY HT (1998) A numerical model of the heat transfer for permafrost regions. J Jpn Soc Hydrol Water Resour 11(4):346–359

Ma X, Fukushima Y, Hiyama T, Hashimoto T, Ohata T (2000) A macro-scale hydrological analysis of the Lena River basin. Hydro Process 14(3):639–651

MacKay MD, Szeto K, Verseghy D, Chan E, Bussieres N (2003a) Mesoscale circulations and surface energy balance during snowmelt in a regional climate model. Nordic Hydrol 34:91–106

MacKay MD, Seglenieks F, Verseghy D, Soulis ED, Sneigrove KR, Walker A, Szeto K (2003b) Modeling Mackenzie Basin surface water balance during CAGES with the Canadian Regional Climate Model. J Hydrometeorol 4:748–767

MacKay MD, Bartlett P, Chan E, Verseghy D, Soulis ED, Seglenieks FR (2007) The MAGS regional climate modeling system: CRCM-MAGS. In: Woo MK (ed) Cold regions atmospheric and hydrologic studies: the Mackenzie GEWEX experience, vol I, Atmospheric dynamics. Springer

Magnuson JJ et al (2000) Historical trends in lake and river ice cover in the Northern Hemisphere. Science 289:1743–1746

Manabe S (1969) Climate and ocean circulation: 1, the atmospheric circulation and the hydrology of the earth's surface. Mon Weather Rev 97:739–805

Mengelkamp H-T, Warrach K, Raschke E (1999) SEWAB-A parameterization of the surface energy and water balance for atmospheric and hydrologic models. Adv Water Resour 23:165–175

Mengelkamp H-T, Warrach K, Ruhe C, Raschke E (2001) Simulation of runoff and streamflow on local and regional scales. Meteorol Atmos Phys 76:107–117

Nijssen B et al (2003) Simulation of high latitude hydrological processes in the Torne-Kalix basin: PILPS phase 2(e): 2. Comparison of model results with observations. Glob Planet Change 38:31–53

Niu G-Y, Yang Z-L (2003) The versatile integrator of surface and atmosphere processes (VISA) Part II: evaluation of three topography-based runoff schemes. Glob Planet Change 38:191–208

Ohta T, Hiyama T, Tanaka H, Kuwada T, Maximov T, Ohata T, Fukushima Y (2001) Seasonal variation in the energy and water exchanges above and below a larch forest in eastern Siberia. Hydrol Process 15:1459–1476

Peterson BJ, Holmes RM, McClelland JW, Vorosmarty CJ, Lammers RB, Shiklomanov AI, Shiklomanov IA, Rahmstorf S (2002) Increasing river discharge to the Arctic Ocean. Science 298:2171–2173

Pietroniro A, Soulis ED (2003) A hydrology modeling framework for the Mackenzie GEWEX program. Hydrol Process 10:1245–1261

Pitman AJ, Slater AG, Desborough CE, Zhao M (1999) Uncertainty in the simulation due to the parameterization of frozen soil moisture using the Global Soil Wetness Project methodology. J Geophys Res 104(D14):16879–16888

Pitman AJ, Xia Y, Leplastrier M, Henderson-Sellers A (2003) The CHAmeleon Surface Model: description and use with the PILPS phase 2(e) forcing data. Glob Planet Change 38:121–135

Pomeroy JW, Li L (2000) Prairie and Arctic areal snow cover mass balance using a blowing snow model. J Geophys Res 105(D21):26619–26634

Pomeroy J, Marsh P, Gray DM (1997) Application of a distributed blowing snow model to the Arctic. Hydrol Process 11:1451–1464

Pomeroy JW, Gray DM, Shook KR, Toth B, Essery RLH, Pietroniro A, Hedstrom N (1998) An evaluation of snow accumulation and ablation processes for land surface modeling. Hydrol Process 12:2339–2367

Pomeroy JW, Gray DM, Hedstrom NR, Janowicz JR (2002) Prediction of seasonal snow accumulation in cold climate forests. Hydrol Process 9:213–228

Prowse TD, Flegg PO (2000) Arctic river flow: a review of contributing areas. In: Levis EL et al (eds) The freshwater budget of the Arctic Ocean. Springer, New York

Raschke et al (2001) A European contribution to investigate the energy and water cycle over a large drainage basin. Bull Am Meteorol Soc 82(11):2389–2413

Rouse WR (2000) Progress in hydrological research in the Mackenzie GEWEX study. Hydrol Process 14:1667–1685

Rouse WR et al (2003) Energy and water cycles in a high-latitude north-flowing river system: summary of results from the Mackenzie GEWEX study-phase 1. Bull Am Meteorol Soc 84:73–87

Rouse WR, Blanken PD, Duguay CR, Oswald CJ, Schertzer WM (2007) Climate-lake interactions. In: Woo MK (ed) Cold regions atmospheric and hydrologic studies: the Mackenzie GEWEX experience, vol II, Hydrologic processes. Springer

Shiklomanov IA, Shiklomanov AI, Lammers RB, Peterson BJ, Vorosmarty CJ (2000) The dynamics of river water inflow to the Arctic Ocean. In: Levis EL et al (eds) The freshwater budget of the Arctic Ocean. Springer, New York

Smith SL, Burgess MM, Riseborough D, Nixon FM (2005) Recent trends from Canadian permafrost thermal monitoring network sites. Permafr Periglacial Process 16:19–30

Snelgrove K, Soulis ED, Seglenieks F, Kouwen N, Pietroniro A (2005) The application of hydrological models in MAGS: lessons learned for PUB. In: Pomeroy J, Pietroniro A, Spence C (eds) Prediction in ungauged basins: approaches for Canada's cold regions. Canadian Water Resources Association

Soulis ED, Seglenieks FR (2007) The MAGS integrated modeling system. In: Woo MK (ed) Cold regions atmospheric and hydrologic studies: the Mackenzie GEWEX experience, vol II, Hydrologic processes. Springer

Soulis ED, Snelgrove KR, Kouwen N, Seglenieks FR, Verseghy DL (2000) Toward closing the vertical water balance in Canadian atmospheric models: coupling the land surface scheme CLASS with the distributed hydrologic model WATFLOOD. Atmosphere-Ocean 38:251–269

Soulis ED, Kouwen N, Pietroniro A, Seglenieks FR, Snelgrove KR, Pellerin P, Shaw DW, Martz LW (2005) A framework for hydrological modeling in MAGS. In: Pomeroy J, Pietroniro A (eds) Prediction in ungauged basins: approaches for Canada's cold regions. Canadian Water Resources Association

Stewart RE et al (1998) The Mackenzie GEWEX study: the water and energy cycles of a major north American river basin. Bull Am Meteorol Soc 79(12):2665–2683

Storck P, Lettenmaier DP (1999) Predicting the effect of a forest canopy on ground snow accumulation and ablation in maritime climates. In: Troendle C (ed) Proceedings of 67th western snow conference. Colo State University, Fort Collins

Su F, Adam JC, Bowling LC, Lettenmaier DP (2005) Streamflow simulations of the terrestrial Arctic domain. J Geophys Res. doi:10.1029/2004JD005518

Van den Hurk B, Viterbo P (2003) The Torne-Kalix PILPS 2(e) experiment as a test bed for modifications to the ECMWF land surface scheme. Glob Planet Change 38:165–173

Verseghy DL (1991) CLASS-A Canadian land surface scheme for GCMs, Part1: soil model. Int J Climatol 11:111–133

Verseghy DL, McFarland NA, Lazare M (1993) CLASS-A Canadian land surface scheme for GCMs, Part 2: vegetation model and coupled runs. Int J Climatol 13:347–370

Vorosmarty CJ, Hinzman LD, Peterson BJ, Bromwich DH, Hamilton LC, Morison J, Romanovsky VE, Sturm M, Webb RS (2001) The hydrologic cycle and its role in Arctic and global environmental change: a rationale and strategy for synthesis study. Fairbanks, Alaska: Arctic Research Consortium of the US, 84 pp

Wang XL, Cho HR (1997) Spatial-temporal structures of trend and oscillatory variabilities of precipitation over northern Eurasia. J Clim 10:2285–2298

Warrach K, Mengelkamp H-T, Raschke E (1999) Runoff parameterization in a SVAT scheme: sensitivity tests with the model SEWAB, regionalization in hydrology. IAHS Publ 254:123–130

Warrach K, Mengelkamp H-T, Raschke E (2001) Treatment of frozen soil and snow cover in the land surface model SEWAB. Theor Appl Climatol 69:23–37

WCRP (1994) Arctic Climate System Study (ACSYS) Initial implementation plan. WCRP-No. 85, WMO/TD-No.627

WCRP (1999) Report on the hydrology models intercomparison planning meeting, Koblenz, 27–29 Mar 1999, WCRP Informal Report No. 12/1999

Woo M-K, Marsh P (2005) Snow, frozen soils and permafrost hydrology in Canada, 1999–2002. Hydrol Process 19:215–229

Woo M-K, Carey SK, Martz LW (1998) Effects of seasonal frost and permafrost on the hydrology of subalpine slopes and drainage basins. In: Strong GS, Wilkinson YML (eds) Proceedings of the 4th scientific workshop for the Mackenzie GEWEX study. Montreal, Quebec

Woo M-K, Marsh P, Pomeroy JW (2000) Snow, frozen soil and permafrost hydrology in Canada, 1995–1998. Hydrol Process 14:1591–1611

Wood EF (1998) The project for intercomparison of land-surface parameterization schemes (PILPS) phase 2(c) Red-Arkansas basin experiment: 1. Experiment description and summary intercomparisons. Glob Planet Change 19:115–135

Xue Y, Shukla J (1993) The influence of landsurface properties on Sahel climate, Part I: desertification. J Climatol 6:2232–2245

Yamazaki T (2001) A one-dimensional land surface model adaptable to intensely cold regions and its application in eastern Siberia. J Meteorol Soc Japan 79:1107–1118

Yamazaki T, Yabuki H, Ishii Y, Ohta T, Ohata T (2004) Water and energy exchanges at forests and a grassland in Eastern Siberia evaluated using a one-dimensional land surface model. J Hydrometeorol 5(3):504–515

Yang D, Kane DL, Hinzman L, Zhang X, Zhang T, Ye H (2002) Siberian Lena River hydrologic regime and recent change. J Geophys Res. doi:10.1029/2002JD002542

Ye H, Ladochy S, Yang D, Zhang T, Zhang X, Ellison M (2004) The impact of climatic conditions on seasonal river discharges in Siberia. J Hydrometeorol 5(2):286–295

Zhang T, Barry RG, Knowles K, Heginbottom JA, Brown J (1999) Statistics and characteristics of permafrost and ground-ice distribution in the Northern Hemisphere. Polar Geogr 23(2):132–154

Zhang T, Heginbottom JA, Brown J (2000) Further statistics on the distribution of permafrost and ground ice in the Northern Hemisphere. Polar Geogr 24(2):126–131

Chapter 10
Sea-Ice–Ocean Modelling

Rüdiger Gerdes and Peter Lemke

Abstract The interaction of sea ice with the underlying ocean is an important component of the climate system. Dynamic and thermodynamic processes determine this interaction. Whereas thermodynamic sea-ice processes have been incorporated in climate models early on, the dynamics of the sea ice has been neglected for quite some time. Therefore, a major activity during the ACSYS project was the optimization of dynamical processes in sea-ice models and the development of coupled sea-ice–ocean general circulation models.

10.1 Introduction

High latitudes have received increasing attention recently because of significant changes in the atmosphere–sea-ice–ocean system. The surface air temperature in the Arctic has increased by 1.1°C over the last 50 years, i.e. more than twice as fast as the global air temperature, which increased by 0.7°C over 100 years. Since 1978, the annual average Arctic sea-ice extent has shrunk by 2.7% per decade, with larger decreases in summer of 7.4% per decade. Submarine-derived data for the Central Arctic indicate that the average sea-ice thickness in the Central Arctic has decreased by up to 1 m from 1987 to 1997. Warming is also observed in the Arctic Ocean at intermediate depths.

These amplified trends are in agreement with warming scenarios performed with coupled climate models, which indicate an amplified response in high latitudes to increased greenhouse gas concentrations (polar amplification). But details

R. Gerdes (✉)
Sea Ice Physics, Alfred Wegener Institute for Polar and Marine Research, Bussestr. 24,
27570 Bremerhaven
e-mail: Ruediger.Gerdes@awi.de

P. Lemke
Alfred Wegener Institute for Polar and Marine Research, Bremerhaven, Germany
e-mail: Peter.Lemke@awi.de

P. Lemke and H.-W. Jacobi (eds.), *Arctic Climate Change: The ACSYS Decade and Beyond,* Atmospheric and Oceanographic Sciences Library 43,
DOI 10.1007/978-94-007-2027-5_10, © Springer Science+Business Media B.V. 2012

of the complex interaction between atmosphere, sea ice and ocean are still only marginally known.

Changes in the sea-ice boundaries are among the most important characteristics of climate fluctuations in the polar regions. Understanding and predicting these changes not only are of interest in view of questions on regional and global climate but also have a practical meaning because the polar regions are increasingly used for commercial purposes.

Sea ice is involved in several climatically relevant feedback processes. A known positive feedback is the ice-albedo feedback: An initial decrease in air temperature causes the sea ice to extend. This enlarges the surface albedo, cools the atmosphere and produces additional sea ice. In the case of an initial increase in air temperature, the opposite happens: The sea ice shrinks, the ocean absorbs more solar radiation, and the atmosphere is heated, melting more sea ice.

A similar positive feedback process acts through evaporation: An increase in atmospheric water-vapour content due to shrinking pack ice results in an intensified absorption of infrared (IR) radiation, which causes warming of the air and a subsequent reduction of sea ice. In the case of an increased sea-ice area, however, the air gets dryer because evaporation is inhibited. This results in a lower absorption of IR radiation and thus cooling of air and further expansion of the sea ice.

The sea-ice boundary is determined by two main processes:

- The thermodynamics, which governs the freezing and melting
- The dynamics, which determines the sea-ice drift

The sea-ice drift is determined by the influence of wind stress, ocean currents, Coriolis force, the sea surface tilt and the deformational behaviour of the sea ice as a plastic material. Melting and freezing, however, are influenced by solar radiation and the heat exchange with the atmosphere and the ocean. All these factors depend on the state of the atmosphere–ice–ocean system, which, in itself, is altered by the moving sea-ice boundary. The difficulty in modelling this closed feedback loop of interactions is due not only to the complexity of the physical processes but also to the fact that the three interacting subsystems are characterized by quite different time scales: the atmosphere – several days; sea ice and oceanic surface layer – several months and the deep ocean – several decades to centuries. Even if all processes influencing the sea-ice variations were fully known, the different time scales would nevertheless cause considerable difficulties in numerical modelling.

The distribution of sea ice and open water in polar latitudes and the seasonally produced or melted sea ice have a considerable impact not only on atmospheric but also on oceanic circulation. The reason is that sea ice significantly alters the radiation balance and the exchange of heat and momentum between the ocean and the atmosphere because of its high albedo and its isolating role.

The changes in the fluxes at the sea surface have a strong impact on the dynamics of the oceanic surface layer. In regions where the oceanic stratification is weak, cooling and the release of salt when sea ice freezes (sea ice has a salt content of only 5 psu on average; the ocean water from which it originates, however, contains 35 psu) cause the formation of deep and bottom water, which influences the circulation in the deep ocean (see Rudels et al., Chap. 4).

10.2   Sea-Ice Models

Changes of sea-ice extent, concentration, thickness and drift are caused by dynamic
and thermodynamic processes. The thermodynamics, i.e. freezing and melting, is controlled by radiative and turbulent, latent and sensible heat fluxes. The dynamics, i.e. the
sea-ice drift and deformation, is determined by atmospheric and oceanic stresses (winds
and currents), Coriolis force, ocean surface tilt and internal forces in the sea ice.

In dynamic sea-ice models, it is generally assumed that the sea ice can be treated as
a two-dimensional continuum, which is characterized by five variables, i.e. the fields
of snow and ice thickness, compactness and the two components of the horizontal
velocity. The prognostic equations are derived from conservation equations for snow
and ice mass and compactness and from a momentum balance, including internal ice
stresses. Dynamic–thermodynamic sea-ice models consist of four major components:

1. A surface energy balance which uses the atmospheric forcing (wind, temperature, humidity, incoming solar radiation) and the heat conduction through snow
   and ice to compute the surface temperature. This energy balance contains
   furthermore parameterizations of incoming and outgoing long-wave radiation
   and sensible and latent heat fluxes.
2. A model which describes the heat conduction through the ice for given surface
   temperature and oceanic heat flux into the ice. The difference between the conductive and the oceanic heat fluxes determines the freezing (or bottom-melting)
   rate. The surface melting is calculated if the surface temperature is above $0°C$.
   Then, the surface temperature is set to zero, and the net energy for this temperature is used to melt snow and ice.
3. A momentum balance, including acceleration, Coriolis force, atmospheric and
   oceanic stress, sea surface tilt and internal ice stress, from which the ice velocity
   is calculated. The internal ice stress is determined from a particular rheology,
   which relates the internal stress to the deformation rate.
4. Balance equations for snow and ice mass and ice concentration. These balance
   equations use the sea-ice velocity and the freezing/melting rate determined from
   the heat conduction model together with the surface energy balance for ice and
   open water to calculate the new ice thickness and concentration in each grid cell.

Although there are some differences in the thermodynamic part (1. and 2.) of the
models, especially in the treatment of internal heat sources and sinks and in the
modification of the sea-ice compactness due to atmospheric and oceanic heat fluxes,
the basic difference between existing dynamic–thermodynamic sea-ice models lies
in the treatment of the internal ice stress.

10.2.1   Thermodynamics

The thermodynamic change of sea-ice thickness (h_1) is given by the difference of
the heat conduction through the ice (Q_c) and the oceanic heat flux into the ice (Q_o).
If $Q_o < Q_c$, then ice is formed, and the latent heat released flows through the ice and

snow layers and is used to close the energy balance at the surface. If $Q_o > Q_c$, then ice is melted at the bottom.

$$\frac{\partial h_1}{\partial t} = \frac{Q_c - Q_o}{\rho L},$$

where ρ is the density of seawater and L is the latent heat of fusion.

The most elaborate thermodynamic model to calculate Q_c originates from Maykut and Untersteiner (1971). In this model, the heat conduction equation is numerically integrated into a two-layer system (snow, ice). The great complexity of this model is due to the fact that density, specific heat and heat conductivity of sea ice are functions of temperature and salt content. These dependencies are caused by brine, trapped in salt pockets, which is in a phase balance with the surrounding ice. A three-layer approximation of this model is presented in Winton (2000).

In a simplified model, introduced by Semtner (1976), it is no longer necessary to solve the elaborate heat conduction equation because linear temperature profiles in snow and ice are assumed. This linear one-resp. two-layer approximation to the heat conduction equation is justified for longer time scales (seasonal cycle and longer). Due to its simple numerics, this thermodynamic sea-ice model is most often used in climate modelling.

Assuming a linear temperature profile in the ice, the heat flux Q_c is given by the heat conductivity of the ice k_I multiplied with the temperature difference between the surface and the bottom of the ice. A similar equation follows for the snow layer.

The temperature at the ice–ocean interface is given by the freezing point of sea water $T_f \approx -2°C$. When internal heat sources and sinks are neglected, the heat flux through the ice equals the heat flux through the snow. With this assumption, the heat flux through the two-layer snow/ice system can be determined as a function of the surface temperature T_s and of the snow and ice thicknesses.

$$Q_c = k_I \frac{T_f - T_s}{h^*},$$

where h^* is the effective thickness of the snow/ice layer determined as

$$h^* = h_1 + h_s \frac{k_I}{k_s},$$

with the snow thickness h_s and the snow conductivity k_s .

Once the heat conduction is given, the surface temperature is determined from an energy balance for the snow or ice surface, which consists of the balance of incoming minus reflected solar radiation $(1 - \alpha)R_s^\downarrow$, incoming minus outgoing long-wave radiation $\left(R_l^\downarrow - \varepsilon \sigma T_s^4\right)$, sensible heat flux Q_s , latent heat flux Q_l and the heat conduction through ice and snow Q_c .

$$(1-\alpha)R_s^{\downarrow} + R_l^{\downarrow} - \varepsilon\sigma T_s^4 + Q_s + Q_l + Q_c = \begin{cases} -\rho_I L \dfrac{\partial h_I}{\partial t} & T_s = 273 \\ 0 & T_s < 273 \end{cases},$$

where ε denotes the emissivity; σ, the Stefan–Boltzmann constant; ρ_I, the sea-ice density; L, the latent heat of fusion and α, the albedo.

10.2.2   Dynamics

Dynamic–thermodynamic sea-ice models can be distinguished by their treatment of the internal ice stress in the momentum balance. Although the sea-ice cover consists of floes of different sizes and thicknesses, it can be described successfully as a two-dimensional continuum. The momentum balance of such a system includes the acceleration term, Coriolis force, wind stress (τ_a), oceanic stress (τ_w), ocean surface tilt and internal forces in the ice (F), which are caused by the deformation.

$$m\frac{Du}{Dt} = -mfk \times u + \tau_a + \tau_w - mg\nabla H + F,$$

where m is the sea-ice mass; u, the sea-ice velocity; f, the Coriolis parameter; g, the gravitational acceleration and H, the ocean surface topography (Hibler 1979; Hibler and Flato 1992). The internal forces are described by the divergence of the stress tensor σ.

$$F = \nabla \cdot \sigma$$

The stress tensor is a function of the tensor of the deformation rate $\dot{\varepsilon}$ via a particular rheology, which also depends on sea-ice thickness h_I and concentration A.

$$\sigma = \sigma(\varepsilon, h_I, A)$$

The choice of $\dot{\varepsilon}$ instead of ε allows the Eulerian description of a geophysical fluid traditionally used in climate modelling. In order to represent sea-ice dynamics realistically, a flow rule is required, which describes the interaction of a system of floes of different sizes and thicknesses under the influence of external forces. Analogies to granular media and viscous fluids have been used. Furthermore, it is found that sea ice is broken or deformed only under shearing and/or compression above certain threshold values (yield strengths). These yield strengths are a function of sea-ice thickness and concentration. Statistical collisions of the different floes also cause an effective pressure in the system, which must be considered.

Observations show the following:

- Large stresses in the ice for convergent motion.
- No internal forces in case of divergent drift.

- Internal forces are nearly independent of the deformation rate (a characteristic of plastic materials).

For this reason, sea ice is best described as a two-dimensional, linear viscous, compressible fluid in the case of small deformation rates and as a plastic material in the case of large deformation rates. The viscous–plastic flow rule, describing a so-called Stokes fluid, is the most appropriate approach. This flow rule includes a shear viscosity η and a bulk viscosity ζ, and a static pressure P (Hibler and Flato 1992).

$$\sigma_{ij} = 2\eta\dot{\varepsilon}_{ij} + [(\zeta-\eta)\dot{\varepsilon}_{kk} - P]\delta_{ij}$$

In the case of a static fluid, it can easily be seen that the gradient of static pressure balances gravity. The static pressure is the only term in the flow rule, which can balance gravity without the fluid being constantly compressed. Therefore, the static pressure is very important for the description of fluids. The term plays a similar role for sea ice. When the wind constantly pushes sea ice against a coast, the ice is compressed and thickened by deformation until it is strong enough to resist further deformation. It then follows that the stress tensor is only given by the pressure P, i.e. the pressure gradient balances the wind stress. It is obvious that the static pressure P has to be described as a function of sea-ice thickness and coverage by some kind of equation of state.

In Hibler's model, the ice resists a deformation up to a limit P, which depends linearly on sea-ice thickness and exponentially on coverage. This empirical approach has been used successfully in many applications.

$$P^* = 2P = P_0 h_1 e^{-c(1-A)},$$

where h_1 represents sea-ice thickness; A, sea-ice coverage and P_0 and C are empirical constants. P^* is the maximum compressive stress above which deformation sets in. When the internal stresses reach the yield stress, the ice becomes plastic, i.e. the ice breaks and rafting occurs without increasing the internal stresses further, and ice gets thicker.

Variations of the ice thickness (h_1) are finally given by a continuity equation, which includes local and advective effects.

$$\dot{h}_1 = -\nabla\cdot(uh_1) + S_h$$

S_h represents the thermodynamic source term, which is calculated separately for the ice-covered and the ice-free portion of each model grid cell. Similar equations can be derived for snow thickness and sea-ice concentration (Owens and Lemke 1990).

10.2.3   Model Simulations

The optimal representation of the sea-ice rheology was investigated in the ACSYS Sea-Ice Model Intercomparison Project (SIMIP) (Lemke et al. 1997; Kreyscher et al. 2000). The rheologies tested included a *viscous–plastic model* (Hibler and Flato 1992),

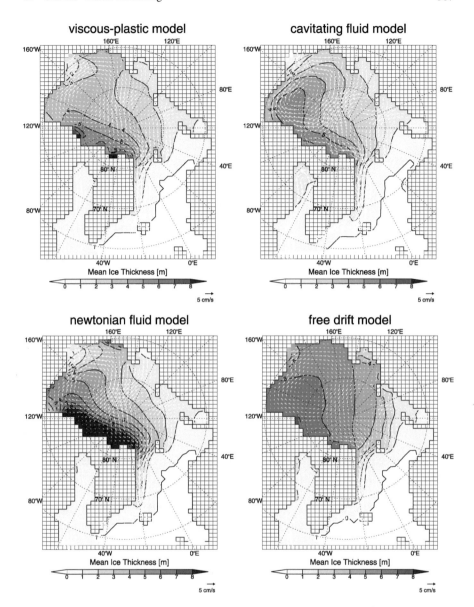

Fig. 10.1 Mean sea-ice motion and thickness (1979–1995) simulated with the viscous–plastic model, compared to other rheologies (adapted from Kreyscher et al. 2000; Hilmer 2001)

a *cavitating-fluid model* without any shear forces (Flato and Hibler 1992), a *Newtonian-fluid model* with constant bulk and shear viscosities and a corrected *free drift model* in which the drift was calculated with zero internal ice stresses but the velocities were set to zero for sea-ice thicknesses above a threshold value of 4 m under convergent flow conditions (Fig. 10.1).

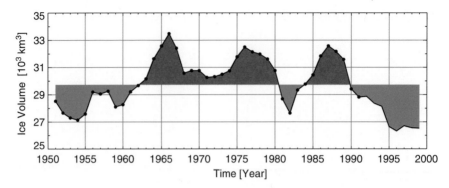

Fig. 10.2 Time series of the simulated annual mean sea-ice volume in the Arctic (Hilmer 2001)

Contrary to purely thermodynamic models, a realistic ice-thickness distribution is obtained with the viscous–plastic model. Of the four tested rheologies, this model matches the observations best. Applying a free drift, i.e. neglecting the internal ice stresses, results in an unrealistic distribution of sea-ice thickness. Due to the missing shear force, the cavitating-fluid model produces too large velocities and thus too small ice thicknesses in the Arctic basin. The Newtonian-fluid model achieves a similar pattern of sea-ice thickness as the viscous–plastic model, but the ice thicknesses are far too large. Further results regarding the model comparison are described in detail in Lemke et al. (1997) and Kreyscher et al. (2000).

Integration of the optimized sea-ice model with atmospheric boundary conditions (wind, air temperature) for the period 1951–1999, as given by the NCEP/NCAR reanalyses, yields a pronounced variability (Hilmer and Lemke 2000; Hilmer 2001). The total Arctic ice volume displays a small long-term trend superimposed with strong decadal variations (Fig. 10.2; see also Sect. 10.3.3).

The simulated thickness and motion fields of the viscous–plastic model correspond well with the observations. The mean drift pattern (Fig. 10.3, left) shows the well-known Beaufort Gyre, the Transpolar Drift Stream and a pronounced southward flow east and west of Greenland through Fram, Denmark and Davis Straits. Because of the Beaufort Gyre, the sea ice is pushed against the north coast of Greenland and Canada and reaches the largest thicknesses in this region (on average, 6 m). In the divergent drift region on the Eurasian shelf, the thickness is only 1–2 m (Fig. 10.3, right).

Because of the pronounced drift pattern, there are regions, especially on the Eurasian shelf, where, on average, more ice is frozen in winter than melted in summer, i.e. the net freezing rate is positive in these divergent drift regions (blue areas in Fig. 10.3, left). Negative net freezing rates are found in the Labrador, Irminger and Greenland Seas. In these areas of the North Atlantic, deep water is generally formed in winter through convection. The melting sea ice stabilizes the oceanic stratification and reduces the deepwater production. These positive or negative net freezing rates represent the basic thermohaline forcing of the ocean circulation in

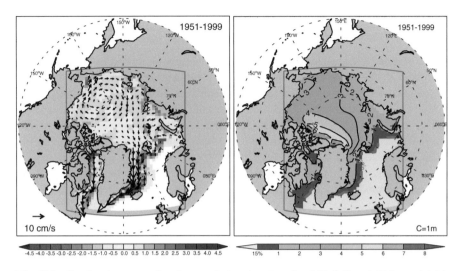

Fig. 10.3 Simulated mean net freezing rate (m/year) and sea-ice drift (*left*) and thickness (*right*) (Hilmer 2001)

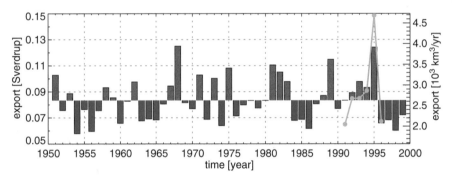

Fig. 10.4 Simulated sea-ice export through Fram Strait (Hilmer 2001). Measurements by Vinje et al. (1998) are indicated as *dots* and *solid line*

higher latitudes. Therefore, in years with small ice export from the Arctic, convection should increase, and in years with strong export, convection should weaken.

Sea-ice transport through Fram Strait is the origin for the large net melting rates in the Greenland Sea. This transport shows a considerable annual cycle and large interannual anomalies (Fig. 10.4) and explains, for example, the Great Salinity Anomaly that appeared in the northern North Atlantic in 1968. The variability in the simulated ice export compares well with the observations taken from upward-looking sonar installed on oceanographic moorings (Vinje et al. 1998). Further details regarding sea-ice export through the Fram Strait are described in Harder et al. (1998), Hilmer (2001) and in Melling (Chap. 3).

10.3 Sea-Ice–Ocean Coupling

In stand-alone sea-ice models, the ocean is generally simulated by a mixed layer with spatially and temporally constant depth, with a given constant heat flux from the deep ocean. Assumed values for this oceanic heat flux range from 2 W/m² in the Arctic Ocean to 20 W/m² in the Southern Ocean (Parkinson and Washington 1979; Hibler 1979). In a study on the sea ice in the Weddell Sea, Hibler (1984) set the vertical oceanic heat flux proportional to the freezing rate. The spatial and temporal mean amounted to 13 W/m².

To examine the interaction between the sea ice and the ocean more accurately, a simple thermodynamic sea-ice model was coupled to a prognostic one-dimensional model for the oceanic mixed layer and the pycnocline below (Lemke 1987). In this model, contrary to earlier studies, the vertical oceanic heat flux was determined prognostically by the entrainment of warm deep water into the cold surface layer. It indicates that the heat flux is temporally and spatially quite variable and certainly cannot be represented by a constant or be set in proportion to the freezing rate.

In comparison to the simulation with a constant mixed layer depth, the prognostic mixed layer model leads to a considerably modified sea-ice distribution because of the spatially and temporally strongly varying oceanic heat flux. It appears that this heat flux is largest (13 W/m²) near the Antarctic continent where the oceanic vertical stratification is weakest. In late winter, when the surface layer has reached its largest depth, and during the retreat phase in spring and summer, the vertical heat flux in the ocean is negligibly small. The seasonal changes of the depth of the mixed layer (i.e. the heat reservoir) and the entrainment heat flux cause a delay in the evolution of the winter sea-ice cover and considerably reduce the maximum sea-ice extent. Too large a sea-ice extent was a known problem in early sea-ice models, which generally was overcome by suitable parameterization of the ice concentration and the albedo. These studies indicate that coupled sea-ice–ocean circulation models are required to realistically describe the observed sea-ice variability.

10.3.1 Development of Regional Ocean–Sea-Ice Simulations

At the beginning of the ACSYS period, ocean–sea-ice modelling was dominated by simplified models and stand-alone sea-ice or ocean component modelling. The majority of existing sea-ice–ocean models were of low spatial resolution and forced with climatological atmospheric fields. Comprehensive coupled ocean–sea-ice models emerged in the second half of the 1990s. An early review of coupled ocean–sea-ice modelling was provided by Mellor and Häkkinen (1994). They stress the requirements for a good mixing parameterization to simulate convection and the need for good estimates of boundary conditions, like surface heat and freshwater fluxes, as well as continental run-off. At that time, the usefulness of some models was compromised by diagnostic elements (restoring towards climatologies of temperature and salinity) in the interior of the ocean. Also, surface forcing taken from individual years or climatologies to force multi-year integrations posed problems in sea-ice distribution and water mass formation. The first fully prognostic ocean–sea-ice models showed improved sea-ice simulations with interannually varying surface forcing (Fleming and Semtner 1991). Häkkinen and Mellor (1992) report excessive sea-ice cover in the Greenland Sea, which was reduced with the adoption of daily wind forcing and the associated strong mixing events that brought sufficient heat from below up to the ice cover. With realistic atmospheric forcing for the period 1955–1975, Häkkinen (1993) was already able to simulate an anomalous sea-ice export from the Arctic for 1968, which delivered as much freshwater as necessary to explain the Great Salinity Anomaly of the late 1960s and the 1970s. With the isopycnal model calculations by Holland et al. (1996), all three major types of vertical coordinate system models were used for Arctic ocean - The meaning is studies of ice and ocean in the Arctic, and a number of sensitivity studies were available. With typical grid sizes of 100–200 km, the lack of resolution was evident in all calculations, and the results relied on the careful adjustment of poorly known parameters in the sea-ice and ocean components.

A few years later, Gerdes (2000) compared several models regarding their representation of the Arctic freshwater balance, the interaction of ocean and sea ice and the exchanges of freshwater between the Arctic Ocean and the subpolar North Atlantic. The reasons for the considerable differences between the model results found were widely unknown, and coordinated model intercomparisons to identify numerical and conceptual differences that led to different responses were requested. The observationally unknown and computationally difficult partitioning of freshwater transports between sea ice and ocean was identified as one of the most severe problems. Also, the transport of liquid and solid freshwater through the different passages connecting the Arctic Ocean with subpolar oceans was resolution dependent and thus represented a major uncertainty.

To a certain degree, these problems are still present in current model simulations. However, due to extensive and long-term observational programmes, the availability of more realistic atmospheric forcing reaching back at least several decades to a century (Kauker et al. 2008), the increase of resolution and the improvement of models, much progress has been made. Combined modelling and observational projects, like ACSYS, were instrumental in this achievement. Model intercomparison projects, especially the Arctic Ocean Model Intercomparison Project (AOMIP; Proshutinsky and Kowalik 2007) that commenced in 2000, provide a strong contribution to identifying model problems and improving parameterizations and numerics.

Drange et al. (2005) briefly reviewed the development of ocean–sea-ice modelling for the high northern latitudes. They also supplied a more in-depth comparison of two typical models, MICOM and NAOSIM, used for hindcasts of the Arctic and subpolar North Atlantic ocean–sea-ice system. Drange et al. found that models have matured such that when forced with realistic atmospheric forcing, they can provide the large-scale context to interpret observational data and overcome their limited scope in space and time. Examples include the spreading of warm anomalies in the Atlantic water of the Arctic Ocean (Gerdes et al. 2003; Polyakov et al. 2005) and the formation and propagation of freshwater anomalies in the Arctic Ocean (Karcher et al. 2005). On the other hand, the NAOSIM and MICOM model systems considered by Drange et al. show differences that would require a well-organized model intercomparison project as AOMIP or the CORE initiative of CLIVAR (Griffies et al. 2009) to clarify. Although resolution is much better than in the earlier models, the results still point at the resolution of boundary currents, passages and other topographic features as the prime reason for model biases and model differences.

Some of these limitations and weaknesses depend on the models' ability to represent details of bottom topography and land geometry, mass and property exchanges through narrow and/or shallow straits, boundary and coastal currents, mesoscale eddies and sea-ice deformations. Global climate models, in particular, have large errors in representing sea-ice conditions (Zhang and Walsh 2006; Gerdes and Köberle 2007), northward fluxes of heat and moisture and export of freshwater into the North Atlantic. These processes control both regional Arctic and, to some degree, global climate variability; hence, their realistic representation is critical to improved future climate predictions. Regional Arctic models use higher resolution compared to global models and provide means to address some of the earlier mentioned limitations.

The focus of the ACSYS modelling programme was the improved representation of polar processes in coupled general circulation models (CGCMs). For ocean–sea-ice processes, this matches the major goal of the Arctic Ocean Modelling Project (AOMIP). The ACSYS programme with its focus on the Arctic region has facilitated the development and increased use of Arctic regional models. Some of the main objectives of Arctic regional ice–ocean modelling are to: (1) study sea-ice and ocean conditions and their variability using coupled ice–ocean models at increasingly high resolution and forced with realistic atmospheric forcing, (2) provide feedback to global ocean and climate models on physical and numerical requirements for adequate modelling of the Arctic Ocean and (3) contribute towards understanding of the role of the Arctic Ocean and sea ice in global climate. A brief summary of advancements made with regional models to address global climate model limitations can be found in Dethloff et al. (Chap. 8).

At the end of the ACSYS period, regional Arctic ocean–sea-ice models are available that rather faithfully simulate the variability of sea ice, hydrography and transports in the Arctic as well as the exchanges between the Arctic and the subpolar oceans. Main developments were fully coupled, comprehensive ocean–sea-ice modelling systems; the availability of high-frequency atmospheric forcing data and the communication between modelling teams that was made possible by projects like the AOMIP. Furthermore, several projects dedicated to a better understanding and quantification of Arctic–sub-Arctic exchanges provided important validation data and ideas concerning the importance of several processes in the Arctic–sub-Arctic climate system.

10.3.2   Dynamical Coupling of Ocean and Sea-Ice Model Components

Earlier approaches of coupling comprehensive sea-ice models with reduced ocean components include a majority of the relevant coupled thermodynamic processes, like the dependence of oceanic heat supply to the ice on turbulence and stratification in the ocean. However, these models do injustice to the fact that sea ice is also dynamically an integral part of the ocean. The transport of sea ice, for example, can constitute a substantial part of the total mass transport in the Ekman layer of the ocean and thus influence the vertical velocity at the bottom of the oceanic Ekman layer. This vertical velocity is the essential vorticity forcing for the large-scale, interior oceanic circulation. Transport of sea ice introduces a spatial separation between freezing and melting regions. The associated net freshwater exchange with the ocean furthermore modifies the buoyancy forcing of the ocean and thus sea level, ocean mixing and circulation. This influences the intermediate water formation on the Arctic shelves and the formation and properties of Arctic water masses. Depending on sea-ice pathways and the fate of the meltwater, sea-ice processes can enhance or suppress oceanic convection. The drag between ocean and sea ice, as well as the surface tilt term (see Sect. 10.2.2), feeds these circulation changes back into the sea-ice drift field. Since sea-ice melting and freezing affect ocean density and thus circulation, feedbacks between sea-ice and ocean processes are possible. These feedbacks are largely unexplored, although Bitz et al. (2006) find a strengthening of high-latitude northward oceanic heat transport as a consequence of increased sea-ice formation over the Siberian shelves. Substantial changes in ocean circulation can especially occur due to fluctuations of the sea-ice supply to regions of massive melting, like the East Greenland Current and the northern sea-ice edge around Antarctica. In the south, the properties and formation rate of Antarctic Intermediate Water are affected by the meltwater input. On large spatial and long temporal scales, density changes associated with sea-ice transport in the Southern Ocean are a major factor affecting the strength of the Antarctic Circumpolar Current (Borowski et al. 2002), while the position of the ACC front determines the northern position of the winter sea-ice edge. In the north, meltwater anomalies (e.g. the Great Salinity Anomaly of the 1970s; Dickson et al. 1988) can affect deepwater formation and the strength of the large-scale oceanic overturning circulation which, in turn, has an essential influence on the sea-ice edge in the Nordic Seas and the Barents Sea (Winton 2003). Sea-ice formation and melting are strongly influenced by the oceanic transport of heat and salt to the sea-ice–ocean interface. The salt transport determines the salinity-dependent freezing point, while oceanic heat transport to the ice is a highly efficient process to melt ice. Despite the small relative contribution of the ocean to the total meridional heat transport at high latitudes, the oceanic heat transport largely determines the position of the sea-ice edge (Winton 2003).

The interior ocean circulation at high latitudes on scales larger than a few kilometres is in geostrophic balance. This implies that vertical velocities at the sloping bottom topography and at the top of the geostrophic interior (the bottom of the surface Ekman layer) are essential drivers for the vertically integrated volume transport of the ocean according to the Sverdrup relationship $\beta V = f(w_E - w_B)$, where f is the Coriolis parameter; β, its meridional derivative; V, the vertically integrated meridional velocity component and w_E and w_B, the vertical velocities at the bottom of the surface Ekman and at the top of the frictional bottom boundary layer, respectively. The divergence of the surface Ekman transports determines the vertical velocity at the top of the geostrophic interior. It is important to realize that the mass transport of sea ice is an integral part of the upper layer transport that determines the interior ocean circulation forcing. Converging sea-ice mass transport will displace ocean water and generate vertical velocity at the base of the Ekman layer. The importance of the sea-ice mass transport for the divergence of the upper ocean flow and the driving of the underlying interior was for the first time fully appreciated and applied in a coupled ocean–sea-ice model by Hibler and Bryan (1987). Their ocean model predicts the average velocity of the mixture of sea ice and ocean in the uppermost level of the model. The momentum forcing of the uppermost level is through the wind stress and the internal ice stresses as they both appear in the vertically integrated momentum balance of the uppermost level. The ocean–sea-ice drag does not appear in this momentum balance because it is internal to the mixture of sea ice and ocean water. As long as the sea ice is thin and of low compactness, the mixture is driven by the wind stress. The wind stress may be modified compared to the open ocean because of the different surface roughness of open ocean and sea ice (Steiner et al. 1999). However, when internal ice stresses become important, the effective momentum forcing can deviate substantially from the wind stress, in the extreme case, the internal ice stress basically opposing the wind stress.

Many coupled ocean–sea-ice models and coupled climate models do not account for the effect of the internal ice stresses on the Ekman pumping velocity. Unfortunately, the systematic effects of the momentum coupling of the sea-ice component to the ocean have not yet thoroughly been evaluated. Another difficulty exists in realizing correct transport velocities in the ocean. Usually, tracers in the ocean – including the dynamically active tracers, temperature and salinity – are advected with the velocity of the sea-ice–water mixture that is calculated in the ocean component. In fact, this is a necessity in order to preserve correct exchanges between kinetic and potential energy and thus to preserve the stability of the ocean model (Bryan 1969). When sea ice occupies a substantial part of the uppermost level in an ocean model, this obviously leads to biases in ocean tracer transports. The choice of thicker uppermost levels is usually not an option as high vertical resolution is needed to capture basin–shelf sea exchanges to properly represent processes in the shallow shelf seas of the Arctic and to incorporate biological and chemical processes that are associated with steep gradients near the ocean surface.

Martin and Gerdes (2007) have analysed AOMIP model results regarding their representation of sea-ice drift speeds. They found two distinct groups of ocean–sea-ice models. The drift speed distribution of one group of models had a mode at drift speeds around 3 cm s^{-1} and a short tail towards higher speeds. The other group showed a more even frequency distribution, with a relatively large probability of drift speeds of 10–20 cm s^{-1}. Observations clearly agree better with the first group of model results. Possible reasons for these differences are manifold, including the discrepancies of wind stress forcing as well as sea-ice model characteristics and sea-ice–ocean coupling. Investigating the drift patterns of anticyclonic and cyclonic wind-driven regimes, Martin and Gerdes (2007) found that all AOMIP models are capable of producing realistic drift pattern variability. The winter of 1994/1995 stands out because of its maximum in Fram Strait ice export. Export estimates of some models agree well with observations, while the corresponding inner Arctic drift pattern is not reproduced. The reason for this is found in the wind forcing and especially in differences in ocean velocities. Besides the coupling mechanism itself, which controls the intensity of the effect that the ocean has on the ice, the different ocean velocities of the models are found to cause some of the differences in sea-ice drift.

10.3.3   Sea-Ice–Ocean Coupling and Arctic Freshwater Balance

The Arctic Ocean freshwater resides in two reservoirs: sea ice (with a residual salinity of a few psu depending on sea-ice age) and liquid freshwater. The latter is defined as the vertical integral of the difference of reference salinity and the actual salinity, scaled by the reference salinity. The reservoirs are of the order of a few 10^4 km^3 for the sea ice to a maximum of 10^5 km^3 for the liquid freshwater (Fig. 10.5). The inter-action of ocean and sea ice with respect to the freshwater export from the Arctic is one of the most important processes for the long-term behaviour of the coupled systems with consequences for sea level as well as ecosystems.

The two reservoirs exchange freshwater through melting and freezing of sea ice. A reduction of the sea-ice volume through increased melting or reduced formation of sea ice within the Arctic will increase the size of the liquid freshwater reservoir. However, external forcing acting in a similar fashion on both reservoirs is often more important for the long-term changes than the internal redistribution of fresh-water between the reservoirs (Köberle and Gerdes 2007).

In a short review, Dickson et al. (2007) compared the liquid freshwater content in the simulations of Karcher et al. (2005), Köberle and Gerdes (2007) and Häkkinen and Proshutinsky (2004). Until very recently, neither ice thickness nor Arctic-wide salinity distributions could be monitored well enough to derive reliable estimates from observations alone (e.g. Swift et al. 2005). The relative lack of validation data, uncertainties in the forcing functions, the implementation of surface freshwater fluxes and resolution issues are important obstacles for present modelling efforts (Gerdes et al. 2008).

For sea ice, the Arctic-wide mass balance consists of the temporal change that is determined by net thermodynamic ice production and by sea-ice export. Köberle and Gerdes (2003) and Rothrock and Zhang (2005) have shown similar results for the simulated Arctic sea-ice volume during the second half of the twentieth century. Sea-ice volume peaked in the mid-1960s. The following decline was punctuated by smaller sea-ice volume maxima in the late 1970s and the late 1980s. While Rothrock

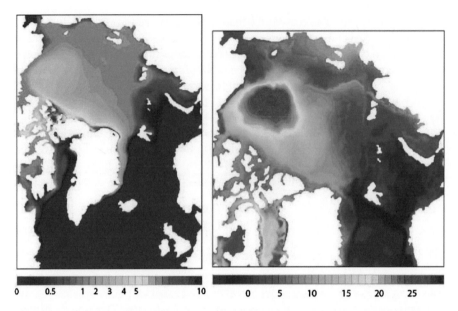

Fig. 10.5 Simulated Arctic annual mean sea-ice thickness in m for 2009 (*left*) and liquid freshwater content in 2009 from the same 9-km-resolution NAOSIM simulation (*right*). Liquid freshwater is defined as the vertical integral of the difference between a reference salinity and actual salinity scaled by the reference salinity. It can be interpreted as the height (in m) of a freshwater column that would have to be added to a water column at the reference salinity to arrive at the actual salinity. The total ice volume is 21,200 km³, while the total liquid freshwater content is 92,500 km³ (Figure courtesy of Cornelia Köberle)

and Zhang simulated a continuous decline over the five decades of the simulation, Köberle and Gerdes found sea-ice volume in the 1950s almost as low as at the end of the century. Both simulations show low sea-ice volume before the large accumulation in the 1960s. Both studies find that the variability in sea-ice volume due to wind forcing and due to thermal forcing is of the same magnitude. However, the wind-forced part has no significant long-term trend, while the thermal-driven part is responsible for most of the long-term decline of the sea-ice volume (Fig. 10.6).

The long-term decline in Arctic sea-ice volume since the mid-1960s through thermodynamic processes implies a freshwater source for the Arctic Ocean. However, the liquid freshwater content has not increased during the same time. Instead, the liquid freshwater content declined similar to that in sea-ice volume. The positive freshwater input anomaly from reduced sea-ice growth was apparently overcompensated by an increase in the export of liquid freshwater such that both reservoirs were depleted. It is conceivable that a common forcing was responsible for this development. This period was characterized by the increasing trend in the strength of the NAO. The NAO has probably contributed to higher temperatures over parts of the Arctic and thus to the reduction in ice volume. A strong positive NAO also leads to enhanced sea-ice export through Fram Strait and an enhanced

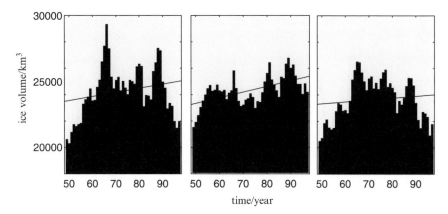

Fig. 10.6 Time series of Arctic sea-ice volume (in km³) from a simulation with NAOSIM for the period 1948–1997. The *left panel* shows the result with the full variability of the atmospheric forcing; the *middle panel*, with variability restricted to the wind forcing and the *right panel*, with variability restricted to the thermal forcing. The straight line indicates the linear trend in sea-ice volume over the integration period (Adapted from Köberle and Gerdes 2003)

cyclonic circulation in the Nordic Seas with extensions into the Arctic Ocean (Karcher et al. 2003). The associated stronger inflow of Atlantic water and stronger outflow of polar water constitute a strong net export of freshwater from the Arctic Ocean. The quantitative effect of increased precipitation that falls over the catchment area of rivers draining into the Arctic during high NAO phases on the Arctic Ocean freshwater balance is probably small (Häkkinen and Proshutinsky 2004). The reversal of the NAO after its maximum in the mid-1990s was correspondingly associated with an increase in the Arctic liquid freshwater content in the second half of the 1990s, as shown in several model simulations (Dickson et al. 2007).

With the general decline of Arctic sea-ice volume, sea-ice thickness near Fram Strait is also decreasing. This has been well established by direct measurements of sea-ice thickness using electromagnetic methods (Haas et al. 2010). These small thicknesses of the southward-flowing sea ice in Fram Strait are consistent with hindcast simulations with a high-resolution version of NAOSIM (Fig. 10.7). There, the average thickness of sea ice drifting southward across 79°N decreases from around 3 m in the early 1990s to around 1.5 m in 2010. Sea-ice drift in Fram Strait in the model has almost no trend – i.e. regarding the sea-ice export, there is little compensation of decreasing ice thickness by faster southward ice drift. The sea-ice export falls from around 0.12 Sv to around 0.09 Sv within the 20-year simulation period. A reduction in sea-ice export by 0.03 Sv over 10 years implies a reduced freshwater loss of 9,500 km³ from the Arctic Ocean. This is consistent with results of Rabe et al. (2011) who used IPY and historical salinity observations to estimate the difference in liquid freshwater content in the Arctic Ocean between the 1990s and 2006–2008 at 8,400 ± 2,000 km³.

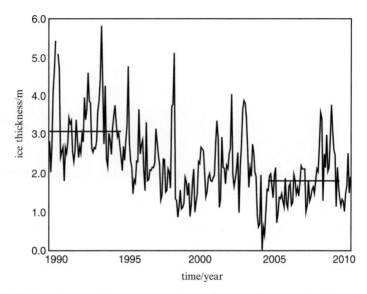

Fig. 10.7 Simulated sea-ice thickness averaged over southward-drifting sea ice in Fram Strait. Ice thicknesses were taken from a simulation with NAOSIM at 9-km horizontal resolution (Figure kindly provided by Cornelia Köberle)

10.3.4 Surface Boundary Conditions

The Arctic-wide freshwater content is determined by net melting of sea ice, run-off, precipitation and exchanges with the Atlantic and Pacific Oceans. In many models, there is an additional source of freshwater through the restoring of model-predicted surface salinities towards some prescribed reference values. Prange and Gerdes (2006) stress the importance of the surface freshwater flux boundary condition in ocean–sea-ice models. Neglecting the volume input associated with river run-off or precipitation (as done with virtual salt fluxes where freshwater fluxes are transformed into salt fluxes by multiplication with a suitable reference salinity) leads to significant salinity increases in the upper Arctic Ocean, compared to an experiment with a fully nonlinear free-surface formulation. This has implications for the density of the polar water leaving the Arctic Ocean via the East Greenland Current and for the large-scale oceanic circulation.

The type of the surface freshwater flux boundary condition is important in other aspects as well. In fully coupled climate models, the freshwater balance adjusts itself into an equilibrium between atmospheric transports; run-off and precipitation and salinity redistribution in the ocean. Global ocean–sea-ice models and especially regional ocean–sea-ice models are much more constrained. Without assimilation of observations – as with restoring of surface salinity – it requires good tuning to achieve realistic salinity distributions with prescribed precipitation, run-off and lateral exchange at the edges of a regional model. A more detailed discussion of the problem can be found in Gerdes et al. (2008). Tuning is always only valid for a specific situation and can lead to unacceptable results in different cases, like for different domains or even different forcing situations. To illustrate this point, we consider the following hypothetical example. A reduction in run-off into the Arctic can lead to a strong salinification of the Arctic and subsequently the subpolar Atlantic. Enhanced deepwater formation might follow with enhanced near-surface flow of saline water of subtropical origin into the deepwater formation region, thus reinforcing the original salinity anomaly. However, evaporation and atmospheric transports in the subtropics will no longer be able to maintain high salinities in the subtropics because the water resides for a reduced time in the positive E-P regime. This means that over the period of several decades to a century, the salinity of the water arriving in subpolar and polar latitudes will be reduced again, bringing the system back to its original state. Further feedbacks involving the reaction of atmospheric moisture transport to SST anomalies (a warm anomaly associated with stronger northward flow of subtropical waters in the above scenario) will also force the system towards its original state. In a regional ocean–sea-ice model, none of the feedbacks can operate because the inflow of subtropical water into the domain is mostly prescribed, as are precipitation and run-off.

From this inherent difficulty of (regional) ocean–sea-ice models arises the need to incorporate a restoring of surface salinity values towards climatological mean values in many models (Steele et al. 2001). Häkkinen et al. (2007) discussed examples of salinity drift in AOMIP models that are run without restoring. Depending on model and freshwater forcing, salinity drifts over several decades can be substantial, obscuring trends that are due to changes in forcing or large-scale oceanic circulation. For the long-term annual mean Arctic freshwater balance in a coarse resolution version of NAOSIM, Köberle and Gerdes (2007) found a contribution from restoring, comparable in magnitude to the largest physical sources of freshwater.

The restoring towards surface salinity climatology also introduces a damping that prevents realistic decadal fluctuations in Arctic freshwater content. Comparing a case with restoring and a case with a constant (in time) surface freshwater flux adjustment derived from a multidecadal run with restoring, Köberle and Gerdes (2007) found an increase in the excursions in freshwater content by at least a factor of two. Processes like the spreading of melt-induced freshwater over long distances are only present in the case without the unrealistic negative feedback that restoring introduces. Although the flux adjustment is found by averaging the restoring term over time, the mean freshwater content in the Arctic Ocean is systematically enhanced with the flux adjustment. This is due to a strong increase in Arctic Ocean freshwater content until the mid-1960s, caused by relatively low freshwater export to lower latitudes.

10.4 Synthesis and Outlook

Understanding of relationships between sea ice and other climate variables on large spatial scales and over long time periods is impossible without reliable numerical models. State-of-the-art regional sea-ice–ocean models are able to provide robust results that rather faithfully reproduce many observations and can help to quantify long-term variability in sea ice and its exchanges with the ocean. Except for the spatial resolution and more sophisticated parameterizations of physical processes, the dedicated regional models are similar to the components that are used in global sea-ice–ocean models as well as in coupled climate models. Thus, these models serve an important function as development platforms for the improvement of climate models. Furthermore, sea-ice–ocean models are increasingly applied as tools, for instance, to calculate the pathways and immissions of contaminants and to calculate environmental conditions important for high-latitude ecosystems. Projects are underway in which sea-ice models are planned to be used for calculating forces acting on offshore structures and for predicting sea-ice conditions to optimize ship routing.

The relatively mature state of ocean–sea-ice models was made possible by improvements in model validation due to increasing availability of observational data for sea-ice concentration, drift, age and, to a somewhat lesser degree, sea-ice thickness. Increasing computer power has made it possible to run ocean–sea-ice models at resolutions that allow the representation of important physical processes without uncertain parameterizations. Similarly important was the availability of realistic atmospheric forcing reaching back several decades through the meteorological reanalysis projects (e.g. Kalnay et al. 1996; Uppala et al. 2005). The prominent role of sea ice in climate change and its detection has led to an increase in the human resources dedicated to the field. Communication and coordination provided by model intercomparison projects, like SIMIP, AOMIP, CORE and CMIP, were instrumental in progress by contributing to the identification of model problems and development of improved parameterizations and numerical methods.

Certainly, sea-ice–ocean models are not perfect. Model intercomparisons have revealed that seemingly similar models can exhibit different behaviours even in well-constrained experiments. Because of the complexity of the models, it is often not easy to isolate the corresponding causes. While spatial resolution has improved substantially over the last years, certain aspects still are not well represented. This concerns especially the ocean components of coupled models. Small-scale processes can be essential for large-scale phenomena. The flows through narrow and shallow passages, like the Bering Strait and the Canadian Arctic Archipelago, probably need higher resolution than what is currently routinely feasible. New techniques like unstructured grids with strong focus on certain regions may be a way to alleviate this problem (Danilov and Schröter 2010). The exchange between the shelf seas and the interior basins requires both high horizontal and vertical resolutions, with high vertical resolution posing conceptional problems in the coupling of sea-ice and ocean components. High-end models now approach horizontal resolutions where the continuum assumption underlying currently used sea-ice rheologies becomes questionable. Alternative approaches that take into account the individual character of ice floes must be considered.

References

Bitz CM, Gent PR, Woodgate RA, Holland MM, Lindsay R (2006) The influence of sea ice on ocean heat uptake in response to increasing CO_2. J Clim 19:2437–2450

Borowski D, Gerdes R, Olbers D (2002) Thermohaline and wind forcing of a circumpolar channel with blocked geostrophic contours. J Phys Oceanogr 32:2520–2540

Bryan K (1969) A numerical method for the study of the circulation of the world oceans. J Comput Phys 4:666–673

Danilov and Schröter (2010) Unstructured meshes in large-scale ocean modeling. In: Nashed MZ, Freeden W (eds) Handbook of geomathematics. Springer, Berlin, pp 371–398

Dickson RR, Meincke J, Malmberg S-A, Lee AJ (1988) The great salinity anomaly in the Northern North Atlantic 1968–82. Prog Oceanogr 20:103–151

Dickson R, Rudels B, Dye S, Karcher M, Meincke J, Yashayaev I (2007) Current estimates of freshwater flux through Arctic and subarctic seas. Prog Oceanogr 73:210–230

Drange H, Gerdes R, Gao Y, Karcher M, Kauker F, Bentsen M (2005) Ocean general circulation modelling of the Nordic Seas. In: Drange H, Dokken T, Furevik T, Gerdes R, Berger W (eds) The Nordic Seas: an integrated perspective, Geophysical monograph 158. AGU, Washington, DC, pp 199–220

Flato GM, Hibler WD (1992) On modelling pack ice as a cavitating fluid. J Phys Oceanogr 22:626–651

Fleming GH, Semtner AJ Jr (1991) A numerical study of interannual ocean forcing on Arctic ice. J Geophys Res 96(C3):4589–4603. doi:10.1029/90JC02268

Gerdes R, Köberle C (2007) Comparison of Arctic sea ice thickness variability in IPCC climate of the 20th century experiments and in ocean–sea ice hindcasts. J Geophys Res 112:C04S13. doi:10.1029/2006JC003616

Gerdes R et al (2000) Modelling the variability of exchanges between the Arctic Ocean and the Nordic Seas. In: Lewis LE (ed) The freshwater budget of the Arctic Ocean. Kluwer Academic, Dordrecht, pp 533–547

Gerdes R, Karcher M, Kauker F, Schauer U (2003) Causes and development of repeated Arctic Ocean warming events. Geophys Res Lett 30(19):1980. doi:10.1029/2003GL018080

Gerdes R, Karcher M, Köberle C, Fieg K (2008) Simulating the long term variability of liquid freshwater export from the Arctic Ocean. In: Dickson B, Meincke J, Rhines P (eds) The role of the Northern Seas in climate. Springer, Dordrecht, pp 405–426

Griffies SM, Biastoch A, Böning C, Bryan F, Danabasoglu G, Chassignet EP, England MH, Gerdes R, Haak H, Hallberg RW, Hazeleger W, Jungclaus J, Large WG, Madec G, Pirani A, Samuels BL, Scheinert M, Sen Gupta A, Severijns CA, Simmons HL, Treguier AM, Winton M, Yeager S, Yin J (2009) Coordinated Ocean-ice Reference Experiments (COREs). Ocean Model 26:1–46. doi:10.1016/j.ocemod.2008.08.007

Haas C, Hendricks S, Eicken H, Herber A (2010) Synoptic airborne thickness surveys reveal state of Arctic sea ice cover. Geophys Res Lett 37:L09501. doi:10.1029/2010GL0426522009

Häkkinen S (1993) An Arctic source for the Great Salinity Anomaly: a simulation of the Arctic ice-ocean system for 1955–1975. J Geophys Res 98:16397–164103

Häkkinen S, Mellor GL (1992) Modeling the seasonal variability of the coupled Arctic ice-ocean system. J Geophys Res 97:20285–20304

Häkkinen S, Proshutinsky A (2004) Freshwater content variability in the Arctic Ocean. J Geophys Res 109:C03051. doi:10.1029/2003JC001940

Häkkinen S, Dupont F, Karcher M, Kauker F, Worthen D, Zhang J (2007) Model simulation of Greenland Sea upper-ocean variability. J Geophys Res 112:C06S90. doi:10.1029/2006JC003687

Harder M, Lemke P, Hilmer M (1998) Simulation of sea ice transport through Fram Strait: natural variability and sensitivity to forcing. J Geophys Res 103:5595–5606

Hibler WD (1979) A dynamic thermodynamic sea ice model. J Phys Oceanogr 9:815–846

Hibler WD III (1984) The role of sea ice dynamics in modeling CO_2 increases. In: Hansen JE, Takahashi T (eds) Climate processes and climate sensitivity, vol 29, Geophysical monograph series. AGU, Washington, DC, pp 238–253

Hibler WD III, Bryan K (1987) A diagnostic ice-ocean model. J Phys Oceanogr 17:987–1015

Hibler WD, Flato GM (1992) Sea ice models. In: Trenberth KE (ed) Climate system modelling. Cambridge University Press, New York, pp 413–436

Hilmer M (2001) A model study of Arctic Sea ice variability, Dissertation, Universität Kiel

Hilmer M, Lemke P (2000) On the decrease of Arctic sea ice volume. Geophys Res Lett 27:3751–3754

Holland DM, Mysak LA, Oberhuber JM (1996) Simulation of the mixed-layer circulation in the Arctic Ocean. J Geophys Res 101(C1):1111–1128. doi:10.1029/95JC02819

Kalnay E et al (1996) The NCEP/NCAR 40-year reanalysis project. Bull Am Meteorol Soc 77(3):437–471

Karcher MJ, Gerdes R, Kauker F, Köberle C (2003) Arctic warming - evolution and spreading of the, warm event in the Nordic Seas and the Arctic Ocean. J Geophys Res 108. doi:10.1029/2001JC001265

Karcher M, Gerdes R, Kauker F, Köberle C, Yashayaev I (2005) Simulation of a 1990s fresh water event in the Nordic Seas and the subpolar North Atlantic. Geophys Res Lett 32:L21606. doi:10.1029/2005GL023861

Kauker F, Köberle C, Gerdes R, Karcher M (2008) Reconstructing atmospheric forcing data for an ocean-sea ice model of the North Atlantic for the period 1900–2003. J Geophys Res 113:C10004. doi:10.1029/2006JC004023

Köberle C, Gerdes R (2003) Mechanisms determining Arctic sea ice conditions and export. J Clim 16:2843–2858

Köberle C, Gerdes R (2007) Simulated variability of the Arctic Ocean fresh water balance 1948–2001. J Phys Oceanogr 37:1628–1644

Kreyscher M, Harder M, Lemke P, Flato GM (2000) Results of the sea-ice model intercomparison project: evaluation of sea ice rheology schemes for use in climate simulations. J Geophys Res 105:11299–11320

Lemke P (1987) A coupled one-dimensional sea ice–ocean model. J Geophys Res 92(C12):13164–13172

Lemke P, Hibler WD, Flato G, Harder M, Kreyscher M (1997) On the improvement of sea ice models for climate simulations: the sea ice model intercomparison project. Ann Glaciol 25:183–187

Martin T, Gerdes R (2007) Sea ice drift variability in Arctic Ocean Model Intercomparison Project models and observations. J Geophys Res 112:C04S10. doi:10.1029/2006JC003617

Maykut GA, Untersteiner N (1971) Some results from a time-dependent thermodynamic model of sea ice. J Geophys Res 76:1550–1575

Mellor GL, Häkkinen S (1994) A review of coupled ice-ocean models. In: The Polar Oceans and their role in shaping the global environment, Geophysical Monograph, vol 85. American Geophysical Union, Washington, DC, pp 21–31

Owens WB, Lemke P (1990) Sensitivity studies with a sea ice-mixed layer-pycnocline model in the Weddell Sea. J Geophys Res 95(C6):9527–9538

Parkinson CL, Washington WM (1979) A large-scale numerical model of sea ice. J Geophys Res 84(C1):311–337

Polyakov IV, Beszczynska A, Carmack EC, Dmitrenko IA, Fahrbach E, Frolov IE, Gerdes R, Hansen E, Holfort J, Ivanov VV, Johnson MA, Karcher M, Kauker F, Morison J, Orvik KA, Schauer U, Simmons HL, Skagseth O, Sokolov VT, Steele M, Tomokhov LA, Walsh D, Walsh JE (2005) One more step towards a warmer Arctic. Geophys Res Lett 32:L17605. doi:10.1029/2005GL023740

Prange M, Gerdes R (2006) The role of surface fresh water flux boundary conditions in prognostic Arctic ocean-sea ice models. Ocean Model 13:25–43

Proshutinsky A, Kowalik Z (2007) Preface to special section on Arctic Ocean Model Intercomparison Project (AOMIP) studies and results. J Geophys Res 112:C04S01. doi:10.1029/2006JC004017

Rabe B, Karcher M, Schauer U, Toole JM, Krishfield RA, Pisarev S, Kauker F, Gerdes R, Kikuchi T (2011) An assessment of Arctic Ocean freshwater content changes from the 1990s to the 2006–2008 period. Deep Sea Research Part I, 58(2), doi:10.1016/j.dsr.2010.12.002

Rothrock DA, Zhang J (2005) Arctic Ocean sea ice volume: what explains its recent depletion? J Geophys Res 110:C01002. doi:10.1029/2004JC002282

Semtner AJ (1976) A model for the thermodynamic growth of sea ice in numerical investigations of climate. J Phys Oceanogr 6:379–389

Steele M, Ermold W, Häkkinen S, Holland D, Holloway G, Karcher M, Kauker F, Maslowski W, Steiner N, Zhang J (2001) Adrift in the Beaufort Gyre: a model intercomparison. Geophys Res Lett 28(15):2935–2938

Steiner N, Harder M, Lemke P (1999) Sea-ice roughness and drag coefficients in a dynamic-thermodynamic sea-ice model for the Arctic. Tellus 51A:964–978

Swift JH, Aagaard K, Timokhov L, Nikiforov EG (2005) Long-term variability of Arctic Ocean waters: evidence from a reanalysis of the EWG data set. J Geophys Res 110:C03012. doi:10.1029/2004JC002312

Uppala SM et al (2005) The ERA-40 re-analysis. Q J R Meteorol Soc 131:2961–3012

Vinje T, Nordlund N, Kvambekk A (1998) Monitoring ice thickness in Fram Strait. J Geophys Res 103:10437–10450

Winton M (2000) A reformulated three-layer sea ice model. J Atmos Ocean Technol 17(4):525–531

Winton M (2003) On the climate impact of ocean circulation. J Clim 16:2875–2889

Zhang X, Walsh J (2006) Toward a seasonally ice-covered Arctic Ocean: scenarios from the IPCC AR4 model simulations. J Clim 19:1730–1746

Chapter 11
Global Climate Models and 20th and 21st Century Arctic Climate Change

Cecilia M. Bitz, Jeff K. Ridley, Marika Holland, and Howard Cattle

Abstract We review the history of global climate model (GCM) development with regard to Arctic climate beginning with the ACSYS era. This was a time of rapid improvement in many models. We focus on those aspects of the Arctic climate system that are most likely to amplify the Arctic response to anthropogenic greenhouse gas forcing in the twentieth and twenty-first centuries. Lessons from past GCM modeling and the most likely near-future model developments are discussed. We present highlights of GCM simulations from two sophisticated climate models that have the highest Arctic amplification among the the models that participated in the World Climate Research Programme's third Coupled Model Intercomparison Project (CMIP3). The two models are the Hadley Center Global Environmental Model (HadGEM1) and the Community Climate System Model version 3 (CCSM3). These two models have considerably larger climate change in the Arctic than the CMIP3 model mean by mid-twenty-first century. Thus, the surface warms by about 50% more on average north of 75 N in HadGEM1 and CCSM3 than in the CMIP3 model mean, which amounts to more than three times the global average warming.

C.M. Bitz (✉)
Atmospheric Sciences, University of Washington, Seattle, WA, USA
e-mail: bitz@atmos.washington.edu

J.K. Ridley
Hadley Centre for Climate Prediction, Met Office, Exeter, UK
e-mail: jeff.ridley@metoffice.gov.uk

M. Holland
National Center for Atmospheric Research, Boulder, CO, USA
e-mail: mholland@ucar.edu

H. Cattle
National Oceanography Centre, Southampton, UK
e-mail: hyc@noc.soton.ac.uk

P. Lemke and H.-W. Jacobi (eds.), *Arctic Climate Change: The ACSYS Decade and Beyond,* Atmospheric and Oceanographic Sciences Library 43, DOI 10.1007/978-94-007-2027-5_11, © Springer Science+Business Media B.V. 2012

The sea ice thins and retreats 50–100% more in HadGEM1 and CCSM3 than in the CMIP3 model mean. Further, the oceanic transport of heat into the Arctic increases much more in HadGEM1 and CCSM3 than in other CMIP3 models and contributes to the larger climate change.

11.1 Introduction

The ACSYS era spanned a period of rapid developments in global climate models (GCMs), especially with regard to polar climates. In 1992, experiments from only four global atmosphere-ocean general circulation models appeared in the Intergovernmental Panel on Climate Change supplementary assessment report (IPCC 1992), while more than 20 different models provided their output for inter-comparison in the most recent IPCC report (IPCC 2007). Models in the earlier IPCC were coarse-resolution, had relatively simple physics, and the majority needed unphysical adjustments to the heat and moisture exchange between the ocean and atmosphere.

Before the ACSYS era, new physics developed for GCMs were usually designed and tested for midlatitude and tropical climate applications. The focus steered clear of the polar regions probably because most modelers thought that too little was known about polar processes and data were scarce (Randall et al. 1998). Programs like ACSYS have helped expand our knowledge and observations of Arctic climate processes, so that GCM developers now pay special attention to the polar regions and high-latitude model physics are improving.

While early climate model development had little emphasis on high-latitude processes, Arctic climate simulated by the models has long attracted scientific attention. Even the earliest coupled atmosphere-ocean energy balance models had an amplified response at the poles when subject to an increase in radiative forcing (Budyko 1969; Sellers 1969). Predictions in the early 1990s of future Arctic climate change were so dire that one of two questions in the ACSYS mission (see http://acsys.npolar.no) read, "Is the Arctic climate system as sensitive to increased greenhouse gas concentrations as climate models suggest?" But when comparing models and observations for the last two decades of the twentieth century, studies find that the multi-model ensemble mean of the most current models agrees well with observed trends in Arctic surface air temperature and sea ice extent (Arzel et al. 2006; Zhang and Walsh 2006; Wang et al. 2007). Now, after repeated record-setting sea ice conditions in recent decades, the question has turned full circle, and studies are asking if models can keep pace with observed trends (e.g., Stroeve et al. 2007).

In this chapter, we describe some of the key GCM developments with regard to Arctic climate since the start of the ACSYS era (Sect. 11.2). We focus on those aspects of the Arctic climate system that are likely the most influential at amplifying the Arctic response to anthropogenic greenhouse gas forcing in the twentieth and twenty-first centuries, including the sea ice component, ocean-atmosphere exchange and ocean mixing, and clouds. Next we present highlights of GCM simulations of the late twentieth century climate and changes at mid-twenty-first century from two models with the highest Arctic amplification of surface warming and compare them to the multi-model ensemble mean from the models that participated in the most recent World Climate Research Programme (WCRP) Coupled Model Intercomparison Project, which is version 3 (CMIP3) (Meehl et al. 2007) (Sect. 11.3). These same models were analyzed for the IPCC fourth assessment report. A summary and future outlook are given at the end (Sect. 11.4).

11.2 GCM Developments Since the Beginning of the ACSYS Era

11.2.1 Sea Ice Component

Prior to the ACSYS era, sea ice in GCMs was treated as a slab of a single thickness that uniformly covered a grid box (e.g., sea ice could not coexist with an ice-free fraction). Usually, any snow that fell on top of sea ice was converted immediately to an equivalent sea ice thickness. Therefore, the thermal insulating capacity of snow was neglected and the surface albedo did not depend explicitly on snow properties. Heat conduction through the sea ice was calculated by assuming a linear temperature profile between the top and bottom surfaces of the ice (as in the Semtner 1976 zero-layer sea ice model); hence, sea ice had zero heat capacity and surface temperature changes lead to no change in stored sensible heat. Surface albedo was highly parameterized to artificially account for leads, snow cover, and melt ponds, usually by varying with surface temperature and ice thickness. If the sea ice moved at all, it was advected with the surface currents – in what is known as "free drift." Once the sea ice thickness reached some threshold (4 m was common), it was then held motionless to prevent the sea ice from building to excess in regions of convergence. Early GCMs that employed such sea ice models are described in Washington and Meehl (1989), Manabe et al. (1991), and McFarlane et al. (1992).

It is now well known that sea ice dynamics has a first-order influence on the sea ice mean state, variability, and sensitivity to radiative forcing. Hibler (1980) showed that a motionless sea ice model would have the thickest ice cover centered on the north pole, while a model with dynamics is needed to simulate the observed thick ice against Greenland and the Canadian Archipelago coasts. Additional studies with uncoupled, sea ice-only models indicated that adding sea ice dynamics to a sea ice component in a GCM would likely reduce the model's sensitivity to radiative forcing (Hibler 1984; Lemke et al. 1990; Holland et al. 1993; Fichefet and Morales Maqueda 1997). The results suggest that sea ice dynamics acts as a negative feedback on sea ice thickness because thinner ice more easily converges and deforms (building thickness dynamically and increasing winter open water formation and ice growth rates), while thicker ice resists dynamical thickening. The association of sea ice dynamics with negative feedback was verified in at least three separate GCMs where studies showed models with dynamics tend to retreat less in response to increasing radiative forcing than the same model with sea ice held motionless (Vavrus 1999; Holland et al. 2001; Hewitt et al. 2001; Vavrus and Harrison 2003).

Several of the first GCMs to include sea ice dynamics adopted the cavitating fluid (CF) sea ice dynamics from Flato and Hibler (1992) (e.g., the early NCAR and CSIRO models, see Pollard and Thompson 1994 and Gordon and O'Farrell 1997) because it offered simplicity and numerical efficiency to describe the ice internal stress over the more comprehensive viscous-plastic (VP) rheology (Hibler 1979). The VP rheology takes into account failure under compression and shear, while the CF physics disregard the influence of shear stress. Both VP and CF treatments assume the amalgam of sea ice floes and leads can be treated as a continuum. The first GCMs to employ the full VP physics were the ECHAM4+OPYC3 model (Oberhuber 1993) and the ECHAM4+HOPE-G (ECHO-G) model (Wolff et al. 1997). Among the models that participated in the first Coupled Model Intercomparison Project (CMIP1, see Meehl et al. 2000), which are contemporaries of these early GCMs with sea ice dynamics, 11 of 18 models had motionless sea ice and another 3 had ice in free drift. The first Sea Ice Model Intercomparison Project (SIMIP1), sponsored by ACSYS, took place roughly at the same time as CMIP1 and compared sea ice models with different dynamics schemes (Lemke et al. 1997). SIMIP1 investigators found that the VP rheology produced a more realistic simulation than CF or free-drift models (Kreyscher et al. 2000). In addition, they noted that the computational cost for the VP scheme was marginal compared to the rest of a typical GCM. A more efficient numerical scheme for ice rheology that could be adapted to parallel computing known as the elastic-viscous-plastic (EVP) soon became available and made sea ice dynamics schemes even more attractive for GCMs (Hunke and Dukowicz 1997; Hunke and Zhang 2000; Zhang and Rothrock 2000). Over half of the CMIP3 models, the most recent Coupled Model Intercomparison Project, have VP or EVP sea ice dynamics.

Sea ice thermodynamics still varies widely across sea ice components of global climate models. An effort to improve model thermodynamics ushered in the second Sea Ice Model Intercomparison Project (SIMIP2), which was a joint initiative of the ACSYS/CliC Numerical Experimentation Group and the GEWEX Cloud System Study, Working Group on Polar Clouds. One study from this project showed that a multi-layer sea ice model that explicitly resolved brine pockets reproduced well the sea ice thickness and temperature measured during the Surface Heat Budget of the Arctic Ocean (SHEBA) experiment (Huwald et al. 2005). Global sea ice models have also shown how sea ice thermodynamics influences the mass, heat, and fresh-water balance of the climate system. One-dimensional (Maykut and Untersteiner 1971; Semtner 1976) and global-scale (Holland et al. 1993; Fichefet and Morales Maqueda 1997; Bitz et al. 2001) models showed that taking into account internal melt in brine pockets in sea ice can shift the seasonal extrema in ice area and volume by up to several weeks. The GCMs also showed the mean distribution of the sea ice and its growth and melt rates were altered substantially. Bitz and Lipscomb (1999) updated the thermodynamic sea ice model of Maykut and Untersteiner (1971) to conserve energy and use faster numerics, and Bitz et al. (2001) noted that implementing this scheme is a small portion of the computational cost of running a full GCM. Nonetheless, it is still common for GCMs to use the Semtner (1976) zero-layer thermodynamics (e.g., MPI ECHAM5 and MRI-CGCM2.3.2, see Marsland et al. 2003; Yukimoto et al. 2006). Still other models use a multi-layer approach that restricts their influence to the upper ice layer after Semtner (1976) or Winton (2000) (e.g., GFDL 2.0 and 2.1 and CSIRO 3.0 and 3.5, see Delworth et al. 2006; O'Farrell 1998). Only two CMIP3 models (NCAR CCSM3 and PCM, see Holland et al. 2006) have adopted multi-layer thermodynamics with explicit brine pocket physics.

Snow cover insulates the underlying sea ice from atmospheric temperature changes. Because snow has a higher albedo than sea ice, snow cover can delay the onset of summer melt. Land surface schemes include complex multi-layer representations of snow that allow freezing of surface melt and metamorphosis of the snow grain size. However, at this time, we know of no GCM that has more than one resolved layer of snow properties on top of sea ice. One practical reason why snow physics in sea ice models has lagged behind its terrestrial counterpart is that each state variable in a sea ice model must be transported with the sea ice motion, and transport schemes with desirable numerical properties (e.g., high order, stable, and conservative) can be expensive. However, a new sea ice transport scheme that uses incremental remapping by Lipscomb and Hunke (2004) can efficiently transport large numbers of sea ice state variables.

The parameterization of melt ponds and radiative transfer in the sea ice and snow has been crude at best in GCMs until now. Heat and freshwater storage in melt ponds was ignored altogether in CMIP3 GCMs (as far as we know). Ponding was only considered to the extent that the surface albedo is typically a function of surface temperature: When melting, the surface albedo of bare sea ice is assigned a value that is meant to represent Arctic-wide conditions with some average pond fraction (e.g., Briegleb et al. 2004). Yet a sophisticated physical treatment of melt ponds was implemented in a single-column sea ice model quite some time ago (Ebert and Curry 1993). A GCM with explicit melt-pond mass and heat balance has been investigated with ECHAM5 (Pedersen et al. 2009). The ponds substantially lowered the summer surface albedo and reduced the ice thickness.

A more consistent treatment of the radiative transfer through melt ponds and sea ice was developed in a one-dimensional sea ice model by Taylor and Feltham (2003). In this case, a two-stream radiative transfer scheme was used to compute the surface albedo; absorption within the snow, ice, and pond; and transmission to the underlying ocean. Another radiative transfer method for sea ice and melt ponds (Briegleb and Light 2007) incorporates a Delta-Eddington, multiple-scattering radiative transfer model to account for multiple scattering from snow grains, bubbles, and brine pockets. We anticipate that these new methods will soon appear in GCMs.

Another important aspect of sea ice physics is its varied distribution of thicknesses that exists on the scale of a typical GCM grid box. In a given region, sea ice thickness is best described by the probability density of ice thicknesses, known as the ice thickness distribution (ITD). The ITD can be considered at the interface of thermodynamics and dynamics, as both class of processes fundamentally alters the ITD time evolution. In models, the ITD is represented by a number of ice thickness categories (or bins), including open water, in each GCM grid cell. Growth and melt processes may shift the ice between categories or create new thin ice, while deformation tends to break up thin ice and raft it or pile it up into ridges, which broadens the probability distribution and creates a long tail of thick ice. An increase in the resolution of the ITD in a model increases the total ice volume (and thickness), and hence the freshwater transport by sea ice is greater (Bitz et al. 2001; Holland et al. 2001, 2006). Several GCMs implemented parameterizations of an ITD in their latest versions (e.g., NCAR CCSM3 and PCM, UKMO HADGEM1, GFDL 2.0 and 2.1, and CNRM-CM3, see Holland et al. 2006; McLaren et al. 2006; Salas-Mélia 2002).

11.2.2 Flux Adjustments, Ocean Parameterizations, and Grids

Flux adjustments were a common feature of models in the early ACSYS era, which have been subsequently eliminated in most GCMs. Flux adjustments are prescribed offsets added to the freshwater and/or heat flux. They are intended to account for deficiencies in the coupled simulation that cause drift in the ocean surface salinity and/or temperature. The offsets usually vary from month to month but repeat year to year, and they are estimated by computing the mismatch in surface fluxes that arise in uncoupled simulations of the atmosphere and ocean with prescribed surface boundary conditions (see e.g., Manabe et al. 1991). Flux adjustment typically can be eliminated and a stable climate simulation can be achieved without them, by raising the ocean component's resolution and improving ocean mixing parameterizations (e.g., Boville and Gent 1998; Gordon et al. 2000). The elimination of flux adjustments is a positive step in improving climate models, as they have been shown to influence climate sensitivity (Gregory and Mitchell 1997).

Many models also now incorporate a representation of the freshwater input to the Arctic Ocean from continental river inflow, which is important for the freshwater balance of the Arctic and the dynamics of the Arctic shelf areas. Schemes are often very simple, with runoff at the land surface as a result of snowmelt and rainfall less evapotranspiration in excess of the needs of the model's soil moisture intake. Runoff is routed into the ocean via defined basins defined by the model's surface topography.

Recent models without flux adjustments simulate the twentieth century sea ice cover or Arctic surface air temperature with about the same fidelity as models with flux adjustments (e.g., Flato 2004; Hu et al. 2004). Among the models that do not have flux adjustments, many use a parameterization of advection by mesoscale eddies from Gent and McWilliams (1990) (GM). Poleward heat transport by ocean mesoscale eddies tends to be large in high southern latitudes, and some studies have found that using the GM parameterizations reduces the modeled sea ice extent in the Southern Hemisphere, but it has little influence in the Northern Hemisphere (Hirst et al. 2000; Gent et al. 2002).

Many ocean models have progressed from the rigid-lid approximation (with zero vertical motion at the surface) to various free-surface formulations in the past decade or so (Griffies et al. 2000). These new formulations permit more realistic exchange of mass, energy, and momentum across the ice–ocean interface. Even with the latest sea ice thermodynamic formulations that include brine pockets, ice thickness distribution and melt ponds, proper conservation is relatively straightforward (Schmidt et al. 2004). Instabilities have been known to arise from the interaction of sea ice dynamics coupled to free surface formulations, but they can also be avoided with relatively simple solutions (Schmidt et al. 2004).

Ocean and sea ice models often share the same grid and many global models in the past discretized their grid in spherical coordinates. The convergence of meridians at the North Pole demanded very small time steps, filtering small-scale variations in the zonal direction near the pole, and/or imposing an artificial island at the pole. Griffies et al. (2000) point out that filtering introduces noise and can destroy geostrophic and thermodynamic balances in ocean models. In sea ice models, filtering can create unphysical negative ice thicknesses and concentrations (Moritz and Bitz 2000). Further the artificial "shadowing" of fluid flow around an artificial island is undesirable. During the ACSYS era, much effort was placed on generalizing models to arbitrary orthogonal curvilinear coordinates, which permit coordinate singularities to be moved onto land. See Griffies et al. (2000) for a review of this practice in ocean models and Hunke and Dukowicz (1997) for an example in a sea ice model. Examples of CMIP3 models that use generalized orthogonal curvilinear coordinates are GFDL CM 2.0 and 2.1 and NCAR CCSM3 (Delworth et al. 2006; Collins et al. 2006).

The Arctic Ocean Model Intercomparison Project (AOMIP; Proshutinsky et al. 2001) has provided a coordinated effort to validate and improve model simulations of the Arctic ocean. This has led to an improved understanding of the processes affecting Arctic ocean conditions and circulation and subsequent recommendations for model improvements. A recent special issue of the *Journal of Geophysical Research – Oceans* (Proshutinsky and Kowalik 2007) highlights many of these studies. As one example, AOMIP studies have shown that tidal effects (which are not typically included in GCMs) can increase ventilation of the Atlantic layer and thereby increase its heat loss with subsequent impacts on the sea ice mass budget (Holloway and Proshutinsky 2007). These studies suggest that Arctic ocean tidal effects have important climate consequences and should be incorporated in future GCM ocean model developments.

11.2.3   Atmospheric Circulation and Clouds

Many GCMs have a systematic bias in the atmospheric surface circulation in the Arctic with a tendency for the mean sea-level pressure to be too high over the Arctic Ocean, except in the Beaufort and Chukchi seas in winter, where it is too low (Walsh and Crane 1992; Walsh et al. 2002; Chapman and Walsh 2007). The across-model variance of sea level pressure in late twentieth-century GCMs is larger in the Arctic than anywhere else in the Northern Hemisphere (Walsh et al. 2002). Bitz et al. (2002) applied biases in the geostrophic winds derived from AMIP1 models to a sea ice model and showed that the sea-level pressure biases created severe errors in the sea ice thickness and ice transport in the Arctic. These sea ice errors in turn had a first-order influence on the freshwater exchange with the ocean surface.

Earlier intercomparison studies proved difficult at attributing biases in the sea level pressure to any particular model parameterization or resolution (Bitz et al. 2002). With higher resolution models available now, deWeaver and Bitz (2006) found that surface winds gave rise to a better sea ice thickness pattern in one model at T85 resolution compared to T42 (about 1.4° compared to 2.8°).

Capturing the true vertical structure of the Arctic circulation in GCMs is also problematic. In at least one model, the Beaufort high in winter was found to have a baroclinic vertical structure, counter to the barotropic vertical structure in atmospheric reanalysis (deWeaver and Bitz 2006). A study comparing synoptic patterns in CMIP3 models found that GCMs tend to have too frequent and too strong anticyclones in the Arctic winter (Cassano et al. 2006).

The summertime Arctic surface circulation is dominated by a polar cyclone. The accompanying surface inflow and rising near the north pole results in a deep (thermally indirect) Ferrel cell north of the well-known polar cell. These summertime features do not appear in most GCMs (Bitz et al. 2002; deWeaver and Bitz 2006). Such biases in the atmospheric circulation aloft are bound to influence the import of heat and moisture from lower latitudes and cloud formation in GCMs.

Clouds play an important role in climate regulation by absorbing and scattering solar and terrestrial radiation. In the Arctic, the role and effect of clouds on climate are more complex owing to the highly reflecting snow-ice surface, low temperatures, variable amounts of water vapor, and the surface-based wintertime temperature inversion (Curry et al. 1996). Observations indicate that Arctic clouds act to warm the surface in winter and cool it for a short period in summer (Shupe and Intrieri 2004).

Early GCM cloud schemes were often purely diagnostic and many adapted methods introduced by Slingo (1987) and Wetherald and Manabe (1988). Cloud fraction parameterizations typically depended on cloud type, which included convective and stratiform clouds. (Sometimes the latter was broken into a number of more specialized types.) Cloud amounts usually depended on the parameterized convective mass flux, temperature and relative humidity profiles, vertical velocity, and atmospheric stability. Cloud optical properties were either based on prescribed fields or they scaled with the vertically integrated water vapor (known as the precipitable water). Precipitation would result from condensation that forms under supersaturated (or nearly supersaturated) conditions. Condensate often fell immediately to the ground; although, it might be reduced somewhat by evaporation along its path. Whether the condensate was converted to snow usually depended on low-level temperature and often the latent heat of fusion was neglected.

In a thorough review of the state of knowledge of Arctic cloud processes, Curry et al. (1996) concluded that too little was known to properly model cloud feedback and that Arctic specific parametizations of clouds were needed in GCMs. Schemes for non-convective cloud schemes, which are the primary challenge in modeling Arctic clouds (Curry et al. 1996), have seen improvements. A major step forward can be realized with the treatment of cloud liquid and ice condensate as prognostic variables – with individual equations that describe their evolution in time. Such schemes permit condensation prior to grid-box wide saturation and allow condensate to spend time within a cloud before converting to precipitation. These features are necessary to simulate ice condensate, which is needed to effectively dissipate moisture in winter. Because ice condensate grows larger and therefore falls faster than liquid condensate, proper mixed-phase cloud schemes are needed to accurately simulate cloud amount and optical properties (Beesley and Moritz 1999). The MPI ECHAM4 model was among the first GCMs to adopt a prognostic cloud water scheme with explicit ice-phase physics (Sundqvist et al. 1989), and the effort returned one of the best simulations of Arctic clouds among the 18 uncoupled atmosphere models analyzed by Tao et al. (1996).

Yet Arctic clouds in most GCMs have been a major source of error for decades. Uncoupled atmosphere models of the early ACSYS era had Arctic average cloud cover ranging from 30% to 90% in winter and 20% to 100% in summer, even though sea ice cover and SST boundary conditions were prescribed from observations (Tao et al. 1996). In addition, Tao et al. (1996) found no association between variations in the across-model winter cloud cover and winter surface temperature. Randall et al. (1998) note that the absence of a positive correlation is counter to observations.

More mature and fully coupled models that participated in CMIP2 had a slightly narrower range in cloud cover, at 40–90% in winter and 40–80% in summer (among the nine models that reported cloud cover, see Holland and Bitz 2003). The most current CMIP3 models have not further narrowed the wintertime range, but the summertime cloud cover in 21 of the 23 models has narrowed to within 10% of the observed cover. Yet CMIP3 models are still puzzling, as there is now a weak but significant *negative* correlation between winter cloud cover and surface temperature across these newer models.

Some have argued that regional climate models (RCMs) offer a good platform for developing and testing parameterizations for GCMs, especially for cloud and radiation processes because the large-scale evolution of atmospheric dynamics is constrained by prescribed lateral boundary conditions (Wyser et al. 2007; Dethloff et al. 2008). Yet at this time, cloud fraction in the Arctic has nearly as large a spread in RCMs as in GCMs. Interestingly, the correlation between observed and modeled surface radiation fluxes individually for longwave and shortwave radiation is much higher than for cloud fraction in RCMs. Wyser et al. (2007) argue that this is because Arctic clouds are frequently very thin, and thus radiation and cloud fraction are not well correlated. Further they found evidence that more work is needed to properly model the correct phase, size distribution, and ice crystal habit of cloud condensate. Models exist that include ice fog and diamond dust (e.g., Girard and Blanchet 2001), but the parameterizations are yet to be included in GCMs. Girard and Blanchet (2001) suggest that diamond dust ought to induce a strong radiative warming at the surface, but based on SHEBA data, Shupe and Intrieri (2004) concluded that diamond dust has very little radiative impact.

11.2.4   Ice Sheet Modeling

A few GCMs have incorporated an ice sheet model (ISM) such that changes to the
global climate in the GCM can interact with the shape and extent of a changing
Greenland ice sheet. The interaction occurs through changes in surface albedo as the
ice sheet retreats or advances over bare soil, elevation-temperature feedbacks, and
through changes in the atmospheric and oceanic circulation (Huybrechts
et al. 2002; Ridley et al. 2005; Driesschaert et al. 2007; Mikolajewicz et al. 2007).
The coupled ISMs show that the Greenland ice sheet declines for almost all future
forcing scenarios, and Greenland's ice melts complete within 1,000–3,000 years in
the fastest warming scenarios. If surface mass balance is considered alone, then it
has been suggested (Gregory and Huybrechts 2006) that a global temperature rise of
3 C could trigger an irreversible decline in the mass of the Greenland ice sheet. The
resolution of the atmospheric component of GCMs is too coarse to resolve the steep
ice sheet margins, a feature which is needed since the surface ablation is highest at
the low elevations of the margins. Consequently, high-resolution (10–20 km) three-
dimensional thermo-mechanical ISMs are coupled to the GCM. The coupling interface allows surface temperature and precipitation to provide the surface mass
balance. The surface runoff combined with ice-berg calving, determined by the ice
dynamics, passes fresh water to the ocean. Surface ablation, combined with ice
dynamics provides a new ice sheet orography for the GCM which influences atmospheric dynamics and surface albedo. Results from coupled GCMs and ISMs show
that even with the fastest warming scenarios, the ice sheet melt water has only a
minor influence on the Atlantic thermohaline overturning circulation and that atmospheric dynamics change after $\sim$ 200 years of ice sheet decline. The inclusion of
ISMs in GCMs allows for their influence on ocean salinity and sea level rise, and it
provides a validation of the carbon cycle and precipitation through comparison of
ISM diagnostics with observed ice cores.

11.3   CMIP3 Model Highlights

In this section, we highlight the simulated Arctic climate in the late twentieth century and mid twenty-first century in two state-of-the-art climate models that are part
of the CMIP3 dataset, the NCAR CCSM3 and UKMO HadGEM1. These models
made great strides in development during the ACSYS era. Here we explore the
extent to which the development effort influences the model results.

We analyze the SRES A1B scenario for the twenty-first century where the rate of
anthropogenic greenhouse gas emissions increases during the first half of the
twenty-first century, and then slowly declines in the second half. We compare
HadGEM1 and CCSM3 to the CMIP3 model mean (which includes the two
models) and to observations, where possible. A list of CMIP3 models is given in
Table 11.1 and much more information about the model physics can be found at
http://www-pcmdi.llnl.gov/ipcc/model_documentation/ipcc_model_

Table 11.1 CMIP3 Models used in this study

Modeling center	Model abbreviations
Bjerknes Centre for Climate Research (Norway)	BCCR BCM2.0
Canadian Centre for Climate Modelling and Analysis (Canada)	CCCMA CGCM3.1 T47, T63
Centre National de Recherches Meteorologiques, Meteo-France (France)	CNRM CM3
Commonwealth Scientific and Industrial Research Organization (Australia)	CSIRO MK3.0, MK3.5
Geophysical Fluid Dynamics Laboratory (USA)	GFDL CM2.0, CM2.1
Goddard Institute for Space Studies (USA)	GISS AOM, EH, ER
Institute for Numerical Mathematics (Russia)	INMCM3.0
Institut Pierre Simon Laplace (France)	IPSL CM4
Center for Climate System Research (Japan)	MIROC3.2 MEDRES, HIRES
University of Bonn (Germany)	MIUB ECHO G
Max-Planck-Institut fuer Meteorologie (Germany)	MPI ECHAM5
Meteorological Research Institute (Japan)	MRI CGCM3.2.2A
National Center for Atmospheric Research (USA)	NCAR PCM, CCSM3
United Kingdom Meteorological Office (UK)	UKMO HADCM3, HADGEM1

documentation.php. Sea ice output from GISS model EH and NCAR PCM was not available in the CMIP3 model archive at the time we wrote this paper. The IPSL CM4 model is excluded from sea ice diagnostics because its sea ice thickness changed abruptly at year 2000 owing to a change in aerosol forcing (S. Denvil, personal communication). The IAP FGOALS model is excluded from our analysis because the sea ice in that model is about twice as extensive as observed and the mean thickness in the Arctic is almost 10 m. A few other CMIP3 models are also known to have severe biases in the Arctic, especially in the sea ice. Nonetheless, we use all available models with equal weights in our multi-model ensemble mean except as noted below. Each model contributes about 5% to the mean, and usually the extreme biases are not of a single sign, so the ensemble mean is not significantly affected by any one model.

The highlighted models, CCSM3 and HadGEM1, stand out as among the most advanced in their sea ice physics, which include the elastic-viscous-plastic rheology and an explicit ice-thickness distribution with deformation and redistribution (Bitz et al. 2001; Lipscomb 2001). CCSM3 has sea ice with explicit brine pocket physics and a vertical temperature profile (Bitz and Lipscomb 1999). At least one study has argued that CCSM3 and HadGEM1 have the most realistic pattern of present-day Arctic sea ice thickness among CMIP3 models (Gerdes and Köberle 2007), and they are the only models that simulate recent Arctic summer ice retreat that is consistent with satellite observations (Stroeve et al. 2007) (although other models might compare favorably if they had run larger ensembles). Both models have free-surface oceans and are free of flux adjustments. They have prognostic ice and liquid condensate cloud physics, and above average horizontal resolution in all components.

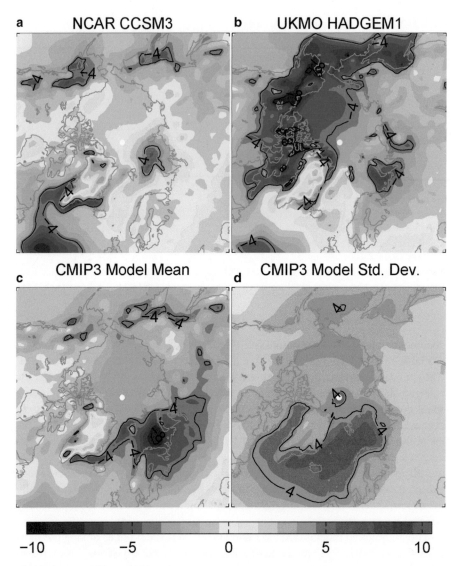

Fig. 11.1 1980–1999 mean bias (**a–c**) and across-model standard deviation (**d**) of the 2 m surface air temperature. The bias is relative to observations from ERA-40

11.3.1   Late 20th Century Climate

We begin by describing the surface air temperature near the end of the twentieth century. Figure 11.1a–c shows the annual mean bias in the models relative to the ECMWF 40-year reanalysis (ERA-40) (Uppala et al. 2005). The bias in CCSM3 in the Arctic is less than 2 °C in most regions, except notably it is too warm by

3–4 °C just north of Novaya Zemlya and too cold by a similar amount on Kamchutka and over southeastern Alaska. CCSM3 is also too cold by about 4–7 °C in the Labrador Sea, around the southern coast of Greenland, and further southward in the North Atlantic drift. HadGEM1 is about 4–8 °C too cold in a large swath over northern Canada and Alask, stretching out over the western central Arctic and over eastern Asia. There are also cold spots over Novaya Zemlya and eastward over northern Russia. HadGEM1 is also cold in the North Atlantic drift, but much less so than CCSM3.

The difference between the two models in the Barents Sea is likely due to the splitting of the West Spitzbergen current around Svalbard where the western branch sinks and flows to the north and the eastern branch encounters the Barents Sea shelf. HadGEM1 sends a greater portion of its warm Atlantic water west of Svalbard, while CCSM3 sends too much to the east, where it cannot sink at first and instead melts too much sea ice (Jochum et al. 2008). This explanation is consistent with the pattern of net upward surface heat flux in the two models (see Fig. 11.2a,b). The net surface flux maps the convergence of ocean heat transport, assuming the change in heat stored in the ocean column is small. Unfortunately, observational climatologies of surface heat fluxes are not reliable enough in the Arctic and subpolar seas to compute biases in Fig. 11.2.

The cold bias in CCSM3 in the Labrador Sea coincides with much too extensive sea ice (see Fig. 11.3a) and 50–150 W m^{-2} lower net surface heat flux than in HadGEM1 (see Fig. 11.2a,b). The net surface heat flux on the southern flank of the Gulf Stream is about 50 W m^{-2} lower in CCSM3 than in HadGEM1, consistent with the more negative surface air temperature bias in this region in CCSM3. In the Pacific sector, the net surface heat flux in HadGEM1 is lower than in CCSM3 by about 25–50 W m^{-2} and the sea ice is more extensive in HadGEM1.

The region with the largest surface air temperature bias in the CMIP3 model mean is in the Barents Sea and along the sea ice edge east of Greenland (see Fig. 11.1c). The sea ice edge is also on average too far south in this region in the CMIP3 models (see Fig. 11.3c). The across-model standard deviation in the annual mean surface air temperature varies most in the Nordic Seas and around Iceland (see Fig. 11.1d). It is also large in the marginal ice zones, especially in the Labrador Sea, Baffin Bay, and Davis Strait. The pattern was similar in CMIP1 models (see Fig. 11.6b of Walsh et al. 2002), but the CMIP1 models had considerably less variability in the Nordic seas. There is no apparent improvement in the model spread in surface air temperature from CMIP1 to CMIP3 when all models are considered from both eras. However, generally the surface air temperature and ice extent biases are lower in this region in HadGEM1 and CCSM3. The net surface flux in Fig. 11.2 along the ice edge between Norway and the southern tip of Greenland in the CMIP3 model mean is much smaller than in HadGEM1 and CCSM3. Although we have not ruled out the role of the atmosphere in our analysis, we note that the greater convergence of heat by the ocean in the subpolar Atlantic in HadGEM1 and CCSM3 is likely a major factor in their higher quality simulations east of Greenland.

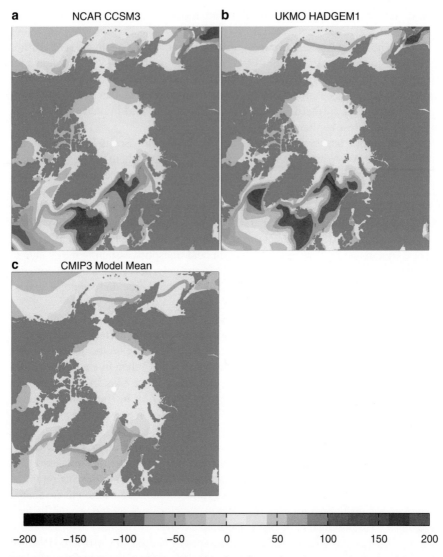

Fig. 11.2 1980–1999 mean net upward surface heat flux and sea ice extent in the models (**a–c**). Ice extent is defined as the 15% concentration contour

Figure 11.3a–c shows the annual mean sea ice thickness. CCSM3 and HadGEM1 have much thicker ice than the CMIP3 model average (averages are given in Table 11.3). Despite colder surface air temperatures, HadGEM1 has thinner ice than CCSM3. Ice thickness is not observed uniformly in space, so we do not include a figure for comparison. However, HadGEM1 and CCSM3 agree more favorably with the sporadic measurements from submarine upward looking sonar (see, e.g., Bourke and Garrett 1987; Rothrock et al. 1999), with values derived from satellite

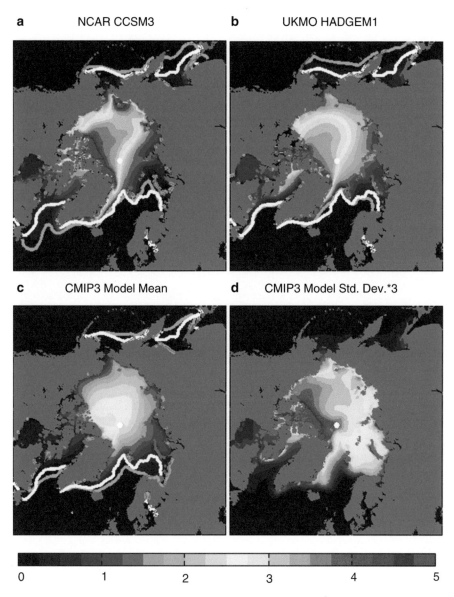

a   NCAR CCSM3   **b**   UKMO HADGEM1

c   CMIP3 Model Mean   **d**   CMIP3 Model Std. Dev.*3

0   1   2   3   4   5

Fig. 11.3 1980–1999 mean sea ice thickness (in m) and annual mean ice extent from the models (*green line*) with observed ice extent (*white line*) (**a–c**) and standard deviation of annual mean sea ice thickness scaled by a factor of 3 (**d**). Observations are from Comiso (1995). Ice extent is defined as the 15% concentration contour

altimetry data (Laxon et al. 2003), and with a hindcast using an ice-ocean model forced with observed atmospheric conditions (Gerdes and Köberle 2007). The buildup of thick ice along the Canadian Archipelago in the HadGEM1 and CCSM3 is most likely an indication of reasonable surface winds in the Arctic.

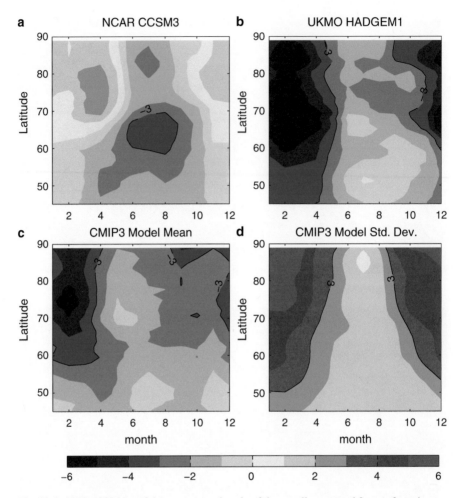

Fig. 11.4 1980–1999 bias of the mean annual cycle of the zonally averaged 2 m surface air temperature (**a–c**) and across-model standard deviation (**d**) in C. The bias is relative to observations from ERA40

Figure 11.4a–c shows the bias in the mean annual cycle of the zonal-mean surface air temperature in the models relative to the ERA40. The warm bias in CCSM3 in the Arctic is mostly a wintertime phenomena, reaching a maximum north of 70°N in late winter. The cold bias in CCSM3 is the worst at about 65°N in summer. There is a cold bias in HadGEM1 nearly year-round, but it is the worst in winter at all latitudes considered. The CMIP3 ensemble mean has a similar bias pattern though with slightly lower magnitude than HadGEM1.

In the across-model standard deviation computed for the zonal means by month (see Fig. 11.4d), the magnitude is largest in winter and at the highest latitudes. The maximum standard deviation is not at the transitions between melt/freeze periods. Instead the maximum variance is likely due to variations in downwelling longwave

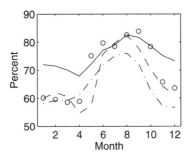

Fig. 11.5 1980–1999 total cloud cover averaged 70–90 N. Lines indicate model results: *dot-dashed* for CCSM3, *dashed* for HADGEM1, and *solid* for CMIP3 Model Mean. *Circles* indicate observations, which are for the period 1954-1997 from Hahn and Warren (2007)

radiation associated with biases in the wintertime clouds and atmospheric heat transport, and associated biases in the ice thickness and/or snow depth.

The seasonal cycle in cloud cover is shown in Fig. 11.5. HadGEM1 is within a few percent of recently observed cloud cover in all months except May and Oct. CCSM3 matches the observations well from Jan–Apr, but its cloud cover is at least 10% too low the rest of the year. Both models simulate the mean annual cycle of cloud cover well compared to the average of the CMIP3 models. These modest cloud biases in HadGEM1 also do not help explain the large cold bias in that model.

We believe HadGEM1 and CCSM3 simulate some cloud properties relatively well compared to other CMIP3 models because they have mixed-phase cloud scheme that independently predict the ice and liquid water content in clouds (Collins et al. 2006; Martin et al. 2006). Yet, cloud cover is only one cloud property of interest. Gorodetskaya et al. (2007) recently showed that despite relatively good agreement with summertime cloud fraction in CCSM3, the cloud liquid water content exceeds observed values, biasing cloud radiative properties. (Gorodetskaya et al. (2007) did not analyze HadGEM1.)

Model intercomparisons usually find that an across-model ensemble mean performs better than any individual model, especially for large-scale performance metrics (e.g., Gleckler et al. 2008). However, the accuracy of many aspects of the Arctic climatology in HadGEM1 and CCSM3 is substantially better than the CMIP3 model mean. It is apparent that the efforts to improve these models have paid off.

11.3.2 Mid-21st Century Climate Change

The pattern of surface warming at mid-twenty-first century in HadGEM1 and CCSM3 and in the CMIP3 model mean is shown in Fig. 11.6a–c. Clearly HadGEM1 and CCSM3 warm much more than the average CMIP3 model. All three panels have the largest warming over the sea ice on the Atlantic side of the Arctic Ocean. There is a complementary local maximum in the change in upward net surface heat flux in the same region (see Fig. 11.7). The magnitude is at least three times larger

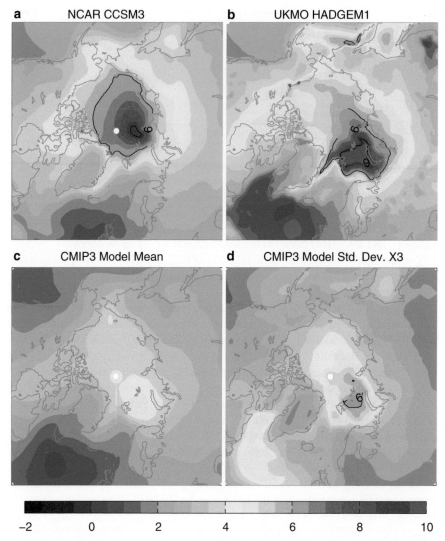

Fig. 11.6 Change in mean surface air temperature at mid-twenty-first century in C (2040–2059 minus 1980–1999) (**a–c**) and across-model standard deviation (**d**) scaled by a factor of 3

in HadGEM1 and CCSM3, consistent with the larger surface warming and sea ice retreat in these models (Fig. 11.7). The center of maximum warming in CCSM3 extends far deeper into the Arctic Ocean compared to HADGEM1. It is here where twentieth-century perennial ice becomes seasonal by mid-twenty-first century in CCSM3. Arzel et al. (2006) speculated that the large ice retreat in the Barents Sea in the twenty-first century on average in the CMIP3 models results from an increase in oceanic heat transport there. Our Fig. 11.7 confirms their suspicion.

There are other large differences in twenty-first-century net surface heat fluxes between CCSM3 and HadGEM1. The net surface heat flux increases by more than

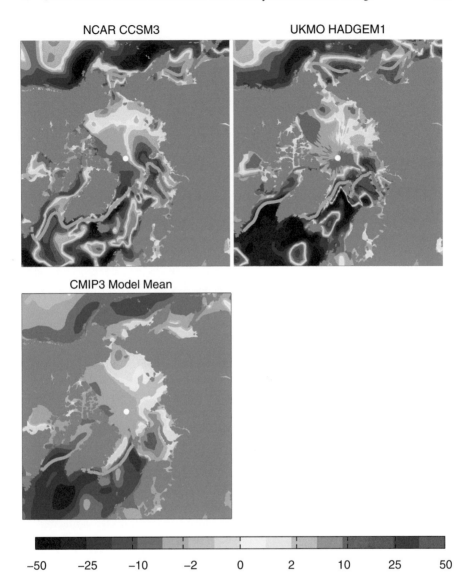

Fig. 11.7 Change in mean net surface heat flux at mid-twenty-first century in W m^{-2} (2040–2059 minus 1980–1999) with 2040–2059 mean sea ice extent. Positive indicates upward. Ice extent is defined as the 15% concentration contour

30 W m^{-2} in the Labrador Sea in the CCSM3 and the sea ice retreats at a high rate there, while the net surface heat flux change has the opposite sign in this region in HadGEM1 (and in the CMIP3 model mean). There is a 10–30 W m^{-2} increase in heat flux along the North Pacific ice edge in HadGEM1, where in CCSM3 the heat flux increase is much less. We suspect these differences arise from major changes in the ocean circulation that are driven by shifts in the midlatitude jets, but we have not analyzed this in depth.

Table 11.2 Change in mean temperature north of 75°N (ΔT_a), change in mean global temperature (ΔT_g), and Polar Amplification (PA=$\Delta T_a/\Delta T_g$)

Model	ΔT_a	ΔT_g	PA
	(°C)	(°C)	
CCSM3	6.07	1.84	3.30
HADGEM1	5.72	1.71	3.35
CMIP3 mean	3.67	1.59	2.35

The change is the mean of 2040–2059 minus 1980–1999

Table 11.3 Northern Hemisphere Sea Ice Extent (SIE) and annual mean Sea Ice Thickness averaged north of 70 N (SIT)

Model	Sep. SIE	Apr. SIE	SIT	Sep. ΔSIE	Apr. ΔSIE	ΔSIT
	10^6 km^2	10^6 km^2	m	10^6 km^2	10^6 km^2	m
CCSM3	7.5	16.8	2.13	−6.2	−2.9	−1.45
HADGEM1	7.3	17.3	2.11	−4.2	−2.8	−1.06
CMIP3 mean	7.7	16.4	1.72	−2.9	−1.6	−0.67
Observations	7.1	15.0				

The first series (without the Δ symbol) are means from 1980–1999. The second series (with the Δ symbol) is the mean of 2040–2059 minus mean of 1980–1999.

Table 11.2 lists the mid-twenty-first century Arctic temperature change (75–90°N), global mean temperature change, and their ratio, which we call the polar amplification. The highlighted models, HadGEM1 and CCSM3, are considerably higher in all three statistics than the CMIP3 model mean. These two models warm about 6°C on average from 75°N to 90°N, which is more than three times the global mean.

The large warming in HadGEM1 and CCSM3 is also associated with relatively high thinning and retreat of sea ice in these models (see Fig. 11.8a–c). Area average sea ice statistics listed in Table 11.3 indicate that thickness and extent changes in HadGEM1 and CCSM3 are 50–100% larger than the CMIP3 model means. The more uniform surface warming over the Arctic Ocean in CCSM3 compared to HadGEM1 (Fig. 11.6) coincides with greater thinning across the Arctic Ocean in CCSM3. In contrast, in HadGEM1 the warming is sharply peaked at the ice edge in the Barents Sea, where there is also a large gradient in the change in upward net surface heat flux in that model.

It had been shown with CMIP2 models that sea ice thickness influences the Arctic response to increasing anthropogenic greenhouse forcing, while the extent has little or no influence (Rind et al. 1995; Holland and Bitz 2003; Walsh and Timlin 2003). Across-model correlation analysis of these variables in the CMIP3 models is given in Table 11.4. With monthly mean output available in the CMIP3 archive, we are able to examine monthly relations. However, thickness anomalies are highly correlated from month to month (and year to year), so monthly thickness data are not needed. Table 11.4 indicates that in across-models for the period 1980–1999, the mean thickness is highly correlated with the mean extent in September but not April. Presumably, this is because the summer surface energy balance has a large influence on ice

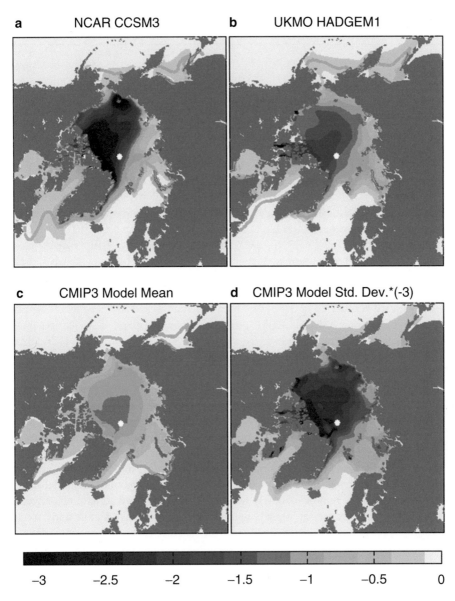

Fig. 11.8 Change in annual mean sea ice thickness at mid-twenty-first century in m (2040–2059 minus 1980-1999) with the 2040–2059 annual mean ice extent (*green line*) (**a–c**) and across-model standard deviation (**d**) scaled by a factor of – 3

thickness and summertime extent, while the winter extent is heavily influenced by wintertime winds and ocean heat fluxes (see Bitz et al. 2005).

However, the positive across-model correlation between thickness and September extent does not carry-over to HadGEM1 and CCSM3. Instead these models have above average thickness but below average September extent (see Table 11.3).

Table 11.4 Across-model correlations of sea ice variables as defined in Table 11.3

	Sep. SIE	Apr. SIE	SIT	Sep. ΔSIE	Apr. ΔSIE	ΔSIT
Sep. SIE				**0.52**	**0.40**	−0.19
Apr. SIE	0.18			−0.02	**−0.42**	0.18
SIT	**0.80**	0.02		0.30	0.17	**−0.58**

MIROC HIRES is eliminated from correlations that involve the change at mid-twenty-first century, because the ice melts away in that model by about 2020, 60 years earlier than in any other model. Numbers in Bold are significant at the 5% confidence level

We believe that CCSM3 and HADGEM1 are unusual because they explicitly resolve the time evolution of the sea ice thickness distribution (ITD) in their sea ice component models. When models resolve the ITD, the total ice volume in the Arctic increases (Bitz et al. 2001; Holland et al. 2001, 2006) and the ice thickness increases on average (over grid cells and larger regions). The ice extent seems to improve as well (Bitz et al. 2001; Salas-Mélia 2002). We suspect that most CMIP3 models have been compromised when tuning: If they had been tuned to be less extensive, they would also have become too thin. This compromise may be reduced when an ITD is included. In addition, the winter surface temperature tends to be slightly warmer when thin ice is resolved in regions with both perennial and first year ice. It is also apparent that HadGEM1 and CCSM3 have a larger annual range of sea ice area than the CMIP3 model mean, which is expected in models with an ITD (Bitz et al. 2001; Holland et al. 2001, 2006).

The across-model correlations in Table 11.4 that relate quantities in the late twentieth century with the changes at mid-twenty-first century are relevant for understanding relative changes in HadGEM1 and CCSM3 compared to the CMIP3 mean. Across CMIP3 models, we find that the September extent correlates significantly with the September retreat, such that models with more extensive ice retreat more slowly. Interestingly, the opposite relation occurs in April, albeit with a weaker correlation. Because the relations are seasonally opposing, the annual mean extent is not well correlated with annual mean retreat. One might imagine that models with thicker ice would retreat more slowly, but there is no significant correlation between these quantities because the models with thicker ice also have significantly more thinning. This unintuitive result stems from the fact that ice growth in winter damps anthropogenic thinning to some extent (Bitz and Roe 2004). The growth rate is inversely related to ice thickness, and hence the thinner the ice, the more strongly damped is its rate of thinning. The argument holds provided the net damping dominates over the positive ice albedo feedback, which must be so when and where sea ice is stable.

Figure 11.9 illustrates the probability density of the fractional sea ice coverage (which depicts the ITD) for the late twentieth century and the mid- and late twenty-first century. The two individual models have a tendency to lose multiyear ice and gain first year ice in the twenty-first century. In the late twentieth century, CCSM3 has more multiyear (a larger thick ice tail) than HadGEM1 (see Fig. 11.9). Yet the multiyear ice disappears sooner in CCSM3 than in HadGEM1, despite similar magnitudes of global and Arctic warming in the models. Also note that in CCSM3

Fig. 11.9 Annual mean sea ice thickness distribution averaged from 75–90N only where the sea ice concentration is 15% or greater in individual years

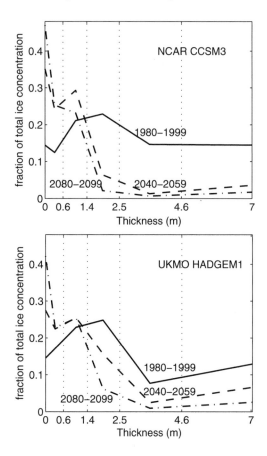

compared to HadGEM1 by mid-twenty-first century, the sea ice retreat in September retreat is about 50% greater (see Table 11.3). Thus, the multiyear sea ice cover appears to be more sensitive to warming in CCSM3 than HadGEM1.

Figure 11.10 shows the mean annual cycle of the change in zonal mean surface air temperature. The season of maximum warming is in early winter (Oct.–Nov.), when the CMIP3 model mean warms on average more than 6°C north of 80°N and HadGEM1 and CCSM3 warm more than 11°C. The warming is about 3–5°C lower in deep winter (Jan–Mar) in HadGEM1. In summer (Jun.–Aug.), the warming is a minimum (at 2°C or less) north of 70°N. Polar amplification is most apparent during the cold season, and it is absent during the melt season, because the temperature is limited to the melting temperature over a substantial portion of the Arctic.

Greater springtime warming in HadGEM1 may result from the larger increase in cloud cover in May (see Figs. 11.4 and 11.11). The larger cloud increase in Fall in CCSM3 could be a factor in the larger Fall surface air warming in CCSM3.

We end our analysis with a discussion of the atmospheric and ocean heat transport into the Arctic. The poles are sometimes referred to as heat sinks for the planet.

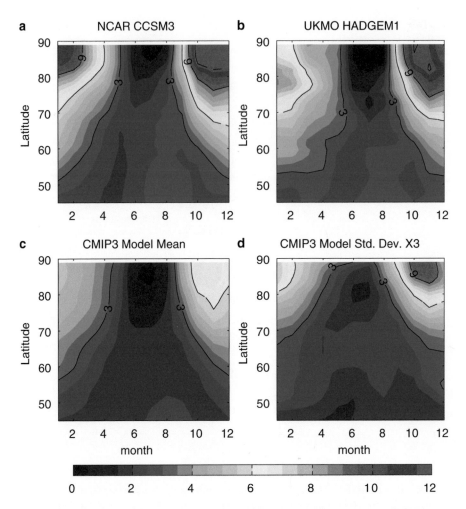

Fig. 11.10 Change in mean annual cycle of the zonally averaged surface air temperature at mid-twenty-first century in °C (2040–2059 minus 1980–1999) (**a–c**) and across-model standard deviation (**d**)

Indeed about 100 W m^{-2} of heat escapes the top of the atmosphere on average north of 70°N over the year (Oort 1974). About two third of the atmospheric heat transport across 70°N is due to sensible heat and potential energy transport, known as the dry static energy (DSE) transport, and about one third is due to latent energy (LE) transport, with kinetic energy making up a near-negligible contribution (Overland and Turet 1994). The DSE transport is thought to depend strongly on the meridional temperature gradients, while the latent heat transport depends mostly on temperature (e.g., see Oort 1974; Held and Soden 2006). In a greenhouse warming climate, one expects the annual mean poleward temperature gradient to decrease on average owing to polar amplification of the warming, thus, the DSE transport into; the Arctic

Fig. 11.11 Change in
percent of total cloud cover
at mid-twenty-first century
(2040–2059 minus 1980–1999)
averaged 70–90 N
in CCSM3 (*dot-dashed*),
HADGEM1 (*dashed*), and
the CMIP3 Model Mean
(*solid*)

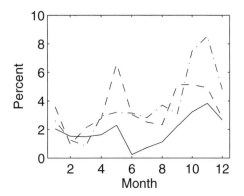

should decrease, giving rise to a negative feedback. At the same time, the rising temperature is expected to increase the LE transport into the Arctic. Figure 11.12 shows this expected behavior in HadGEM1, CCSM3, and the CMIP3 model mean. The sum of the two components in the CMIP3 model mean is near zero (see Fig. 11.12c). In contrast, the decrease in northward DSE transport is greater than the increase in northward LE transport in HadGEM1 and CCSM3, consistent with the very large polar amplifications in these two models.

Figure 11.12 also shows the change in northward oceanic heat transport. This quantity increases slightly north of 60°N in the CMIP3 model mean, and it increases a relatively much larger amount in HadGEM1 and CCSM3. An increase in the oceanic heat transport into the Arctic was found in the majority of CMIP2 models as well (Holland and Bitz 2003). Bitz et al. (2006) analyzed this increase in CCSM3 and posited a positive feedback between ocean heat import into the Arctic and sea ice retreat. Thus, the poleward ocean heat transport likely contributes to polar amplification in the Arctic surface warming at mid-twenty-first century and may even be part of another positive feedback.

11.4 Summary and Future Outlook

A number of modeling centers have devoted a considerable amount of energy to improve high-latitude climate physics in their models during the ACSYS era. In the best models, the sea ice components now take into account the ice rheology, ice thickness distribution, and multiple vertical layers. For the ocean component of coupled models generally, flux adjustments have been eliminated, terrestrial runoff schemes have been adopted, and vertical mixing schemes have been updated. There is some evidence that modeled atmospheric circulation benefits from higher resolution, and many models have implemented improved schemes for treating clouds.

Stimulated by the immense and immensely valuable CMIP3 archive, IPCC 2007 initiated a new paradigm in research with GCMs (Meehl et al. 2007). With the

Fig. 11.12 Change in atmospheric and oceanic heat transport (*thick dashed* and *solid lines*, resp.) at mid-twenty-first century in PW (2040–2059 minus 1980–1999). The atmospheric heat transport is broken into latent energy and dry static energy components (*thin dot-dashed* and *dashed*, resp)

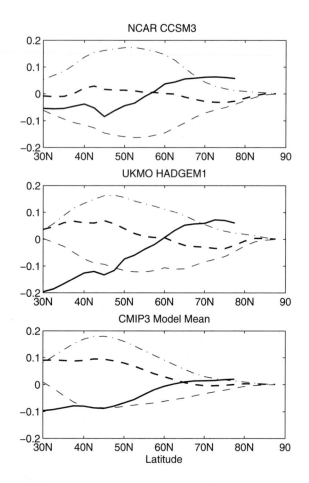

model output archived substantially in advance, analysis included in the assessment could test hypothesis with a variety of GCMs at once. In addition, a more thorough model intercomparison was possible. In this Chapter, we have reviewed numerous valuable studies that analyzed the Arctic climate in the CMIP3 models. These studies have had a significant impact on understanding model behavior and will steer the development of new model physics.

In Sect. 11.3, we featured results from two climate models that saw substantial improvements during the ACSYS era. The pay-off is clear. The large-scale pattern of sea ice thickness in the Arctic is well represented, and the sea ice edge east of Greenland is positioned fairly well. Cloud cover is within about 10% of observations when averaged north of 70°N. In some ways HadGEM1 and CCSM3 capture the late twentieth century Arctic climate better than the multi-model ensemble mean of the CMIP3 models. These two models appear to have reduced sea ice albedo tuning compromises by including better sea ice physics. Thus, the multi-model ensemble mean may not be the best forecast in an area such as the Arctic, where model physics in some models lag severely behind the best models.

It is unfortunate that we were unable to examine each CMIP3 model with the same level of detail that we gave to HadGEM1 and CCSM3. We do not claim that these two models are the best models. Indeed different studies have found superior behavior in other CMIP3 models (e.g., Chapman and Walsh 2007). They were chosen because we participated in their development.

These two models have considerably larger climate change in the Arctic than the multi-model ensemble mean by mid-twenty-first century compared to the late twentieth century. The surface warms by about $6\,^{\circ}$C on average north of $75\,^{\circ}$N in HadGEM1 and CCSM3, which is more than three times the global average. In contrast the surface warming in the CMIP3 model mean is less than $4\,^{\circ}$C, which is closer to twice the global average. The September sea ice extent declines by 4.2 and 6.2×10^6 km^2 in HadGem1 and CCSM3, respectively, compared to just 2.9×10^6 km^2 in the CMIP3 model mean. The reduction in annual mean thickness north of $70\,^{\circ}$N is similarly higher at 1.06 and 1.45m in HadGem1 and CCSM3, respectively, compared to just 0.67m in the CMIP3 model mean. Generally, the changes across models can be attributed somewhat to the late-twentieth century mean state (see Table 11.4). Models with below average September extent and above average thickness (as in HadGEM1 and CCSM3) tend to also have larger sea ice changes. The twentieth-century ice thickness and September extent in HadGEM1 and CCSM3 are a good match to observations, which gives us some confidence that their large future changes are plausible. Compared to the CMIP3 model mean, HadGEM1 and CCSM3 also have above average increases in ocean heat transport into the Arctic, which appears to contribute to the large climate change in the two models.

We expect it will not be long before almost every GCM has a dynamical sea ice component with an ice thickness distribution that builds ridges under compression and shear. Non-continuum, or discrete element, sea ice models that split floes based on the theory of fracture mechanics are still on the distant horizon for GCMs. Solving a multi-layer thermodynamics scheme for the vertical temperature profile and subsequently for sea ice growth and melt is not computationally expensive compared to sea ice dynamics and transport schemes. Brine pocket energy storage adds only a minor complication. New physically based methods for treating melt ponds and radiative transfer in sea ice are well developed. New efficient sea ice transport schemes are making more sophisticated treatment of sea ice and snow thermodynamics and the sea ice thickness distribution (and the associated expanding lists of state variables) feasible in GCMs. Owing to the importance of these sea ice physics, which was discussed in Sect. 11.2, we expect many models will adopt them soon.

Expanding computing capacity within the next decade will permit much higher resolution GCMs. In the Arctic, this is likely to be important for improving atmospheric circulation and the representation of ocean eddies. Simulated oceanic heat and freshwater transport could benefit a great deal. Many GCMs will soon have the capacity to run high-resolution regional components embedded within them, which should be useful for further investigating the role of tides and eddies in Arctic climate.

Cloud models are beginning to resolve size distributions of ice and liquid cloud condensate as well as a variety of ice crystal habits. Their development is part of the continued long path toward higher-quality Arctic cloud simulations.

Another area that will see rapid development in the next decade is the treatment of ice sheets. So far ice sheets in GCMs have only dealt with the behavior of deep, cold land ice. The important roles of ice shelves and their grounding line, ice streams, and calving have not yet been considered. New ice sheet models are needed to incorporate their behavior. The potential critical influence of ice sheet decay on sea level rise and ocean circulation has called the attention of many modelers to this important new work.

GCMs are evolving into Earth System Models that couple physical and biogeo-chemical systems to model Earth's cycles of carbon and aerosols. Many of the problems of interest for Earth System Modeling involve the polar regions, so the continued development of new physics and new capability must not be carried out without special emphasis on their operation in the polar regions.

Acknowledgments The authors gratefully acknowledge the support of the National Science Foundation through grants ATM0304662 and OPP0454843 (CMB) and OPP0084273 (MMH). We acknowledge the modeling groups, the Program for Climate Model Diagnosis and Intercomparison (PCMDI) and the WCRP's Working Group on Coupled Modelling (WGCM) for their roles in making available the WCRP CMIP3 multi-model dataset. Support for this dataset is provided by the Office of Science, U.S. Department of Energy.

References

Arzel O, Fichefet T, Goosse H (2006) Sea ice evolution over the 20th and 21st centuries as simulated by the current AOGCMs. Ocean Model 12:401–415

Beesley JA, Moritz RE (1999) Toward an explanation of the annul cycle of cloudiness over the arctic ocean. J Clim 12:395–415

Bitz CM, Lipscomb WH (1999) An energy-conserving thermodynamic model of sea ice. J Geophys Res 104:15,669–15,677

Bitz CM, Roe GH (2004) A mechanism for the high rate of sea-ice thinning in the arctic ocean. J Clim 18:3622–3631

Bitz CM, Holland MM, Weaver AJ, Eby M (2001) Simulating the ice-thickness distribution in a coupled climate model. J Geophys Res 106:2441–2464

Bitz CM, Fyfe JC, Flato GM (2002) Sea ice response to wind forcing from amip models. J Clim 15:522–536

Bitz CM, Holland MM, Hunke EC, Moritz RE (2005) On the maintenance of the sea-ice edge. J Clim 18:2903–2921

Bitz CM, Gent PR, Woodgate RA, Holland MM, Lindsay R (2006) The influence of sea ice on ocean heat uptake in response to increasing CO_2. J Clim 19:2437–2450

Bourke RH, Garrett RP (1987) Sea ice thickness distribution in the arctic ocean. Cold Reg Sci Tech 13:259–280

Boville BA, Gent PR (1998) The NCAR climate system model, version one. J Clim 11:1115–1130

Briegleb BP, Light B (2007) A Delta-Eddington multiple scattering parameterization for solar radiation in the sea ice component of the community climate system model. NCAR/TN-472+STR

Briegleb BP, Bitz CM, Hunke EC, Lipscomb WH, Schramm JL (2004) Scientific Description of the sea ice component in the community climate system model, version 3. NCAR/TN-463+STR

Budyko MI (1969) The effect of solar radiatin variations on the climate of the earth. Tellus 21:611–619

Cassano JJ, Uotila P, Lynch AH (2006) Changes in synoptic weather patterns in the polar regions in the 20th and 21st centuries. Part 1. Arctic. Int J Climatol 26. doi:10.1002/JOC.1306

Chapman WL, Walsh JE (2007) Simulation of arctic temperature and pressure by global climate models. J Clim 20:609–632

Collins WD et al (2006) The community climate system model, Version 3. J Clim 19:2122–2143

Comiso JC (1995) SSM/I concentrations using the Bootstrap Algorithm. Tech Rep RP 1380, 40 pp, NASA, Technical Report

Curry JA, Rossow WB, Randall D, Schramm JL (1996) Overview of arctic cloud and radiation charactereistics. J Clim 9:1731–1764

Delworth TL et al (2006) CM2 global coupled climate models – part 1: formulation and simulation characteristics. J Clim 19:675–697

Dethloff K, Rinke A, Lynch A, Dorn W, Saha S, Handorf D (2008) Chapter 8: Arctic regional climate models. In: Arctic climate change — The ACSYS decade and beyond, this volume

deWeaver E, Bitz CM (2006) Atmospheric circulation and Arctic sea ice in CCSM3 at medium and high resolution. J Clim 19:2415–2436

Driesschaert E, Fichefet T, Goosse H, Huybrechts P, Janssens L, Mouchet A, Munhove G, Brovkin V, Weber SL (2007) Modeling the influence of greenland ice sheet melting on the atlantic meridional overturning circulation during the next millennia. Geophys Res Lett 34:L10707

Ebert EE, Curry JA (1993) An intermediate one-dimensional thermodynamic sea ice model for investigating ice-atmosphere interactions. J Geophys Res 98:10,085–10,109

Fichefet T, Morales Maqueda M (1997) Sensitivity of a global sea ice model to the treatment of ice thermodynamics and dynamics. J Geophys Res 102:12,609–12,646

Flato GM (2004) Sea-ice climate and sensitivity as simulated by several global climate models. Clim Dyn 23:229–241

Flato GM, Hibler WD (1992) Modeling pack ice as a cavitating fluid. J Phys Oceanogr 22:626–651

Gent PR, McWilliams JC (1990) Isopycnal mixing in ocean circulation models. J Phys Oceanogr 20:150–155

Gent PR, Craig AP, Bitz CM, Weatherly JW (2002) Parameterization improvements in an eddy-permitting ocean model. J Clim 13:1447–1459

Gerdes R, Köberle C (2007) Comparison of arctic sea ice thickness variability in IPCC climate of the 20th century experiments and in ocean sea ice hindcasts. J Geophys Res 112:C04S13. doi:10.1029/2006JC003,616

Girard E, Blanchet JP (2001) Simulation of arctic diamond dust, ice fog, and thin stratus using an explicit aerosol-cloud-radiation model. J Atmos Sci 58:1199–1221

Gleckler PJ, Taylor KE, Doutriaux C (2008) Performance metrics for climate models. J Geophys Res 113:D06,104. doi:10.1029/2007JD008,972

Gordon C, Cooper C, Senior CA, Banks HT, Gregory JM, Johns TC, Mitchell JFB, Wood RA (2000) The simulation of SST, sea ice extents and ocean heat transports in a version of the Hadley Centre coupled model without flux adjustments. Clim Dyn 16:147–168

Gordon HB, O'Farrell SP (1997) Transient climate change in the CSIRO coupled model with dynamics sea ice. Mon Weather Rev 125:875–907

Gorodetskaya IV, Tremblay L-B, Lipert B, Cane MA, Cullather RI (2007) Modification of the arctic ocean short-wave radiation budget due to cloud and sea ice properties in coupled models and observations. J Climate 21:866–882

Gregory JM, Huybrechts P (2006) Ice-sheet contributions to future sea-level change. Philos Trans R Soc A 364:1709–1731

Gregory JM, Mitchell JFB (1997) The climate response to CO_2 of the Hadley Centre coupled aogcm with and without flux adjustment. Geophys Res Lett 24:1943–1946

Griffies SM, Böning C, Bryan FO, Chassingnet EP, Gerdes R, Hasumi H, Hirst A, Treguier A-M, Webb D (2000) Developments in ocean climate modeling. Ocean Model 2:123–190

Hahn CJ, Warren SG (2007) A gridded climatology of clouds over land (1971–96) and ocean (1954–97) from surface observations worldwide, Tech. Rep. Documentation, 70pp, Carbon Dioxide Information Analysis Center (CDIAC), Department of Energy, Oak Ridge, Tennessee

Held IM, Soden BJ (2006) Robust responses of the hydrological cycle to global warming. J Clim 19:5686–5699

Hewitt CD, Senior CS, Mitchell J (2001) The impact of dynamic sea-ice on the climate sensitivity of a GCM: a study of past, present, and future climates. Clim Dyn 17:655–668

Hibler WD (1979) A dynamic thermodynamic sea ice model. J Phys Oceanogr 9:815–846

Hibler WD (1980) Modeling a variable thickness ice cover. Mon Weather Rev 108:1943–1973

Hibler WD (1984) The role of sea ice dynamics in modeling CO_2 increases. In: Hansen JE, Takahashi T (eds) Climate processes and climate sensitivity. Geophysical monograph 29, vol 5. American Geophysical Union, Washington, DC, pp 238–253

Hirst AC, O'Farrell SP, Gordon HB (2000) Comparison of a coupled oceanatmosphere model with and without oceanic eddy-induced advection. Part I: Ocean spinup and control integrations. J Clim 13:139–163

Holland DM, Mysak LA, Manak DK, Oberhuber JM (1993) Sensitivity study of a dynamic thermodynamic sea ice model. J Geophys Res 98:2561–2586

Holland MM, Bitz CM (2003) Polar amplification of climate change in the coupled model intercomparison project. Clim Dyn 21:221–232

Holland MM, Bitz C, Weaver A (2001) The influence of sea ice physics on simulations of climate change. J Geophys Res 106:2441–2464

Holland MM, Bitz CM, Hunke EC, Lipscomb WH, Schramm JL (2006) Influence of the sea ice thickness distribution on polar climate in CCSM3. J Clim 19:2398–2414

Holloway G, Proshutinsky A (2007) Role of tides in Arctic ocean/ice climate. J Geophys Res 112:C04S06. doi:10.1029/2006JC003,643

Hu Z-Z, Kuzmina SI, Bengtsson L, Holland DM (2004) Sea-ice change and its connection with climate change in the arctic in CMIP2 simulations. J Geophys Res 109:D10,106. doi:10.1029/2003JD004,454

Hunke EC, Dukowicz JK (1997) An elastic-viscous-plastic model for sea ice dynamics. J Phys Oceanogr 27:1849–1867

Hunke EC, Zhang Y (2000) Comparison of sea ice dynamics models at high resolution. Mon Weather Rev 127:396–408

Huwald H, Tremblay L-B, Blatter H (2005) A multilayer sigma-coordinate thermodynamic sea ice model: Validation against Surface Heat Budget of the Arctic Ocean (SHEBA)/Sea Ice Model Intercomparison Project Part 2 (SIMIP2) data. J Geophys Res 110:C05,010. doi:10.1029/2004JC002,328

Huybrechts P, Janssens I, Pocin C, Fichefet T (2002) The response of the greenland ice sheet to climate changes in the 21st century by interactive coupling of an aogcm with a thermomechanical ice-sheet model. Ann Glaciol 34:408–415

IPCC (1992) Climate change 1992: the IPCC scientific assembly supplementary report. Cambridge University Press, Cambridge, 198pp

IPCC (2007) Climate change 2007 the physical science basis. Contribution of Working Group I to the fourth assessment report of the intergovernmental panel on climate change. Cambridge University Press, Cambridge, 996pp

Jochum M, Danabasoglu G, Holland MM, Kwon Y, Large W (2008) Ocean viscosity and climate. J Geophys Res 113:C06017. doi:10.1029/2007JC004,515

Kreyscher M, Harder M, Lemke P, Flato GM (2000) Results of the sea ice model intercomparison project: evaluation of sea ice rheology schemes for use in climate simulations. J Geophys Res 105:11,299–11,320

Laxon S, Peacock N, Smith D (2003) High interannual variability of sea ice thickness in the Arctic region. Nature 425:947–950

Lemke P, Owens W, Hibler W (1990) A coupled sea ice-mixed layer-pycnocline model for the weddell sea. J Geophys Res 95:9513–9525

Lemke P, Hibler W, Flato G, Harder M, Kreyscher M (1997) On the improvement of sea ice models for climate simulations: the sea ice model intercomparison project. Ann Glaciol 25:183–187

Lipscomb WH (2001) Remapping the thickness distribution in sea ice models. J Geophys Res 106:13,989–14,000

Lipscomb WH, Hunke EC (2004) Modeling sea ice transport using incremental remapping. Mon Weather Rev 132:1341–1354

Manabe S, Stouffer RJ, Spellman MJ, Bryan K (1991) Transient responses of a coupled ocean-atmosphere model to gradual changes of atmospheric CO_2. Part I. Annual mean response. J Clim 4:785–818

Marsland SJ, Haak H, Jungclaus JH, Latif M, Roeske F (2003) The max-planck-institute global ocean/sea ice model with orthogonal curvilinear coordinates. Ocean Model 5:91–127

Martin G, Ringer M, Pope V, Jones A, Dearden C, Hinton T (2006) The physical properties of the atmosphere in the new Hadley Centre Global Environmental Model, HadGEM1. Part 1: Model description and global climatology. J Clim 19:147–168

Maykut GA, Untersteiner N (1971) Some results from a time-dependent thermodynamic model of sea ice. J Geophys Res 76:1550–1575

McFarlane NA, Boer GJ, Blanchet J-P, Lazare M (1992) The Canadian Climate Centre second-generation general circulation model and its equilibrium climate. J Clim 5:1013–1044

McLaren AJ et al (2006) Evaluation of the sea ice simulation in a new coupled atmosphere-ocean climate model (HadGEM1). J Geophys Res 111:C12,014, doi:10.1029/2005JC003,033

Meehl GA, Boer G, Covey C, Latif M, Stouffer R (2000) Coupled model intercomparison project. Bull Am Meteorol Soc 81:313–318

Meehl GA, Covey C, Delworth T, Latif M, McAvaney B, Mitchell JFB, Stouffer RJ, Taylor KE (2007) The WCRP CMIP3 multimodel dataset: A new era in climate change research. Bull Am Meteorol Soc. doi:10.1175/BAMS-88-9-1383

Mikolajewicz U, Vizcaino M, Jungclaus J, Schurgers G (2007) Effect of ice sheet interactions in anthropogenic climate change simulations. Geophys Res Lett 34:L18706

Moritz RE, Bitz CM (2000) Climate model underestimates natural variability of Northern Hemisphere sea ice extent. Science 288:927a

Oberhuber JM (1993) Simulation of the Atlantic circulation with a coupled sea-ice-mixed layer-isopycnical general circulation model. Part I: model description. J Phys Oceanogr 23:808–829

O'Farrell SP (1998) Investigation of the dynamic sea-ice component of a coupled atmosphere sea-ice general circulation model. J Geophys Res 103:15,751–15,782

Oort AH (1974) Year-to-year variations in the energy balance of the arctic atmosphere. J Geophys Res 79:1253–1260

Overland JE, Turet P (1994) Variability of the atmospheric energy flux across 70 N computed from the GFDL data set. In: Johannessen OM, Muench RD, Overland JE (eds) Polar oceans and their role in shaping the global environment. Geophysics Monograph 85. American Geophysical Union, Washington, DC

Pedersen CA, Roeckner E, Lüthje M, Winther J-G (2009) A new sea ice albedo scheme including melt ponds for ECHAM5 general circulation model. J Geophys Res 114. doi:10.1029/2008JD010,440

Pollard D, Thompson SL (1994) Sea-ice dynamics and co_2 sensitivity in a global climate model. Atmos Ocean 32:449–467

Proshutinsky A, Kowalik Z (2001) Preface to special section on Arctic Ocean Model Intercomparison Project (AOMIP) studies and results. J Geophys Res 112:C04S01. doi:10.1029/2006JC004,017

Proshutinsky A et al (2001) The Arctic Ocean Model Intercomparison Project (AOMIP). EOS Trans Am Geophys Union 82(51):637–644

Randall D et al (1998) Status of and outlook for large-scale modeling of atmosphere-ice-ocean interactions in the arctic. Bull Am Meteorol Soc 79:197–219

Ridley JK, Huybrechts P, Gregory JM, Lowe JA (2005) Elimination of the greenland ice sheet in a high co2 climate. J Clim 18:3409–3427

Rind D, Healy R, Parkinson C, Martinson D (1995) The role of sea ice in $2XCO_2$ climate model sensitivity. 1. The total influence of sea ice thickness and extent. J Clim 8:449–463

Rothrock DA, Yu Y, Maykut GA (1999) Thinning of the arctic sea ice cover. Geophys Res Lett 26:3469–3472

Salas-Mélia D (2002) A global coupled sea ice-ocean model. Ocean Model 4:137–172

Schmidt GA, Bitz CM, Mikolajewicz U, Tremblay LB (2004) Ice-ocean boundary conditions for coupled models. Ocean Model 7:59–74. doi:10.1016/S1463–5003(03)00,030–1

Sellers WD (1969) A global climate model based on the energy balance of the earth-atmosphere system. J Appl Meteorol 8:392–400

Semtner AJ (1976) A model for the thermodynamic growth of sea ice in numerical investigaions of climate. J Phys Oceanogr 6:379–389

Shupe MD, Intrieri JM (2004) Cloud radiative forcing of the arctic surface: the influence of cloud properties, surface albedo, and solar zenith angle. J Clim 17:616–628

Slingo JM (1987) The development and verification of a cloud prediction scheme for the ECMWF model. Q J R Meteorol Soc 113:899–927

Stroeve J, Holland MM, Meier W, Scambos T, Serreze M (2007) Arctic sea ice decline: faster than forecast. Geophys Res Lett 34. doi:10.1029/2007GL029,703

Sundqvist H, Berge E, Kristjánsson JE (1989) Condensation and cloud parameterization studies with a mesoscale numerical weather prediction model. Mon Weather Rev 117:1641–1657

Tao X, Walsh JE, Chapman WL (1996) An assessment of global climate model simulations of arctic air temperatures. J Clim 9:1060–1075

Taylor PD, Feltham DL (2003) A model of melt pond evolution on sea ice. J Geophys Res 109:C12,007. doi:10.1029/2004JC002,361

Uppala S et al (2005) The ERA-40 re-analysis. Q J R Meteorol Soc 131:2961–3012

Vavrus SJ (1999) The response of the coupled Arctic sea ice-atmosphere system to orbital forcing and ice motion at 6 ka and 115 ka BP. J Clim 12:873–896

Vavrus SJ, Harrison S (2003) The impact of sea-ice dynamics on the Arctic climate system. Clim Dyn 20:741–757

Walsh J, Timlin M (2003) Northern Hemisphere sea ice simulations by global climate models. Polar Res 22:75–82

Walsh JE, RG Crane (1992) A comparison of gcm simulations of Arctic climate. Geophys Res Lett 19:29–32

Walsh JE, Kattsov VM, Chapman WL, Govorkova V, Pavlova T (2002) Comparison of arctic climate simultations by uncoupled and coupled global models. J Clim 15:1429–1446

Wang M, Overland JE, Kattsov V, Walsh JE, Zhang X, Pavlova T (2007) Intrinsic versus forced variation in coupled climate model simulations over the arctic during the twentieth century. J Clim 20:1093–1107. doi:10.1175/JCLI4043.1

Washington WM, Meehl GA (1989) Climate sensitivity due to increased co_2: Experiments with a coupled atmosphere and ocean general circulation model. Clim Dyn 8:211–223

Wetherald R, Manabe S (1988) Cloud feedback processes in a general circulation model. J Atmos Sci 45:1397–1415

Winton M (2000) A reformulated three-layer sea ice model. J Atmos Ocean Technol 17:525–531

Wolff J-O, Maier-Reimer E, legutke S (1997) The Hamburg ocean primitive equation model. Tech. Rep., No. 13, German Climate Computer Center (DKRZ), Hamburg, 98pp

Wyser K et al (2007) An evaluation of arctic cloud and radiation processes during the SHEBA year: Simulation results from eight arctic regional climate models. Clim Dyn 22. doi:10.1007/s00,382–007–0286–1

Yukimoto S, Noda A, Uchiyama T, Kusunoki S (2006) Climate change of the twentieth through twenty-first centuries simulated by the MRI-CGCM2.3. Pap Meteorol Geophys 56:9–24

Zhang J, Rothrock D (2000) Modeling arctic sea ice with an efficient plastic solution. J Geophys Res 108:3325–3338

Zhang X, Walsh JE (2006) Toward a seasonally ice-covered Arctic Ocean: Scenarios from the IPCC AR4 model simulations. J Clim 19:1730–1747

Chapter 12
ACSYS: A Scientific Foundation for the Climate and Cryosphere (CliC) Project

Konrad Steffen, Daqing Yang, Vladimir Ryabinin, and Ghassem Asrar

Abstract The cryosphere, derived from the Greek word cryo for "cold", is the term which collectively describes the portions of the Earth's surface where water is in solid form, including sea ice, lake ice, river ice, snow cover, glaciers, ice caps and ice sheets and permafrost and seasonally frozen ground. Thus, there is a wide overlap with the hydrosphere. The cryosphere is an integral part of the global climate system with important linkages and feedbacks generated through its influence on surface energy and moisture fluxes, clouds, precipitation, hydrology, atmospheric and oceanic circulation. Through these feedback processes, the cryosphere plays a significant role in global climate and in climate model response to global change. The World Climate Research Programme (WCRP) established the Climate and Cryosphere (CliC) Project in 2000 as an evolution from the Arctic Climate System Study (ACSYS) with a global focus on the cryosphere and all its components in the Earth system. The CliC Project coordinates and enables research on (a) terrestrial cryosphere and hydroclimatology of cold regions with special focus on the carbon budget and permafrost, (b) ice masses and sea level which includes ice sheets, ice caps and glaciers, (c) the marine cryosphere and climate which includes all forms of sea ice and (d) the global predictions and the cryosphere to improve the prediction for regional climate models with the inclusion of cryospheric components.

K. Steffen (✉)
University of Colorado, CIRES, Campus Box 216, Boulder, CO 80309, USA
e-mail: Konrad.steffen@colorado.edu

D. Yang
CliC International Project Office, Norwegian Polar Institute,
Polarmiljøsenteret, NO-9296, Tromso, Norway
e-mail: daqing.yang@npolar.no

V. Ryabinin • G. Asrar
World Climate Research Programme, c/o WMO Secretariat, 7bis, Avenue de la Paix,
CP2300, Geneva 2, CH-1211, Switzerland
e-mail: VRyabinin@wmo.int; GAsrar@wmo.int

P. Lemke and H.-W. Jacobi (eds.), *Arctic Climate Change: The ACSYS Decade and Beyond,* Atmospheric and Oceanographic Sciences Library 43, DOI 10.1007/978-94-007-2027-5_12, © Springer Science+Business Media B.V. 2012

12.1   Introduction

The term "cryosphere" collectively describes parts of the Earth system containing water in its frozen state. It includes snow cover and solid precipitation, sea ice, ice shelves, icebergs, lake and river ice, glaciers, ice caps, ice sheets, permafrost and seasonally frozen ground (Fig. 12.1). The cryosphere affects the Earth's climate system, and it is also sensitive to the changing climate. Consequently, its state parameters provide informative indicators of change in the climate system (CliC Project 2001).

Through its influence on surface energy and moisture fluxes, clouds, precipitation, hydrology and atmospheric and oceanic circulation, the cryosphere plays a significant role in global climate system. Changes in the cryosphere are involved in the strong positive feedback of a warming climate associated with a decreasing surface albedo as snow and ice melt (Fig. 12.2). Intensified melting of glaciers, ice caps and ice sheets in the 1990s is now considered to be the main cause of the increased pace of global sea-level variability and change during that period of time (Steffen et al. 2008). The potential for accelerated melting of various parts of the Greenland and Antarctic ice sheets, which could lead to significant amount of sea-level rise, is a matter of great societal concern (Velicogna 2009).

In 1997, during ACSYS implementation phase, the WCRP Conference on its Achievements, Benefits and Challenges (Geneva, 26–28 August 1997) was analysing the course corrections required for the WCRP to align to the science requirements for the twenty-first century. In a report on behalf of ACSYS to the Conference (Goodison et al. 1998a, b), it was stated that WCRP could rely on ACSYS in meeting the initial needs and interests of the cryosphere/climate community because

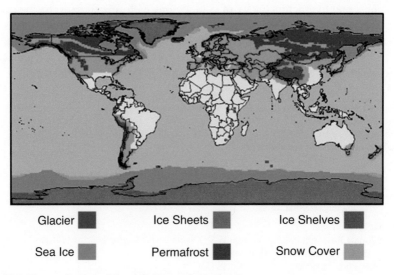

Fig. 12.1 Type and extent of the global cryospheric elements

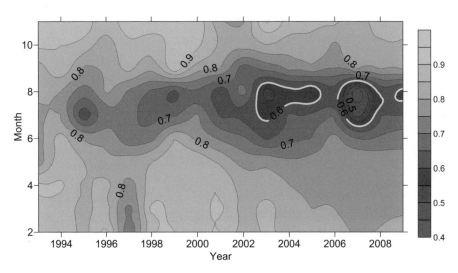

Fig. 12.2 Surface albedo time series at the Swiss Camp, equilibrium line altitude (1,150 m) on the western slope of the Greenland ice sheet (69° 34′ N, 49° 20′ W); the albedo during the melt season (month 7 and 8) decreased markedly from 2001 through 2009 (*Source: K. Steffen, CIRES/University of Colorado*)

ACSYS was focused on the most important aspects of the cryospheric research, despite the fact that its focus was primarily local to regional, and not global. The participants in the conference recommended to expand the WCRP activities in the field of cryospheric research to examine factors determining the extent and variability of the cryosphere, its feedbacks to the global climate system and its role in global climate variability and change and in sea-level rise. Thus, WCRP agreed to start developing a new global project and to have an overlap period with ACSYS, which had its 'sunset' planned for 2003. This led to the development of the CliC initiative, whose science plan was prepared in early 2000 (CliC Project 2001). CliC was formally endorsed as a new WCRP Project by the WCRP Joint Scientific Committee at its twenty-first Session in March 2000, in Tokyo, Japan (WMO Report 2000). It is still the youngest Project in the family of WCRP four core projects.

Until 2003, the development of both ACSYS and CliC projects was coordinated by the ACSYS/CliC Scientific Steering Group (SSG). In 2004, at the first session of the CliC SSG, a decision was made to hold the first CliC Science Conference and start developing an implementation strategy for the project. In the same year, the Scientific Committee on Antarctic Research (SCAR) joined WCRP as a cosponsor of CliC. In 2005, the China Meteorological Administration hosted a very successful 1st CliC International Science Conference entitled, *"Cryosphere, the frozen frontier of climate science – theory, observations and practical applications"* (Beijing, China, 11–15 April 2005). In 2008, the International Arctic Science Committee (IASC) became another cosponsor of CliC.

12.2   Project Goals and Objectives

The principal goal of CliC is to assess and quantify the impacts of climatic variability and change on components of the cryosphere and their consequences for the climate system, and to determine the stability of the global cryosphere.

CliC intends to achieve its principal goal by working, independently and with partners, to provide:

- Enhanced observation and monitoring of the cryosphere and the climate of cold regions in support of process studies, model evaluation, change detection and other applications;
- Improved understanding of the physical processes and feedbacks through which the cryosphere interacts within the climate system;
- Improved representation of cryospheric processes in models to reduce uncertainties in simulations of climate and predictions of climate variability and change; and
- Facilitation and support to scientific assessments of changes in the cryosphere and their impacts, in particular to the IPCC fifth Assessment Report.

WCRP affiliated scientists have identified several key questions for CliC to address:

- What are the magnitudes, patterns and rates of change in the terrestrial cryosphere on seasonal-to-century timescales? What are the associated changes in the water cycle?
- What will be the contribution of the cryosphere to changes in global sea level on decadal-to-century timescales?
- What will be the nature of changes in sea-ice distribution and mass balance in both polar regions in response to climate change and variability?
- What will be the impacts of the changes in cryosphere on atmospheric and oceanic circulation? What is the likelihood of abrupt climate and/or Earth system changes resulting from processes involving the cryosphere?
- How do monitored changes in the cryosphere reflect the variability and change in the climate system? How can these monitored changes be combined with proxy or paleo-records and the results of modelling studies to improve our understanding of climate change?
- What elements of climate predictability over a range of timescales involve the cryosphere and associated processes and how can we increase the predictive skill of long-range forecasting techniques through the use of cryospheric data, information and modelling?

12.3   The CliC Main Scientific Themes

At the meeting of the ACSYS/CliC SSG in October 2002, in Beijing, China, it was decided that implementation of CliC should be organised through four broad themes.

The first Theme is *"The Terrestrial Cryosphere and Hydroclimatology of Cold Regions"* (TCHM). The terrestrial cryosphere includes land areas where snow cover, lake and river ice, glaciers and ice caps, permafrost and seasonally frozen ground and solid precipitation occur. The main task of this theme is to improve estimates and quantify the uncertainty of water balance and related energy flux components in cold climate regions. This includes research on observations and processes that control precipitation (both solid and liquid) distribution, properties of snow, snow melt, river and lake ice, evapotranspiration, sublimation, water movement through frozen and unfrozen ground, its storage in watersheds and river runoff.

The main scientific questions to be addressed by this Theme are:

- What are the magnitudes, patterns and rates of change in terrestrial cryosphere regimes on seasonal-to-century timescales? What are the associated changes in the water cycle?
- What is the role of terrestrial cryospheric processes in the spatial and temporal variability of the water, energy and carbon cycles of cold climate regions, and how can they be parameterised in models?
- What are the interactions and feedbacks between the terrestrial cryosphere and atmosphere/ocean systems and current climate? How variable are these interactions and how will they change in the future?

CliC will facilitate the development and validation of physically based land-atmosphere-cryosphere process models, including permafrost-hydrology and carbon cycle interactions, with appropriate complexity, for their inclusion into coupled models at a range of scales. The proper scaling and verification of the scaling procedure will be essential in the development of land surface process models and their linking to hydrological models and for incorporation into climate models. Improvement of regional climate models will be associated with interactive climate-cryosphere-hydrosphere schemes allowing assessments of impacts on the terrestrial cryosphere of future climate change, and the effects of changes in the cryosphere on regional and local climates on the scale of mountain and watershed systems. There is a need to account and include river-ice processes in watershed hydrology in the cold regions, particularly in the Arctic for the regions of large rivers with thick-ice cover. Modelling and monitoring studies should proceed concurrently.

Expected outcomes of this Theme include:

- Daily/monthly cryospheric data products for cold regions.
- Establishment of CliC 'Super Study Sites' for process studies and model validation.
- Advanced and validated remote sensing algorithms.
- Improved radiative transfer and surface process models for cryospheric regions.
- Enhanced information for water resource management.
- Improved estimation of carbon emissions from permafrost and frozen ground regions.
- Improved prediction of changes in the cryosphere that will have socio-economic impacts.

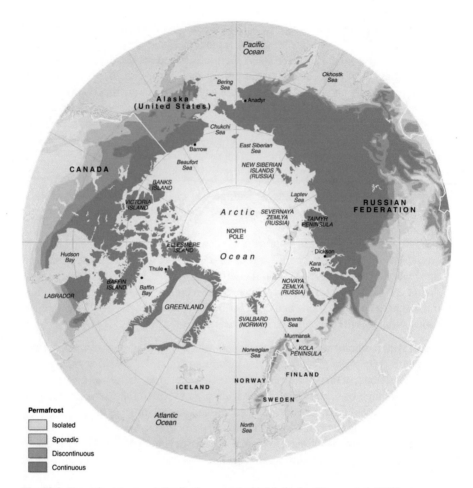

Fig. 12.3 Permafrost type and distribution over the high latitudes (Brown et al. 1998)

A new CliC and AIMES (IGBP) activity called CAPER (CArbon and PERmafrost) is being initiated to estimate the unknown magnitude of the positive feedback between the warming climate and emission of greenhouse gases into the atmosphere from natural sources, particularly from thawing permafrost. The controlling processes are further complicated by insufficient understanding of the interactions between the terrestrial cryosphere, hydrology and vegetation in the northern high latitudes in the warming climate. CAPER will promote regional to global observations, modelling and analyses on a range of scales to improve the understanding and quantification of the carbon cycle and permafrost dynamics (Fig. 12.3). Some estimates will be made of the amount and form of carbon stored in the ground and its release rates to the atmosphere or surface and ground water in a warming climate.

Specifically, CAPER will develop:

- An observations and measurements strategy based on collaboration with international observing networks and programmes, relevant regional science projects such as NEESPI (Northern Eurasia Earth Science Partnership Initiative) and global climate modelling projects such as C4MIP.
- Coordinated modelling framework for the northern high latitudes to quantify key vulnerabilities and thresholds of the coupled carbon-climate system and improve representation of major feedbacks between biogeochemistry, hydrology and vegetation in RCMs and ESMs for high latitudes.

The WMO sponsored Solid Precipitation Measurement Intercomparison Project (Goodison et al. 1998a, b) revealed significant systematic errors (biases) that exist in the gauge-measured precipitation records. These biases must be documented and corrected to obtain a compatible, accurate data set for hydrological and climatic investigations over a wide range of time and space scales. CliC and GEWEX, particularly its Coordinated Energy and water cycle Observation Project (CEOP), and its high elevations and cold region components, with other partner research activities in cryospheric sciences, will work to address biases of precipitation measurements in the cold regions and generate an accurate precipitation database and its improved climatology. These will enable the development and validation of satellite retrievals, initialization of numerical models, and the validation of climate and hydrological model output. Value-added precipitation products will be especially relevant to studies of climate variability and change, as well as water and energy balance estimates in the cold and high altitude regions. They will help to answer the following questions:

- What is the distribution and variability of precipitation in cold climate regions?
- What is the accuracy in measurement and model prediction of solid precipitation in high latitudes and high elevations, and how can their uncertainties be reduced?
- What are the accuracy requirements for solid precipitation products in cold climate regions for use in water budget analyses and change detection over monthly-to-decadal timescales?

The second CliC Theme is called '*Ice Masses and Sea Level*' (IMSL). It was proposed to develop estimates of the mass balance of the Antarctic and Greenland ice sheets, ice caps and glaciers, and their contribution to sea-level changes. It is also imperative to develop an enhanced capability to estimate the past and predict future changes in ice mass.

Increased melting of the large polar ice sheets contribute to the observed increase in sea level. Observations of the area of the Greenland ice sheet that has been at the melting point temperature at least one day during the summer period shows a 50% increase during the period 1979–2008 (Fig. 12.4). The Greenland region experienced an extremely warm summer in 2007. The whole area of south Greenland reached the melting temperatures during that summer, and the melt season began 10–20 days earlier and lasted up to 60 days longer in South Greenland.

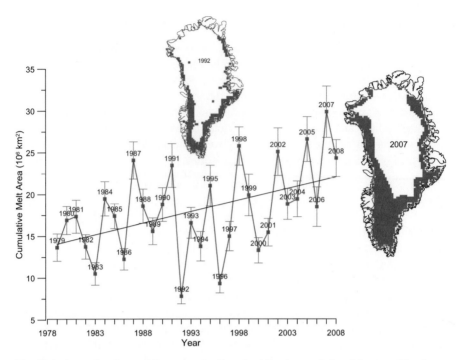

Fig. 12.4 Area of surface melting across the Greenland ice sheet as inferred from satellite observations (Steffen et al. 2008, updated)

In addition to melting, the large polar ice sheets lose mass by ice discharge, which is also sensitive to regional temperature. Satellite measurements of very small changes in gravity have revolutionised the ability to estimate loss of mass from these processes. Figure 12.5 shows that the Greenland ice sheet has been losing mass at a rate of 180 Gt/year since 2003 according to the GRACE satellite measurements. Similar ice loss is also observed for West Antarctic and the Antarctic Peninsula.

The main scientific questions to be addressed include:

- What is the contribution of glaciers, ice caps and ice sheets to changes in global sea level on decadal-to-century timescales?
- What are the key sensitive regions for rapid changes in ice volume (i.e. ice masses grounded below sea level, over-deepened channels of outlet glaciers) and what are their potential contributions to sea level in a warming climate?
- How stable are the Antarctic ice shelves given the recent increase in the ocean temperatures along the ice margins?
- How can we reduce the uncertainty of these estimates?

New technologies, particularly satellite remote sensing instrumentation, will be essential to provide an observational network to monitor elevation and gravitational

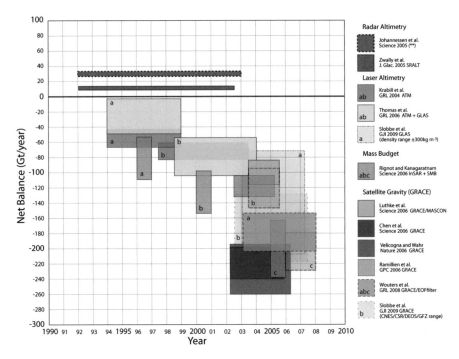

Fig. 12.5 Summarised results of total mass balance measurements for the Greenland ice sheet based on radar altimetry, mass budget and satellite gravity estimates. The heights of the boxes cover the published error bars or ranges in mass change rate over those intervals (Dahl-Jensen et al. 2009)

changes of the major ice sheets. This will be a major contribution to enable us to assess the mass balance of the bodies and thereby their contribution to sea-level changes. To improve estimates of mass loss from ice sheets, observations are required of the ice stream behaviour, ice velocities and thickness, basal melt, ablation processes and the role of ice shelves in restraining ice streams and glaciers. Both long-term observations and intensive process studies will be needed. In addition, improved observations of solid precipitation are needed to provide comprehensive data on accumulation on glaciers and ice sheets and to assist with development of models and evaluation and calibration of satellite measurements. Methods for measuring the influence of sublimation on their mass balance need also to be improved.

This CliC theme will generate and facilitate:

- Improved long-term ice-sheet, ice-cap and glacier monitoring systems (including more measurements in data-sparse, thick-ice areas and also relevant atmospheric parameters), together with an inventory and related databases for assessing their mass balance and its uncertainties.

- Mass balance records for a selection of large glaciers and ice caps representative of different climatic regions.
- Improved observations and representation in models of spatial and temporal variability of surface mass budget of ice sheets, ice caps and glaciers, which are expected to be significantly contributing to sea-level change and where mass balance evolution may be sensitive to sea level.
- Realistic treatment of ice deformation, basal sliding and/or deforming basal till, and appropriate stress configurations in ice stream models, and of ice sheet/ice stream/ice shelf/ocean interaction and coupling.
- Records of continuous ice velocities for a selection of sensitive regions to determine the dynamic response of ice sheets to climate perturbation on seasonal and longer timescales.
- Assessment of the Greenland and West Antarctic ice-sheet stability and vulnerability to climate change including sudden and potentially irreversible changes.
- Facilitate and organise summer schools that will bring current and future ice-sheet scientists together to develop better models for the projection of future sea-level rise

Recognising the need to consolidate the efforts of the science community on substantiating estimates of all factors contributing to the mean sea-level change and variability, the important role of the cryosphere among these factors such as the key uncertainties associated with the ice-sheet dynamics and the unique contributions of WCRP to address the challenges facing the science community. CliC is undertaking this crosscutting activity in collaboration with partners, such as CLIVAR, GEWEX, and the International Association for Cryospheric Sciences (IACS), SCAR, and a number of ice modelling groups.

The third Theme is called '*The Marine Cryosphere and Climate*' (MarC). The marine cryosphere includes all forms of sea ice, as well as the snowfall and snow cover on sea ice. Sea ice is one of the most sensitive climate parameters, yet one of the least understood. Sea ice changes fast on seasonal scale and often exhibits large interannual variability and is dependent on drift (except for landfast sea ice). The work of the theme will focus on understanding the ocean-ice-atmosphere system in the polar regions through enhanced observations, process studies and models. This includes the dynamics and thermodynamics of sea ice and its snow cover, changes in the thickness distribution and characteristics of these media, polynya processes, the impact of sea ice on water mass modification (addition of brine during sea-ice formation, freshening during melt) and interactions of sea ice and snow cover with the atmosphere, including the surface albedo feedback process. Ice shelf–ocean interactions, the role of ice shelves and icebergs in the ocean freshwater budget, and the impact of changes in sea ice and ice shelf properties that may affect thermohaline circulation will also be investigated.

The main scientific questions to be addressed by this Theme include:

- What are the present mean states, natural variability and recent trends in sea-ice characteristics in both hemispheres, and what are the physical processes that determine these?

Fig. 12.6 Ice thickness experiment in the Bellingshausen Sea of Southern Ocean as part of the ASPeCt experiment (*Photo credit: K. Steffen, CIRES/University of Colorado*)

- How will sea ice respond in future to a changing climate; e.g. will the Arctic Ocean be ice-free in summer and if yes, how soon?
- How will a changing sea-ice cover affect climate through its interactions with the atmosphere and ocean?
- How do processes of ice-ocean interaction – including basal melt and marine ice accretion – affect the mass balance and stability of ice shelves?
- What is the role of the seasonal production of sea ice in the neighbourhood of ice shelves in driving currents beneath the shelves and thereby influencing basal melting and refreezing?
- What is the distribution and variability of fresh water input to the oceans from ice shelves, icebergs and ice-sheet runoff, and what role does this have on ocean circulation (e.g. maintenance of global thermohaline circulation)?

The Antarctic Sea Ice Processes and Climate (ASPeCt) study, originally sponsored under SCAR, and now jointly sponsored with CliC, is a coordinated research activity focused on Antarctic sea-ice observations and process studies (Fig. 12.6). The ASPeCt community is very active and has recently been complemented in the Arctic by the establishment of a CliC Arctic Sea Ice Working Group. This working group is initially focussed on improving the coordination of surface-based sea ice and snow observations, establishing protocols for standardising and archiving data across the different national and international activities, and linking partners to ensure that functional, sustained observing networks are established for long-term observation and monitoring of these regions. Progress in observations should be continued in the area of assimilation of the variety of sea-ice observations into coupled atmosphere/ocean general circulation models. This will contribute to advances in the weather and climate forecasting on a variety of timescales, and it will create a basis for global ocean and climate system reanalysis.

Predicting the future of Arctic and Antarctic sea ice will be a high priority for MarC. Of major concern is the inability of models to reproduce adequately the observed loss of summer sea ice from the Arctic Ocean. Further work will focus on identifying all relevant positive and negative feedbacks related to the marine cryosphere and its interactions with the oceans and atmosphere. There is also a need for improved models of the atmosphere and oceans in these regions. Better numerical modelling of the three-dimensional ocean circulation and thermohaline circulation in sub-ice shelf cavities is essential in improving our knowledge of the exchange of ice and heat at the base of ice shelves, which are expected to respond sensitively to changes in ocean temperature. Such models must ultimately be coupled to sea-ice models in order to adequately simulate the seasonal variations of temperature and salinity in the adjoining ocean.

This theme will develop:

- Improved capability to measure sea ice thickness on a regional scale, for development of long-term monitoring programmes and calibration and validation of CryoSat-2 and other remotely sensed data.
- Information on the structure and volume of ice including sea ice ridges and the effects of basal melt on the ice thickness distribution.
- Realistic treatment of the ice thickness distribution in climate models, particularly in relation to the movement of sea ice between thickness classes in response to dynamic processes, and the response of the ice thickness distribution to basal melt. This will allow improved estimates of sea ice response to global warming in the both hemispheres.
- Improved parameterisation of sea ice advection and ice-ocean processes in the upper water column in coupled models.
- Improved understanding of ice-ocean processes beneath ice shelves, including basal melt, marine ice accretion and iceberg calving, and the implications for the mass balance and stability of ice shelves.
- Information on the spatial and temporal distribution of icebergs (and their variability), in particular, their drift tracks and impact on fresh water input to the oceans and effects on ocean circulation (e.g. maintenance of global thermohaline circulation).
- Improved understanding of snow processes on sea ice, particularly in the context of improving the interpretation of airborne and space-borne laser and radar altimetry data.
- Maintenance and expansion of existing networks to monitor sea-ice drift such as the WCRP/SCAR International Program on Antarctic Buoys (IPAB) and regional changes in fast ice properties (Antarctic Fast Ice Network).

In both the Arctic and Southern Ocean, CliC will facilitate the development of the regional components of the Global Ocean Observing System (GOOS), Arctic GOOS and Southern Ocean Observing Systems, which would cover not only the domain of physical oceanographic observations but also marine component of the global carbon cycle and other crucial variables. In addition to sea-ice observations, CliC will promote observations of oceanographic conditions/parameters under the sea-ice cover. CliC will support novel research into in situ observations of polar oceans and the development, testing, calibration of new observing technologies and their conversion into operations. Given the remoteness of most cryospheric marine regions, properly evaluated and calibrated satellite remote sensing will be vital to provide broad coverage of marine cryospheric observations. CliC will promote and support continuation and enhancement of satellite cryospheric missions such as ICESat-2 and CryoSat-2. To measure the ocean circulation under the ice shelves, enhanced observations in front of, and through, ice shelves are required. Similarly, remote sensing of iceberg population provides information on drift tracks, but additional research is required for improved understanding of iceberg calving processes, dissolution rates, effects of grounded icebergs on ocean circulation and sea-ice dynamics and impacts on ecosystems. In addition, the presence of refrozen seawater ('jade ice') in icebergs may serve as a useful indicator of areas of the parent ice shelf where refreezing processes occur. CliC will address, jointly with other partner programmes/agencies including space agencies, the development of observations of precipitation over the polar oceans.

The Arctic is moving towards a new state, with the seasonal loss of sea ice and other significant impacts on cryospheric elements. Pivotal to such changes are the terrestrial cryosphere and the associated hydrologic system. Correspondingly, there is a concern about how changing cryosphere and hydrology will affect (1) global feedbacks and impacts, (2) biological productivity and diversity and (3) human and economic systems. Impacts will lead to intra-Arctic changes in bio-geophysical and socio-economic systems, and extra-Arctic effects that will have global consequences. CliC will initiate an assessment of the freshwater balance of the Arctic and the role of cryosphere in it. This work will be later expanded to the Southern Ocean to improve our understanding of the cryospheric elements that affect the freshwater balance of the Southern Ocean through cooperation with the CLIVAR/CliC/SCAR Southern Ocean Region Implementation Panel.

This CliC cross-cut theme will study and provide a scientific assessment of the following issues:

- Snow cover melt and retreat impacting surface radiative balance, regional and basin hydrology.
- Melt of ice sheets contributing to sea-level rise and freshwater distribution over the polar oceans, with potential effects on thermohaline circulation and global climate.
- Retreat of sea ice leading to increased radiative absorption and reduced salt flux during growth season.
- Permafrost thaw changing geomorphic and geochemical processes and fluxes. As permafrost degrades methane production increases. Potential drying of wetlands

may result in CO_2 production increase due to the oxidation of organic material. Both processes may involve significant positive climate feedbacks.

- Shrinking river and lake ice cover and changes in the magnitude and timing of snowmelt runoff and river-ice processes such as ice-jam flooding.
- Retreat of small glaciers and ice caps. This initially causes increased run-off but lower flows will eventually result as ice masses diminish and disappear.

The assessment will enable a number of important applications in several areas of societal benefit and will facilitate related studies, e.g. in aquatic ecosystems.

The fourth CliC Theme 'Global Predictions and the Cryosphere' (GPC) is designed as an integrator of the previous three themes. Within this theme, the key initial research question for CliC is how to improve prediction of the cryosphere from the regional climate modelling (RCM) perspective. Validation and development of polar RCMs that will help to achieve this objective will focus on:

- Forcing of cryospheric models with high-resolution atmospheric data
- Improvement and evaluation of polar RCMs
- Complementing GCMs by RCMs
- Polar process studies with RCM simulations
- Prediction experiments for the high latitudes
- Polar climate change projections at high resolution

Following the initial stage of developing and testing RCMs, a range of issues linking the cryosphere and the rest of the Earth's climate system will be addressed with the general aim to enable a range of predictions and assessments across time and space.

The major outcomes of this Theme and, hence, the entire CliC project will include:

- Improved knowledge of cryosphere-related feedbacks to the rest of the climate system.
- Understanding of the cryospheric processes that underlie global teleconnectivity in climate.
- Determination of the pathways of moisture transport that control the mass balance of ice sheets.
- Better knowledge of abrupt climate changes associated with impacts of the cryosphere on the rest of the global climate system.
- Better understanding and quantification of the role of cryosphere in global water, energy and carbon cycles.

Predictability of the atmospheric general circulation is much higher in the tropics throughout the year than it is in mid- and high latitudes. However, there is some predictability in the Northern extratropical winter atmospheric circulation through some patterns of teleconnection. Predictability of the general circulation at the polar regions has still remained as a "cold" topic. Nevertheless, based on a study of predictability with a general circulation model, it is found that the SST-related predictability of the Southern winter lower atmospheric circulation in Antarctica is reasonably high. Key questions are how the low frequency forcing agents like sea ice, SST, snow cover, stratospheric conditions and oceanic heat anomalies affect the predictability of polar climate on seasonal to longer timescales. More research of this subject based on data analysis and model simulations is needed.

It is known that the average position of the Polar front in the Northern Hemisphere is largely determined by the geographical extent of sea ice and snow. The role of sea ice and snow cover as a controlling mechanism for seasonal developments is only starting to be established. A few teleconnections are known, such as the influence of Tibetan snow cover on the Asian monsoon and a possible influence of late winter East Asian snow cover on North American spring conditions. CliC will continue to develop quality control and assessment of data available for the initialisation of sea-ice and snow conditions in seasonal and longer-range forecasts, and cryospheric components of the models used for that purpose. In terms of decadal predictability, CliC will undertake an activity in cooperation with CLIVAR and SPARC towards better resolving modes of atmospheric variability in the polar regions, the role of atmospheric and ocean circulation in forming prevailing patterns of interannual change in the Arctic and Southern Ocean, and their likely evolution in a warmer climate with less ice-covered Arctic Ocean.

12.4   CliC and Cryospheric Observations

In order to gain a better understanding of the role of the cryosphere in the global climate system, a comprehensive, coordinated cryospheric observation system is needed. A number of international programmes – such as the Global Climate Observing System (GCOS) – address the cryosphere in part, but none cover in a comprehensive and integrated manner what is required. ACSYS and CliC have made significant contributions in identifying the critical observations required to characterise, understand, model and ultimately predict the changes taking place in the Earth's cryosphere. In early 2004, a concept paper was prepared by the CliC project in collaboration with the Scientific Committee on Antarctic Research (SCAR) and in consultation with several Global Observing Strategy (IGOS) Partners, and presented as a proposal for an IGOS Cryosphere Theme. The Cryosphere Theme proposal was approved by the IGOS Partnership in May 2004.

The IGOS Cryosphere Theme Report (IGOS 2007) was developed based on input from approximately 100 scientists from 17 countries. It is a robust compilation of observing system capabilities, needs and shortcomings, with separate chapters covering each major element of the cryosphere: sea ice, land- and river ice, ice sheets, glaciers and ice caps, surface temperature and albedo, permafrost and seasonally frozen ground and solid precipitation. In November 2007, the report was published with support from the World Meteorological Organization. It was first 'released' at the Group on Earth Observations (GEO) Plenary Meeting in Cape Town, South Africa, and has since been distributed widely.

The implementation of the IGOS Cryospheric Observing System (CryOS) was proposed to be phased over three time intervals: 2007–2009 (the IPY period), 2010–2015 (the period during which available satellite missions were known) and beyond 2015 (the period during which satellite missions are yet to be proposed or require further elaboration).

CryOS has been a major development and achieved all its goals for the early period by:

- Creating a framework for improved coordination of cryospheric observations conducted by research, long-term scientific monitoring and operational programmes
- Improving availability and accessibility of data and information needed for both operational services and research
- Strengthening national and international institutional structures responsible for cryospheric observations and increased resources for ensuring the transition of research-based cryosphere observing projects to sustained observations, e.g. through the ESA projects – GlobSnow, GlobIce and GlobGlacier.

On every occasion, the IGOS Cryosphere Report has proved its usefulness as an authoritative source of agreed requirements in cryospheric observations and recommendations on means to establish them. The report solidly put cryospheric observations on the agenda of WMO Space Programme, space agencies (CSA, ESA, JAXA, and NASA), CEOS and GEO.

Numerous space-based cryospheric data sets or products have already been developed, and more are under development from present and near-future satellite/sensor systems. These new systems will complement the current and future systems including Special Sensor Microwave Imager (SSM/I), Advanced Microwave Scanning Radiometer (AMSR) and Synthetic Aperture Radars (SARs) on Canadian, Japanese and European satellites, which provide valuable information on snow and ice resources. The recent availability of missions such as Japan ALOS, Canada Radarsat-2, German TerraSAR-X and Italian Cosmo-Skymed has already opened the door to new exciting scientific research, enhanced observations and novel applications. In the near future, the advent of the ESA's CryoSat-2 and the Sentinel series will further enhance this global observation capability. The full exploitation of this multi-mission capability by the scientific community and operational users will require coordinated research and development efforts to develop and validate novel products and applications.

In order to support existing scientific efforts carried out by CliC, the European Space Agency (ESA) and the CliC Project Office plan to launch a dedicated joint activity to address some of the key scientific priorities in Cryosphere science where earth observation (EO) technology and ESA missions may contribute. In particular, the planned activity aims at:

- Advancing the development and validation of novel advanced EO-based multi-mission-based products, improved data sets and enhanced applications that may respond directly to the specific scientific requirements of the CliC and cryosphere scientific community;

- Exploiting novel EO approaches to improve the observation, understanding and prediction of cryosphere processes with special attention to CliC priorities. In particular, two major areas of interest have been identified;
- Developing a Scientific Roadmap as a basis for further ESA scientific support activities dedicated to the CliC community. The project will be funded by the new ESA's Support to Science Element;
- Identifying the main geographical areas of interest and existing data sets that may contribute to the project.

12.5   CliC's Role in IPY

With more than 50,000 participants from around 60 countries, the International Polar Year 2007–2008 (IPY) was one of the biggest scientific environmental campaigns in the history of science. WCRP affiliated experts supported and participated in the early discussions on the concept of IPY and helped to focus IPY on achieving a few practical goals rather than being a mere collection of individual projects. Due to efforts of CliC and partners, the climate variability and change science was at the centre of IPY agenda. The CliC Project led the overall coordination of WCRP input to IPY. The State and Fate of the Polar Cryosphere project managed by the CliC SSG was the IPY umbrella project for the cryosphere. All WCRP affiliated projects proposed one or several activities for consideration by the IPY Joint Committee, and more than 20 such activities were approved as major research clusters.

The CliC-led IGOS Cryosphere Theme made several recommendations on the development of the polar and cryospheric observing systems and identified the IPY time frame as the first phase of their implementation. One of the recommendations formulated by the cryosphere Theme was implemented through the IPY Project "Global Interagency IPY Polar Snapshot Year" (GIIPSY). Proposed and coordinated jointly with leading satellite agencies, this project contributed to strengthening cooperation among space agencies in polar observations, and led to the implementation of the first virtual constellation of satellites for polar regions. The resulting data archive, especially of the Synthetic Aperture Radar (SAR) data, represents a unique snapshot of the polar cryosphere and provides a rich source of data for future cryospheric studies. Working with and through partners, WCRP made marked advances in the establishment of the integrated Arctic Ocean Observing System (iAOOS) and Southern Ocean Observing System (SOOS). In ship cruises conducted before and during the IPY, a significant warming of the Southern Ocean was detected and documented. This warming will likely play a significant role in determining the pace of the future sea-level variability and change.

Current observational activities of WCRP contribute to preserving and further developing the IPY legacy. These activities will be coordinated under the Sustaining Arctic Observing Networks Initiative (SAON), the Pan-Antarctic Observing System (PAntOS), and the WMO Global Cryosphere Watch (GCW) initiative. Building on these improved observations and enhanced knowledge of the polar processes, WCRP will continue to cooperate with the World Climate Programme in establishing a Polar Climate Outlook Forum (PCOF). The goal of this forum is to disseminate the best available scientific knowledge regarding the polar regions to support policy and management decisions that will sustain the polar ecosystems and environment.

12.6   Polar Research Planning (ICARP II)

The successful conclusion of ACSYS after a decade led to the birth of CliC, and the importance of cryosphere in climate system has thus stayed in the forefront of research and development across the world. CliC affiliated experts, following the successful legacy of ACSYS, continue to make significant contribution to the preparation and outcomes of the seminal Second International Conference on Arctic Research Planning (ICARP II, November 2005) (Bowden et al. 2005). CliC organised the development of two science plans on Terrestrial Cryospheric and Hydrologic Processes and Systems, and on Modelling and Predicting Arctic Weather and Climate, as well as a special session on global observations and the impact on Arctic programmes. CliC leads implementation of the ICARP recommendations on terrestrial cryospheric and hydrologic systems and co-ordination of Arctic weather and climate modelling activities with other Arctic organisations. Following the conference, the European Polar Board decided to initiate a transnational programme in polar science. It was also acknowledged that co-ordination by funding agencies to support diverse Arctic science programmes was essential. A further outcome was a co-ordinated plan to develop a Sustaining Arctic Observing Network, engaging research and operational entities. CliC with IASC, AMAP and NSF, were founding members of the initiative, which has grown into a major initiative of the Arctic Council.

The ICARP group that worked on the Terrestrial Cryospheric, Hydrologic Processes and Systems has proposed an approach to developing observations in both high latitude and high altitude regions, which has been developed in partnership with the IGOS Theme on Cryosphere. The cryospheric observing systems are normally sparse and concentrated primarily in coastal and valley locations; however, some of the observing networks have declined and continue to decline in most of these regions. For example, mapping of mountain glaciers using historical sources combined with recent remotely sensed imagery is a pressing need for over half of the world's glaciers that are not yet included in the World Glacier Inventory. Rapidly shrinking tropical glaciers and ice caps merit special attention. A proposed sampling strategy is to integrate validated in situ and satellite data to provide temporally and spatially consistent data coverage on the terrestrial cryosphere and replace the loss of data from traditional stations. (However, efforts must also be made to prevent further decline and indeed to reverse the trend in in situ observational networks.) The establishment of transects and long-term and intensive multidisciplinary observatories, in other words "Super Sites," spanning a range of terrain settings, will be initiated to provide data for improved understanding of relevant terrestrial processes and their temporal variations, testing parameterisations suitable for larger scale models for future projections and development and validation of remotely sensed cryospheric information.

12.7   Assessments

CliC – affiliated scientists led the preparation of the cryosphere-dedicated chapter in the IPCC AR4, which presented the first holistic look at the simultaneous changes taking place in various parts of the cryosphere (Lemke et al. 2007). The AR4 analysis of the state of the cryosphere was further developed under the Global Outlook for Ice and Snow published by UNEP in 2007 (UNEP 2007), which also included key contributions of many experts associated with CliC. The expected contributions of cryospheric changes to the Earth's climate system and, in turn, the impact of climate variability and changes on the cryosphere are expected to be very significant in the future. For example, melting of mountain glaciers and polar ice sheets contribute to the increased pace of the global sea-level change, while accelerated thawing of permafrost alters the land surface in northern latitudes and affects the hydrological regime and carbon cycles.

After the seminal Arctic Climate Impact Assessment (ACIA) of 2005 (ACIA 2005), a new level of understanding on the changes in the Arctic climate has been obtained and documented in publications such as the World Wide Fund for Nature (WWF) report (WWF 2008) which is an update since ACIA in 2008.

The new level of knowledge on the state of and changes in the cryosphere made it possible for the cryosphere science community to embark on a new assessment entitled "Climate Change and Cryosphere: Snow, Water, Ice, and Permafrost in the Arctic" (SWIPA), which has been commissioned by the Arctic Council and is co-sponsored by the AMAP, IASC, CliC and IPY. In this activity not only observed changes in the cryosphere will be reviewed but also a prediction will be given for their evolution under a warming climate, based on the climate predictions available in the WCRP CMIP3 archive. SWIPA is comprised of three major components: (1) Arctic sea ice in a changing climate, (2) The Greenland ice sheet in a changing climate and (3) Climate change the terrestrial cryosphere.

12.8   WMO Global Cryosphere Watch

In 2007, the 15th WMO Congress welcomed a proposal by Canada to establish a WMO Global Cryosphere Watch (GCW) as an IPY legacy. The proposal largely followed the recommendations of the IGOS Cryosphere Theme. The GCW will provide an international mechanism for supporting all key cryospheric in situ and space-based observations. It will also include cryospheric monitoring, assessment, product development, prediction and related research. It aims to provide authoritative, clear, understandable and useable information on the past, current and future state of the cryosphere for use by the media, public and decision of policy makers. GCW will also directly benefit society at large. In tandem with its partners, WCRP will continue to support the development of the GCW concept and planning of its initial activities.

12.9   Polar and Cryospheric Reanalysis

The Arctic System Reanalysis (ASR), which can be viewed as a blend of modelling and observations, will provide a high-resolution description in space (10–20 km) and time (3 h) of the atmosphere-sea ice-land surface system of the Arctic. The ASR, approved as an IPY proposal under the international CARE/ASR (Climate of the Arctic and its Role for Europe/ASR) activity, will ingest historical data streams along with measurements of the physical components of the Arctic Observing Network. Gridded fields from the ASR, such as temperature, radiation and winds, will also serve as drivers for coupled ice-ocean, land surface and other models, and will offer a focal point for coordinated model intercomparison efforts. The ASR will permit reconstructions of the Arctic system's state, thereby serving as a state-of-the-art synthesis tool for assessing Arctic climate variability and monitoring Arctic change. A similar approach for an Antarctic System Reanalysis is of high priority. Polar reanalyses, synthesised cryospheric datasets will eventually contribute to preparation of a climate system reanalysis.

12.10   Regional and National Developments

Active CliC programmes, national CliC committees or large scale national projects are conducted in China, Russia, Japan, Canada and in several other countries. Only few examples of such activities follow.

A major national project, affiliated with CliC, is the Improved Processes and Parameterisation for Prediction in Cold Regions (IP3), a research network that studies water resources and cryosphere in mountain and northern Canada. The IP3 Network includes researchers from across Canada, Europe and the USA. With recent reports of greater snowfall, melting glaciers and thawing permafrost, Canada's changing mountain and northern cold regions are presenting challenges for management of the nation's water supply, infrastructure and ecosystem health. IP3 research is currently focused on obtaining a better understanding of cold region processes, parameterising these processes and testing and validating these new parameterisations in cold regions models.

National Chinese CliC activities are not only contributing nationally to the understanding of global change. They have proven to be very important for the sustainable development of the national social economy. The focus is on two main regions: the high Asia and the polar regions. Main project areas include:

- The impacts of cryosphere changes on the sustainable development of western China. The cryospheric changes and water cycle
- Global change and the occurrence of extreme climatic events and natural disasters in cryosphere
- Understanding the interactions between cryospheric processes and Asian monsoon, the relationships between climate of coldness and aridity, and the regional climatic sensitivity
- Understanding the atmospheric chemistry and their relationships with cryospheric changes
- Standardisation of the monitoring system
- Simulation of the cryospheric processes at different spatial scales.
- Construction of a cryosphere database
- Establishment of a platform for data sharing
- Ice mass balance and sea-level change
- Climate records from shallow ice cores
- Biological dynamics and climate change in the Southern Ocean

To celebrate the 1st and 2nd International Polar Year as well as the 50th anniversary of International Geophysics Year, a 5–10-year research programme entitled "International High Asia Cryosphere Years" (IHACYs) is proposed. The expected outcomes of IHACYs are:

- Enhanced observing, data exchange and management system for regions with cold climate including observations of glacier and snow cover variability, seasonally frozen ground and permafrost and hydrological variables such as fresh water ice, cold climate meteorological variables and relevant biological parameters.
- Data sets of observations and (re-) analyses.
- Improved knowledge and ability to model related processes and contribute to studies of predictability of regional climate at several timescales. Advances of knowledge of interactions in the system due to increased scope of interdisciplinary research.
- Potentially, substantiated scenarios of the impacts of atmospheric GHG concentration growth rates on degradation of frozen ground, changes in snow cover, retreat of the glaciers and other aspects.
- Outlooks of changes in cryospheric parameters of significance for socio-economic activities, including permafrost degradation and changes in ecosystem.
- Information for water management between mountain and plateau areas and the arid lowlands.

CliC is attracting attention also in countries located far away from polar regions. Ocean-Atmosphere-Cryosphere Interactions are one of foci of the Brazilian High Latitudes Oceanography Group. Interest in cooperation with CliC has been expressed by scientists from India, Pakistan and the Central Asia.

12.11 Data Management and Data Services

Under the umbrella of the ACSYS and CliC projects, a variety of data sets have been produced. They include sea-ice edge positions dating back to the sixteenth century, novel datasets on the state of the Greenland ice sheet and many others. Efforts aimed at evaluation, calibration and adaptation of algorithms, and reprocessing of data to generate new fundamental climate data records of better quality for all cryospheric elements will continue.

12.12 Way Forward

Long-term interests of CliC will reflect the future priorities of WCRP (WCRP 2009) and will likely be associated with:

- Targeted multidisciplinary sea-level variability and change research.
- Emerging research of understudied cryospheric elements, such as ice shelves, with a focus on the climate change and variability.
- Synthesising activity/assessment on cryospheric contribution to water and carbon balance.
- Scientific support to cryospheric observations and generation of data products, the IGOS Cryosphere Theme, GCW, GCOS and other relevant systems and initiatives.
- Collaboration with space agencies on mission strategy, calibration and validation and data/products applications for global change research.
- Collaboration and interaction with developing countries on cryospheric issues, in particular, for regions of significant ice and snow masses.
- Modelling and prediction of all cryospheric elements, for a broad range of applications, especially climate prediction from months to centuries.
- Jointly with other WCRP projects and partners, developing a science theme on Polar predictability and modes of variability in the changing climate.
- Developing "CliC-certified" predictions of all cryospheric elements that are essential for regional and national decision making.
- Encouraging and promoting additional regional activities (i.e. in South America) that will stimulate related research and provide strategic regional planning on cryospheric issues to carry the legacy of CliC for future decades.
- Enhancing collaboration with IGBP, IHDP and other groups to include cryospheric change in studies of societal impact and mitigation.
- Efforts aimed at developing an ability to study the role of cryosphere in the possible abrupt climate changes.
- Support of early career scientists through APECS and PYRN networks.

References

ACIA (2005) Climate impact assessment. Cambridge University Press, New York, p 1042

Bowden S, Corell RW, Hassol SJ, Symonl C (2005) 2nd International conference on Arctic Research Planning, Copenhagen, 10–12 Nov 2005

Brown J, Ferrians OJ Jr, Heginbottom JA, Melnikov ES (1998) Circum- Arctic map of permafrost and ground-ice conditions. National Snow and Ice Data Center/World Data Center for Glaciology, Boulder

Climate and Cryosphere (CliC) Project (2001) Science and Co-ordination plan, Version 1 WMO/TD-No. 1053. WCRP series report no. 114, p 99

Dahl-Jensen D, Bamber J, Bøggild CE et al (2009) AMAP 2009. The Greenland ice sheet in a changing climate: snow, ice and permafrost (SWIPS), 2009. Arctic Monitoring and Assessment Programme (AMAP), Oslo. ISBN 13 978 82 7971 052 3

Goodison BE, Ye Frolov I, Kotlyakov VM (1998a) The role of cryosphere in the climate system. In: Proceedings of the conference on the world climate research programme: achievements, benefits and challenges, WMO TD-No. 904, Geneva

Goodison BE, Louie PYT, Yang D (1998b) WMO solid precipitation measurement intercomparison, final report. WMO/TD-No.872, Geneva

IGOS (2007) Integrated Global Observing Strategy Cryosphere Theme Report – for the monitoring of our environment from space and from earth. WMO/TD-No. 1405, World Meteorological Organization, Geneva

Lemke P, Ren J, Alley RB et al (2007) Observations: changes in snow, ice and frozen ground. In: Solomon S, Qin D, Manning M, Chen Z, Marquis M, Averyt KB, Tignor M, Miller HL (eds) Climate change 2007: the physical science basis. Contribution of Working Group I to the Fourth Assessment Report of the Intergovernmental Panel on Climate Change. Cambridge University Press, Cambridge/New York

Steffen K, Clark PU, Cogley JG et al (2008) Rapid changes in glaciers and ice sheets and their impacts on sea level. In: Abrupt climate change. A report by the U.S. climate change science program and the subcommittee on global change research. U.S. Geological Survey, Reston

UNEP (2007) Global outlook for ice & snow. United Nations Environmental Programme, Nairobi

Velicogna I (2009) Increasing rates of ice mass loss from the Greenland and Antarctic ice sheets revealed by GRACE. Geophys Res Lett. doi:10.1029/2009GL040222

WCRP (2009) Implementation plan 2010–2015, WMO/TD-No. 1503

WMO Report (2000) Annual review of the world climate research programme and report of the 21st Session of the Joint Scientific Committee, Tokyo, 13–17 Mar 2000, WMO TD-No. 1049

WWF (2008) Arctic climate impact science – an update since ACIA, WWF, Oslo, Norway

Index

P. Lemke and H.-W. Jacobi (eds.), *Arctic Climate Change: The ACSYS Decade
and Beyond,* Atmospheric and Oceanographic Sciences Library 43,
DOI 10.1007/978-94-007-2027-5, © Springer Science+Business Media B.V. 2012

Printed in France by Amazon
Brétigny-sur-Orge, FR